ADENOVIRUS DNA

DEVELOPMENTS IN MOLECULAR VIROLOGY

Yechiel Becker, Series Editor

Julia Hadar, Managing Editor

Other books in the series:

Becker, Y. (ed.), *Herpesvirus DNA* (1981).
Becker, Y. (ed.), *Replication of Viral and Cellular Genomes* (1983).
Kohn, A. and Fuchs, P. (eds.), *Mechanisms of Viral Pathogenesis: From Gene to Pathogen* (1984).
Becker, Y. (ed.), *Antiviral Drugs and Interferon: The Molecular Basis of Their Activity* (1984).
Becker, Y. (ed.), *Recombinant DNA Research and Viruses: Cloning and Expression of Viral Genes* (1985).
Feitelson, M., *Molecular Components of Hepatitis B Virus* (1985).
Becker, Y. (ed.), *Viral Messenger RNA: Transcription, Processing, Splicing and Molecular Structure* (1985).

ADENOVIRUS DNA

The Viral Genome and Its Expression

edited by

Walter Doerfler
University of Cologne
Federal Republic of Germany

Martinus Nijhoff Publishing
a member of the Kluwer Academic Publishers Group
Boston / Dordrecht / Lancaster

Distributors for North America:
Kluwer Academic Publishers
190 Old Derby Street
Hingham, Massachusetts 02043, USA

Distributors for the UK and Ireland:
Kluwer Academic Publishers
MTP Press Limited
Falcon House, Queen Square
Lancaster LA1 1RN, UNITED KINGDOM

Distributors for all other countries:
Kluwer Academic Publishers Group
Distribution Centre
Post Office Box 322
3300 AH Dordrecht, THE NETHERLANDS

The figure on the cover also appears on page 5 of this book and is used with permission.

Library of Congress Cataloging-in-Publication Data
Main entry under title:

Adenovirus DNA.

(Developments in molecular virology)
Bibliography: p.
Includes index.
1. Adenoviruses. 2. Viral genetics. 3. Transfection. 4. Deoxyribonucleic acid. I. Doerfler, Walter, 1933- . II. Series.
QR396.A33 1985 616.9′2507 85-18899
ISBN 0-89838-758-2

Printed in the United States of America

CONTENTS

CONTRIBUTORS

THOMAS ADRIAN, Institute of Medical Microbiology, University of the Saar, 6650 Homburg, FEDERAL REPUBLIC OF GERMANY

GÖRAN AKUSJÄRVI, University of Uppsala, Biomedical Center, 75123 Uppsala, SWEDEN

Peter ALESTRÖM, University of Uppsala, Biomedical Center, 75123 Uppsala, SWEDEN

WALTER DOERFLER, Institute of Genetics, University of Cologne, 5000 Cologne 41, FEDERAL REPUBLIC OF GERMANY

Alexander J. van der Eb, University of Leiden, Department of Medical Biochemistry, 2300 RA Leiden, THE NETHERLANDS

HELMUT ESCHE, Institut für Molekularbiologie, Universität Essen, 4300 Essen 1, FEDERAL REPUBLIC OF GERMANY

JANE S. FLINT, Department of Biochemical Sciences, Princeton University, Princeton, N.J. 08544, USA

RICHARD E. GELINAS, Fred Hutchinson Cancer Research Center, 1124 Columbia Street, Seattle, WA 98104, USA

THOMAS R. GINGERAS, Salk Institute Biotechnology/Industrial Associates, Inc., P.O.Box 85200, San Diego, CA 92138-9216, USA

MICHAEL J. IMPERIALE, The Rockefeller University, New York, N.Y. 10021, USA

NICHOLAS C. JONES, Department of Biological Sciences, Purdue University, West Lafayette, Indiana 47907, USA

DAGMAR KNEBEL, Institute of Genetics, University of Cologne, 5000 Cologne 41, FEDERAL REPUBLIC OF GERMANY

BIRGITT KNUST-KRON, Institute of Genetics, University of Cologne, 5000 Cologne 41, FEDERAL REPUBLIC OF GERMANY

KLAUS-DIETER LANGNER, Institute of Genetics, University of Cologne, 5000 Cologne 41, FEDERAL REPUBLIC OF GERMANY

HERMANN LÜBBERT, California Institute of Technology, Department of Chemistry, Pasadena, California 91125, USA

STANLEY MAK, Department of Biology, McMaster University, Hamilton, Ontario L8S 4K1, CANADA

IRENE MAK, Department of Biology, McMaster University, Hamilton, Ontario L8S 4K1, CANADA

JOSEPH R. NEVINS, The Rockefeller University, New York, N.Y. 10021, USA

Bernard A. OOSTRA, University of Leiden, Department of Medical Biochemistry, 2300 RA Leiden, THE NETHERLANDS

ULF PETTERSSON, University of Uppsala, Biomedical Center, 75123 Uppsala, SWEDEN

RICHARD J. ROBERTS, Cold Spring Harbor Laboratory, Cold Spring Harbor, N.Y. 11724, USA

DANIELA SCIAKY, Biology Department, Brookhaven National Laboratories, Upton N.Y. 11973, USA

FUYUHIKO TAMANOI, Department of Biochemistry and Molecular Biology, University of Chicago, 920 East 58th Street, Chicago, IL 60637

REINHARD WIGAND, Institute of Medical Microbiology, University of the Saar, 6650 Homburg, FEDERAL REPUBLIC OF GERMANY

JIM F. WILLIAMS, Department of Biological Sciences, Carnegie-Mellon University, Pittsburgh, Pennsylvania 15213, USA

PREFACE

Although adenoviruses have been established for quite some time as one of the most pliable tools to study the molecular biology of mammalian cells, rapid progress continues to be made with this virus system. The adenoviral genome introduced into the nucleus of cells as a Trojan Horse, as it were, is now facilitating investigators to uncover details of cellular mechanisms. In this volume twelve chapters have been collected that summarize important current research on different mechanisms in adenovirus-infected and -transformed cells. It has become increasingly apparent that some of these mechanisms do not only pertain to highly specialized strategies of the viral genome and its expression or replication, but may simultaneously shed light on events indigenous to the cell.

"Adenovirus DNA: The Viral Genome and its Expression" highlights the first publication of the entire annotated sequence of 35.937 nucleotide pairs of adenovirus type 2 DNA by the Cold Spring Harbor and Uppsala groups (chapter 1). Göran Akusjärvi, Ulf Pettersson (Uppsala University) and Rich Roberts (Cold Spring Harbor) present a survey on the structure and function of the adenovirus-2 genome (chapter 2). A considerable amount of information has been collected on the biochemical mechanisms involved in adenovirus DNA replication in human cells. Although highly specialized in the way of initiating DNA replication, adenovirus DNA replication represents probably one of the best studied systems in work on eukaryotic DNA replication. Fuyuhiko Tamanoi's (Cold Spring Harbor) chapter deals with recent progress in work on this system (chapter 3). Studies on the regulation of gene expression - both at the transcriptional and post-transcriptional levels - have assumed a central role in molecular biology and many biological systems reveal interesting solutions to this pivotal control mechanism. Michael Imperiale's and Joseph Nevins' (The Rockefeller University) chapter summarizes the current understanding of the transcriptional and post-transcriptional regulation of adenovirus gene expression (chapter 4). Genetic elements governing adenoviral gene expression have been investigated in great detail by Nicholas Jones (Purdue

University) (chapter 5). Walter Doerfler and colleagues (University of Cologne) summarize their work that helped elucidate the role that sequence-specific promoter methylations can play in the shut-off of eukaryotic genes (chapter 6). Helmut Esche (University of Essen) has studied in detail many of the early adenovirus-coded proteins and has set out to study the functions of some of these proteins in infected and transformed cells (chapter 7). The elucidation of specific functions of viral polypeptides largely hinges on the genetic analysis of the viral genome. Progress in the analysis of viral DNA replication represents a convincing example. Jim Williams (Carnegie-Mellon University) has compiled and analyzed in a functional context the genetics of this viral system (chapter 8).

Ever since the discovery of the oncogenic capacity of adenoviruses in rodents (Trentin et al., Science 137, 835-841, 1962), adenoviruses have been considered to be useful instruments in investigating the mechanism of viral oncogenesis. There are three contributions covering this intriguing topic. Jane Flint (Princeton University) has concentrated on work studying "Alterations in Cellular Functions in Adenovirus-Infected and -Transformed Cells" (chapter 9). Alex van der Eb and Bernard Oostra (Leiden University) have pursued their discovery that the expression of genes coding for class I antigens in the major histocompatibility complex (MHC) can be drastically affected in adenovirus-transformed cells (chapter 10). Stanley and Irene Mak (McMaster University) have reviewed the role which the E1 region of adenoviruses can play in viral transformation and oncogenesis (chapter 11). Lastly, Reinhard Wigand and Thomas Adrian (University of the Saar) present an up-to-date survey on the classification and epidemiology of adenoviruses (chapter 12).

This volume will hopefully prove useful to the long-time adenovirus afficionado in that it presents recent results and summaries of current research, as well as to novices seeking to familiarize themselves with progress in this active field of research in molecular biology.

I should like to thank all the authors for their contributions. I am particularly indebted to Julia Hadar for critically reading the manuscripts and Petra Böhm for expert editorial assistance.

Walter Doerfler
Institute of Genetics
Cologne

ADENOVIRUS DNA

1

A CONSENSUS SEQUENCE FOR THE ADENOVIRUS-2 GENOME

R.J. Roberts, G. Akusjarvi[1], P. Alestrom[1*], R.E. Gelinas[2], T.R. Gingeras[3], D. Sciaky[4] and U. Pettersson[1]

Cold Spring Harbor Laboratory, Cold Spring Harbor, New York 11724

ABSTRACT

The complete sequence of 35,937 nucleotides for the Adenovirus-2 genome is presented in a fully annotated form. This sequence represents a consensus derived by combining sequence data from many different laboratories.

INTRODUCTION

As a result of sequencing efforts in many different laboratories (Table 1) it is now possible to derive a complete sequence for the Adenovirus-2 genome. The final sequence shown in Fig. 1 is 35,937 base pairs in length, although because of heterogeneity at several positions this length may vary by ± 9 nucleotides (1). This heterogeneity is present in viral stocks grown from a single plaque and is not to be confused with the strain variations or sequence errors that are listed in Table 2. A diagrammatic representation of the open reading frames present in the sequence is shown in Fig. 2 and a listing of the locations of known features is presented in Table 3.

[1] Department of Medical Genetics, Biomedical Center Box 589, S-751 23 Uppsala, Sweden

[2] Fred Hutchinson Cancer Research Center, 1124 Columbia Street, Seattle, WA 98104

[3] Salk Institute Biotechnology/Industrial Associates, Inc., P.O. Box 85200, San Diego, CA 92138-9216

[4] Biology Department, Brookhaven National Laboratories, Upton, NY 11973

[*] Present Address: Dept. of Medical Biochemistry, P. O. Box 1112, Blindern, Oslo 3, Norway

Table 1. Published adenovirus-2 sequences.

Location	Reference	Location	Reference
1 to 110	2	18616 to 19233	18
1 to 156	3	18778 to 18918	19
1 to 11600	1	18802 to 18861	13
459 to 608	4	18838 to 19143	20
1226 to 1609	5	18838 to 21744	21
1517 to 1696	6	19555 to 19622	20
3504 to 4109	7	19663 to 19743	20
3932 to 4112	6	19789 to 19816	20
5778 to 11560	8	19823 to 19932	20
5817 to 6051	9	20341 to 20414	20
5848 to 6578	10	20425 to 20490	20
5909 to 6178	11	20674 to 20724	20
5986 to 6236	12	21238 to 21744	20
6039 to 6079	13	21607 to 21816	22
6039 to 6079	14	21607 to 22770	23
6039 to 6140	15	22309 to 22598	24
7023 to 7212	11	22467 to 24123	25
7055 to 7215	15	23909 to 24088	26
7101 to 7172	13	23924 to 25638	27
7101 to 7172	14	24715 to 24791	26
7869 to 8420	16	25634 to 27376	28
7929 to 8423	17	26975 to 27176	29
9452 to 9836	11	27025 to 27092	26
9604 to 9753	15	27373 to 30050	39
9634 to 9723	13	27609 to 27980	40
9635 to 9723	14	28142 to 28259	6
10514 to 10680	30	28328 to 28622	15
10514 to 11066	31	28376 to 28559	14
10610 to 10766	32	28376 to 29792	40
10681 to 10813	33	30047 to 32266	41
11601 to 15726	27	30977 to 31080	15
13898 to 14231	34	31030 to 31095	14
15033 to 18316	35	32092 to 35937	1
15821 to 16495	36	32263 to 35937	42

3

17878	to	18918	37	35358	to	35707	9
17880	to	17975	6	35358	to	35937	43
17962	to	18177	38	35835	to	35937	2
				35804	to	35937	3

Table 2. Sequence variations between the consensus sequence shown in Fig. 1 and other published results.

Base #	Explanation
8	A insertion after base 8. This is a known strain variation within the inverted terminal repetition. eg ref 2.
460	C is given as T in reference 4.
6443	C is given as Y in reference 10.
6575	C is deleted in reference 10. sequence error.
7213	G is deleted in reference 15. sequence error corrected by authors.
9315-9316	CG is inverted to GC in reference 8.
9634	G is deleted in reference 14. sequence error.
10716	C is deleted in reference 33. sequence error.
11062	T is C in reference 31. Probable strain difference.
15856	G is a T in reference 36.
15914	C is a T in reference 36.
15998	G is a C in reference 36.
16207-16209	CGA are deleted in reference 36.
16437	G is a C in reference 36.
17964	G is a C in references 38 and 44.
18915	C is deleted in references 37, 19, and 20. error corrected by authors.
18919	C is given as NN in reference 20. error corrected by authors.
19617	T is a C in reference 20. error corrected by authors.
19666	T is a C in reference 20. error corrected by authors.

19823	A is a G in reference 20. error corrected by authors.
20427	A is a G in reference 20. error corrected by authors.
20487	C is a T in reference 20. error corrected by authors.
22524	T in the final sequence and a C in reference 24. The more recent determination is used. This may be a strain variation.
28339	C insertion after base 28339 in reference 15. error corrected by authors.
28350	A insertion after base 28350 in reference 15. error corrected by authors.
28359	A insertion after base 28359 in reference 15. error corrected by authors.
28466	C is deleted in references 15, 14. sequence error corrected by authors.
28496-28497	TG are deleted in references 15, 14. sequence error corrected by authors.
30981	T is deleted in reference 15. sequence error corrected by authors.
34344-34357	14 T residues. Only 13 T residues in reference 42. It is known that heterogeneity is found at this position and the 13 T residues reflect the use of cloned DNA. 14 T residues occur in the predominant form of the sequence.
35134-35142	9 A residues. 10 residues in reference 42. These lie in an apparently untranslated region between two open reading frames and probably represent a strain difference.
35447	T. But a C in reference 43. T has been found 3 times independently. Probable sequencing error.
35503	A. But a G in reference 43. A has been found 3 times independently. Probable sequencing error.
35930	T insertion after base 35930 in most strains. eg. references 42, 2, 43. This is a known strain variation within the inverted terminal repetition.

```
                                                                                                   100
CATCATCATAATATACCTTATTTTGGATTGAAGCCAATATGATAATGAGGGGGTGGAGTTTGTGACGTGGCGCGGGGCGTGGGAACGGGGCGGGTGACGT
GTAGTAGTATTATATGGAATAAAACCTAACTTCGGTTATACTATTACTCCCCCACCTCAAACACTGCACCGCGCCCCGCACCCTTGCCCCGCCCACTGCA

                                                                                                   200
AGTAGTGTGGCGGAAGTGTGATGTTGCAAGTGTGGCGGAACACATGTAAGCGCCGGATGTGGTAAAAGTGACGTTTTTGGTGTGCGCCGGTGTATACGGG
TCATCACACCGCCTTCACACTACAACGTTCACACCGCCTTGTGTACATTCGCGGCCTACACCATTTTCACTGCAAAAACCACACGCGGCCACATATGCCC
*
0.28% end of inverted terminal repetition                        0.74% Bal I
                                                                 *                                 300
AAGTGACAATTTTCGCGCGGTTTTAGGCGGATGTTGTAGTAAATTTGGGCGTAACCAAGTAATGTTTGGCCATTTTCGCGGGAAAACTGAATAAGAGGAA
TTCACTGTTAAAAGCGCGCCAAAATCCGCCTACAACATCATTTAAACCCGCATTGGTTCATTACAAACCGGTAAAAGCGCCCTTTTGACTTATTCTCCTT

                                                                                                   400
GTGAAATCTGAATAATTCTGTGTTACTCATAGCGCGTAATATTTGTCTAGGGCCGCGGGGACTTTGACCGTTTACGTGGAGACTCGCCCAGGTGTTTTTC
CACTTTAGACTTATTAAGACACAATGAGTATCGCGCATTATAAACAGATCCCGGCGCCCCTGAAACTGGCAAATGCACCTCTGAGCGGGTCCACAAAAAG

                                                                                  1.38% Cap site for EIa mRNAs
                                                                                                  *
TCAGGTGTTTTCCGCGTTCCGGGTCAAAGTTGGCGTTTTATTATTATAGTCAGCTGACGCGCAGTGTATTTATACCCGGTGAGTTCCTCAAGAGGCCACT
AGTCCACAAAAGGCGCAAGGCCCAGTTTCAACCGCAAAATAATAATATCAGTCGACTGCGCGTCACATAAATATGGGCCACTCAAGGAGTTCTCCGGTGA

                                                                 1.55% Initiator AUG for E1a proteins
                                                                 *                                 600
CTTGAGTGCCAGCGAGTAGAGTTTTCTCCTCCGAGCCGCTCCGACACCGGGACTGAAAATGAGACATATTATCTGCCACGGAGGTGTTATTACCGAAGAA
GAACTCACGGTCGCTCATCTCAAAAGAGGAGGCTCGGCGAGGCTGTGGCCCTGACTTTTACTCTGTATAATAGACGGTGCCTCCACAATAATGGCTTCTT

                                    1.76% DONOR splice site for E1a 9s mRNA
                                    *                                                              700
ATGGCCGCCAGTCTTTTGGACCAGCTGATCGAAGAGGTACTGGCTGATAATCTTCCACCTCCTAGCCATTTTGAACCACCTACCCTTCACGAACTGTATG
TACCGGCGGTCAGAAAACCTGGTCGACTAGCTTCTCCATGACCGACTATTAGAAGGTGGAGGATCGGTAAAACTTGGTGGATGGGAAGTGCTTGACATAC

                                                                                                   800
ATTTAGACGTGACGGCCCCCGAAGATCCCAACGAGGAGGCGGTTTCGCAGATTTTTCCCGAGTCTGTAATGTTGGCGGTGCAGGAAGGGATTGACTTATT
TAAATCTGCACTGCCGGGGGCTTCTAGGGTTGCTCCTCCGCCAAAGCGTCTAAAAAGGGCTCAGACATTACAACCGCCACGTCCTTCCCTAACTGAATAA

                                                                                                   900
CACTTTTCCGCCGGCGCCCGGTTCTCCGGAGCCGCCTCACCTTTCCCGGCAGCCCGAGCAGCCGGAGCAGAGAGCCTTGGGTCCGGTTTCTATGCCAAAC
GTGAAAAGGCGGCCGCGGGCCAAGAGGCCTCGGCGGAGTGGAAAGGGCCGTCGGGCTCGTCGGCCTCGTCTCTCGGAACCCAGGCCAAAGATACGGTTTG

                                                                             2.70% DONOR splice site for E1a 12s mRNA
                                                                             *                    1000
CTTGTGCCGGAGGTGATCGATCTTACCTGCCACGAGGCTGGCTTTCCACCCAGTGACGACGAGGATGAAGAGGGTGAGGAGTTTGTGTTAGATTATGTGG
GAACACGGCCTCCACTAGCTAGAATGGACGGTGCTCCGACCGAAAGGTGGGTCACTGCTGCTCCTACTTCTCCCACTCCTCAAACACAATCTAATACACC

     2.79% Xma I
     *                                                                                            1100
AGCACCCCGGGCACGGTTGCAGGTCTTGTCATTATCACCGGAGGAATACGGGGGACCCAGATATTATGTGTTCGCTTTGCTATATGAGGACCTGTGGCAT
TCGTGGGGCCCGTGCCAACGTCCAGAACAGTAATAGTGGCCTCCTTATGCCCCCTGGGTCTATAATACACAAGCGAAACGATATACTCCTGGACACCGTA
```

FIGURE 1: Complete Nucleotide Sequence of Adenovirus-2 Genome

```
      3.09% DONOR splice site for E1a 13s mRNA
          *
                                                                                                1200
GTTTGTCTACAGTAAGTGAAAATTATGGGCAGTCGGTGATAGAGTGGTGGGTTTGGTGTGGTAATTTTTTTTTAATTTTTACAGTTTTGTGGTTTAAAGA
CAAACAGATGTCATTCACTTTTAATACCCGTCAGCCACTATCTCACCACCCAAACCACACCATTAAAAAAAAATTAAAAATGTCAAAACACCAAATTTCT

                         3.41% ACCEPTOR splice site for all E1a mRNAs
                         *
                                                                                                1300
ATTTTGTATTGTGATTTTTTAAAAGGTCCTGTGTCTGAACCTGAGCCTGAGCCCGAGCCAGAACCGGAGCCTGCAAGACCTACCCGGCGTCCTAAATTGG
TAAAACATAACACTAAAAAATTTTCCAGGACACAGACTTGGACTCGGACTCGGGCTCGGTCTTGGCCTCGGACGTTCTGGATGGGCCGCAGGATTTAACC

                                   3.71% Xba I
            3.65% Terminator UGA for E1a 6K polypeptide (9s mRNA)
            *                      *
                                                                                                1400
TGCCTGCTATCCTGAGACGCCCGACATCACCTGTGTCTAGAGAATGCAATAGTAGTACGGATAGCTGTGACTCCGGTCCTTCTAACACACCTCCTGAGAT
ACGGACGATAGGACTCTGCGGGCTGTAGTGGACACAGATCTCTTACGTTATCATCATGCCTATCGACACTGAGGCCAGGAAGATTGTGTGGAGGACTCTA

                                                                                                1500
ACACCCGGTGGTCCCGCTGTGCCCCATTAAACCAGTTGCCGTGAGAGTTGGTGGGCGTCGCCAGGCTGTGGAATGTATCGAGGACTTGCTTAACGAGTCT
TGTGGGCCACCAGGGCGACACGGGGTAATTTGGTCAACGGCACTCTCAACCACCCGCAGCGGTCCGACACCTTACATAGCTCCTGAACGAATTGCTCAGA

                        4.28% Terminator UAA for E1a polypeptides   4.36% Hpa I
                                       *                            *
                                                                                                1600
GGGCAACCTTTGGACTTGAGCTGTAAACGCCCCAGGCCATAAGGTGTAAACCTGTGATTGCGTGTGTGGTTAACGCCTTTGTTTGCTGAATGAGTTGATG
CCCGTTGGAAACCTGAACTCGACATTTGCGGGGTCCGGTATTCCACATTTGGACACTAACGCACACACCAATTGCGGAAACAAACGACTTACTCAACTAC

 4.47% AAUAAA for E1a mRNAs 4.53% Poly-A addition site for E1a mRNAs                      4.72% Cap site for E1b mRNAs
       *                     *                                                                    *
TAAGTTTAATAAAGGGTGAGATAATGTTTAACTTGCATGGCGTGTTAAATGGGGCGGGGCTTAAAGGGTATATAATGCGCCGTGGGCTAATCTTGGTTAC
ATTCAAATTATTTCCCACTCTATTACAAATTGAACGTACCGCACAATTTACCCCGCCCCGAATTTCCCATATATTACGCGGCACCCGATTAGAACCAATG

          4.76% Initiator AUG for E1b 15K polypeptide                    4.91% Sac I
          *                                                             *
                                                                                                1800
ATCTGACCTCATGGAGGCTTGGGAGTGTTTGGAAGATTTTTCTGCTGTGCGTAACTTGCTGGAACAGAGCTCTAACAGTACCTCTTGGTTTTGGAGGTTT
TAGACTGGAGTACCTCCGAACCCTCACAAACCTTCTAAAAAGACGACACGCATTGAACGACCTTGTCTCGAGATTGTCATGGAGAACCAAAACCTCCAAA

                                                                                                1900
CTGTGGGGCTCCTCCCAGGCAAAGTTAGTCTGCAGAATTAAGGAGGATTACAAGTGGGAATTTGAAGAGCTTTTGAAATCCTGTGGTGAGCTGTTTGATT
GACACCCCGAGGAGGGTCCGTTTCAATCAGACGTCTTAATTCCTCCTAATGTTCACCCTTAAACTTCTCGAAAACTTTAGGACACCACTCGACAAACTAA

                                                                                                2000
CTTTGAATCTGGGTCACCAGGCGCTTTTCCAAGAGAAGGTCATCAAGACTTTGGATTTTTCCACACCGGGGCGCGCTGCGGCTGCTGTTGCTTTTTTGAG
GAAACTTAGACCCAGTGGTCCGCGAAAAGGTTCTCTTCCAGTAGTTCTGAAACCTAAAAAGGTGTGGCCCCGCGCGACGCCGACGACAACGAAAAAACTC

5.60% Initiator AUG for E1b 57K polypeptide 5.69% Kpn I       5.74% Bal I
               *                            *                 *
                                                                                                2100
TTTTATAAAGGATAAATGGAGCGAAGAAACCCATCTGAGCGGGGGGTACCTGCTGGATTTTCTGGCCATGCATCTGTGGAGAGCGGTGGTGAGACACAAG
AAAATATTTCCTATTTACCTCGCTTCTTTGGGTAGACTCGCCCCCCATGGACGACCTAAAAGACCGGTACGTAGACACCTCTCGCCACCACTCTGTGTTC
```

```
                                                                                                   2200
AATCGCCTGCTACTGTTGTCTTCCGTCCGCCCGGCAATAATACCGACGGAGGAGCAACAGCAGGAGGAAGCCAGGCGGCGGCGGCAGGAGCAGAGCC
TTAGCGGACGATGACAACAGAAGGCAGGCGGGCCGTTATTATGGCTGCCTCCTCGTTGTCGTCCTCCTTCGGTCCGCCGCCGCCGTCCTCGTCTCGG

  6.22% Terminator UGA for E1b 15K polypeptide  6.25% DONOR splice site for E1b 13s, 1.26 & 1.31kb mRNAs
                                         *            *                                            2300
CATGGAACCCGAGAGCCGGCCTGGACCCTCGGGAATGAATGTTGTACAGGTGGCTGAACTGTTTCCAGAACTGAGACGCATTTTAACCATTAACGAGGAT
GTACCTTGGGCTCTCGGCCGGACCTGGGAGCCCTTACTTACAACATGTCCACCGACTTGACAAAGGTCTTGACTCTGCGTAAAATTGGTAATTGCTCCTA

                                                                                                   2400
GGGCAGGGGCTAAAGGGGGTAAAGAGGGAGCGGGGGGCTTCTGAGGCTACAGAGGAGGCTAGGAATCTAACTTTTAGCTTAATGACCAGACACCGTCCTG
CCCGTCCCCGATTTCCCCCATTTCTCCCTCGCCCCCCGAAGACTCCGATGTCTCCTCCGATCCTTAGATTGAAAATCGAATTACTGGTCTGTGGCAGGAC

                                                                                                   2500
AGTGTGTTACTTTTCAGCAGATTAAGGATAATTGCGCTAATGAGCTTGATCTGCTGGCGCAGAAGTATTCCATAGAGCAGCTGACCACTTACTGGCTGCA
TCACACAATGAAAAGTCGTCTAATTCCTATTAACGCGATTACTCGAACTAGACGACCGCGTCTTCATAAGGTATCTCGTCGACTGGTGAATGACCGACGT

                                                                                                   2600
GCCAGGGGATGATTTTGAGGAGGCTATTAGGGTATATGCAAAGGTGGCACTTAGGCCAGATTGCAAGTACAAGATTAGCAAACTTGTAAATATCAGGAAT
CGGTCCCCTACTAAAACTCCTCCGATAATCCCATATACGTTTCCACCGTGAATCCGGTCTAACGTTCATGTTCTAATCGTTTGAACATTTATAGTCCTTA

                                                                                                   2700
TGTTGCTACATTTCTGGGAACGGGGCCGAGGTGGAGATAGATACGGAGGATAGGGTGGCCTTTAGATGTAGCATGATAAATATGTGGCCGGGGGTGCTTG
ACAACGATGTAAAGACCCTTGCCCCGGCTCCACCTCTATCTATGCCTCCTATCCCACCGGAAATCTACATCGTACTATTTATACACCGGCCCCCACGAAC

                                                                    7.69% Bal I                   7.78% Hind III
                                                                    *                             *
GCATGGACGGGGTGGTTATTATGAATGTGAGGTTTACTGGTCCCAATTTTAGCGGTACGGTTTTCCTGGCCAATACCAATCTTATCCTACACGGTGTAAG
CGTACCTGCCCCACCAATAATACTTACACTCCAAATGACCAGGGTTAAAATCGCCATGCCAAAAGGACCGGTTATGGTTAGAATAGGATGTGCCACATTC

                                                                                                   2900
CTTCTATGGGTTTAACAATACCTGTGTGGAAGCCTGGACCGATGTAAGGGTTCGGGGCTGTGCCTTTTACTGCTGCTGGAAGGGGGTGGTGTGTCGCCCC
GAAGATACCCAAATTGTTATGGACACACCTTCGGACCTGGCTACATTCCCAAGCCCCGACACGGAAAATGACGACGACCTTCCCCCACCACACAGCGGGG

                                                                                                   3000
AAAAGCAGGGCTTCAATTAAGAAATGCCTGTTTGAAAGGTGTACCTTGGGTATCCTGTCTGAGGGTAACTCCAGGGTGCGCCACAATGTGGCCTCCGACT
TTTTCGTCCCGAAGTTAATTCTTTACGGACAAACTTTCCACATGGAACCCATAGGACAGACTCCCATTGAGGTCCCACGCGGTGTTACACCGGAGGCTGA

                                                                                                   3100
GTGGTTGCTTCATGCTAGTGAAAAGCGTGGCTGTGATTAAGCATAACATGGTGTGTGGCAACTGCGAGGACAGGGCCTCTCAGATGCTGACCTGCTCGGA
CACCAACGAAGTACGATCACTTTTCGCACCGACACTAATTCGTATTGTACCACACACCGTTGACGCTCCTGTCCCGGAGAGTCTACGACTGGACGAGCCT

                                                           8.78% Bal I
                                                           *                                       3200
CGGCAACTGTCACTTGCTGAAGACCATTCACGTAGCCAGCCACTCTCGCAAGGCCTGGCCAGTGTTTGAGCACAACATACTGACCCGCTGTTCCTTGCAT
GCCGTTGACAGTGAACGACTTCTGGTAAGTGCATCGGTCGGTGAGAGCGTTCCGGACCGGTCACAAACTCGTGTTGTATGACTGGGCGACAAGGAACGTA
```

```
                                                   9.05% Terminator UAA for E1b 1.31kb mRNA
   8.93% ACCEPTOR splice site for E1b 1.31kb mRNA                9.09% ACCEPTOR splice site for E1b 1.26kb mRNA
             *                                             *         *                               3300
TTGGGTAACAGGAGGGGGGTGTTCCTACCTTACCAATGCAATTTGAGTCACACTAAGATATTGCTTGAGCCCGAGAGCATGTCCAAGGTGAACCTGAACG
AACCCATTGTCCTCCCCCCACAAGGATGGAATGGTTACGTTAAACTCAGTGTGATTCTATAACGAACTCGGGCTCTCGTACAGGTTCCACTTGGACTTGC

                    9.24% Bgl II
                    *                                                                            3400
GGGTGTTTGACATGACCATGAAGATCTGGAAGGTGCTGAGGTACGATGAGACCCGCACCAGGTGCAGACCCTGCGAGTGTGGCGGTAAACATATTAGGAA
CCCACAAACTGTACTGGTACTTCTAGACCTTCCACGACTCCATGCTACTCTGGGCGTGGTCCACGTCTGGGACGCTCACACCGCCATTTGTATAATCCTT

                                                                                                 3500
CCAGCCTGTGATGCTGGATGTGACCGAGGAGCTGAGGCCCGATCACTTGGTGCTGGCCTGCACCCGCGCTGAGTTTGGCTCTAGCGATGAAGATACAGAT
GGTCGGACACTACGACCTACACTGGCTCCTCGACTCCGGGCTAGTGAACCACGACCGGACGTGGGCGCGACTCAAACCGAGATCGCTACTTCTATGTCTA

                                                                        10.01% Initiator AUG for polypeptide I
   9.75% DONOR splice site for E1b 22s, 1.26 & 1.31kb mRNAs           9.98% ACCEPTOR splice site for all E1b mRNAs
9.74% Terminator UGA for E1b 57K polypeptide    9.95% Cap site for polypeptide IX mRNA
*  *                                                                         *         *        *
TGAGGTACTGAAATGTGTGGGCGTGGCTTAAGGGTGGGAAAGAATATATAAGGTGGGGGTCTCATGTAGTTTTGTATCTGTTTTGCAGCAGCCGCCGCCA
ACTCCATGACTTTACACACCCGCACCGAATTCCCACCCTTTCTTATATATTCCACCCCCAGAGTACATCAAAACATAGACAAAACGTCGTCGGCGGCGGT

                            10.10% Sac I
                            *                                                                    3700
TGAGCGCCAACTCGTTTGATGGAAGCATTGTGAGCTCATATTTGACAACGCGCATGCCCCCATGGGCCGGGGTGCGTCAGAATGTGATGGGCTCCAGCAT
ACTCGCGGTTGAGCAAACTACCTTCGTAACACTCGAGTATAAACTGTTGCGCGTACGGGGGTACCCGGCCCCACGCAGTCTTACACTACCCGAGGTCGTA

                                                                                                 3800
TGATGGTCGCCCCGTCCTGCCCGCAAACTCTACTACCTTGACCTACGAGACCGTGTCTGGAACGCCGTTGGAGACTGCAGCCTCCGCCGCCGCTTCAGCC
ACTACCAGCGGGGCAGGACGGGCGTTTGAGATGATGGAACTGGATGCTCTGGCACAGACCTTGCGGCAACCTCTGACGTCGGAGGCGGCGGCGAAGTCGG

                                                                                                 3900
GCTGCAGCCACCGCCCGCGGGATTGTGACTGACTTTGCTTTCCTGAGCCCGCTTGCAAGCAGTGCAGCTTCCCGTTCATCCGCCCGCGATGACAAGTTGA
CGACGTCGGTGGCGGGCGCCCTAACACTGACTGAAACGAAAGGACTCGGGCGAACGTTCGTCACGTCGAAGGGCAAGTAGGCGGGCGCTACTGTTCAACT

                           10.93% Xma I
                           *                                                                     4000
CGGCTCTTTTGGCACAATTGGATTCTTTGACCCGGGAACTTAATGTCGTTTCTCAGCAGCTGTTGGATCTGCGCCAGCAGGTTTCTGCCCTGAAGGCTTC
GCCGAGAAAACCGTGTTAACCTAAGAAACTGGGCCCTTGAATTACAGCAAAGAGTCGTCGACAACCTAGACGCGGTCGTCCAAAGACGGGACTTCCGAAG
```

```
                                                        11.30% Poly-A addition site for E1b and polypeptide IX mRNAs
                              11.21% AAUAAA for E1b and polypeptide IX mRNAs
          11.18% Terminator UAA for polypeptide IX      11.28% Bcl I
                     *         *                           *     *                                     4100
CTCCCCTCCCAATGCGGTTTAAAACATAAATAAAAACCAGACTCTGTTTGGATTTTGATCAAGCAAGTGTCTTGCTGTCTTTATTTAGGGGTTTTGCGCG
GAGGGGAGGGTTACGCCAAATTTTGTATTTATTTTTGGTCTGAGACAAACCTAAAACTAGTTCGTTCACAGAACGACAGAAATAAATCCCCAAAACGCGC
                                                   *                                 *  *
                            11.26% Poly-A addition site for IVa2 mRNA                   11.36% AAUAAA for IVa2 mRNA
          11.43% Xma I                                                        11.36% Terminator UAA for IVa2 polypeptide
          *                                                                                            4200
CGCGGTAGGCCCGGGACCAGCGGTCTCGGTCGTTGAGGGTCCTGTGTATTTTTTCCAGGACGTGGTAAAGGTGACTCTGGATGTTCAGATACATGGGCAT
GCGCCATCCGGGCCCTGGTCGCCAGAGCCAGCAACTCCCAGGACACATAAAAAAGGTCCTGCACCATTTCCACTGAGACCTACAAGTCTATGTACCCGTA

                                                                                                    4300
AAGCCCGTCTCTGGGGTGGAGGTAGCACCACTGCAGAGCTTCATGCTGCGGGGTGGTGTTGTAGATGATCCAGTCGTAGCAGGAGCGCTGGGCGTGGTGC
TTCGGGCAGAGACCCCACCTCCATCGTGGTGACGTCTCGAAGTACGACGCCCCACCACAACATCTACTAGGTCAGCATCGTCCTCGCGACCCGCACCACG

                                                                                                    4400
CTAAAAATGTCTTTCAGTAGCAAGCTGATTGCCAGGGGCAGGCCCTTGGTGTAAGTGTTTACAAAGCGGTTAAGCTGGGATGGGTGCATACGTGGGGATA
GATTTTTACAGAAAGTCATCGTTCGACTAACGGTCCCCGTCCGGGAACCACATTCACAAATGTTTCGCCAATTCGACCCTACCCACGTATGCACCCCTAT

                                                                                                    4500
TGAGATGCATCTTGGACTGTATTTTTAGGTTGGCTATGTTCCCAGCCATATCCCTCCGGGGATTCATGTTGTGCAGAACCACCAGCACAGTGTATCCGGT
ACTCTACGTAGAACCTGACATAAAAATCCAACCGATACAAGGGTCGGTATAGGGAGGCCCCTAAGTACAACACGTCTTGGTGGTCGTGTCACATAGGCCA

                                                                                                    4600
GCACTTGGGAAATTTGTCATGTAGCTTAGAAGGAAATGCGTGGAAGAACTTGGAGACGCCCTTGTGACCTCCGAGATTTTCCATGCATTCGTCCATAATG
CGTGAACCCTTTAAACAGTACATCGAATCTTCCTTTACGCACCTTCTTGAACCTCTGCGGGAACACTGGAGGCTCTAAAAGGTACGTAAGCAGGTATTAC

                                                                                                    4700
ATGGCAATGGGCCCACGGGCGGCGGCCTGGGCGAAGATATTTCTGGGATCACTAACGTCATAGTTGTGTTCCAGGATGAGATCGTCATAGGCCATTTTTA
TACCGTTACCCGGGTGCCCGCCGCCGGACCCGCTTCTATAAAGACCCTAGTGATTGCAGTATCAACACAAGGTCCTACTCTAGCAGTATCCGGTAAAAAT

                                                                                                    4800
CAAAGCGCGGGCGGAGGGTGCCAGACTGCGGTATAATGGTTCCATCCGGCCCAGGGGCGTAGTTACCCTCACAGATTTGCATTTCCCACGCTTTGAGTTC
GTTTCGCGCCCGCCTCCCACGGTCTGACGCCATATTACCAAGGTAGGCCGGGTCCCCGCATCAATGGGAGTGTCTAAACGTAAAGGGTGCGAAACTCAAG

                                                                                                    4900
AGATGGGGGGATCATGTCTACCTGCGGGGCGATGAAGAAAACCGTTTCCGGGGTAGGGGAGATCAGCTGGGAAGAAAGCAGGTTCCTGAGCAGCTGCGAC
TCTACCCCCCTAGTACAGATGGACGCCCCGCTACTTCTTTTGGCAAAGGCCCCATCCCCTCTAGTCGACCCTTCTTTCGTCCAAGGACTCGTCGACGCTG

                                                                                                    5000
TTACCGCAGCCGGTGGGCCCGTAAATCACACCTATTACCGGCTGCAACTGGTAGTTAAGAGAGCTGCAGCTGCCGTCATCCCTGAGCAGGGGGGCCACTT
AATGGCGTCGGCCACCCGGGCATTTAGTGTGGATAATGGCCGACGTTGACCATCAATTCTCTCGACGTCGACGGCAGTAGGGACTCGTCCCCCCGGTGAA

                                                                                                    5100
CGTTAAGCATGTCCCTGACTTGCATGTTTTCCCTGACCAAATGCGCCAGAAGGCGCTCGCCGCCCAGCGATAGCAGTTCTTGCAAGGAAGCAAAGTTTTT
GCAATTCGTACAGGGACTGAACGTACAAAAGGGACTGGTTTACGCGGTCTTCCGCGAGCGGCGGGTCGCTATCGTCAAGAACGTTCCTTCGTTTCAAAAA
```

```
                                                                                                   5200
CAACGGTTTGAGGCCGTCCGCCGTAGGCATGCTTTTGAGCGTTTGACCAAGCAGTTCCAGGCGGTCCCACAGCTCGGTCACGTGCTCTACGGCATCTCGA
GTTGCCAAACTCCGGCAGGCGGCATCCGTACGAAAACTCGCAAACTGGTTCGTCAAGGTCCGCCAGGGTGTCGAGCCAGTGCACGAGATGCCGTAGAGCT
                                                                                         *
                                                    14.43% Terminator UAG for DNA polymerase
                                                                                                   5300
TCCAGCATATCTCCTCGTTTCGCGGGTTGGGGCGGCTTTCGCTGTACGGCAGTAGTCGGTGCTCGTCCAGACGGGCCAGGGTCATGTCTTTCCACGGGCG
AGGTCGTATAGAGGAGCAAAGCGCCCAACCCCGCCGAAAGCGACATGCCGTCATCAGCCACGAGCAGGTCTGCCCGGTCCCAGTACAGAAAGGTGCCCGC

                                                           14.91% Bal I
                                                           *                                       5400
CAGGGTCCTCGTCAGCGTAGTCTGGGTCACGGTGAAGGGGTGCGCTCCGGGCTGCGCGCTGGCCAGGGTGCGCTTGAGGCTGGTCCTGCTGGTGCTGAAG
GTCCCAGGAGCAGTCGCATCAGACCCAGTGCCACTTCCCCACGCGAGGCCCGACGCGCGACCGGTCCCACGCGAACTCCGACCAGGACGACCACGACTTC

                                                                                                   5500
CGCTGCCGGTCTTCGCCCTGCGCGTCGGCCAGGTAGCATTTGACCATGGTGTCATAGTCCAGCCCCTCCGCGGCGTGGCCCTTGGCGCGCAGCTTGCCCT
GCGACGGCCAGAAGCGGGACGCGCAGCCGGTCCATCGTAAACTGGTACCACAGTATCAGGTCGGGGAGGCGCCGCACCGGGAACCGCGCGTCGAACGGGA
            *
            15.07% ACCEPTOR splice site for IVa2 mRNA
                                                                                                   5600
TGGAGGAGGCGCCGCACGAGGGGCAGTGCAGACTTTTAAGGGCGTAGAGCTTGGGCGCGAGAAATACCGATTCCGGGGAGTAGGCATCCGCGCCGCAGGC
ACCTCCTCCGCGGCGTGCTCCCCGTCACGTCTGAAAATTCCCGCATCTCGAACCCGCGCTCTTTATGGCTAAGGCCCCTCATCCGTAGGCGCGGCGTCCG

                                15.67% Sac I
                                *                                                                  5700
CCCGCAGACGGTCTCGCATTCCACGAGCCAGGTGAGCTCTGGCCGTTCGGGGTCAAAAACCAGGTTTCCCCCATGCTTTTTGATGCGTTTCTTACCTCTG
GGGCGTCTGCCAGAGCGTAAGGTGCTCGGTCCACTCGAGACCGGCAAGCCCCAGTTTTTGGTCCAAAGGGGGTACGAAAAACTACGCAAAGAATGGAGAC
                                                                                               *
                                                         15.84% DONOR splice site for IVa2 mRNA
                                                                          16.07% Xho I
                                                                          *                        5800
GTTTCCATGAGCCGGTGTCCACGCTCGGTGACGAAAAGGCTGTCCGTGTCCCCGTATACAGACTTGAGAGGCCTGTCCTCGAGCGGTGTTCCGCGGTCCT
CAAAGGTACTCGGCCACAGGTGCGAGCCACTGCTTTTCCGACAGGCACAGGGGCATATGTCTGAACTCTCCGGACAGGAGCTCGCCACAAGGCGCCAGGA
      *
      15.88% Initiator AUG for IVa2 polypeptide
                                                                                                   5900
CCTCGTATAGAAACTCGGACCACTCTGAGACGAAGGCTCGCGTCCAGGCCAGCACGAAGGAGGCTAAGTGGGAGGGGTAGCGGTCGTTGTCCACTAGGGG
GGAGCATATCTTTGAGCCTGGTGAGACTCTGCTTCCGAGCGCAGGTCCGGTCGTGCTTCCTCCGATTCACCCTCCCCATCGCCAGCAACAGGTGATCCCC
                       *
                       16.21% Cap site for IVa2 mRNA
                                                                                                   6000
GTCCACTCGCTCCAGGGTGTGAAGACACATGTCGCCCTCTTCGGCATCAAGGAAGGTGATTGGTTTATAGGTGTAGGCCACGTGACCGGGTGTTCCTGAA
CAGGTGAGCGAGGTCCCACACTTCTGTGTACAGCGGGAGAAGCCGTAGTTCCTTCCACTAACCAAATATCCACATCCGGTGCACTGGCCCACAAGGACTT

                  16.80% Cap site for major late transcript         16.91% DONOR splice site for 1st late leader
                                   *                                        *                      6100
GGGGGGCTATAAAAGGGGGTGGGGGCGCGTTCGTCCTCACTCTCTTCCGCATCGCTGTCTGCGAGGGCCAGCTGTTGGGGTGAGTACTCCCTCTCAAAAG
CCCCCCGATATTTTCCCCCACCCCCGCGCAAGCAGGAGTGAGAGAAGGCGTAGCGACAGACGCTCCCGGTCGACAACCCCACTCATGAGGGAGAGTTTTC
```

```
                                                                                                  6200
CGGGCATGACTTCTGCGCTAAGATTGTCAGTTTCCAAAAACGAGGAGGATTTGATATTCACCTGGCCCGCGGTGATGCCTTTGAGGGTGGCCGCGTCCAT
GCCCGTACTGAAGACGCGATTCTAACAGTCAAAGGTTTTTGCTCCTCCTAAACTATAAGTGGACCGGGCGCCACTACGGAAACTCCCACCGGCGCAGGTA

                        17.33% Hind III
                        *                                                                         6300
CTGGTCAGAAAAGACAATCTTTTTGTTGTCAAGCTTGGTGGCAAACGACCCGTAGAGGGCGTTGGACAGCAACTTGGCGATGGAGCGCAGGTTTGGTTT
GACCAGTCTTTTCTGTTAGAAAAACAACAGTTCGAACCACCGTTTGCTGGGCATCTCCCGCAACCTGTCGTTGAACCGCTACCTCGCGTCCCAAACCAAA

                                                                                                  6400
TTGTCGCGATCGGCGCGCTCCTTGGCCGCGATGTTTAGCTGCACGTATTCGCGCGCAACGCACCGCCATTCGGGAAAGACGGTGGTGCGCTCGTCGGGCA
AACAGCGCTAGCCGCGCGAGGAACCGGCGCTACAAATCGACGTGCATAAGCGCGCGTTGCGTGGCGGTAAGCCCTTTCTGCCACCACGCGAGCAGCCCGT

                                                                                                  6500
CCAGGTGCACGCGCCAACCGCGGTTGTGCAGGGTGACAAGGTCAACGCTGGTGGCTACCTCTCCGCGTAGGCGCTCGTTGGTCCAGCAGAGGCGGCCGCC
GGTCCACGTGCGCGGTTGGCGCCAACACGTCCCACTGTTCCAGTTGCGACCACCGATGGAGAGGCGCATCCGCGAGCAACCAGGTCGTCTCCGCCGGCGG

                                                                          18.29% Xma I
                                                                          *
CTTGCGCGAACAGAATGGCGGTAGTGGGTCTAGCTGCGTCTCGTCCGGGGGGTCTGCGTCCACGGTAAAGACCCCGGGCAGCAGGCGCGCGTCGAAGTAG
GAACGCGCTTGTCTTACCGCCATCACCCAGATCGACGCAGAGCAGGCCCCCCAGACGCAGGTGCCATTTCTGGGGCCCGTCGTCCGCGCGCAGCTTCATC

                                                                                                  6700
TCTATCTTGCATCCTTGCAAGTCTAGCGCCTGCTGCCATGCGCGGGCGGCAAGCGCGCGCTCGTATGGGTTGAGTGGGGGACCCCATGGCATGGGGTGGG
AGATAGAACGTAGGAACGTTCAGATCGCGGACGACGGTACGCGCCCGCCGTTCGCGCGCGAGCATACCCAACTCACCCCCTGGGGTACCGTACCCCACCC

                                                                                                  6800
TGAGCGCGGAGGCGTACATGCCGCAAATGTCGTAAACGTAGAGGGGCTCTCTGAGTATTCCAAGATATGTAGGGTAGCATCTTCCACCGCGGATGCTGGC
ACTCGCGCCTCCGCATGTACGGCGTTTACAGCATTTGCATCTCCCCGAGAGACTCATAAGGTTCTATACATCCCATCGTAGAAGGTGGCGCCTACGACCG

                                                                                                  6900
GCGCACGTAATCGTATAGTTCGTGCGAGGGAGCGAGGAGGTCGGGACCGAGGTTGCTACGGGCGGGCTGCTCTGCTCGGAAGACTATCTGCCTGAAGATG
CGCGTGCATTAGCATATCAAGCACGCTCCCTCGCTCCTCCAGCCCTGGCTCCAACGATGCCCGCCCGACGAGACGAGCCTTCTGATAGACGGACTTCTAC

                                                                                                  7000
GCATGTGAGTTGGATGATATGGTTGGACGCTGGAAGACGTTGAAGCTGGCGTCTGTGAGACCTACCGCGTCACGCACGAAGGAGGCGTAGGAGTCGCGCA
CGTACACTCAACCTACTATACCAACCTGCGACCTTCTGCAACTTCGACCGCAGACACTCTGGATGGCGCAGTGCGTGCTTCCTCCGCATCCTCAGCGCGT

                                                                                                  7100
GCTTGTTGACCAGCTCGGCGGTGACCTGCACGTCTAGGGCGCAGTAGTCCAGGGTTTCCTTGATGATGTCATACTTATCCTGTCCCTTTTTTTTCCACAG
CGAACAACTGGTCGAGCCGCCACTGGACGTGCAGATCCCGCGTCATCAGGTCCCAAAGGAACTACTACAGTATGAATAGGACAGGGAAAAAAAAGGTGTC

19.75% ACCEPTOR splice site for 2nd late leader                            19.95% DONOR splice site for 2nd late leader
*                                                                         *                       7200
CTCGCGGTTGAGGACAAACTCTTCGCGGTCTTTCCAGTACTCTTGGATCGGAAACCCGTCGGCCTCCGAACGGTAAGAGCCTAGCATGTAGAACTGGTTG
GAGCGCCAACTCCTGTTTGAGAAGCGCCAGAAAGGTCATGAGAACCTAGCCTTTGGGCAGCCGGAGGCTTGCCATTCTCGGATCGTACATCTTGACCAAC
```

```
                                                                                                        7300
ACGGCCTGGTAGGCGCAGCATCCCTTTTCTACGGGTAGCGCGTATGCCTGCGCGGCCTTCCGGAGCGAGGTGTGGGTGAGCGCAAAGGTGTCCCTAACCA
TGCCGGACCATCCGCGTCGTAGGGAAAAGATGCCCATCGCGCATACGGACGCGCCGGAAGGCCTCGCTCCACACCCACTCGCGTTTCCACAGGGATTGGT

                                                                                                        7400
TGACTTTGAGGTACTGGTATTTGAAGTCAGTGTCGTCGCATCCGCCCTGCTCCCAGAGCAAAAAGTCCGTGCGCTTTTTGGAACGCGGGTTTGGCAGGGC
ACTGAAACTCCATGACCATAAACTTCAGTCACAGCAGCGTAGGCGGGACGAGGGTCTCGTTTTTCAGGCACGCGAAAAACCTTGCGCCCAAACCGTCCCG

                                                                                                        7500
GAAGGTGACATCGTTGAAAAGTATCTTTCCCGCGCGAGGCATAAAGTTGCGTGTGATGCGGAAGGGTCCCGGCACCTCGGAACGGTTGTTAATTACCTGG
CTTCCACTGTAGCAACTTTTCATAGAAAGGGCGCGCTCCGTATTTCAACGCACACTACGCCTTCCCAGGGCCGTGGAGCCTTGCCAACAATTAATGGACC

                                                                                                        7600
GCGGCGAGCACGATCTCGTCGAAGCCGTTGATGTTGTGGCCCACGATGTAAAGTTCCAAGAAGCGCGGGGTGCCCTTGATGGAGGGCAATTTTTTAAGTT
CGCCGCTCGTGCTAGAGCAGCTTCGGCAACTACAACACCGGGTGCTACATTTCAAGGTTCTTCGCGCCCCACGGGAACTACCTCCCGTTAAAAAATTCAA

          21.17% Sac I
          *                                                                                21.38% Sac I
                                                                                           *    7700
CCTCGTAGGTGAGCTCCTCAGGGGAGCTGAGCCCGTGTTCTGACAGGGCCCAGTCTGCAAGATGAGGGTTGGAAGCGACGAATGAGCTCCACAGGTCACG
GGAGCATCCACTCGAGGAGTCCCCTCGACTCGGGCACAAGACTGTCCCGGGTCAGACGTTCTACTCCCAACCTTCGCTGCTTACTCGAGGTGTCCAGTGC

                                                    21.56% Bal I
                                                    *                                           7800
GGCCATTAGCATTTGCAGGTGGTCGCGAAAGGTCCTAAACTGGCGACCTATGGCCATTTTTTCTGGGGTGATGCAGTAGAAGGTAAGCGGGTCTTGTTCC
CCGGTAATCGTAAACGTCCACCAGCGCTTTCCAGGATTTGACCGCTGGATACCGGTAAAAAAGACCCCACTACGTCATCTTCCATTCGCCCAGAACAAGG

                                                                                                        7900
CAGCGGTCCCATCCAAGGTCCACGGCTAGGTCTCGCGCGGCGGTCACCAGAGGCTCATCTCCGCCGAACTTCATAACCAGCATGAAGGGCACGAGCTGCT
GTCGCCAGGGTAGGTTCCAGGTGCCGATCCAGAGCGCGCCGCCAGTGGTCTCCGAGTAGAGGCGGCTTGAAGTATTGGTCGTACTTCCCGTGCTCGACGA

                    22.09% ACCEPTOR splice site for 'i' leader    22.17% Initiator AUG for 13.6K polypeptide
                                          *                                *                          8000
TCCCAAAGGCCCCCATCCAAGTATAGGTCTCTACATCGTAGGTGACAAAGAGACGCTCGGTGCGAGGATGCGAGCCGATCGGGAAGAACTGGATCTCCCG
AGGGTTTCCGGGGGTAGGTTCATATCCAGAGATGTAGCATCCACTGTTTCTCTGCGAGCCACGCTCCTACGCTCGGCTAGCCCTTCTTGACCTAGAGGGC

                                                                                                        8100
CCACCAGTTGGAGGAGTGGCTGTTGATGTGGTGAAAGTAGAAGTCCCTGCGACGGGCCGAACACTCGTGCTGGCTTTTGTAAAAACGTGCGCAGTACTGG
GGTGGTCAACCTCCTCACCGACAACTACACCACTTTCATCTTCAGGGACGCTGCCCGGCTTGTGAGCACGACCGAAAACATTTTTGCACGCGTCATGACC

                                                                                                        8200
CAGCGGTGCACGGGCTGTACATCCTGCACGAGGTTGACCTGACGACCGCGCACAAGGAAGCAGAGTGGGAATTTGAGCCCCTCGCCTGGCGGGTTTGGCT
GTCGCCACGTGCCCGACATGTAGGACGTGCTCCAACTGGACTGCTGGCGCGTGTTCCTTCGTCTCACCCTTAAACTCGGGGAGCGGACCGCCCAAACCGA

                                               22.94% Xho I
                                               *                                                8300
GGTGGTCTTCTACTTCGGCTGCTTGTCCTTGACCGTCTGGCTGCTCGAGGGGAGTTATGGTGGATCGGACCACCACGCCGCGCGAGCCCAAAGTCCAGAT
CCACCAGAAGATGAAGCCGACGAACAGGAACTGGCAGACCGACGAGCTCCCCTCAATACCACCTAGCCTGGTGGTGCGGCGCGCTCGGGTTTCAGGTCTA
```

23.32% DONOR splice site for 'i' leader
23.27% Sac I
23.34% Sac I
8400
GTCCGCGCGCGGCGGTCGGAGCTTGATGACAACATCGCGCAGATGGGAGCTGTCCATGGTCTGGAGCTCCCGCGGCGACAGGTCAGGCGGGAGCTCCTGC
CAGGCGCGCGCCGCCAGCCTCGAACTACTGTTGTAGCGCGTCTACCCTCGACAGGTACCAGACCTCGAGGGCGCCGCTGTCCAGTCCGCCCTCGAGGACG

23.41% Terminator UAG for 13.6K polypeptide
23.25% 1st AUG in DNA polymerase reading frame
8500
AGGTTTACCTCGCATAGCCGGGTCAGGGCGCGGGCTAGGTCCAGGTGATACCTGATTTCCAGGGGCTGGTTGGTGGCGGCGTCGATGACTTGCAAGAGGC
TCCAAATGGAGCGTATCGGCCCAGTCCCGCGCCCGATCCAGGTCCACTATGGACTAAAGGTCCCCGACCAACCACCGCCGCAGCTACTGAACGTTCTCCG

23.71% Kpn I
8600
CGCATCCCCGCGGCGCGACTACGGTACCGCGCGGCGGGCGGTGGGCCGCGGGGGTGTCCTTGGATGATGCATCTAAAAGCGGTGACGCGGGCGGCCCCC
GCGTAGGGGCGCCGCGCTGATGCCATGGCGCGCCGCCCGCCACCCGGCGCCCCCACAGGAACCTACTACGTAGATTTTCGCCACTGCGCCCGCCCGGGGG
23.86% Terminator UAG for Terminal protein

8700
GGAGGTAGGGGGGGCTCGGGACCCGCCGGGAGAGGGGGCAGGGGCACGTCGGCGCCGCGCGCGGGCAGGAGCTGGTGCTGCGCGCGGAGGTTGCTGGCGA
CCTCCATCCCCCCCGAGCCCTGGGCGGCCCTCTCCCCCGTCCCCGTGCAGCCGCGGCGCGCGCCCGTCCTCGACCACGACGCGCGCCTCCAACGACCGCT

8800
ACGCGACGACGCGGCGGTTGATCTCCTGAATCTGGCGCCTCTGCGTGAAGACGACGGGCCCGGTGAGCTTGAACCTGAAAGAGAGTTCGACAGAATCAAT
TGCGCTGCTGCGCCGCCAACTAGAGGACTTAGACCGCGGAGACGCACTTCTGCTGCCCGGGCCACTCGAACTTGGACTTTCTCTCAAGCTGTCTTAGTTA

8900
TTCGGTGTCGTTGACGGCGGCCTGGCGCAAAATCTCCTGCACGTCTCCTGAGTTGTCTTGATAGGCGATTTCGGCCATGAACTGCTCGATCTCTTCCTCC
AAGCCACAGCAACTGCCGCCGGACCGCGTTTTAGAGGACGTGCAGAGGACTCAACAGAACTATCCGCTAAAGCCGGTACTTGACGAGCTAGAGAAGGAGG

24.77% Bgl II
9000
TGGAGATCTCCGCGTCCGGCTCGCTCCACGGTGGCGGCGAGGTCGTTGGAGATGCGGGCCATGAGCTGCGAGAAGGCGTTGAGGCCTCCCTCGTTCCAGA
ACCTCTAGAGGCGCAGGCCGAGCGAGGTGCCACCGCCGCTCCAGCAACCTCTACGCCCGGTACTCGACGCTCTTCCGCAACTCCGGAGGGAGCAAGGTCT

25.21% Sac I
9100
CGCGGCTGTAGACCACGCCCCCTTCGGCATCGCGGGCGCGCATGACCACCTGCGCGAGATTGAGCTCCACGTGCCGGGCGAAGACGGCGTAGTTTCGCAG
GCGCCGACATCTGGTGCGGGGGAAGCCGTAGCGCCCGCGCGTACTGGTGGACGCGCTCTAACTCGAGGTGCACGGCCCGCTTCTGCCGCATCAAAGCGTC

9200
GCGCTGAAAGAGGTAGTTGAGGGTGGTGGCGGTGTGTTCTGCCACGAAGAAGTACATAACCCAGCGTCGCAACGTGGATTCGTTGATATCCCCCAAGGCC
CGCGACTTTCTCCATCAACTCCCACCACCGCCACACAAGACGGTGCTTCTTCATGTATTGGGTCGCAGCGTTGCACCTAAGCAACTATAGGGGGTTCCGG

25.79% Hpa I
25.86% Sac I
TCAAGGCGCTCCATGGCCTCGTAGAAGTCCACGGCGAAGTTGAAAAACTGGGAGTTGCGCGCCGACACGGTTAACTCCTCCTCCAGAAGACGGATGAGCT
AGTTCCGCGAGGTACCGGAGCATCTTCAGGTGCCGCTTCAACTTTTTGACCCTCAACGCGCGGCTGTGCCAATTGAGGAGGAGGTCTTCTGCCTACTCGA

```
                                                                                                    9400
CGGCGACAGTGTCGCGCACCTCGCGCTCAAAGGCTACAGGGGCCTCTTCTTCTTCAATCTCCTCTTCCATAAGGGCCTCCCCTTCTTCTTCTTCTTCTGG
GCCGCTGTCACAGCGCGTGGAGCGCGAGTTTCCGATGTCCCCGGAGAAGAAGAAGTTAGAGGAGAAGGTATTCCCGGAGGGGAAGAAGAAGAAGAAGACC

                                                   26.30% Sal I
                                                   *                                                9500
CGGCGGTGGGGGAGGGGGGACACGGCGGCGACGACGGCGCACCGGGAGGCGGTCGACAAAGCGCTCGATCATCTCCCCGCGGCGACGGCGCATGGTCTCG
GCCGCCACCCCCTCCCCCCTGTGCCGCCGCTGCTGCCGCGTGGCCCTCCGCCAGCTGTTTCGCGAGCTAGTAGAGGGGCGCCGCTGCCGCGTACCAGAGC

                                                                                                    9600
GTGACGGCGCGGCCGTTCTCGCGGGGGCGCAGTTGGAAGACGCCGCCCGTCATGTCCCGGTTATGGGTTGGCGGGGGGCTGCCGTGCGGCAGGGATACGG
CACTGCCGCGCCGGCAAGAGCGCCCCCGCGTCAACCTTCTGCGGCGGGCAGTACAGGGCCAATACCCAACCGCCCCCCGACGGCACGCCGTCCCTATGCC

                                    26.80% ACCEPTOR splice site for 3rd late leader       26.96% Xho I
                                    *                                                       *       9700
CGCTAACGATGCATCTCAACAATTGTTGTGTAGGTACTCCGCCACCGAGGGACCTGAGCGAGTCCGCATCGACCGGATCGGAAAACCTCTCGAGAAAGGC
GCGATTGCTACGTAGAGTTGTTAACAACACATCCATGAGGCGGTGGCTCCCTGGACTCGCTCAGGCGTAGCTGGCCTAGCCTTTTGGAGAGCTCTTTCCG

                         27.05% DONOR splice site for 3rd late leader
                         *
GTCTAACCAGTCACAGTCGCAAGGTAGGCTGAGCACCGTGGCGGGCGGCAGCGGGTGGCGGTCGGGGTTGTTTCTGGCGGAGGTGCTGCTGATGATGTAA
CAGATTGGTCAGTGTCAGCGTTCCATCCGACTCGTGGCACCGCCCGCCGTCGCCCACCGCCAGCCCCAACAAAGACCGCCTCCACGACGACTACTACATT

                              27.35% Sal I
                              *                                                                     9900
TTAAAGTAGGCGGTCTTGAGACGGCGGATGGTCGACAGAAGCACCATGTCCTTGGGTCCGGCCTGCTGAATGCGCAGGCGGTCGGCCATGCCCCAGGCTT
AATTTCATCCGCCAGAACTCTGCCGCCTACCAGCTGTCTTCGTGGTACAGGAACCCAGGCCGGACGACTTACGCGTCCGCCAGCCGGTACGGGGTCCGAA

                                                                                                   10000
CGTTTTGACATCGGCGCAGGTCTTTGTAGTAGTCTTGCATGAGCCTTTCTACCGGCACTTCTTCTTCTCCTTCCTCTTGTCCTGCATCTCTTGCATCTAT
GCAAAACTGTAGCCGCGTCCAGAAACATCATCAGAACGTACTCGGAAAGATGGCCGTGAAGAAGAAGAGGAAGGAGAACAGGACGTAGAGAACGTAGATA

                                                                                                   10100
CGCTACGGCGGCGGCGGAGTTTGGCCGTAGGTGGCGCCCTCTTCCTCCCATGCGTGTGACCCCGAAGCCCCTCATCGGCTGAAGCAGGGCCAGGTCGGCG
GCGATGCCGCCGCCGCCTCAAACCGGCATCCACCGCGGGAGAAGGAGGGTACGCACACTGGGGCTTCGGGGAGTAGCCGACTTCGTCCCGGTCCAGCCGC

                                                                                                   10200
ACAACGCGCTCGGCTAATATGGCCTGCTGCACCTGCGTGAGGGTAGACTGGAAGTCATCCATGTCCACAAAGCGGTGGTATGCGCCCGTGTTGATGGTGT
TGTTGCGCGAGCCGATTATACCGGACGACGTGGACGCACTCCCATCTGACCTTCAGTAGGTACAGGTGTTTCGCCACCATACGCGGGCACAACTACCACA

        28.41% Bal I     28.45% Hpa I                 28.53% Sac I
        *                *                            *
                                                                                                   10300
AAGTGCAGTTGGCCATAACGGACCAGTTAACGGTCTGGTGACCCGGCTGCGAGAGCTCGGTGTACCTGAGACGCGAGTAAGCCCTTGAGTCAAAGACGTA
TTCACGTCAACCGGTATTGCCTGGTCAATTGCCAGACCACTGGGCCGACGCTCTCGAGCCACATGGACTCTGCGCTCATTCGGGAACTCAGTTTCTGCAT
```

```
                                                                                                   10400
GTCGTTGCAAGTCCGCACCAGGTACTGATATCCCACCAAAAAGTGCGGCGGCGGCTGGCGGTAGAGGGGCCAGCGTAGGGTGGCCGGGGCTCCGGGGGCG
CAGCAACGTTCAGGCGTGGTCCATGACTATAGGGTGGTTTTTCACGCCGCCGCCGACCGCCATCTCCCCGGTCGCATCCCACCGGCCCCGAGGCCCCCGC

                                                                                                   10500
AGGTCTTCCAACATAAGGCGATGATATCCGTAGATGTACCTGGACATCCAGGTGATGCCGGCGGCGGTGGTGGAGGCGCGCGGAAAGTCGCGGACGCGGT
TCCAGAAGGTTGTATTCCGCTACTATAGGCATCTACATGGACCTGTAGGTCCACTACGGCCGCCGCCACCACCTCCGCGCGCCTTTCAGCGCCTGCGCCA

                                                                                  29.43% Xba I
                                                                                  *                10600
TCCAGATGTTGCGCAGCGGCAAAAAGTGCTCCATGGTCGGGACGCTCTGGCCGGTGAGGCGTGCGCAGTCGTTGACGCTCTAGACCGTGCAAAAGGAGAG
AGGTCTACAACGCGTCGCCGTTTTTCACGAGGTACCAGCCCTGCGAGACCGGCCACTCCGCACGCGTCAGCAACTGCGAGATCTGGCACGTTTTCCTCTC
                                *
                         29.31% 1st AUG in terminal protein reading frame
29.52% 5' end of VA I RNA                                                          29.71% BamH I
          *                                                                        *               10700
CCTGTAAGCGGGCACTCTTCCGTGGTCTGGTGGATAAATTCGCAAGGGTATCATGGCGGACGACCGGGGTTCGAACCCCGGATCCGGCCGTCCGCCGTGA
GGACATTCGCCCGTGAGAAGGCACCAGACCACCTATTTAAGCGTTCCCATAGTACCGCCTGCTGGCCCCAAGCTTGGGGCCTAGGCCGGCAGGCGGCACT

                                                                         29.95% 3' end of VA I RNA
                                                                         *                         10800
TCCATGCGGTTACCGCCCGCGTGTCGAACCCAGGTGTGCGACGTCAGACAACGGGGGAGCGCTCCTTTTGGCTTCCTTCCAGGCGCGGCGGCTGCTGCGC
AGGTACGCCAATGGCGGGCGCACAGCTTGGGTCCACACGCTGCAGTCTGTTGCCCCCTCGCGAGGAAAACCGAAGGAAGGTCCGCGCCGCCGACGACGCG


  30.08% Bal I                                                           30.23% 5' end of VA II RNA
         *                                                               *                         10900
TAGCTTTTTTGGCCACTGGCCGCGCGCGGCGTAAGCGGTTAGGCTGGAAAGCGAAAGCATTAAGTGGCTCGCTCCCTGTAGCCGGAGGGTTATTTTCCAA
ATCGAAAAAACCGGTGACCGGCGCGCGCCGCATTCGCCAATCCGACCTTTCGCTTTCGTAATTCACCGAGCGAGGGACATCGGCCTCCCAATAAAAGGTT

                                                                                                   11000
GGGTTGAGTCGCAGGACCCCCGGTTCGAGTCTCGGGCCGGCCGGACTGCGGCGAACGGGGGTTTGCCTCCCCGTCATGCAAGACCCCGCTTGCAAATTCC
CCCAACTCAGCGTCCTGGGGGCCAAGCTCAGAGCCCGGCCGGCCTGACGCCGCTTGCCCCCAAACGGAGGGGCAGTACGTTCTGGGGCGAACGTTTAAGG

                                             30.72% ACCEPTOR splice site for 52,55K mRNA
            30.67% 3' end of VA II RNA       30.72% Initiator AUG for 52,55K polypeptide
                       *                     *                                                     11100
TCCGGAAACAGGGACGAGCCCCTTTTTTGCTTTTCCCAGATGCATCCGGTGCTGCGGCAGATGCGCCCCCCTCCTCAGCAGCGGCAAGAGCAAGAGCAGC
AGGCCTTTGTCCCTGCTCGGGGAAAAAACGAAAAGGGTCTACGTAGGCCACGACGCCGTCTACGCGGGGGGAGGAGTCGTCGCCGTTCTCGTTCTCGTCG

                                                                                                   11200
GGCAGACATGCAGGGCACCCTCCCCTTCTCCTACCGCGTCAGGAGGGGCAACATCCGCGGCTGACGCGGCGGCAGATGGTGATTACGAACCCCCGCGGCG
CCGTCTGTACGTCCCGTGGGAGGGGAAGAGGATGGCGCAGTCCTCCCCGTTGTAGGCGCCGACTGCGCCGCCGTCTACCACTAATGCTTGGGGGCGCCGC

                                                                                                   11300
CCGGGCCCGGCACTACCTGGACTTGGAGGAGGGCGAGGGCCTGGCGCGGCTAGGAGCGCCCTCTCCTGAGCGACACCCAAGGGTGCAGCTGAAGCGTGAC
GGCCCGGGCCGTGATGGACCTGAACCTCCTCCCGCTCCCGGACCGCGCCGATCCTCGCGGGAGAGGACTCGCTGTGGGTTCCCACGTCGACTTCGCACTG
```

```
                                                                                               11400
ACGCGCGAGGCGTACGTGCCGCGGCAGAACCTGTTTCGCGACCGCGAGGGAGAGGAGCCCGAGGAGATGCGGGATCGAAAGTTCCACGCAGGGCGCGAGT
TGCGCGCTCCGCATGCACGGCGCCGTCTTGGACAAAGCGCTGGCGCTCCCTCTCCTCGGGCTCCTCTACGCCCTAGCTTTCAAGGTGCGTCCCGCGCTCA

                                                                                               11500
TGCGGCATGGCCTGAACCGCGAGCGGTTGCTGCGCGAGGAGGACTTTGAGCCCGACGCGCGGACCGGGATTAGTCCCGCGCGCGCACACGTGGCGGCCGC
ACGCCGTACCGGACTTGGCGCTCGCCAACGACGCGCTCCTCCTGAAACTCGGGCTGCGCGCCTGGCCCTAATCAGGGCGCGCGCGTGTGCACCGCCGGCG

                                                        32.15% Hind III
                                                        *                                      11600
CGACCTGGTAACCGCGTACGAGCAGACGGTGAACCAGGAGATTAACTTTCAAAAAAGCTTTAACAACCACGTGCGCACGCTTGTGGCGCGCGAGGAGGTG
GCTGGACCATTGGCGCATGCTCGTCTGCCACTTGGTCCTCTAATTGAAAGTTTTTTCGAAATTGTTGGTGCACGCGTGCGAACACCGCGCGCTCCTCCAC

                                                                                               11700
GCTATAGGACTGATGCATCTGTGGGACTTTGTAAGCGCGCTGGAGCAAAACCCAAATAGCAAGCCGCTCATGGCGCAGCTGTTCCTTATAGTGCAGCACA
CGATATCCTGACTACGTAGACACCCTGAAACATTCGCGCGACCTCGTTTTGGGTTTATCGTTCGGCGAGTACCGCGTCGACAAGGAATATCACGTCGTGT

                                                                                               11800
GCAGGGACAACGAGGCATTCAGGGATGCGCTGCTAAACATAGTAGAGCCCGAGGGCCGCTGGCTGCTCGATTTGATAAACATTCTGCAGAGCATAGTGGT
CGTCCCTGTTGCTCCGTAAGTCCCTACGCGACGATTTGTATCATCTCGGGCTCCCGGCGACCGACGAGCTAAACTATTTGTAAGACGTCTCGTATCACCA

                                                                                               11900
GCAGGAGCGCAGCTTGAGCCTGGCTGACAAGGTGGCCGCCATTAACTATTCCATGCTCAGTCTGGGCAAGTTTTACGCCCGCAAGATATACCATACCCCT
CGTCCTCGCGTCGAACTCGGACCGACTGTTCCACCGGCGGTAATTGATAAGGTACGAGTCAGACCCGTTCAAAATGCGGGCGTTCTATATGGTATGGGGA

                                                                                               12000
TACGTTCCCATAGACAAGGAGGTAAAGATCGAGGGGTTCTACATGCGCATGGCGTTGAAGGTGCTTACCTTGAGCGACGACCTGGGCGTTTATCGCAACG
ATGCAAGGGTATCTGTTCCTCCATTTCTAGCTCCCCAAGATGTACGCGTACCGCAACTTCCACGAATGGAACTCGCTGCTGGACCCGCAAATAGCGTTGC

                                   33.50% Sac I
                                   *                                                           12100
AGCGCATCCACAAGGCCGTGAGCGTGAGCCGGCGGCGCGAGCTCAGCGACCGCGAGCTGATGCACAGCCTGCAAAGGGCCCTGGCTGGCACGGGCAGCGG
TCGCGTAGGTGTTCCGGCACTCGCACTCGGCCGCCGCGCTCGAGTCGCTGGCGCTCGACTACGTGTCGGACGTTTCCCGGGACCGACCGTGCCCGTCGCC

                                                                                               12200
CGATAGAGAGGCCGAGTCCTACTTTGACGCGGGCGCTGACCTGCGCTGGGCCCCAAGCCGACGCGCCCTGGAGGCAGCTGGGGCCGGACCTGGGCTGGCG
GCTATCTCTCCGGCTCAGGATGAAACTGCGCCCGCGACTGGACGCGACCCGGGGTTCGGCTGCGCGGGACCTCCGTCGACCCCGGCCTGGACCCGACCGC

                                                     34.18% Terminator UAA for 52,55K polypeptide
                                                                                   *           12300
GTGGCACCCGCGCGCGCTGGCAACGTCGGCGGCGTGGAGGAATATGACGAGGACGATGAGTACGAGCCAGAGGACGGCGAGTACTAAGCGGTGATGTTTC
CACCGTGGGCGCGCGCGACCGTTGCAGCCGCCGCACCTCCTTATACTGCTCCTGCTACTCATGCTCGGTCTCCTGCCGCTCATGATTCGCCACTACAAAG
```

```
      34.24% 1st AUG in Polypeptide IIIa reading frame
34.22% Bcl I
*         *                                                                                          12400
TGATCAGATGATGCAAGACGCAACGGACCCGGCGGTGCGGGCGGCGCTGCAGAGCCAGCCGTCCGGCCTTAACTCCACGGACGACTGGCGCCAGGTCATG
ACTAGTCTACTACGTTCTGCGTTGCCTGGGCCGCCACGCCCGCCGCGACGTCTCGGTCGGCAGGCCGGAATTGAGGTGCCTGCTGACCGCGGTCCAGTAC

                                                                                                   12500
GACCGCATCATGTCGCTGACTGCGCGTAACCCTGACGCGTTCCGGCAGCAGCCGCAGGCCAACCGGCTCTCCGCAATTCTGGAAGCGGTGGTCCCGGCGC
CTGGCGTAGTACAGCGACTGACGCGCATTGGGACTGCGCAAGGCCGTCGTCGGCGTCCGGTTGGCCGAGAGGCGTTAAGACCTTCGCCACCAGGGCCGCG

                                                                                                   12600
GCGCAAACCCCACGCACGAGAAGGTGCTGGCGATCGTAAACGCGCTGGCCGAAAACAGGGCCATCCGGCCCGATGAGGCCGGCCTGGTCTACGACGCGCT
CGCGTTTGGGGTGCGTGCTCTTCCACGACCGCTAGCATTTGCGCGACCGGCTTTTGTCCCGGTAGGCCGGGCTACTCCGGCCGGACCAGATGCTGCGCGA

                                                                                                   12700
GCTTCAGCGCGTGGCTCGTTACAACAGCGGCAACGTGCAGACCAACCTGGACCGGCTGGTGGGGGATGTGCGCGAGGCCGTGGCGCAGCGTGAGCGCGCG
CGAAGTCGCGCACCGAGCAATGTTGTCGCCGTTGCACGTCTGGTTGGACCTGGCCGACCACCCCCTACACGCGCTCCGGCACCGCGTCGCACTCGCGCGC

                                                                                                   12800
CAGCAGCAGGGCAACCTGGGCTCCATGGTTGCACTAAACGCCTTCCTGAGTACACAGCCCGCCAACGTGCCGCGGGGACAGGAGGACTACACCAACTTTG
GTCGTCGTCCCGTTGGACCCGAGGTACCAACGTGATTTGCGGAAGGACTCATGTGTCGGGCGGTTGCACGGCGCCCCTGTCCTCCTGATGTGGTTGAAAC

                                                                                                   12900
TGAGCGCACTGCGGCTAATGGTGACTGAGACACCGCAAAGTGAGGTGTACCAGTCCGGGCCAGACTATTTTTTCCAGACCAGTAGACAAGGCCTGCAGAC
ACTCGCGTGACGCCGATTACCACTGACTCTGTGGCGTTTCACTCCACATGGTCAGGCCCGGTCTGATAAAAAAGGTCTGGTCATCTGTTCCGGACGTCTG

                                                                                                   13000
CGTAAACCTGAGCCAGGCTTTCAAGAACTTGCAGGGGCTGTGGGGGGTGCGGGCTCCCACAGGCGACCGCGCGACCGTGTCTAGCTTGCTGACGCCCAAC
GCATTTGGACTCGGTCCGAAAGTTCTTGAACGTCCCCGACACCCCCCACGCCCGAGGGTGTCCGCTGGCGCGCTGGCACAGATCGAACGACTGCGGGTTG

                                                     36.32% Xma I
                                                     *                                             13100
TCGCGCCTGTTGCTGCTGCTAATAGCGCCCTTCACGGACAGTGGCAGCGTGTCCCGGGACACATACCTAGGTCACTTGCTGACACTGTACCGCGAGGCCA
AGCGCGGACAACGACGACGATTATCGCGGGAAGTGCCTGTCACCGTCGCACAGGGCCCTGTGTATGGATCCAGTGAACGACTGTGACATGGCGCTCCGGT

                                                                                                   13200
TAGGTCAGGCGCATGTGGACGAGCATACTTTCCAGGAGATTACAAGTGTCAGCCGCGCGCTGGGGCAGGAGGACACGGGCAGCCTGGAGGCAACCCTGAA
ATCCAGTCCGCGTACACCTGCTCGTATGAAAGGTCCTCTAATGTTCACAGTCGGCGCGCGACCCCGTCCTCCTGTGCCCGTCGGACCTCCGTTGGGACTT

                                                                                                   13300
CTACCTGCTGACCAACCGGCGGCAGAAGATCCCCTCGTTGCACAGTTTAAACAGCGAGGAGGAGCGCATCTTGCGCTATGTGCAGCAGAGCGTGAGCCTT
GATGGACGACTGGTTGGCCGCCGTCTTCTAGGGGAGCAACGTGTCAAATTTGTCGCTCCTCCTCGCGTAGAACGCGATACACGTCGTCTCGCACTCGGAA

                                                                                                   13400
AACCTGATGCGCGACGGGGTAACGCCCAGCGTGGCGCTGGACATGACCGCGCGCAACATGGAACCGGGCATGTATGCCTCAAACCGGCCGTTTATCAATC
TTGGACTACGCGCTGCCCCATTGCGGGTCGCACCGCGACCTGTACTGGCGCGCGTTGTACCTTGGCCCGTACATACGGAGTTTGGCCGGCAAATAGTTAG
```

```
                                                                                               13500
GCCTAATGGACTACTTGCATCGCGCGGCCGCCGTGAACCCCGAGTATTTCACCAATGCCATCTTGAACCCGCACTGGCTACCGCCCCCTGGTTTCTACAC
CGGATTACCTGATGAACGTAGCGCGCCGGCGGCACTTGGGGCTCATAAAGTGGTTACGGTAGAACTTGGGCGTGACCGATGGCGGGGGACCAAAGATGTG

                                                                                               13600
CGGGGGATTTGAGGTGCCCGAGGGTAACGATGGATTCCTCTGGGACGACATAGACGACAGCGTGTTTTCCCCGCAACCGCAGACCCTGCTAGAGTTGCAA
GCCCCCTAAACTCCACGGGCTCCCATTGCTACCTAAGGAGACCCTGCTGTATCTGCTGTCGCACAAAAGGGGCGTTGGCGTCTGGGACGATCTCAACGTT

                                   37.94% Hind III
                                   *                                                           13700
CAGCGCGAGCAGGCAGAGGCGGCGCTGCGAAAGGAAAGCTTCCGCAGGCCAAGCAGCTTGTCCGATCTAGGCGCTGCGGCCCCGCGGTCAGATGCGAGTA
GTCGCGCTCGTCCGTCTCCGCCGCGACGCTTTCCTTTCGAAGGCGTCCGGTTCGTCGAACAGGCTAGATCCGCGACGCCGGGGCGCCAGTCTACGCTCAT

          38.15% Hind III
          *                                                                                    13800
GCCCATTTCCAAGCTTGATAGGGTCTTTTACCAGCACTCGCACCACCCGCCCGCGCCTGCTGGGCGAGGAGGAGTACCTAAACAACTCGCTGCTGCAGCC
CGGGTAAAGGTTCGAACTATCCCAGAAAATGGTCGTGAGCGTGGTGGGCGGGCGCGGACGACCCGCTCCTCCTCATGGATTTGTTGAGCGACGACGTCGG

                                                                                               13900
GCAGCGCGAAAAGAACCTGCCTCCGGCATTTCCCAACAACGGGATAGAGAGCCTAGTGGACAAGATGAGTAGATGGAAGACGTATGCGCAGGAGCACAGG
CGTCGCGCTTTTCTTGGACGGAGGCCGTAAAGGGTTGTTGCCCTATCTCTCGGATCACCTGTTCTACTCATCTACCTTCTGCATACGCGTCCTCGTGTCC

                                                                                               14000
GATGTGCCCGGCCCGCGCCCGCCCACCCGTCGTCAAAGGCACGACCGTCAGCGGGGTCTGGTGTGGGAGGACGATGACTCGGCAGACGACAGCAGCGTCC
CTACACGGGCCGGGCGCGGGCGGGTGGGCAGCAGTTTCCGTGCTGGCAGTCGCCCCAGACCACACCCTCCTGCTACTGAGCCGTCTGCTGTCGTCGCAGG

                                              39.13% Terminator UAA for IIIa          39.21% AAUAAA for L1 mRNAs
                                                            *                           *     14100
TGGATTTGGGAGGGAGTGGCAACCCGTTTGCGCACCTTCGCCCCAGGCTGGGGAGAATGTTTTAAAAAAAAAAAAAAAAAAGCATGATGCAAAATAAAAAA
ACCTAAACCCTCCCTCACCGTTGGGCAAACGCGTGGAAGCGGGGTCCGACCCCTCTTACAAAATTTTTTTTTTTTTTTTTTCGTACTACGTTTTATTTTTT

                                                 39.37% Initiator AUG for penton
 39.27% Poly-A addition site for L1 mRNAs       39.37% ACCEPTOR splice site for penton mRNA
            *                                   **                                             14200
CTCACCAAGGCCATGGCACCGAGCGTTGGTTTTCTTGTATTCCCCTTAGTATGCAGCGCGCGGCGATGTATGAGGAAGGTCCTCCTCCCTCCTACGAGAG
GAGTGGTTCCGGTACCGTGGCTCGCAACCAAAAGAACATAAGGGGAATCATACGTCGCGCGCCGCTACATACTCCTTCCAGGAGGAGGGAGGATGCTCTC

                                                                                  39.73% Kpn I
                                                                                  *            14300
CGTGGTGAGCGCGGCGCCAGTGGCGGCGGCGCTGGGTTCCCCCTTCGATGCTCCCCTGGACCCGCCGTTTGTGCCTCCGCGGTACCTGCGGCCTACCGGG
GCACCACTCGCGCCGCGGTCACCGCCGCCGCGACCCAAGGGGGAAGCTACGAGGGGACCTGGGCGGCAAACACGGAGGCGCCATGGACGCCGGATGGCCC

                                                                                               14400
GGGAGAAACAGCATCCGTTACTCTGAGTTGGCACCCCTATTCGACACCACCCGTGTGTACCTTGTGGACAACAAGTCAACGGATGTGGCATCCCTGAACT
CCCTCTTTGTCGTAGGCAATGAGACTCAACCGTGGGGATAAGCTGTGGTGGGCACACATGGAACACCTGTTGTTCAGTTGCCTACACCGTAGGGACTTGA
```

```
                                                  40.22% Xma I
                                                  *                                                14500
ACCAGAACGACCACAGCAACTTTCTAACCACGGTCATTCAAAACAATGACTACAGCCCGGGGGAGGCAAGCACACAGACCATCAATCTTGACGACCGTTC
TGGTCTTGCTGGTGTCGTTGAAAGATTGGTGCCAGTAAGTTTTGTTACTGATGTCGGGCCCCCTCCGTTCGTGTGTCTGGTAGTTAGAACTGCTGGCAAG

                                                                                                   14600
GCACTGGGGCGGCGACCTGAAAACCATCCTGCATACCAACATGCCAAATGTGAACGAGTTCATGTTTACCAATAAGTTTAAGGCGCGGGTGATGGTGTCG
CGTGACCCCGCCGCTGGACTTTTGGTAGGACGTATGGTTGTACGGTTTACACTTGCTCAAGTACAAATGGTTATTCAAATTCCGCGCCCACTACCACAGC

                                                                                                   14700
CGCTCGCTTACTAAGGACAAACAGGTGGAGCTGAAATATGAGTGGGTGGAGTTCACGCTGCCCGAGGGCAACTACTCCGAGACCATGACCATAGACCTTA
GCGAGCGAATGATTCCTGTTTGTCCACCTCGACTTTATACTCACCCACCTCAAGTGCGACGGGCTCCCGTTGATGAGGCTCTGGTACTGGTATCTGGAAT

                                                                                                   14800
TGAACAACGCGATCGTGGAGCACTACTTGAAAGTGGGCAGGCAGAACGGGGTTCTGGAAAGCGACATCGGGGTAAAGTTTGACACCCGCAACTTCAGACT
ACTTGTTGCGCTAGCACCTCGTGATGAACTTTCACCCGTCCGTCTTGCCCCAAGACCTTTCGCTGTAGCCCCATTTCAAACTGTGGGCGTTGAAGTCTGA

                                                                                                   14900
GGGGTTTGACCCAGTCACTGGTCTTGTCATGCCTGGGGTATATACAAACGAAGCCTTCCATCCAGACATCATTTTGCTGCCAGGATGCGGGGTGGACTTC
CCCCAAACTGGGTCAGTGACCAGAACAGTACGGACCCCATATATGTTTGCTTCGGAAGGTAGGTCTGTAGTAAAACGACGGTCCTACGCCCCACCTGAAG

                                                                                                   15000
ACCCACAGCCGCCTGAGCAACTTGTTGGGCATCCGCAAGCGGCAACCCTTCCAGGAGGGCTTTAGGATCACCTACGATGACCTGGAGGGTGGTAACATTC
TGGGTGTCGGCGGACTCGTTGAACAACCCGTAGGCGTTCGCCGTTGGGAAGGTCCTCCCGAAATCCTAGTGGATGCTACTGGACCTCCCACCATTGTAAG

                            41.82% Hind III
                            *                                                                      15100
CCGCACTGTTGGATGTGGACGCCTACCAGGCAAGCTTAAAAGATGACACCGAACAGGGCGGGGATGGCGCAGGCGGCGGCAACAACAGTGGCAGCGGCGC
GGCGTGACAACCTACACCTGCGGATGGTCCGTTCGAATTTTCTACTGTGGCTTGTCCCGCCCCTACCGCGTCCGCCGCCGTTGTTGTCACCGTCGCCGCG

                                                                                                   15200
GGAAGAGAACTCCAACGCGGCAGCCGCGGCAATGCAGCCGGTGGAGGACATGAACGATCATGCCATTCGCGGCGACACCTTTGCCACACGGGCGGAGGAG
CCTTCTCTTGAGGTTGCGCCGTCGGCGCCGTTACGTCGGCCACCTCCTGTACTTGCTAGTACGGTAAGCGCCGCTGTGGAAACGGTGTGCCCGCCTCCTC

                                                                               42.52% Bcl I
                                                                               *                   15300
AAGCGCGCTGAGGCCGAGGCAGCGGCAGAAGCTGCCGCCCCCGCTGCGCAACCCGAGGTCGAGAAGCCTCAGAAGAAACCGGTGATCAAACCCCTGACAG
TTCGCGCGACTCCGGCTCCGTCGCCGTCTTCGACGGCGGGGGCGACGCGTTGGGCTCCAGCTCTTCGGAGTCTTCTTTGGCCACTAGTTTGGGGACTGTC

                                                                   42.76% Kpn I
                                                                   *                               15400
AGGACAGCAAGAAACGCAGTTACAACCTAATAAGCAATGACAGCACCTTCACCCAGTACCGCAGCTGGTACCTTGCATACAACTACGGCGACCCTCAGAC
TCCTGTCGTTCTTTGCGTCAATGTTGGATTATTCGTTACTGTCGTGGAAGTGGGTCATGGCGTCGACCATGGAACGTATGTTGATGCCGCTGGGAGTCTG

  42.86% BamH I
  *                                                                                                15500
CGGGATCCGCTCATGGACCCTCCTTTGCACTCCTGACGTAACCTGCGGCTCGGAGCAGGTCTACTGGTCGTTGCCAGACATGATGCAAGACCCCGTGACC
GCCCTAGGCGAGTACCTGGGAGGAAACGTGAGGACTGCATTGGACGCCGAGCCTCGTCCAGATGACCAGCAACGGTCTGTACTACGTTCTGGGGCACTGG
```

```
                                                                                               15600
TTCCGCTCCACGAGCCAGATCAGCAACTTTCCGGTGGTGGGCGCCGAGCTGTTGCCCGTGCACTCCAAGAGCTTCTACAACGACCAGGCCGTCTACTCCC
AAGGCGAGGTGCTCGGTCTAGTCGTTGAAAGGCCACCACCCGCGGCTCGACAACGGGCACGTGAGGTTCTCGAAGATGTTGCTGGTCCGGCAGATGAGGG

                                                                                               15700
AGCTCATCCGCCAGTTTACCTCTCTGACCCACGTGTTCAATCGCTTTCCCGAGAACCAGATTTTGGCGCGCCCGCCAGCCCCCACCATCACCACCGTCAG
TCGAGTAGGCGGTCAAATGGAGAGACTGGGTGCACAAGTTAGCGAAAGGGCTCTTGGTCTAAAACCGCGCGGGCGGTCGGGGGTGGTAGTGGTGGCAGTC

                                                                                               15800
TGAAAACGTTCCTGCTCTCACAGATCACGGGACGCTACCGCTGCGCAACAGCATCGGAGGAGTCCAGCGAGTGACCATTACTGACGCCAGACGCCGCACC
ACTTTTGCAAGGACGAGAGTGTCTAGTGCCCTGCGATGGCGACGCGTTGTCGTAGCCTCCTCAGGTCGCTCACTGGTAATGACTGCGGTCTGCGGCGTGG

                                44.14% Terminator UGA for Penton        44.16% Initiator AUG for Pro-VII
                                                                *       *                      15900
TGCCCCTACGTTTACAAGGCCCTGGGCATAGTCTCGCCGCGCGTCCTATCGAGCCGCACTTTTTGAGCAAACATGTCCATCCTTATATCGCCCAGCAATA
ACGGGGATGCAAATGTTCCGGGACCCGTATCAGAGCGGCGCGCAGGATAGCTCGGCGTGAAAAACTCGTTTGTACAGGTAGGAATATAGCGGGTCGTTAT

                                            44.36% Gly/Ala cleavage site for Pro-VII to VII
                                            *                                                  16000
ACACAGGCTGGGGCCTGCGCTTCCCAAGCAAGATGTTTGGCGGGGCAAAGAAGCGCTCCGACCAACACCCAGTGCGCGTGCGCGGGCACTACCGCGCGCC
TGTGTCCGACCCCGGACGCGAAGGGTTCGTTCTACAAACCGCCCCGTTTCTTCGCGAGGCTGGTTGTGGGTCACGCGCACGCGCCCGTGATGGCGCGCGG

                                                                                               16100
CTGGGGCGCGCACAAACGCGGCCGCACTGGGCGCACCACCGTCGATGACGCCATTGACGCGGTGGTGGAGGAGGCGCGCAACTACACGCCCACGCCGCCA
GACCCCGCGCGTGTTTGCGCCGGCGTGACCCGCGTGGTGGCAGCTACTGCGGTAACTGCGCCACCACCTCCTCCGCGCGTTGATGTGCGGGTGCGGCGGT

                                                                                               16200
CCAGTGTCCACAGTGGACGCGGCCATTCAGACCGTGGTGCGCGGAGCCCGGCGTTATGCTAAAATGAAGAGACGGCGGAGGCGCGTAGCACGTCGCCACC
GGTCACAGGTGTCACCTGCGCCGGTAAGTCTGGCACCACGCGCCTCGGGCCGCAATACGATTTTACTTCTCTGCCGCCTCCGCGCATCGTGCAGCGGTGG

                                                                                               16300
GCCGCCGACCCGGCACTGCCGCCCAACGCGCGGCGGCGGCCCTGCTTAACCGCGCACGTCGCACCGGCCGACGGGCGGCCATGCGGGCCGCTCGAAGGCT
CGGCGGCTGGGCCGTGACGGCGGGTTGCGCGCCGCCGCCGGGACGAATTGGCGCGTGCAGCGTGGCCGGCTGCCCGCCGGTACGCCCGGCGAGCTTCCGA

                                                                                               16400
GGCCGCGGGTATTGTCACTGTGCCCCCCAGGTCCAGGCGACGAGCGGCCGCCGCAGCAGCCGCGGCCATTAGTGCTATGACTCAGGGTCGCAGGGGCAAC
CCGGCGCCCATAACAGTGACACGGGGGGTCCAGGTCCGCTGCTCGCCGGCGGCGTCGTCGGCGCCGGTAATCACGATACTGAGTCCCAGCGTCCCCGTTG

                                                                  45.82% Terminator UAG for Pro-VII
                                                                  *                            16500
GTGTACTGGGTGCGCGACTCGGTTAGCGGCCTGCGCGTGCCCGTGCGCACCCGCCCCCCGCGCAACTAGATTGCAAGAAAAAACTACTTAGACTCGTACT
CACATGACCCACGCGCTGAGCCAATCGCCGGACGCGCACGGGCACGCGTGGGCGGGGGGCGCGTTGATCTAACGTTCTTTTTTGATGAATCTGAGCATGA

                                      46.02% Initiator AUG for polypeptide V
               45.95% ACCEPTOR splice site for protein V mRNA                                46.15% Bgl II
               *                      *                                                *       16600
GTTGTATGTATCCAGCGGCGGCGGCGCGCAACGAAGCTATGTCCAAGCGCAAAATCAAAGAAGAGATGCTCCAGGTCATCGCGCCGGAGATCTATGGCCC
CAACATACATAGGTCGCCGCCGCCGCGCGTTGCTTCGATACAGGTTCGCGTTTTAGTTTCTTCTCTACGAGGTCCAGTAGCGCGGCCTCTAGATACCGGG
```

```
                                                                                                   16700
CCCGAAGAAGGAAGAGCAGGATTACAAGCCCCGAAAGCTAAAGCGGGTCAAAAAGAAAAAGAAAGATGATGATGATGATGAACTTGACGACGAGGTGGAA
GGGCTTCTTCCTTCTCGTCCTAATGTTCGGGGCTTTCGATTTCGCCCAGTTTTTCTTTTTCTTTCTACTACTACTACTACTTGAACTGCTGCTCCACCTT

                                            46.59% Sal I
                                            *                                                      16800
CTGCTGCACGCAACCGCGCCCAGGCGGCGGGTACAGTGGAAAGGTCGACGCGTAAGACGTGTTTTGCGACCCGGCACCACCGTAGTTTTTACGCCCGGTG
GACGACGTGCGTTGGCGCGGGTCCGCCGCCCATGTCACCTTTCCAGCTGCGCATTCTGCACAAAACGCTGGGCCGTGGTGGCATCAAAAATGCGGGCCAC

                                                                                                   16900
AGCGCTCCACCCGCACCTACAAGCGCGTGTATGATGAGGTGTACGGCGACGAGGACCTGCTTGAGCAGGCCAACGAGCGCCTCGGGGAGTTTGCCTACGG
TCGCGAGGTGGGCGTGGATGTTCGCGCACATACTACTCCACATGCCGCTGCTCCTGGACGAACTCGTCCGGTTGCTCGCGGAGCCCCTCAAACGGATGCC

                                                                                                   17000
AAAGCGGCATAAGGACATGTTGGCGTTGCCGCTGGACGAGGGCAACCCAACACCTAGCCTAAAGCCCGTGACACTGCAGCAGGTGCTGCCCACGCTTGCA
TTTCGCCGTATTCCTGTACAACCGCAACGGCGACCTGCTCCCGTTGGGTTGTGGATCGGATTTCGGGCACTGTGACGTCGTCCACGACGGGTGCGAACGT

                                                                   47.48% Kpn I
                                                                   *                               17100
CCGTCCGAAGAAAAGCGCGGCCTAAAGCGCGAGTCTGGTGACTTGGCACCCACCGTGCAGCTGATGGTACCCAAGCGCCAGCGACTGGAAGATGTCTTGG
GGCAGGCTTCTTTTCGCGCCGGATTTCGCGCTCAGACCACTGAACCGTGGGTGGCACGTCGACTACCATGGGTTCGCGGTCGCTGACCTTCTACAGAACC

                                                                                                   17200
AAAAAATGACCGTGGAGCCTGGGCTGGAGCCCGAGGTCCGCGTGCGGCCAATCAAGCAGGTGGCACCGGGACTGGGCGTGCAGACCGTGGACGTTCAGAT
TTTTTTACTGGCACCTCGGACCCGACCTCGGGCTCCAGGCGCACGCCGGTTAGTTCGTCCACCGTGGCCCTGACCCGCACGTCTGGCACCTGCAAGTCTA

                                                                                                   17300
ACCCACCACCAGTAGCACTAGTATTGCCACTGCCACAGAGGGCATGGAGACACAAACGTCCCCGGTTGCCTCGGCGGTGGCAGATGCCGCGGTGCAGGCG
TGGGTGGTGGTCATCGTGATCATAACGGTGACGGTGTCTCCCGTACCTCTGTGTTTGCAGGGGCCAACGGAGCCGCCACCGTCTACGGCGCCACGTCCGC

                                                                                                   17400
GCCGCTGCGGCCGCGTCCAAAACCTCTACGGAGGTGCAAACGGACCCGTGGATGTTTCGCGTTTCAGCCCCCCGGCGCCCGCGCCGTTCCAGGAAGTACG
CGGCGACGCCGGCGCAGGTTTTGGAGATGCCTCCACGTTTGCCTGGGCACCTACAAAGCGCAAAGTCGGGGGGCCGCGGGCGCGGCAAGGTCCTTCATGC

                                                                                                   17500
GCACCGCCAGCGCACTACTGCCCGAATATGCCCTACATCCTTCCATCGCGCCTACCCCCGGCTATCGTGGCTACACCTACCGCCCCAGAAGACGAGCGAC
CGTGGCGGTCGCGTGATGACGGGCTTATACGGGATGTAGGAAGGTAGCGCGGATGGGGGCCGATAGCACCGATGTGGATGGCGGGGTCTTCTGCTCGCTG

                                                                                                   17600
TACCCGACGCCGAACCACCACTGGAACCCGCCGCCGCCGTCGCCGTCGCCAGCCCGTGCTGGCCCCGATTTCCGTGCGCAGGGTGGCTCGCGAAGGAGGC
ATGGGCTGCGGCTTGGTGGTGACCTTGGGCGGCGGCGGCAGCGGCAGCGGTCGGGCACGACCGGGGCTAAAGGCACGCGTCCCACCGAGCGCTTCCTCCG

                                             49.10% Terminator UAA for polypeptide V
                                             *                                                     17700
AGGACCCTGGTGCTGCCAACAGCGCGCTACCACCCCAGCATCGTTTAAAAGCCGGTCTTTGTGGTTCTTGCAGATATGGCCCTCACCTGCCGCCTCCGTT
TCCTGGGACCACGACGGTTGTCGCGCGATGGTGGGGTCGTAGCAAATTTTCGGCCAGAAACACCAAGAACGTCTATACCGGGAGTGGACGGCGGAGGCAA
```

```
                                                                                               17800
TCCCGGTGCCGGGATTCCGAGGAAGAATGCACCGTAGGAGGGGCATGGCCGGCCACGGCCTGACGGGCGGCATGCGTCGTGCGCACCACCGGCGGCGGCG
AGGGCCACGGCCCTAAGGCTCCTTCTTACGTGGCATCCTCCCCGTACCGGCCGGTGCCGGACTGCCCGCCGTACGCAGCACGCGTGGTGGCCGCCGCCGC

                                                                                               17900
CGCGTCGCACCGTCGCATGCGCGGCGGTATCCTGCCCCTCCTTATTCCACTGATCGCCGCGGCGATTGGCGCCGTGCCCGGAATTGCATCCGTGGCCTTG
GCGCAGCGTGGCAGCGTACGCGCCGCCATAGGACGGGGAGGAATAAGGTGACTAGCGGCGCCGCTAACCGCGGCACGGGCCTTAACGTAGGCACCGGAAC

                                                                                   50.08% ACCEPTOR splice site for pVI
                      49.94% AAUAAA for L2 mRNAs      50.00% Poly-A addition site for L2 mRNAs
                                                 *                    *                           *
CAGGCGCAGAGACACTGATTAAAAACAAGTTGCATGTGGAAAAATCAAAATAAAAAGTCTGGAGTCTCACGCTCGCTTGGTCCTGTAACTATTTTGTAGA
GTCCGCGTCTCTGTGACTAATTTTTGTTCAACGTACACCTTTTTAGTTTTATTTTTCAGACCTCAGAGTGCGAGCGAACCAGGACATTGATAAAACATCT

50.09% Initiator AUG for polypeptide VI
*                                                                                              18100
ATGGAAGACATCAACTTTGCGTCTCTGGCCCCGCGACACGGCTCGCGCCCGTTCATGGGAAACTGGCAAGATATCGGCACCAGCAATATGAGCGGTGGCG
TACCTTCTGTAGTTGAAACGCAGAGACCGGGGCGCTGTGCCGAGCGCGGGCAAGTACCCTTTGACCGTTCTATAGCCGTGGTCGTTATACTCGCCACCGC

                                                                                               18200
CCTTCAGCTGGGGCTCGCTGTGGAGCGGCATTAAAAATTTCGGTTCCACCATTAAGAACTATGGCAGCAAGGCCTGGAACAGCAGCACAGGCCAGATGCT
GGAAGTCGACCCCGAGCGACACCTCGCCGTAATTTTTAAAGCCAAGGTGGTAATTCTTGATACCGTCGTTCCGGACCTTGTCGTCGTGTCCGGTCTACGA

                                                                                  50.86% Bal I
                                                                                  *            18300
GAGGGACAAGTTGAAAGAGCAAAATTTCCAACAAAAGGTGGTAGATGGCCTGGCCTCTGGCATTAGCGGGGTGGTGGACCTGGCCAACCAGGCAGTGCAA
CTCCCTGTTCAACTTTCTCGTTTTAAAGGTTGTTTTCCACCATCTACCGGACCGGAGACCGTAATCGCCCCACCACCTGGACCGGTTGGTCCGTCACGTT

               50.96% Hind III
               *                                                                               18400
AATAAGATTAACAGTAAGCTTGATCCCCGCCCTCCCGTAGAGGAGCCTCCACCGGCCGTGGAGACAGTGTCTCCAGAGGGGCGTGGCGAAAAGCGTCCGC
TTATTCTAATTGTCATTCGAACTAGGGGCGGGAGGGCATCTCCTCGGAGGTGGCCGGCACCTCTGTCACAGAGGTCTCCCCGCACCGCTTTTCGCAGGCG

                                                                                               18500
GGCCCGACAGGGAAGAAACTCTGGTGACGCAAATAGATGAGCCTCCCTCGTACGAGGAGGCACTAAAGCAAGGCCTGCCCACCACCCGTCCCATCGCGCC
CCGGGCTGTCCCTTCTTTGAGACCACTGCGTTTATCTACTCGGAGGGAGCATGCTCCTCCGTGATTTCGTTCCGGACGGGTGGTGGGCAGGGTAGCGCGG

                                                                                               18600
CATGGCTACCGGAGTGCTGGGCCAGCACACACCTGTAACGCTGGACCTGCCTCCCCCCGCTGACACCCAGCAGAAACCTGTGCTGCCAGGGCCGTCCGCC
GTACCGATGGCCTCACGACCCGGTCGTGTGTGGACATTGCGACCTGGACGGAGGGGGGCGACTGTGGGTCGTCTTTGGACACGACGGTCCCGGCAGGCGG

                                                                                               18700
GTTGTTGTAACCCGCCCTAGCCGCGCGTCCCTGCGCCGTGCCGCCAGCGGTCCGCGATCGATGCGGCCCGTAGCCAGTGGCAACTGGCAAAGCACACTGA
CAACAACATTGGGCGGGATCGGCGCGCAGGGACGCGGCACGGCGGTCGCCAGGCGCTAGCTACGCCGGGCATCGGTCACCGTTGACCGTTTCGTGTGACT

                                                  52.17% Terminator UAA for polypeptide VI
                                                  *                                            18800
ACAGCATCGTGGGTCTGGGGGTGCAATCCCTGAAGCGCCGACGATGCTTCTAAATAGCTAACGTGTCGTATGTGTCATGTATGCGTCCATGTCGCCGCCA
TGTCGTAGCACCCAGACCCCCACGTTAGGGACTTCGCGGCTGCTACGAAGATTTATCGATTGCACAGCATACACAGTACATACGCAGGTACAGCGGCGGT
```

```
                                                52.41% Initiator AUG for Hexon
52.31% ACCEPTOR splice site for Hexon mRNA
*                                                *                                                   18900
GAGGAGCTGCTGAGCCGCCGTGCGCCCGCTTTCCAAGATGGCTACCCCTTCGATGATGCCGCAGTGGTCTTACATGCACATCTCGGGCCAGGACGCCTCG
CTCCTCGACGACTCGGCGGCACGCGGGCGAAAGGTTCTACCGATGGGGAAGCTACTACGGCGTCACCAGAATGTACGTGTAGAGCCCGGTCCTGCGGAGC

              52.62% Xma I
              *                                                                                      19000
GAGTACCTGAGCCCCGGGCTGGTGCAGTTTGCCCGCGCCACCGAGACGTACTTCAGCCTGAATAACAAGTTTAGAAACCCCACGGTGGCACCTACGCACG
CTCATGGACTCGGGGCCCGACCACGTCAAACGGGCGCGGTGGCTCTGCATGAAGTCGGACTTATTGTTCAAATCTTTGGGGTGCCACCGTGGATGCGTGC

                                                                                                     19100
ACGTAACCACAGACCGGTCCCAGCGTTTGACGCTGCGGTTCATCCCTGTGGACCGCGAGGATACCGCGTACTCGTACAAAGCGCGGTTCACCCTGGCTGT
TGCATTGGTGTCTGGCCAGGGTCGCAAACTGCGACGCCAAGTAGGGACACCTGGCGCTCCTATGGCGCATGAGCATGTTTCGCGCCAAGTGGGACCGACA

                                                                                                     19200
GGGTGACAACCGTGTGCTTGATATGGCTTCCACGTACTTTGACATCCGCGGCGTGCTGGACAGGGGGCCTACTTTTAAGCCCTACTCCGGCACTGCCTAC
CCCACTGTTGGCACACGAACTATACCGAAGGTGCATGAAACTGTAGGCGCCGCACGACCTGTCCCCCGGATGAAAATTCGGGATGAGGCCGTGACGGATG

                                                                                                     19300
AACGCTCTAGCTCCCAAGGGCGCTCCTAACTCCTGTGAGTGGGAACAAACCGAAGATAGCGGCCGGGCAGTTGCCGAGGATGAAGAAGAGGAAGATGAAG
TTGCGAGATCGAGGGTTCCCGCGAGGATTGAGGACACTCACCCTTGTTTGGCTTCTATCGCCGGCCCGTCAACGGCTCCTACTTCTTCTCCTTCTACTTC

                                   53.79% Xho I
                                   *                                                                 19400
ATGAAGAAGAGGAAGAAGAAGAGCAAAACGCTCGAGATCAGGCTACTAAGAAAACACATGTCTATGCCCAGGCTCCTTTGTCTGGAGAAACAATTACAAA
TACTTCTTCTCCTTCTTCTTCTCGTTTTGCGAGCTCTAGTCCGATGATTCTTTTGTGTACAGATACGGGTCCGAGGAAACAGACCTCTTTGTTAATGTTT

                                                                                                     19500
AAGCGGGCTACAAATAGGATCAGACAATGCAGAAACACAAGCTAAACCTGTATACGCAGATCCTTCCTATCAACCAGAACCTCAAATTGGCGAATCTCAG
TTCGCCCGATGTTTATCCTAGTCTGTTACGTCTTTGTGTTCGATTTGGACATATGCGTCTAGGAAGGATAGTTGGTCTTGGAGTTTAACCGCTTAGAGTC

                                                                                                     19600
TGGAACGAAGCTGATGCTAATGCGGCAGGAGGGAGAGTGCTTAAAAAAACAACTCCCATGAAACCATGCTATGGATCTTATGCCAGGCCTACAAATCCTT
ACCTTGCTTCGACTACGATTACGCCGTCCTCCCTCTCACGAATTTTTTTGTTGAGGGTACTTTGGTACGATACCTAGAATACGGTCCGGATGTTTAGGAA

                                                                                                     19700
TTGGTGGTCAATCCGTTCTGGTTCCGGATGAAAAAGGGGTGCCTCTTCCAAAGGTTGACTTGCAATTCTTCTCAAATACTACCTCTTTGAACGACCGGCA
AACCACCAGTTAGGCAAGACCAAGGCCTACTTTTTCCCCACGGAGAAGGTTTCCAACTGAACGTTAAGAAGAGTTTATGATGGAGAAACTTGCTGGCCGT

                                                                                                     19800
AGGCAATGCTACTAAACCAAAAGTGGTTTTGTACAGTGAAGATGTAAATATGGAAACCCCAGACACACATCTGTCTTACAAACCTGGAAAAGGTGATGAA
TCCGTTACGATGATTTGGTTTTCACCAAAACATGTCACTTCTACATTTATACCTTTGGGGTCTGTGTGTAGACAGAATGTTTGGACCTTTTCCACTACTT

                                                                                                     19900
AATTCTAAAGCTATGTTGGGTCAACAATCTATGCCAAACAGACCCAATTACATTGCTTTCAGGGACAATTTTATTGGCCTAATGTATTATAACAGCACTG
TTAAGATTTCGATACAACCCAGTTGTTAGATACGGTTTGTCTGGGTTAATGTAACGAAAGTCCCTGTTAAAATAACCGGATTACATAATATTGTCGTGAC
```

```
                                                                                                   20000
GCAACATGGGTGTTCTTGCTGGTCAGGCATCGCAGCTAAATGCCGTGGTAGATTTGCAAGACAGAAACACAGAGCTGTCCTATCAACTCTTGCTTGATTC
CGTTGTACCCACAAGAACGACCAGTCCGTAGCGTCGATTTACGGCACCATCTAAACGTTCTGTCTTTGTGTCTCGACAGGATAGTTGAGAACGAACTAAG

                                                                                                   20100
CATAGGTGATAGAACCAGATATTTTTCTATGTGGAATCAGGCTGTAGACAGCTATGATCCAGATGTTAGAATCATTGAAAACCATGGAACTGAGGATGAA
GTATCCACTATCTTGGTCTATAAAAAGATACACCTTAGTCCGACATCTGTCGATACTAGGTCTACAATCTTAGTAACTTTTGGTACCTTGACTCCTACTT

                                                                                                   20200
TTGCCAAATTATTGTTTTCCTCTTGGGGGTATTGGGGTAACTGACACCTATCAAGCTATTAAGGCTAATGGCAATGGCTCAGGCGATAATGGAGATACTA
AACGGTTTAATAACAAAAGGAGAACCCCCATAACCCCATTGACTGTGGATAGTTCGATAATTCCGATTACCGTTACCGAGTCCGCTATTACCTCTATGAT

                                                                                                   20300
CATGGACAAAAGATGAAACTTTTGCAACACGTAATGAAATAGGAGTGGGTAACAACTTTGCCATGGAAATTAACCTAAATGCCAACCTATGGAGAAATTT
GTACCTGTTTTCTACTTTGAAAACGTTGTGCATTACTTTATCCTCACCCATTGTTGAAACGGTACCTTTAATTGGATTTACGGTTGGATACCTCTTTAAA

                                                                                                   20400
CCTTTACTCCAATATTGCGCTGTACCTGCCAGACAAGCTAAAATACAACCCCACCAATGTGGAAATATCTGACAACCCCAACACCTACGACTACATGAAC
GGAAATGAGGTTATAACGCGACATGGACGGTCTGTTCGATTTTATGTTGGGGTGGTTACACCTTTATAGACTGTTGGGGTTGTGGATGCTGATGTACTTG

               56.81% Xma I
               *                                                                                   20500
AAGCGAGTGGTGGCTCCCGGGCTTGTAGACTGCTACATTAACCTTGGGGCGCGCTGGTCTCTGGACTACATGGACAACGTTAATCCCTTTAACCACCACC
TTCGCTCACCACCGAGGGCCCGAACATCTGACGATGTAATTGGAACCCCGCGCGACCAGAGACCTGATGTACCTGTTGCAATTAGGGAAATTGGTGGTGG


                                                                                                   20600
GCAATGCGGGCCTCCGTTATCGCTCCATGTTGTTGGGAAACGGCCGCTACGTGCCCTTTCACATTCAGGTGCCCCAAAAGTTTTTTGCCATTAAAAACCT
CGTTACGCCCGGAGGCAATAGCGAGGTACAACAACCCTTTGCCGGCGATGCACGGGAAAGTGTAAGTCCACGGGGTTTTCAAAAAACGGTAATTTTTGGA

                                                      57.46% Hpa I        57.50% Sac I
                                                      *                   *                        20700
CCTCCTCCTGCCAGGCTCATATACATATGAATGGAACTTCAGGAAGGATGTTAACATGGTTCTGCAGAGCTCTCTGGGAAACGATCTTAGAGTTGACGGG
GGAGGAGGACGGTCCGAGTATATGTATACTTACCTTGAAGTCCTTCCTACAATTGTACCAAGACGTCTCGAGAGACCCTTTGCTAGAATCTCAACTGCCC

                                                                                                   20800
GCTAGCATTAAGTTTGACAGCATTTGTCTTTACGCCACCTTCTTCCCCATGGCCCACAACACGGCCTCCACGCTGGAAGCCATGCTCAGAAATGACACCA
CGATCGTAATTCAAACTGTCGTAAACAGAAATGCGGTGGAAGAAGGGGTACCGGGTGTTGTGCCGGAGGTGCGACCTTCGGTACGAGTCTTTACTGTGGT

                                                                                                   20900
ACGACCAGTCCTTTAATGACTACCTTTCCGCCGCCAACATGCTATACCCCATACCCGCCAACGCCACCAACGTGCCCATCTCCATCCCATCGCGCAACTG
TGCTGGTCAGGAAATTACTGATGGAAAGGCGGCGGTTGTACGATATGGGGTATGGGCGGTTGCGGTGGTTGCACGGGTAGAGGTAGGGTAGCGCGTTGAC


                                                                                                   21000
GGCAGCATTTCGCGGTTGGGCCTTCACACGCTTGAAGACAAAGGAAACCCCTTCCCTGGGATCAGGCTACGACCCTTACTACACCTACTCTGGCTCCATA
CCGTCGTAAAGCGCCAACCCGGAAGTGTGCGAACTTCTGTTTCCTTTGGGGAAGGGACCCTAGTCCGATGCTGGGAATGATGTGGATGAGACCGAGGTAT
```

```
                                                      58.56% Bal I
                                                      *                                              21100
CCATACCTTGACGGAACCTTCTATCTTAATCACACCTTTAAGAAGGTGGCCATTACCTTTGACTCTTCTGTTAGCTGGCCGGGCAACGACCGCCTGCTTA
GGTATGGAACTGCCTTGGAAGATAGAATTAGTGTGGAAATTCTTCCACCGGTAATGGAAACTGAGAAGACAATCGACCGGCCCGTTGCTGGCGGACGAAT

                                                                                                  58.98% Bal I
                                                                                                  *
CTCCCAATGAGTTTGAGATTAAACGCTCAGTTGACGGGGAGGGCTACAACGTAGCTCAGTGCAACATGACCAAGGACTGGTTCCTGGTGCAGATGTTGGC
GAGGGTTACTCAAACTCTAATTTGCGAGTCAACTGCCCCTCCCGATGTTGCATCGAGTCACGTTGTACTGGTTCCTGACCAAGGACCACGTCTACAACCG

                                                                                                 21300
CAACTACAATATTGGCTACCAGGGCTTCTACATTCCAGAAAGCTACAAGGACCGCATGTACTCGTTCTTCAGAAACTTCCAGCCCATGAGCCGGCAAGTG
GTTGATGTTATAACCGATGGTCCCGAAGATGTAAGGTCTTTCGATGTTCCTGGCGTACATGAGCAAGAAGTCTTTGAAGGTCGGGTACTCGGCCGTTCAC

                                             59.37% EcoR I
                                             *                                                   21400
GTTGACGATACTAAATACAAGGAGTATCAGCAGGTTGGAATTCTTCACCAGCATAACAACTCAGGATTCGTAGGCTACCTCGCTCCCACCATGCGCGAGG
CAACTGCTATGATTTATGTTCCTCATAGTCGTCCAACCTTAAGAAGTGGTCGTATTGTTGAGTCCTAAGCATCCGATGGAGCGAGGGTGGTACGCGCTCC

                                                                                                 21500
GACAGGCTTACCCCGCCAACGTGCCCTACCCACTAATAGGCAAAACCGCGGTTGACAGTATTACCCAGAAAAAGTTTCTTTGCGATCGCACCCTTTGGCG
CTGTCCGAATGGGGCGGTTGCACGGGATGGGTGATTATCCGTTTTGGCGCCAACTGTCATAATGGGTCTTTTTCAAAGAAACGCTAGCGTGGGAAACCGC

                                                                                                 21600
CATCCCATTCTCCAGTAACTTTATGTCCATGGGCGCACTCACAGACCTGGGCCAAAACCTTCTCTACGCCAACTCCGCCCACGCGCTAGACATGACTTTT
GTAGGGTAAGAGGTCATTGAAATACAGGTACCCGCGTGAGTGTCTGGACCCGGTTTTGGAAGAGATGCGGTTGAGGCGGGTGCGCGATCTGTACTGAAAA

    60.12% BamH I                                        60.24% ACCEPTOR splice site for 23K mRNA
    *                                                    *                                       21700
GAGGTGGATCCCATGGACGAGCCCACCCTTCTTTATGTTTTGTTTGAAGTCTTTGACGTGGTCCGTGTGCACCAGCCGCACCGCGGCGTCATCGAGACCG
CTCCACCTAGGGTACCTGCTCGGGTGGGAAGAAATACAAAACAAACTTCAGAAACTGCACCAGGCACACGTGGTCGGCGTGGCGCCGCAGTAGCTCTGGC

                                             60.50% Terminator UAA for Hexon    60.60% Initiator AUG for 23K polypeptide
                                             *                                  *                21800
TGTACCTGCGCACGCCCTTCTCGGCCGGCAACGCCACAACATAAAAGAAGCAAGCAACATCAACAACAGCTGCCGCCATGGGCTCCAGTGAGCAGGAACT
ACATGGACGCGTGCGGGAAGAGCCGGCCGTTGCGGTGTTGTATTTTCTTCGTTCGTTGTAGTTGTTGTCGACGGCGGTACCCGAGGTCACTCGTCCTTGA

          60.70% Bgl II
          *                                                                                      21900
GAAAGCCATTGTCAAAGATCTTGGTTGTGGGCCATATTTTTTGGGCACCTATGACAAGCGCTTTCCAGGCTTTGTTTCTCCACACAAGCTCGCCTGCGCC
CTTTCGGTAACAGTTTCTAGAACCAACACCCGGTATAAAAAACCCGTGGATACTGTTCGCGAAAGGTCCGAAACAAAGAGGTGTGTTCGAGCGGACGCGG

                                                                                                 22000
ATAGTCAATACGGCCGGTCGCGAGACTGGGGGCGTACACTGGATGGCCTTTGCCTGGAACCCGCGCTCAAAAACATGCTACCTCTTTGAGCCCTTTGGCT
TATCAGTTATGCCGGCCAGCGCTCTGACCCCCGCATGTGACCTACCGGAAACGGACCTTGGGCGCGAGTTTTTGTACGATGGAGAAACTCGGGAAACCGA

                                                                                                 22100
TTTCTGACCAACGACTCAAGCAGGTTTACCAGTTTGAGTACGAGTCACTCCTGCGCCGTAGCGCCATTGCTTCTTCCCCCGACCGCTGTATAACGCTGGA
AAAGACTGGTTGCTGAGTTCGTCCAAATGGTCAAACTCATGCTCAGTGAGGACGCGGCATCGCGGTAACGAAGAAGGGGGCTGGCGACATATTGCGACCT
```

```
                                                                                               22200
AAAGTCCACCCAAAGCGTGCAGGGGCCCAACTCGGCCGCCTGTGGACTATTCTGCTGCATGTTTCTCCACGCCTTTGCCAACTGGCCCCAAACTCCCATG
TTTCAGGTGGGTTTCGCACGTCCCCGGGTTGAGCCGGCGGACACCTGATAAGACGACGTACAAAGAGGTGCGGAAACGGTTGACCGGGGTTTGAGGGTAC

                        61.86% Kpn I
                        *                                                                      22300
GATCACAACCCCACCATGAACCTTATTACCGGGGTACCCAACTCCATGCTTAACAGTCCCCAGGTACAGCCCACCCTGCGTCGCAACCAGGAACAGCTCT
CTAGTGTTGGGGTGGTACTTGGAATAATGGCCCCATGGGTTGAGGTACGAATTGTCAGGGGTCCATGTCGGGTGGGACGCAGCGTTGGTCCTTGTCGAGA

                                                      62.30% Terminator UAA for 23K polypeptide
                                                                                    *          22400
ACAGCTTCCTGGAGCGCCACTCGCCCTACTTCCGCAGCCACAGTGCGCAGATTAGGAGCGCCACTTCTTTTTGTCACTTGAAAAACATGTAAAAATAATG
TGTCGAAGGACCTCGCGGTGAGCGGGATGAAGGCGTCGGTGTCACGCGTCTAATCCTCGCGGTGAAGAAAAACAGTGAACTTTTTGTACATTTTTATTAC

    62.38% AAUAAA for L3 mRNAs             62.45% Poly-A addition site for L3 mRNAs
                *                          *                                                   22500
TACTAGGAGACACTTTCAATAAAGGCAAATGTTTTTATTTGTACACTCTCGGGTGATTATTTACCCCCCACCCTTGCCGTCTGCGCCGTTTAAAAATCAA
ATGATCCTCTGTGAAAGTTATTTCCGTTTACAAAAATAAACATGTGAGAGCCCACTAATAAATGGGGGGTGGGAACGGCAGACGCGGCAAATTTTTAGTT
                  *                       *                                             *
                                          62.43% AAUAAA for E2a mRNAs          62.58% Terminator UAA for 72K protein
                  62.38% Poly-A addition site for E2a mRNAs
                                                                                               22600
AGGGGTTCTGCCGCGCATCGCTATGCGCCACTGGCAGGGACACGTTGCGATACTGGTGTTTAGTGCTCCACTTAAACTCAGGCACAACCATCCGCGGCAG
TCCCCAAGACGGCGCGTAGCGATACGCGGTGACCGTCCCTGTGCAACGCTATGACCACAAATCACGAGGTGAATTTGAGTCCGTGTTGGTAGGCGCCGTC

                                                                                               22700
CTCGGTGAAGTTTTCACTCCACAGGCTGCGCACCATCACCAACGCGTTTAGCAGGTCGGGCGCCGATATCTTGAAGTCGCAGTTGGGGCCTCCGCCCTGC
GAGCCACTTCAAAAGTGAGGTGTCCGACGCGTGGTAGTGGTTGCGCAAATCGTCCAGCCCGCGGCTATAGAACTTCAGCGTCAACCCCGGAGGCGGGACG

                                                               63.34% Bal I
                                                               *                               22800
GCGCGCGAGTTGCGATACACAGGGTTGCAGCACTGGAACACTATCAGCGCCGGGTGGTGCACGCTGGCCAGCACGCTCTTGTCGGAGATCAGATCCGCGT
CGCGCGCTCAACGCTATGTGTCCCAACGTCGTGACCTTGTGATAGTCGCGGCCCACCACGTGCGACCGGTCGTGCGAGAACAGCCTCTAGTCTAGGCGCA

                                                                                               22900
CCAGGTCCTCCGCGTTGCTCAGGGCGAACGGAGTCAACTTTGGTAGCTGCCTTCCCAAAAAGGGTGCATGCCCAGGCTTTGAGTTGCACTCGCACCGTAG
GGTCCAGGAGGCGCAACGAGTCCCGCTTGCCTCAGTTGAAACCATCGACGGAAGGGTTTTTCCCACGTACGGTCCGAAACTCAACGTGAGCGTGGCATC

                                                                                               23000
TGGCATCAGAAGGTGACCGTGCCCGGTCTGGGCGTTAGGATACAGCGCCTGCATGAAAGCCTTGATCTGCTTAAAAGCCACCTGAGCCTTTGCGCCTTCA
ACCGTAGTCTTCCACTGGCACGGGCCAGACCCGCAATCCTATGTCGCGGACGTACTTTCGGAACTAGACGAATTTTCGGTGGACTCGGAAACGCGGAAGT

                                                                                  64.24% Bgl II
                                                                                  *            23100
GAGAAGAACATGCCGCAAGACTTGCCGGAAAACTGATTGGCCGGACAGGCCGCGTCATGCACGCAGCACCTTGCGTCGGTGTTGGAGATCTGCACCACAT
CTCTTCTTGTACGGCGTTCTGAACGGCCTTTTGACTAACCGGCCTGTCCGGCGCAGTACGTGCGTCGTGGAACGCAGCCACAACCTCTAGACGTGGTGTA
```

```
                                                                                                   23200
TTCGGCCCCACCGGTTCTTCACGATCTTGGCCTTGCTAGACTGCTCCTTCAGCGCGCGCTGCCCGTTTTCGCTCGTCACATCCATTTCAATCACGTGCTC
AAGCCGGGGTGGCCAAGAAGTGCTAGAACCGGAACGATCTGACGAGGAAGTCGCGCGCGACGGGCAAAAGCGAGCAGTGTAGGTAAAGTTAGTGCACGAG

                                                                                                   23300
CTTATTTATCATAATGCTCCCGTGTAGACACTTAAGCTCGCCTTCGATCTCAGCGCAGCGGTGCAGCCACAACGCGCAGCCCGTGGGCTCGTGGTGCTTG
GAATAAATAGTATTACGAGGGCACATCTGTGAATTCGAGCGGAAGCTAGAGTCGCGTCGCCACGTCGGTGTTGCGCGTCGGGCACCCGAGCACCACGAAC

                                                                                                   23400
TAGGTTACCTCTGCAAACGACTGCAGGTACGCCTGCAGGAATCGCCCCATCATCGTCACAAAGGTCTTGTTGCTGGTGAAGGTCAGCTGCAACCCGCGGT
ATCCAATGGAGACGTTTGCTGACGTCCATGCGGACGTCCTTAGCGGGGTAGTAGCAGTGTTTCCAGAACAACGACCACTTCCAGTCGACGTTGGGCGCCA

                                                                                                   23500
GCTCCTCGTTTAGCCAGGTCTTGCATACGGCCGCCAGAGCTTCCACTTGGTCAGGCAGTAGCTTGAAGTTTGCCTTTAGATCGTTATCCACGTGGTACTT
CGAGGAGCAAATCGGTCCAGAACGTATGCCGGCGGTCTCGAAGGTGAACCAGTCCGTCATCGAACTTCAAACGGAAATCTAGCAATAGGTGCACCATGAA

                         65.46% 1st AUG in 100K reading frame
                         *                                                                         23600
GTCCATCAACGCGCGCGCAGCCTCCATGCCCTTCTCCCACGCAGACACGATCGGCAGGCTCAGCGGGTTTATCACCGTGCTTTCACTTTCCGCTTCACTG
CAGGTAGTTGCGCGCGCGTCGGAGGTACGGGAAGAGGGTGCGTCTGTGCTAGCCGTCCGAGTCGCCCAAATAGTGGCACGAAAGTGAAAGGCGAAGTGAC

                                                                                                   23700
GACTCTTCCTTTTCCTCTTGCGTCCGCATACCCCGCGCCACTGGGTCGTCTTCATTCAGCCGCCGCACCGTGCGCTTACCTCCCTTGCCGTGCTTGATTA
CTGAGAAGGAAAAGGAGAACGCAGGCGTATGGGGCGCGGTGACCCAGCAGAAGTAAGTCGGCGGCGTGGCACGCGAATGGAGGGAACGGCACGAACTAAT

                                                                                                   23800
GCACCGGTGGGTTGCTGAAACCCACCATTTGTAGCGCCACATCTTCTCTTTCTTCCTCGCTGTCCACGATCACCTCTGGGGATGGCGGGCGCTCGGGCTT
CGTGGCCACCCAACGACTTTGGGTGGTAAACATCGCGGTGTAGAAGAGAAAGAAGGAGCGACAGGTGCTAGTGGAGACCCCTACCGCCCGCGAGCCCGAA

                                   66.32% Bal I
                                   *                                                               23900
GGGAGAGGGGCGCTTCTTTTTCTTTTTGGACGCAATGGCCAAATCCGCCGTCGAGGTCGATGGCCGCGGGCTGGGTGTGCGCGGCACCAGCGCATCTTGT
CCCTCTCCCCGCGAAGAAAAAGAAAAACCTGCGTTACCGGTTTAGGCGGCAGCTCCAGCTACCGGCGCCCGACCCACACGCGCCGTGGTCGCGTAGAACA

                      66.57% Xho I
                      *                                                                            24000
GACGAGTCTTCTTCGTCCTCGGACTCGAGACGCCGCCTCAGCCGCTTTTTTGGGGGCGCGCGGGGAGGCGGCGGCGACGGCGACGGGGACGACACGTCCT
CTGCTCAGAAGAAGCAGGAGCCTGAGCTCTGCGGCGGAGTCGGCGAAAAAACCCCCGCGCGCCCCTCCGCCGCCGCTGCCGCTGCCCCTGCTGTGCAGGA

                                                         66.98% Bal I   67.04% ACCEPTOR splice site for 100K mRNA
                                                                    *                           *
CCATGGTTGGTGGACGTCGCGCCGCACCGCGTCCGCGCTCGGGGGTGGTTTCGCGCTGCTCCTCTTCCCGACTGGCCATTTCCTTCTCCTATAGGCAGAA
GGTACCAACCACCTGCAGCGCGGCGTGGCGCAGGCGCGAGCCCCCACCAAAGCGCGACGAGGAGAAGGGCTGACCGGTAAAGGAAGAGGATATCCGTCTT
                                                                               *        *
                                          67.00% Initiator AUG for 72K protein  67.02% ACCEPTOR splice site
    67.08% Initiator AUG for 100K polypeptide                                            for 72K mRNA
    *                                                                                              24200
AAAGATCATGGAGTCAGTCGAGAAGGAGGACAGCCTAACCGCCCCCTTTGAGTTCGCCACCACCGCCTCCACCGATGCCGCCAACGCGCCTACCACCTTC
TTTCTAGTACCTCAGTCAGCTCTTCCTCCTGTCGGATTGGCGGGGGAAACTCAAGCGGTGGTGGCGGAGGTGGCTACGGCGGTTGCGCGGATGGTGGAAG
```

```
                                                                                                      24300
CCCGTCGAGGCACCCCCGCTTGAGGAGGAGGAAGTGATTATCGAGCAGGACCCAGGTTTTGTAAGCGAAGACGACGAGGATCGCTCAGTACCAACAGAGG
GGGCAGCTCCGTGGGGGCGAACTCCTCCTCCTTCACTAATAGCTCGTCCTGGGTCCAAAACATTCGCTTCTGCTGCTCCTAGCGAGTCATGGTTGTCTCC

                                                                                                      24400
ATAAAAAGCAAGACCAGGACGACGCAGAGGCAAACGAGGAACAAGTCGGGCGGGGGGACCAAAGGCATGGCGACTACCTAGATGTGGGAGACGACGTGCT
TATTTTTCGTTCTGGTCCTGCTGCGTCTCCGTTTGCTCCTTGTTCAGCCCGCCCCCCTGGTTTCCGTACCGCTGATGGATCTACACCCTCTGCTGCACGA

                                                                                                      24500
GTTGAAGCATCTGCAGCGCCAGTGCGCCATTATCTGCGACGCGTTGCAAGAGCGCAGCGATGTGCCCCTCGCCATAGCGGATGTCAGCCTTGCCTACGAA
CAACTTCGTAGACGTCGCGGTCACGCGGTAATAGACGCTGCGCAACGTTCTCGCGTCGCTACACGGGGAGCGGTATCGCCTACAGTCGGAACGGATGCTT

                                                                                                      24600
CGCCACCTGTTCTCACCGCGCGTACCCCCCAAACGCCAAGAAAACGGCACATGCGAGCCCAACCCGCGCCTCAACTTCTACCCCGTATTTGCCGTGCCAG
GCGGTGGACAAGAGTGGCGCGCATGGGGGGTTTGCGGTTCTTTTGCCGTGTACGCTCGGGTTGGGCGCGGAGTTGAAGATGGGGCATAAACGGCACGGTC

                                                                                                      24700
AGGTGCTTGCCACCTATCACATCTTTTTCCAAAACTGCAAGATACCCCTATCCTGCCGTGCCAACCGCAGCCGAGCGGACAAGCAGCTGGCCTTGCGGCA
TCCACGAACGGTGGATAGTGTAGAAAAAGGTTTTGACGTTCTATGGGGATAGGACGGCACGGTTGGCGTCGGCTCGCCTGTTCGTCGACCGGAACGCCGT

                                                                                                      24800
GGGCGCTGTCATACCTGATATCGCCTCGCTCGACGAAGTGCCAAAAATCTTTGAGGGTCTTGGACGCGACGAGAAACGCGCGGCAAACGCTCTGCAACAA
CCCGCGACAGTATGGACTATAGCGGAGCGAGCTGCTTCACGGTTTTTAGAAACTCCCAGAACCTGCGCTGCTCTTTGCGCGCCGTTTGCGAGACGTTGT
          *                                                                                  *
   68.77% DONOR splice site for 2nd leader in 72K mRNA          68.98% ACCEPTOR splice site for 2nd leader in 72K mRNA
                                                                                                      24900
GAAAACAGCGAAAATGAAAGTCACTGTGGAGTGCTGGTGGAACTTGAGGGTGACAACGCGCGCCTAGCCGTGCTGAAACGCAGCATCGAGGTCACCCACT
CTTTTGTCGCTTTTACTTTCAGTGACACCTCACGACCACCTTGAACTCCCACTGTTGCGCGCGGATCGGCACGACTTTGCGTCGTAGCTCCAGTGGGTGA

                                                                                                      25000
TTGCCTACCCGGCACTTAACCTACCCCCCAAGGTTATGAGCACAGTCATGAGCGAGCTGATCGTGCGCCGTGCACGACCCCTGGAGAGGGATGCAAACTT
AACGGATGGGCCGTGAATTGGATGGGGGGTTCCAATACTCGTGTCAGTACTCGCTCGACTAGCACGCGGCACGTGCTGGGGACCTCTCCCTACGTTTGAA

                                                                                                      25100
GCAAGAACAAACCGAGGAGGGCCTACCCGCAGTTGGCGATGAGCAGCTGGCGCGCTGGCTTGAGACGCGCGAGCCTGCCGACTTGGAGGAGCGACGCAAG
CGTTCTTGTTTGGCTCCTCCCGGATGGGCGTCAACCGCTACTCGTCGACCGCGCGACCGAACTCTGCGCGCTCGGACGGCTGAACCTCCTCGCTGCGTTC

                                                                                                      25200
CTAATGATGGCCGCAGTGCTTGTTACCGTGGAGCTTGAGTGCATGCAGCGGTTCTTTGCTGACCCGGAGATGCAGCGCAAGCTAGAGGAAACGTTGCACT
GATTACTACCGGCGTCACGAACAATGGCACCTCGAACTCACGTACGTCGCCAAGAAACGACTGGGCCTCTACGTCGCGTTCGATCTCCTTTGCAACGTGA

                                                      70.26% Sac I
                                                      *                                               25300
ACACCTTTCGCCAGGGCTACGTGCGCCAGGCCTGCAAAATTTCCAACGTGGAGCTCTGCAACCTGGTCTCCTACCTTGGAATTTTGCACGAAAACCGCCT
TGTGGAAAGCGGTCCCGATGCACGCGGTCCGGACGTTTTAAAGGTTGCACCTCGAGACGTTGGACCAGAGGATGGAACCTTAAAACGTGCTTTTGGCGGA

                                                                                                      25400
CGGGCAAAACGTGCTTCATTCCACGCTCAAGGGCGAGGCGCGCCGCGACTACGTCCGCGACTGCGTTTACTTATTTCTGTGCTACACCTGGCAAACGGCC
GCCCGTTTTGCACGAAGTAAGGTGCGAGTTCCCGCTCCGCGCGGCGCTGATGCAGGCGCTGACGCAAATGAATAAAGACACGATGTGGACCGTTTGCCGG
```

```
                                                                                                    25500
ATGGGCGTGTGGCAGCAATGCCTGGAGGAGCGCAACCTAAAGGAGCTGCAGAAGCTGCTAAAGCAAAACTTGAAGGACCTATGGACGGCCTTCAACGAGC
TACCCGCACACCGTCGTTACGGACCTCCTCGCGTTGGATTTCCTCGACGTCTTCGACGATTTCGTTTTGAACTTCCTGGATACCTGCCGGAAGTTGCTCG

                                                                                                    25600
GCTCCGTGGCCGCGCACCTGGCGGACATTATCTTCCCCGAACGCCTGCTTAAAACCCTGCAACAGGGTCTGCCAGACTTCACCAGTCAAAGCATGTTGCA
CGAGGCACCGGCGCGTGGACCGCCTGTAATAGAAGGGGCTTGCGGACGAATTTTGGGACGTTGTCCCAGACGGTCTGAAGTGGTCAGTTTCGTACAACGT

                                71.32% EcoR I
                                *                                                                   25700
AAACTTTAGGAACTTTATCCTAGAGCGTTCAGGAATTCTGCCCGCCACCTGCTGTGCGCTTCCTAGCGACTTTGTGCCCATTAAGTACCGTGAATGCCCT
TTTGAAATCCTTGAAATAGGATCTCGCAAGTCCTTAAGACGGGCGGTGGACGACACGCGAAGGATCGCTGAAACACGGGTAATTCATGGCACTTACGGGA

                                                                                                    25800
CCGCCGCTTTGGGGTCACTGCTACCTTCTGCAGCTAGCCAACTACCTTGCCTACCACTCCGACATCATGGAAGACGTGAGCGGTGACGGCCTACTGGAGT
GGCGGCGAAACCCCAGTGACGATGGAAGACGTCGATCGGTTGATGGAACGGATGGTGAGGCTGTAGTACCTTCTGCACTCGCCACTGCCGGATGACCTCA

                                                                                  72.01% Kpn I
                                                                                  *                 25900
GTCACTGTCGCTGCAACCTATGCACCCCGCACCGCTCCCTGGTCTGCAATTCGCAACTGCTTAGCGAAAGTCAAATTATCGGTACCTTTGAGCTGCAGGG
CAGTGACAGCGACGTTGGATACGTGGGGCGTGGCGAGGGACCAGACGTTAAGCGTTGACGAATCGCTTTCAGTTTAATAGCCATGGAAACTCGACGTCCC
                                                                                      *
                                72.03% DONOR splice site for 1st leader of 72K mRNA late
                                                                                                    26000
TCCCTCGCCTGACGAAAAGTCCGCGGCTCCGGGGTTGAAACTCACTCCGGGGCTGTGGACGTCGGCTTACCTTCGCAAATTTGTACCTGAGGACTACCAC
AGGGAGCGGACTGCTTTTCAGGCGCCGAGGCCCCAACTTTGAGTGAGGCCCCGACACCTGCAGCCGAATGGAAGCGTTTAAACATGGACTCCTGATGGTG
                                                  * *
                                                  72.22% Alternative cap sites for 72K mRNA late
                                                                                   72.59% Bal I
                                                                                   *                26100
GCCCACGAGATTAGGTTCTACGAAGACCAATCCCGCCCGCCAAATGCGGAGCTTACCGCCTGCGTCATTACCCAGGGCCACATCCTTGGCCAATTGCAAG
CGGGTGCTCTAATCCAAGATGCTTCTGGTTAGGGCGGGCGGTTTACGCCTCGAATGGCGGACGCAGTAATGGGTCCCGGTGTAGGAACCGGTTAACGTTC

                                                                            72.83% Sac I
                                                                            *                       26200
CCATCAACAAAGCCCGCCAAGAGTTTCTGCTACGAAAGGACGGGGGGTTTACCTGGACCCCCAGTCCGGCGAGGAGCTCAACCCAATCCCCCCGCCGCC
GGTAGTTGTTTCGGGCGGTTCTCAAAGACGATGCTTTCCCTGCCCCCCAAATGGACCTGGGGGTCAGGCCGCTCCTCGAGTTGGGTTAGGGGGGCGGCGG

                                      73.01% Initiator AUG for 33K polypeptide
                                      *                                                             26300
GCAGCCCTATCAGCAGCCGCGGGCCCTTGCTTCCCAGGATGGCACCCAAAAAGAAGCTGCAGCTGCCGCCGCCGCCACCCACGGACGAGGAGGAATACTG
CGTCGGGATAGTCGTCGGCGCCCGGGAACGAAGGGTCCTACCGTGGGTTTTTCTTCGACGTCGACGGCGGCGGCGGTGGGTGCCTGCTCCTCCTTATGAC

                                                                       73.37% Hind III
                                                                       *                            26400
GGACAGTCAGGCAGAGGAGGTTTTGGACGAGGAGGAGGAGATGATGGAAGACTGGGACAGCCTAGACGAAGCTTCCGAGGCCGAAGAGGTGTCAGACGAA
CCTGTCAGTCCGTCTCCTCCAAAACCTGCTCCTCCTCCTCTACTACCTTCTGACCCTGTCGGATCTGCTTCGAAGGCTCCGGCTTCTCCACAGTCTGCTT
```

```
                                                                                                   26500
ACACCGTCACCCTCGGTCGCATTCCCCTCGCCGGCGCCCCAGAAATTGGCAACCGTTCCCAGCATCGCTACAACCTCCGCTCCTCAGGCGCCGCCGGCAC
TGTGGCAGTGGGAGCCAGCGTAAGGGGAGCGGCCGCGGGGTCTTTAACCGTTGGCAAGGGTCGTAGCGATGTTGGAGGCGAGGAGTCCGCGGCGGCCGTG

   73.80% Terminator UAG for 100K polypeptide     73.88% Possible DONOR splice site for 33K mRNA
                      *                            *                                               26600
TGCCTGTTCGCCGACCCAACCGTAGATGGGACACCACTGGAACCAGGGCCGGTAAGTCTAAGCAGCCGCCGCCGTTAGCCCAAGAGCAACAACAGCGCCA
ACGGACAAGCGGCTGGGTTGGCATCTACCCTGTGGTGACCTTGGTCCCGGCCATTCAGATTCGTCGGCGGCGGCAATCGGGTTCTCGTTGTTGTCGCGGT

                                                                                                   26700
AGGCTACCGCTCGTGGCGCGGGCACAAGAACGCCATAGTTGCTTGCTTGCAAGACTGTGGGGGCAACATCTCCTTCGCCCGCCGCTTTCTTCTCTACCAT
TCCGATGGCGAGCACCGCGCCCGTGTTCTTGCGGTATCAACGAACGAACGTTCTGACACCCCCGTTGTAGAGGAAGCGGGCGGCGAAAGAAGAGATGGTA

                                                     74.44% possible ACCEPTOR splice site for 33K mRNA
                                                      *                                            26800
CACGGCGTGGCCTTCCCCCGTAACATCCTGCATTACTACCGTCATCTCTACAGCCCCTACTGCACCGGCGGCAGCGGCAGCGGCAGCAACAGCAGCGGTC
GTGCCGCACCGGAAGGGGGCATTGTAGGACGTAATGATGGCAGTAGAGATGTCGGGGATGACGTGGCCGCCGTCGCCGTCGCCGTCGTTGTCGTCGCCAG

                                                                                                   26900
ACACAGAAGCAAAGGCGACCGGATAGCAAGACTCTGACAAAGCCCAAGAAATCCACAGCGGCGGCAGCAGCAGGAGGAGGAGCGCTGCGTCTGGCGCCCA
TGTGTCTTCGTTTCCGCTGGCCTATCGTTCTGAGACTGTTTCGGGTTCTTTAGGTGTCGCCGCCGTCGTCGTCCTCCTCCTCGCGACGCAGACCGCGGGT

                                                                                                   27000
ACGAACCCGTATCGACCCGCGAGCTTAGAAATAGGATTTTTCCCACTCTGTATGCTATATTTCAACAAAGCAGGGGCCAAGAACAAGAGCTGAAAATAAA
TGCTTGGGCATAGCTGGGCGCTCGAATCTTTATCCTAAAAAGGGTGAGACATACGATATAAAGTTGTTTCGTCCCCGGTTCTTGTTCTCGACTTTTATTT

                                                                                                   27100
AAACAGGTCTCTGCGCTCCCTCACCCGCAGCTGCCTGTATCACAAAAGCGAAGATCAGCTTCGGCGCACGCTGGAAGACGCGGAGGCTCTCTTCAGCAAA
TTTGTCCAGAGACGCGAGGGAGTGGGCGTCGACGGACATAGTGTTTTCGCTTCTAGTCGAAGCCGCGTGCGACCTTCTGCGCCTCCGAGAGAAGTCGTTT
                    *                                                                       **
       75.20% DONOR splice site for 1st leader of 72K mRNA          75.38% Cap sites for 72K mRNA early

                   75.47% Terminator UAG for 33K polypeptide
                    *                                                                              27200
TACTGCGCGCTGACTCTTAAGGACTAGTTTCGCGCCCTTTCTCAAATTTAAGCGCGAAAACTACGTCATCTCCAGCGGCCACACCCGGCGCCAGCACCTG
ATGACGCGCGACTGAGAATTCCTGATCAAAGCGCGGGAAAGAGTTTAAATTCGCGCTTTTGATGCAGTAGAGGTCGCCGGTGTGGGCCGCGGTCGTGGAC

          75.72% Initiator AUG for pVIII
          *                                                                                        27300
TCGTCAGCGCCATTATGAGCAAGGAAATTCCCACGCCCTACATGTGGAGTTACCAGCCACAAATGGGACTTGCGGCTGGAGCTGCCCAAGACTACTCAAC
AGCAGTCGCGGTAATACTCGTTCCTTTAAGGGTGCGGGATGTACACCTCAATGGTCGGTGTTTACCCTGAACGCCGACCTCGACGGGTTCTGATGAGTTG

                                        76.07% Xma I                       76.16% EcoR I
                                        *                                  *                       27400
CCGAATAAACTACATGAGCGCGGGACCCCACATGATATCCCGGGTCAACGGAATCCGCGCCCACCGAAACCGAATTCTCCTCGAACAGGCGGCTATTACC
GGCTTATTTGATGTACTCGCGCCCTGGGGTGTACTATAGGGCCCAGTTGCCTTAGGCGCGGGTGGCTTTGGCTTAAGAGGAGCTTGTCCGCCGATAATGG
```

```
                                                                                                     27500
ACCACACCTCGTAATAACCTTAATCCCCGTAGTTGGCCCGCTGCCCTGGTGTACCAGGAAAGTCCCGCTCCCACCACTGTGGTACTTCCCAGAGACGCCC
TGGTGTGGAGCATTATTGGAATTAGGGGCATCAACCGGGCGACGGGACCACATGGTCCTTTCAGGGCGAGGGTGGTGACACCATGAAGGGTCTCTGCGGG

                                                                        76.71% Xma I
                                                                        *                            27600
AGGCCGAAGTTCAGATGACTAACTCAGGGGCGCAGCTTGCGGGCGGCTTTCGTCACAGGGTGCGGTCGCCCGGGCAGGGTATAACTCACCTGAAAATCAG
TCCGGCTTCAAGTCTACTGATTGAGTCCCCGCGTCGAACGCCCGCCGAAAGCAGTGTCCCACGCCAGCGGGCCCGTCCCATATTGAGTGGACTTTTAGTC

76.82% Cap site for E3 mRNAs     76.89% Sac I
     *                           *                                                                   27700
AGGGCGAGGTATTCAGCTCAACGACGAGTCGGTGAGCTCCTCTCTTGGTCTCCGTCCGGACGGGACATTTCAGATCGGCGGCGCTGGCCGCTCTTCATTT
TCCCGCTCCATAAGTCGAGTTGCTGCTCAGCCACTCGAGGAGAGAACCAGAGGCAGGCCTGCCCTGTAAAGTCTAGCCGCCGCGACCGGCGAGAAGTAAA

                                                                                                     27800
ACGCCCCGTCAGGCGATCCTAACTCTGCAGACCTCGTCCTCGGAGCCGCGCTCCGGAGGCATTGGAACTCTACAATTTATTGAGGAGTTCGTGCCTTCGG
TGCGGGGCAGTCCGCTAGGATTGAGACGTCTGGAGCAGGAGCCTCGGCGCGAGGCCTCCGTAACCTTGAGATGTTAAATAACTCCTCAAGCACGGAAGCC

                                                                      77.62% Terminator UGA for pVIII
                                                                                                *
TTTACTTCAACCCCTTTTCTGGACCTCCCGGCCACTACCCGGACCAGTTTATTCCCAACTTTGACGCGGTGAAAGACTCGGCGGACGGCTACGACTGAAT
AAATGAAGTTGGGGAAAAGACCTGGAGGGCCGGTGATGGGCCTGGTCAAATAAGGGTTGAAACTGCGCCACTTTCTGAGCCGCCTGCCGATGCTGACTTA

                                                                                  77.85% DONOR splice site for 'x' leader
                                                                                  *                  28000
GACCAGTGGAGAGGCAGAGCGACTGCGCCTGACACACCTCGACCACTGCCGCCGCCACAAGTGCTTTGCCCGCGGCTCCGGTGAGTTTTGTTACTTTGAA
CTGGTCACCTCTCCGTCTCGCTGACGCGGACTGTGTGGAGCTGGTGACGGCGGCGGTGTTCACGAAACGGGCGCCGAGGCCACTCAAAACAATGAAACTT

                                                                                                     28100
TTGCCCGAAGAGCATATCGAGGGCCCGGCGCACGGCGTCCGGCTCACCACCCAGGTAGAGCTTACACGTAGCCTGATTCGGGAGTTTACCAAGCGCCCCC
AACGGGCTTCTCGTATAGCTCCCGGGCCGCGTGCCGCAGGCCGAGTGGTGGGTCCATCTCGAATGTGCATCGGACTAAGCCCTCAAATGGTTCGCGGGGG

                                                                            78.39% Bgl II
                                                                            *                        28200
TGCTAGTGGAGCGGGAGCGGGGTCCCTGTGTTCTGACCGTGGTTTGCAACTGTCCTAACCCTGGATTACATCAAGATCTTTGTTGTCATCTCTGTGCTGA
ACGATCACCTCGCCCTCGCCCCAGGGACACAAGACTGGCACCAAACGTTGACAGGATTGGGACCTAATGTAGTTCTAGAAACAACAGTAGAGACACGACT

                   78.53% Poly-A addition site for L4 mRNAs
    78.48% AAUAAA for L4 mRNAs
    *                *                                                                               28300
GTATAATAAATACAGAAATTAGAATCTACTGGGGCTCCTGTCGCCATCCTGTGAACGCCACCGTTTTTACCCACCCAAAGCAGACCAAAGCAAACCTCAC
CATATTATTTATGTCTTTAATCTTAGATGACCCCGAGGACAGCGGTAGGACACTTGCGGTGGCAAAAATGGGTGGGTTTCGTCTGGTTTCGTTTGGAGTG

                                                                               78.96% ACCEPTOR splice site for 'y' leader
                                                                               *                     28400
CTCCGGTTTGCACAAGCGGGCCAATAAGTACCTTACCTGGTACTTTAACGGCTCTTCATTTGTAATTTACAACAGTTTCCAGCGAGACGAAGTAAGTTTG
GAGGCCAAACGTGTTCGCCCGGTTATTCATGGAATGGACCATGAAATTGCCGAGAAGTAAACATTAAATGTTGTCAAAGGTCGCTCTGCTTCATTCAAAC
```

```
                                                                                               28500
CCACACAACCTTCTCGGCTTCAACTACACCGTCAAGAAAAACACCACCACCACCACCCTCCTCACCTGCCGGGAACGTACGAGTGCGTCACCGGTTGCTG
GGTGTGTTGGAAGAGCCGAAGTTGATGTGGCAGTTCTTTTTGTGGTGGTGGTGGTGGGAGGAGTGGACGGCCCTTGCATGCTCACGCAGTGGCCAACGAC

                    79.46% DONOR splice site for 'y' leader  79.47% Sac I
                                                          *  *                                 28600
CGCCCACACCTACAGCCTGAGCGTAACCAGACATTACTCCCATTTTTCCAAAACAGGAGGTGAGCTCAACTCCCGGAACTCAGGTCAAAAAAGCATTTTG
GCGGGTGTGGATGTCGGACTCGCATTGGTCTGTAATGAGGGTAAAAAGGTTTTGTCCTCCACTCGAGTTGAGGGCCTTGAGTCCAGTTTTTTCGTAAAAC

                                                    79.73% Hind III
                                                    *                                          28700
CGGGGTGCTGGGATTTTTTAATTAAGTATATGAGCAATTCAAGTAACTCTACAAGCTTGTCTAATTTTTCTGGAATTGGGGTCGGGGTTATCCTTACTCT
GCCCCACGACCCTAAAAAATTAATTCATATACTCGTTAAGTTCATTGAGATGTTCGAACAGATTAAAAAGACCTTAACCCCAGCCCCAATAGGAATGAGA

                                                                                               28800
TGTAATTCTGTTTATTCTTATACTAGCACTTCTGTGCCTTAGGGTTGCCGCCTGCTGCACGCACGTTTGTACCTATTGTCAGCTTTTTAAACGCTGGGGG
ACATTAAGACAAATAAGAATATGATCGTGAAGACACGGAATCCCAACGGCGGACGACGTGCGTGCAAACATGGATAACAGTCGAAAAATTTGCGACCCCC

           80.17% Initiator AUG for E3 19K glycoprotein
           *                                                                                   28900
CAACATCCAAGATGAGGTACATGATTTTAGGCTTGCTCGCCCTTGCGGCAGTCTGCAGCGCTGCCAAAAAGGTTGAGTTTAAGGAACCAGCTTGCAATGT
GTTGTAGGTTCTACTCCATGTACTAAAATCCGAACGAGCGGGAACGCCGTCAGACGTCGCGACGGTTTTTCCAACTCAAATTCCTTGGTCGAACGTTACA

                                                             80.59% Hind III
                                                             *                                 29000
TACATTTAAATCAGAAGCTAATGAATGCACTACTCTTATAAAATGCACCACAGAACATGAAAAGCTTATTATTCGCCACAAAGACAAAATTGGCAAGTAT
ATGTAAATTTAGTCTTCGATTACTTACGTGATGAGAATATTTTACGTGGTGTCTTGTACTTTTCGAATAATAAGCGGTGTTTCTGTTTTAACCGTTCATA

                                                                                               29100
GCTGTATATGCTATTTGGCAGCCAGGTGACACTAACGACTATAATGTCACAGTCTTCCAAGGTGAAAATCGTAAAACTTTTATGTATAAATTTCCATTTT
CGACATATACGATAAACCGTCGGTCCACTGTGATTGCTGATATTACAGTGTCAGAAGGTTCCACTTTTAGCATTTTGAAAATACATATTTAAAGGTAAAA

                                                                                               29200
ATGAAATGTGCGATATTACCATGTACATGAGCAAACAGTACAAGTTGTGGCCCCCACAAAAGTGTTTAGAGAACACTGGCACCTTTTGTTCCACCGCTCT
TACTTTACACGCTATAATGGTACATGTACTCGTTTGTCATGTTCAACACCGGGGGTGTTTTCACAAATCTCTTGTGACCGTGGAAAACAAGGTGGCGAGA

                                                                    81.50% Terminator UGA for E3 19K glycoprotein
                                                                                        *      29300
GCTTATTACAGCGCTTGCTTTGGTATGTACCTTACTTTATCTCAAATACAAAAGCAGACGCAGTTTTATTGATGAAAAGAAAATGCCTTGATTTTCCGCT
CGAATAATGTCGCGAACGAAACCATACATGGAATGAAATAGAGTTTATGTTTTCGTCTGCGTCAAAATAACTACTTTTCTTTTACGGAACTAAAAGGCGA

                                                                                               29400
TGCTTGTATTCCCCTGGACAATTTACTCTATGTGGGATATGCTCCAGGCGGGCAAGATTATACCCACAACCTTCAAATCAAACTTTCCTGGACGTTAGCG
ACGAACATAAGGGGACCTGTTAAATGAGATACACCCTATACGAGGTCCGCCCGTTCTAATATGGGTGTTGGAAGTTTAGTTTGAAAGGACCTGCAATCGC
```

```
                                81.90% Bcl I                         82.00% 1st AUG in E3 11.6K reading frame
                                *                                         *                               29500
CCTGATTTCTGCCAGCGCCTGCACTGCAAATTTGATCAAACCCAGCTTCAGCTTGCCTGCTCCAGAGATGACCGGCTCAACCATCGCGCCCACAACGGAC
GGACTAAAGACGGTCGCGGACGTGACGTTTAAACTAGTTTGGGTCGAAGTCGAACGGACGAGGTCTCTACTGGCCGAGTTGGTAGCGCGGGTGTTGCCTG

                                                                                                    29600
TATCGCAACACCACTGCTACCGGACTAACATCTGCCCTAAATTTACCCCAAGTTCATGCCTTTGTCAATGACTGGGCGAGCTTGGACATGTGGTGGTTTT
ATAGCGTTGTGGTGACGATGGCCTGATTGTAGACGGGATTTAAATGGGGTTCAAGTACGGAAACAGTTACTGACCCGCTCGAACCTGTACACCACCAAAA

                                                                                                    29700
CCATAGCGCTTATGTTTGTTTGCCTTATTATTATGTGGCTTATTTGTTGCCTAAAGCGCAGACGCGCCAGACCCCCCATCTATAGGCCTATCATTGTGCT
GGTATCGCGAATACAAACAAACGGAATAATAATACACCGAATAAACAACGGATTTCGCGTCTGCGCGGTCTGGGGGGTAGATATCCGGATAGTAACACGA

                                                      82.92% Poly-A addition site for E3-1 mRNAs (UP)
                                               82.90% Poly-A addition site for E3-1 mRNAs (SZ)
                                Terminator UAA for E3 11.6K polypeptide     82.88% Xho I
                                                                        *           *  *     *
CAACCCACACAATGAAAAAATTCATAGATTGGACGGTCTGAAACCATGTTCTCTTCTTTTACAGTATGATTAAATGAGACATGATTCCTCGAGTTCTTAT
GTTGGGTGTGTTACTTTTTTAAGTATCTAACCTGCCAGACTTTGGTACAAGAGAAGAAAATGTCATACTAATTTACTCTGTACTAAGGAGCTCAAGAATA

   82.93% Poly-A addition site for E3-1 mRNAs (SZ)
82.92% Poly-A addition site for E3-1 mRNAs (SZ)
*  *                                                                                                29900
ATTATTGACCCTTGTTGCGCTTTTCTGTGCGTGCTCTACATTGGCCGCGGTCGCTCACATCGAAGTAGATTGCATCCCACCTTTCACAGTTTACCTGCTT
TAATAACTGGGAACAACGCGAAAAGACACGCACGAGATGTAACCGGCGCCAGCGAGTGTAGCTTCATCTAACGTAGGGTGGAAAGTGTCAAATGGACGAA

                                                                                                    30000
TACGGATTTGTCACCCTTATCCTCATCTGCAGCCTCGTCACTGTAGTCATCGCCTTCATTCAGTTCATTGACTGGGTTTGTGTGCGCATTGCGTACCTCA
ATGCCTAAACAGTGGGAATAGGAGTAGACGTCGGAGCAGTGACATCAGTAGCGGAAGTAAGTCAAGTAACTGACCCAAACACACGCGTAACGCATGGAGT

                                          83.60% EcoR I       83.64% 1st AUG in 14.5K URF
                                                *                *                                  30100
GGCACCATCCGCAATACAGAGACAGGACTATAGCTGATCTTCTCAGAATTCTTTAATTATGAAACGGAGTGTCATTTTTGTTTTGCTGATTTTTTGCGCC
CCGTGGTAGGCGTTATGTCTCTGTCCTGATATCGACTAGAAGAGTCTTAAGAAATTAATACTTTGCCTCACAGTAAAAACAAAACGACTAAAAAACGCGG

                                                                                                    30200
CTACCTGTGCTTTGCTCCCAAACCTCAGCGCCTCCCAAAAGACATATTTCCTGCAGATTCACTCAAATATGGAACATTCCCAGCTGCTACAACAAACAGA
GATGGACACGAAACGAGGGTTTGGAGTCGCGGAGGGTTTTCTGTATAAAGGACGTCTAAGTGAGTTTATACCTTGTAAGGGTCGACGATGTTGTTTGTCT

                                                                                                    30300
GCGATTTGTCAGAAGCCTGGTTATACGCCATCATCTCTGTCATGGTTTTTTGCAGTACCATTTTTGCCCTAGCCATATATCCATACCTTGACATTGGCTG
CGCTAAACAGTCTTCGGACCAATATGCGGTAGTAGAGACAGTACCAAAAAACGTCATGGTAAAAACGGGATCGGTATATAGGTATGGAACTGTAACCGAC

                                                                                                    30400
GAATGCCATAGATGCCATGAACCACCCTACTTTCCCAGTGCCCGCTGTCATACCACTGCAACAGGTTATTGCCCCAATCAATCAGCCTCGCCCCCCTTCT
CTTACGGTATCTACGGTACTTGGTGGGATGAAAGGGTCACGGGCGACAGTATGGTGACGTTGTCCAATAACGGGGTTAGTTAGTCGGAGCGGGGGGAAGA
```

```
                                                   84.76% Xba I
                                                 84.75% Bgl II
              Terminator UGA for 14.5K URF     84.74% Xba I
                                        84.71% 1st AUG in 14.7K URF
                                    84.69% ACCEPTOR splice site for 'z' leader
                                     *   *   *   * * *                      30500
CCCACCCCCACTGAGATTAGCTACTTTAATTTGACAGGTGGAGATGACTGAATCTCTAGATCTAGAATTGGATGGAATTAACACCGAACAGCGCCTACTA
GGGTGGGGGTGACTCTAATCGATGAAATTAAACTGTCCACCTCTACTGACTTAGAGATCTAGATCTTAACCTACCTTAATTGTGGCTTGTCGCGGATGAT

                                                     85.03% Hpa I        85.09% DONOR splice site for 'z' leader
                                                     *                 *    30600
GAAAGGCGCAAGGCGGCGTCCGAGCGAGAACGCCTAAAACAAGAAGTTGAAGACATGGTTAACCTACACCAGTGTAAAAGAGGTATCTTTTGTGTGGTCA
CTTTCCGCGTTCCGCCGCAGGCTCGCTCTTGCGGATTTTGTTCTTCAACTTCTGTACCAATTGGATGTGGTCACATTTTCTCCATAGAAAACACACCAGT

                                                                            30700
AGCAGGCCAAACTTACCTACGAAAAAACCACTACCGGCAACCGCCTCAGCTACAAGCTACCCACCCAGCGCCAAAAACTGGTGCTTATGGTGGGAGAAAA
TCGTCCGGTTTGAATGGATGCTTTTTTGGTGATGGCCGTTGGCGGAGTCGATGTTCGATGGGTGGGTCGCGGTTTTTGACCACGAATACCACCCTCTTTT

                                                                            30800
ACCTATCACCGTCACCCAGCACTCGGCAGAAACAGAGGGCTGCCTGCACTTCCCCTATCAGGGTCCAGAGGACCTCTGCACTCTTATTAAAACCATGTGT
TGGATAGTGGCAGTGGGTCGTGAGCCGTCTTTGTCTCCCGACGGACGTGAAGGGGATAGTCCCAGGTCTCCTGGAGACGTGAGAATAATTTTGGTACACA

                                                        85.88% Poly-A addition site for E3-2 mRNAs
                                        85.82% AAUAAA for E3 mRNAs
     85.73% Bgl II        85.78% Terminator for 14.7K URF
     *                    *               *             *                   30900
GGTATTAGAGATCTTATTCCATTCAACTAACATAAACACACAATAAATTACTTACTTAAAATCAGTCAGCAAATCTTTGTCCAGCTTATTCAGCATCACC
CCATAATCTCTAGAATAAGGTAAGTTGATTGTATTTGTGTGTTATTTAATGAATGAATTTTAGTCAGTCGTTTAGAAACAGGTCGAATAAGTCGTAGTGG

                                                                            31000
TCCTTTCCTTCCTCCCAACTCTGGTATCTCAGCCGCCTTTTAGCTGCAAACTTTCTCCAAAGTTTAAATGGGATGTCAAATTCCTCATGTTCTTGTCCCT
AGGAAAGGAAGGAGGGTTGAGACCATAGAGTCGGCGGAAAATCGACGTTTGAAAGAGGTTTCAAATTTACCCTACAGTTTAAGGAGTACAAGAACAGGGA

                          86.34% ACCEPTOR splice site for fiber mRNA
                          86.34% Initiator AUG for fiber polypeptide
                          *                                                 31100
CCGCACCCACTATCTTCATATTGTTGCAGATGAAACGCGCCAGACCGTCTGAAGACACCTTCAACCCCGTGTATCCATATGACACAGAAACCGGGCCTCC
GGCGTGGGTGATAGAAGTATAACAACGTCTACTTTGCGCGGTCTGGCAGACTTCTGTGGAAGTTGGGGCACATAGGTATACTGTGTCTTTGGCCCGGAGG

                                                                            31200
AACTGTGCCCTTTCTTACCCCTCCATTTGTTTCACCCAATGGTTTCCAAGAAAGTCCCCCTGGAGTTCTCTCTCTACGCGTCTCCGAACCTTTGGACACC
TTGACACGGGAAAGAATGGGGAGGTAAACAAAGTGGGTTACCAAAGGTTCTTTCAGGGGGACCTCAAGAGAGAGATGCGCAGAGGCTTGGAAACCTGTGG

                                                                            31300
TCCCACGGCATGCTTGCGCTTAAAATGGGCAGCGGTCTTACCCTAGACAAGGCCGGAAACCTCACCTCCCAAAATGTAACCACTGTTACTCAGCCACTTA
AGGGTGCCGTACGAACGCGAATTTTACCCGTCGCCAGAATGGGATCTGTTCCGGCCTTTGGAGTGGAGGGTTTTACATTGGTGACAATGAGTCGGTGAAT
```

```
                                                                                                   31400
AAAAAACAAAGTCAAACATAAGTTTGGACACCTCCGCACCACTTACAATTACCTCAGGCGCCCTAACAGTGGCAACCACCGCTCCTCTGATAGTTACTAG
TTTTTTGTTTCAGTTTGTATTCAAACCTGTGGAGGCGTGGTGAATGTTAATGGAGTCCGCGGGATTGTCACCGTTGGTGGCGAGGAGACTATCAATGATC

                                                                                                   31500
CGGCGCTCTTAGCGTACAGTCACAAGCCCCACTGACCGTGCAAGACTCCAAACTAAGCATTGCTACTAAAGGGCCCATTACAGTGTCAGATGGAAAGCTA
GCCGCGAGAATCGCATGTCAGTGTTCGGGGTGACTGGCACGTTCTGAGGTTTGATTCGTAACGATGATTTCCCGGGTAATGTCACAGTCTACCTTTCGAT

                                                                                                   31600
GCCCTGCAAACATCAGCCCCCCTCTCTGGCAGTGACAGCGACACCCTTACTGTAACTGCATCACCCCCGCTAACTACTGCCACGGGTAGCTTGGGCATTA
CGGGACGTTTGTAGTCGGGGGGAGAGACCGTCACTGTCGCTGTGGGAATGACATTGACGTAGTGGGGGCGATTGATGACGGTGCCCATCGAACCCGTAAT

                                                                                                   31700
ACATGGAAGATCCTATTTATGTAAATAATGGAAAAATAGGAATTAAAATAAGCGGTCCTTTGCAAGTAGCACAAAACTCCGATACACTAACAGTAGTTAC
TGTACCTTCTAGGATAAATACATTTATTACCTTTTTATCCTTAATTTTATTCGCCAGGAAACGTTCATCGTGTTTTGAGGCTATGTGATTGTCATCAATG

                                                                                                   31800
TGGACCAGGTGTCACCGTTGAACAAAACTCCCTTAGAACCAAAGTTGCAGGAGCTATTGGTTATGATTCATCAAACAACATGGAAATTAAAACGGGCGGT
ACCTGGTCCACAGTGGCAACTTGTTTTGAGGGAATCTTGGTTTCAACGTCCTCGATAACCAATACTAAGTAGTTTGTTGTACCTTTAATTTTGCCCGCCA

                             88.57% Xba I
                             *                                                                     31900
GGCATGCGTATAAATAACAACTTGTTAATTCTAGATGTGGATTACCCATTTGATGCTCAAACAAAACTACGTCTTAAACTGGGGCAGGGACCCCTGTATA
CCGTACGCATATTTATTGTTGAACAATTAAGATCTACACCTAATGGGTAAACTACGAGTTTGTTTTGATGCAGAATTTGACCCCGTCCCTGGGGACATAT

                                                                                                   32000
TTAATGCATCTCATAACTTGGACATAAACTATAACAGAGGCCTATACCTTTTTAATGCATCAAACAATACTAAAAAACTGGAAGTTAGCATAAAAAAATC
AATTACGTAGAGTATTGAACCTGTATTTGATATTGTCTCCGGATATGGAAAAATTACGTAGTTTGTTATGATTTTTTGACCTTCAATCGTATTTTTTTAG

                                                                                                   32100
CAGTGGACTAAACTTTGATAATACTGCCATAGCTATAAATGCAGGAAAGGGTCTGGAGTTTGATACAAACACATCTGAGTCTCCAGATATCAACCCAATA
GTCACCTGATTTGAAACTATTATGACGGTATCGATATTTACGTCCTTTCCCAGACCTCAAACTATGTTTGTGTAGACTCAGAGGTCTATAGTTGGGTTAT

                                                                              89.52% Hind III
                                                                              *                    32200
AAAACTAAAATTGGCTCTGGCATTGATTACAATGAAAACGGTGCCATGATTACTAAACTTGGAGCGGGTTTAAGCTTTGACAACTCAGGGGCCATTACAA
TTTTGATTTTAACCGAGACCGTAACTAATGTTACTTTTGCCACGGTACTAATGATTTGAACCTCGCCCAAATTCGAAACTGTTGAGTCCCCGGTAATGTT

                                                                      89.77% EcoR I
                                                                      *                            32300
TAGGAAACAAAAATGATGACAAACTTACCCTGTGGACAACCCCAGACCCATCTCCTAACTGCAGAATTCATTCAGATAATGACTGCAAATTTACTTTGGT
ATCCTTTGTTTTTACTACTGTTTGAATGGGACACCTGTTGGGGTCTGGGTAGAGGATTGACGTCTTAAGTAAGTCTATTACTGACGTTTAAATGAAACCA

                                                                90.04% Bgl II
                                                                *                                  32400
TCTTACAAAATGTGGGAGTCAAGTACTAGCTACTGTAGCTGCTTTGGCTGTATCTGGAGATCTTTCATCCATGACAGGCACCGTTGCAAGTGTTAGTATA
AGAATGTTTTACACCCTCAGTTCATGATCGATGACATCGACGAAACCGACATAGACCTCTAGAAAGTAGGTACTGTCCGTGGCAACGTTCACAATCATAT
```

```
                                                                                                 32500
TTCCTTAGATTTGACCAAAACGGTGTTCTAATGGAGAACTCCTCACTTAAAAAACATTACTGGAACTTTAGAAATGGGAACTCAACTAATGCAAATCCAT
AAGGAATCTAAACTGGTTTTGCCACAAGATTACCTCTTGAGGAGTGAATTTTTTGTAATGACCTTGAAATCTTTACCCTTGAGTTGATTACGTTTAGGTA

                                                                                                 32600
ACACAAATGCAGTTGGATTTATGCCTAACCTTCTAGCCTATCCAAAAACCCAAAGTCAAACTGCTAAAAATAACATTGTCAGTCAAGTTTACTTGCATGG
TGTGTTTACGTCAACCTAAATACGGATTGGAAGATCGGATAGGTTTTTGGGTTTCAGTTTGACGATTTTTATTGTAACAGTCAGTTCAAATGAACGTACC

                                                                                                 32700
TGATAAAACTAAACCTATGATACTTACCATTACACTTAATGGCACTAGTGAATCCACAGAAACTAGCGAGGTAAGCACTTACTCTATGTCTTTTACATGG
ACTATTTTGATTTGGATACTATGAATGGTAATGTGAATTACCGTGATCACTTAGGTGTCTTTGATCGCTCCATTCGTGAATGAGATACAGAAAATGTACC

                                 91.20% Terminator UAA for fiber polypeptide
                                                 91.19% AAUAAA for L5 mRNAs     91.26% Poly-A addition site for L5 mRNA
                                                                      * *                   *
TCCTGGGAAAGTGGAAAATACACCACTGAAACTTTTGCTACCAACTCTTACACCTTCTCCTACATTGCCCAGGAATAAAGAATCGTGAACCTGTTGCATG
AGGACCCTTTCACCTTTTATGTGGTGACTTTGAAAACGATGGTTGAGAATGTGGAAGAGGATGTAACGGGTCCTTATTTCTTAGCACTTGGACAACGTAC

                                                                                        91.52% Bcl I
                                                                                        *    32900
TTATGTTTCAACGTGTTTATTTTTCAATTGCAGAAAATTTCAAGTCATTTTTCATTCAGTAGTATAGCCCCACCACCACATAGCTTATATTGATCACCGT
AATACAAAGTTGCACAAATAAAAAGTTAACGTCTTTTAAAGTTCAGTAAAAAGTAAGTCATCATATCGGGGTGGTGGTGTATCGAATATAACTAGTGGCA
 *                  *
                    91.32% AAUAAA for E4 mRNAs
 91.27% Poly-A addition site for E4 mRNAs

                                                                                                 33000
ACCTTAATCAAACTCACAGAACCCTAGTATTCAACCTGCCACCTCCCTCCCAACACACAGAGTACACAGTCCTTTCTCCCCGGCTGGCCTTAAAAAGCAT
TGGAATTAGTTTGAGTGTCTTGGGATCATAAGTTGGACGGTGGAGGGAGGGTTGTGTGTCTCATGTGTCAGGAAAGAGGGGCCGACCGGAATTTTTCGTA
          *
          91.59% Final terminator UGA in E4                                     92.08% Xma I
                                                                                        *    33100
CATATCATGGGTAACAGACATATTCTTAGGTGTTATATTCCACACGGTTTCCTGTCGAGCCAAACGCTCATCAGTGATATTAATAAACTCCCCGGGCAGC
GTATAGTACCCATTGTCTGTATAAGAATCCACAATATAAGGTGTGCCAAAGGACAGCTCGGTTTGCGAGTAGTCACTATAATTATTTGAGGGGCCCGTCG

                                                                                                 33200
TCGCTTAAGTTCATGTCGCTGTCCAGCTGCTGAGCCACAGGCTGCTGTCCAACTTGCGGTTGCTCAACGGGCGGCGAAGGGGAAGTCCACGCCTACATGG
AGCGAATTCAAGTACAGCGACAGGTCGACGACTCGGTGTCCGACGACAGGTTGAACGCCAACGAGTTGCCCGCCGCTTCCCCTTCAGGTGCGGATGTACC
                                                                                         *
                                                                  92.36% ACCEPTOR splice site in E4
                                                                                                 33300
GGGTAGAGTCATAATCGTGCATCAGGATAGGGCGGTGGTGCTGCAGCAGCGCGCGAATAAACTGCTGCCGCCGCCGCTCCGTCCTGCAGGAATACAACAT
CCCATCTCAGTATTAGCACGTAGTCCTATCCCGCCACCACGACGTCGTCGCGCGCTTATTTGACGACGGCGGCGGCGAGGCAGGACGTCCTTATGTTGTA
                                                                              *
                                                         92.61% ACCEPTOR splice site in E4
                                                                                                 33400
GGCAGTGGTCTCCTCAGCGATGATTCGCACCGCCCGCAGCATGAGACGCCTTGTCCTCCGGGCACAGCAGCGCACCCTGATCTCACTTAAATCAGCACAG
CCGTCACCAGAGGAGTCGCTACTAAGCGTGGCGGGCGTCGTACTCTGCGGAACAGGAGGCCCGTGTCGTCGCGTGGGACTAGAGTGAATTTAGTCGTGTC
                                                                          *
                                                                          92.87% ACCEPTOR splice site in E4
```

```
                                                                                            93.19% Bal I
                                                                                            *    33500
TAACTGCAGCACAGCACCACAATATTGTTCAAAATCCCACAGTGCAAGGCGCTGTATCCAAAGCTCATGGCGGGGACCACAGAACCCACGTGGCCATCAT
ATTGACGTCGTGTCGTGGTGTTATAACAAGTTTTAGGGTGTCACGTTCCGCGACATAGGTTTCGAGTACCGCCCCTGGTGTCTTGGGTGCACCGGTAGTA
  *                                             *
  92.94% ACCEPTOR splice site in E4             93.08% ACCEPTOR splice site in E4
                                                                                               93.48% Kpn I
                                                                                               * 33600
ACCACAAGCGCAGGTAGATTAAGTGGCGACCCCTCATAAACACGCTGGACATAAACATTACCTCTTTTGGCATGTTGTAATTCACCACCTCCCGGTACCA
TGGTGTTCGCGTCCATCTAATTCACCGCTGGGGAGTATTTGTGCGACCTGTATTTGTAATGGAGAAAACCGTACAACATTAAGTGGTGGAGGGCCATGGT

                                                93.63% Bal I
                                                *                                                   33700
TATAAACCTCTGATTAAACATGGCGCCATCCACCACCATCCTAAACCAGCTGGCCAAAACCTGCCCGCCGGCTATGCACTGCAGGGAACCGGGACTGGAA
ATATTTGGAGACTAATTTGTACCGCGGTAGGTGGTGGTAGGATTTGGTCGACCGGTTTTGGACGGGCGGCCGATACGTGACGTCCCTTGGCCCTGACCTT
      *                                                                   *
      93.52% ACCEPTOR splice site in E4                                   93.71% ACCEPTOR splice site in E4

                                                                                                    33800
CAATGACAGTGGAGAGCCCAGGACTCGTAACCATGGATCATCATGCTCGTCATGATATCAATGTTGGCACAACACAGGCACACGTGCATACACTTCCTCA
GTTACTGTCACCTCTCGGGTCCTGAGCATTGGTACCTAGTAGTACGAGCAGTACTATAGTTACAACCGTGTTGTGTCCGTGTGCACGTATGTGAAGGAGT

                                                                                                    33900
GGATTACAAGCTCCTCCCGCGTCAGAACCATATCCCAGGGAACAACCCATTCCTGAATCAGCGTAAATCCCACACTGCAGGGAAGACCTCGCACGTAACT
CCTAATGTTCGAGGAGGGCGCAGTCTTGGTATAGGGTCCCTTGTTGGGTAAGGACTTAGTCGCATTTAGGGTGTGACGTCCCTTCTGGAGCGTGCATTGA
                                                                          *
                                                  94.25% ACCEPTOR splice site in E4
                                                                                                    34000
CACGTTGTGCATTGTCAAAGTGTTACATTCGGGCAGCAGCGGATGATCCTCCAGTATGGTAGCGCGGGTCTCTGTCTCAAAAGGAGGTAGGCGATCCCTA
GTGCAACACGTAACAGTTTCACAATGTAAGCCCGTCGTCGCCTACTAGGAGGTCATACCATCGCGCCCAGAGACAGAGTTTTCCTCCATCCGCTAGGGAT
   *
   94.34% DONOR splice site in E4
                                                                                                    34100
CTGTACGGAGTGCGCCGAGACAACCGAGATCGTGTTGGTCGTAGTGTCATGCCAAATGGAACGCCGGACGTAGTCATATTTCCTGAAGCAAAACCAGGTG
GACATGCCTCACGCGGCTCTGTTGGCTCTAGCACAACCAGCATCACAGTACGGTTTACCTTGCGGCCTGCATCAGTATAAAGGACTTCGTTTTGGTCCAC
                                                                                    *
             94.93% Bgl II                                         94.83% ACCEPTOR splice site in E4
             *                                                                                      34200
CGGGCGTGACAAACAGATCTGCGTCTCCGGTCTCGTCGCTTAGCTCGCTCTGTGTAGTAGTTGTAGTATATCCACTCTCTCAAAGCATCCAGGCGCCCCC
GCCCGCACTGTTTGTCTAGACGCAGAGGCCAGAGCAGCGAATCGAGCGAGACACATCATCAACATCATATAGGTGAGAGAGTTTCGTAGGTCCGCGGGGG

                                                                                                    34300
TGGCTTCGGGTTCTATGTAAACTCCTTCATGCGCCGCTGCCCTGATAACATCCACCACCGCAGAATAAGCCACACCCAGCCAACCTACACATTCGTTCTG
ACCGAAGCCCAAGATACATTTGAGGAAGTACGCGGCGACGGGACTATTGTAGGTGGTGGCGTCTTATTCGGTGTGGGTCGGTTGGATGTGTAAGCAAGAC
                                           *                                              *
              95.28% ACCEPTOR splice site in E4            95.41% DONOR splice site in E4
```

```
                                                                                         95.69% Bgl II
                                                                                         *     34400
CGAGTCACACACGGGAGGAGCGGGAAGAGCTGGAAGAACCATGTTTTTTTTTTTTTTATTCCAAAAGATTATCCAAAACCTCAAAATGAAGATCTATTAA
GCTCAGTGTGTGCCCTCCTCGCCCTTCTCGACCTTCTTGGTACAAAAAAAAAAAAAAATAAGGTTTTCTAATAGGTTTTGGAGTTTTACTTCTAGATAATT
                            *                           **                    *
                            95.52% ACCEPTOR splice site in E4                 95.66% ACCEPTOR splice site in E4
                                                         95.60% Terminator UAA for E4 11K protein
                                                        95.60% variable run of As ( 13-16 ) in E4
                                                                                               34500
GTGAACGCGCTCCCCTCCGGTGGCGTGGTCAAACTCTACAGCCAAAGAACAGATAATGGCATTTGTAAGATGTTGCACAATGGCTTCCAAAAGGCAAACT
CACTTGCGCGAGGGGAGGCCACCGCACCAGTTTGAGATGTCGGTTTCTTGTCTATTACCGTAAACATTCTACAACGTGTTACCGAAGGTTTTCCGTTTGA
                                  *
                                  95.82% ACCEPTOR splice site in E4
                                                                                               34600
GCCCTCACGTCCAAGTGGACGTAAAGGCTAAACCCTTCAGGGTGAATCTCCTCTATAAACATTCCAGCACCTTCAACCATGCCCAAATAATTTTCATCTC
CGGGAGTGCAGGTTCACCTGCATTTCCGATTTGGGAAGTCCCACTTAGAGGAGATATTTGTAAGGTCGTGGAAGTTGGTACGGGTTTATTAAAAGTAGAG

                                                                                               34700
GCCACCTTATCAATATGTCTCTAAGCAAATCCCGAATATTAAGTCCGGCCATTGTAAAAATCTGCTCCAGAGCGCCCTCCACCTTCAGCCTCAAGCAGCG
CGGTGGAATAGTTATACAGAGATTCGTTTAGGGCTTATAATTCAGGCCGGTAACATTTTTAGACGAGGTCTCGCGGGAGGTGGAAGTCGGAGTTCGTCGC
     *
     96.29% DONOR splice site in E4
                                                                                               34800
AATCATGATTGCAAAAATTCAGGTTCCTCACAGACCTGTATAAGATTCAAAAGCGGAACATTAACAAAAATACCGCGATCCCGTAGGTCCCTTCGCAGGG
TTAGTACTAACGTTTTTAAGTCCAAGGAGTGTCTGGACATATTCTAAGTTTTCGCCTTGTAATTGTTTTTATGGCGCTAGGGCATCCAGGGAAGCGTCCC
     *                            *
     96.57% Initiator AUG for E4 11K protein
                                  96.65% ACCEPTOR splice site in E4
                                                                                               34900
CCAGCTGAACATAATCGTGCAGGTCTGCACGGACCAGCGCGGCCACTTCCCCGCCAGGAACCATGACAAAAGAACCCACACTGATTATGACACGCATACT
GGTCGACTTGTATTAGCACGTCCAGACGTGCCTGGTCGCGCCGGTGAAGGGGCGGTCCTTGGTACTGTTTTCTTGGGTGTGACTAATACTGTGCGTATGA

                                97.20% Hind III
                                *                                                              35000
CGGAGCTATGCTAACCAGCGTAGCCCCTATGTAAGCTTGTTGCATGGGCGGCGATATAAAATGCAAGGTGCTGCTCAAAAAATCAGGCAAAGCCTCGCGC
GCCTCGATACGATTGGTCGCATCGGGGATACATTCGAACAACGTACCCGCCGCTATATTTTACGTTCCACGACGAGTTTTTTAGTCCGTTTCGGAGCGCG

                                                                                               35100
AAAAAAGCAAGCACATCGTAGTCATGCTCATGCAGATAAAGGCAGGTAAGTTCCGGAACCACCACAGAAAAAGACACCATTTTTCTCTCAAACATGTCTG
TTTTTTCGTTCGTGTAGCATCAGTACGAGTACGTCTATTTCCGTCCATTCAAGGCCTTGGTGGTGTCTTTTTCTGTGGTAAAAAGAGAGTTTGTACAGAC

                                                                                               35200
CGGGTTCCTGCATTAAACACAAAATAAAATAACAAAAAAAAACATTTAAACATTAGAAGCCTGTCTTACAACAGGAAAAACAACCCTTATAAGCATAAGA
GCCCAAGGACGTAATTTGTGTTTTATTTTATTGTTTTTTTTTGTAAATTTGTAATCTTCGGACAGAATGTTGTCCTTTTTGTTGGGAATATTCGTATTCT
      *
      97.69% ACCEPTOR splice site in E4
```

```
                                                                                                   35300
CGGACTACGGCCATGCCGGCGTGACCGTAAAAAAACTGGTCACCGTGATTAAAAAGCACCACCGACAGTTCCTCGGTCATGTCCGGAGTCATAATGTAAG
GCCTGATGCCGGTACGGCCGCACTGGCATTTTTTTGACCAGTGGCACTAATTTTTCGTGGTGGCTGTCAAGGAGCCAGTACAGGCCTCAGTATTACATTC

                      98.29% Hpa I                         98.39% Xma I
                      *                                    *                                       35400
ACTCGGTAAACACATCAGGTTGGTTAACATCGGTCAGTGCTAAAAAGCGACCGAAATAGCCCGGGGGAATACATACCCGCAGGCGTAGAGACAACATTAC
TGAGCCATTTGTGTAGTCCAACCAATTGTAGCCAGTCACGATTTTTCGCTGGCTTTATCGGGCCCCCTTATGTATGGGCGTCCGCATCTCTGTTGTAATG

                                                                                                   35500
AGCCCCCATAGGAGGTATAACAAAATTAATAGGAGAGAAAAACACATAAACACCTGAAAAACCCTCCTGCCTAGGCAAAATAGCACCCTCCCGCTCCAGA
TCGGGGGTATCCTCCATATTGTTTTAATTATCCTCTCTTTTTGTGTATTTGTGGACTTTTTGGGAGGACGGATCCGTTTTATCGTGGGAGGGCGAGGTCT

                                                                                                   35600
ACAACATACAGCGCTTCCACAGCGGCAGCCATAACAGTCAGCCTTACCAGTAAAAAAACCTATTAAAAAACACCACTCGACACGGCACCAGCTCAATCAG
TGTTGTATGTCGCGAAGGTGTCGCCGTCGGTATTGTCAGTCGGAATGGTCATTTTTTTGGATAATTTTTTGTGGTGAGCTGTGCCGTGGTCGAGTTAGTC
                                                    *
                                                    98.91% DONOR splice site for 1st leader of E4 mRNAs

                                                                                                   35700
TCACAGTGTAAAAAGGGCCAAGTACAGAGCGAGTATATATAGGACTAAAAAATGACGTAACGGTTAAAGTCCACAAAAAACACCCAGAAAACCGCACGCG
AGTGTCACATTTTTCCCGGTTCATGTCTCGCTCATATATATCCTGATTTTTTACTGCATTGCCAATTTCAGGTGTTTTTTGTGGGTCTTTTGGCGTGCGC
             *
             99.10% Cap site (Ts and A) for E4 mRNAs
                                                                                                   35800
AACCTACGCCCAGAAACGAAAGCCAAAAAACCCACAACTTCCTCAAATCTTCACTTCCGTTTTCCCACGATACGTCACTTCCCATTTTAAAAAAACTACA
TTGGATGCGGGTCTTTGCTTTCGGTTTTTTGGGTGTTGAAGGAGTTTAGAAGTGAAGGCAAAAGGGTGCTATGCAGTGAAGGGTAAAATTTTTTTGATGT

                                    99.71% end of inverted terminal repetition
                                    *                                                              35900
ATTCCCAATACATGCAAGTTACTCCGCCCTAAAACCTACGTCACCCGCCCCGTTCCCACGCCCCGCGCCACGTCACAAACTCCACCCCCTCATTATCATA
TAAGGGTTATGTACGTTCAATGAGGCGGGATTTTGGATGCAGTGGGCGGGGCAAGGGTGCGGGGCGCGGTGCAGTGTTTGAGGTGGGGGAGTAATAGTAT

                                  35937
TTGGCTTCAATCCAAAATAAGGTATATTATGATGATG
AACCGAAGTTAGGTTTTATTCCATATAATACTACTAC
```

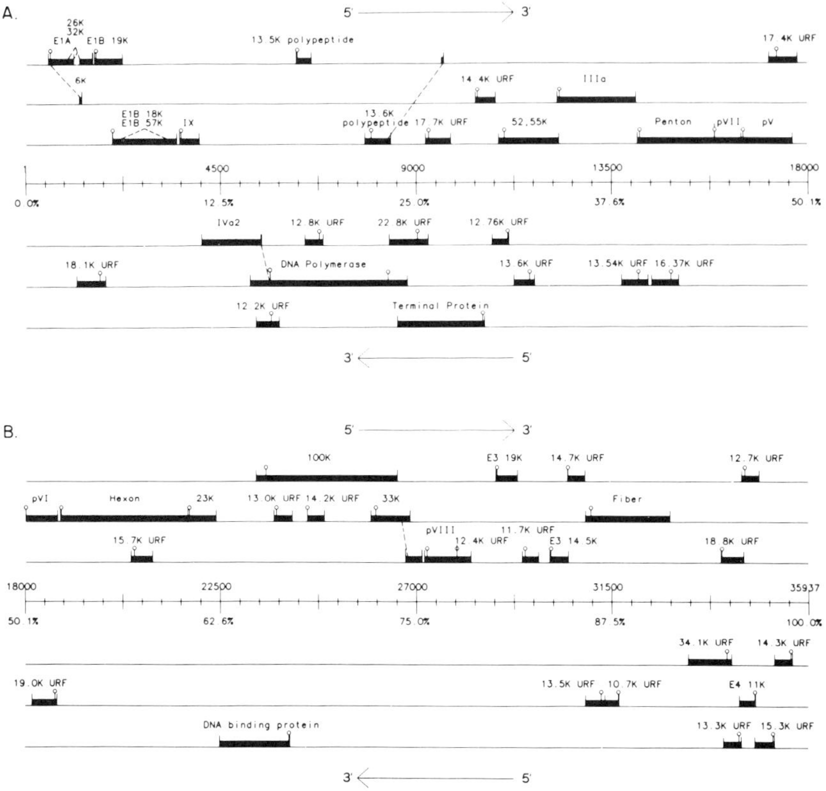

Figure 2 : Open reading frames, containing at least 100 amino-acids between the first AUG and the final terminator, present on the Ad2 genome. The complete reading frame from the preceding terminator to the final terminator is shaded. Within each reading frame the flag shows the position of the first AUG codon. The scale shows base number (upper) and percent coordinates (lower). The direction of transcription is indicated. Panel A 0-50%. Panel B 50-100%.

Table 3. Master Coordinates.

Feature	Base #	Coord	Reference
end of inverted terminal repetition	102	0.28%	3,2
*Bal*I	267	0.74%	-
Cap site for E1a mRNAs	498	1.38%	9,44
Initiator AUG for E1a proteins	559	1.55%	45 (Ad5)
DONOR splice site for E1a 9s mRNA	636	1.76%	46
DONOR splice site for E1a 12s mRNA	973	2.70%	5
*Xma*I	1006	2.79%	-
DONOR splice site for E1a 13s mRNA	1111	3.09%	5
Terminator UAA for 18.1K URF	1196	3.32%	-
ACCEPTOR splice site for all E1a mRNAs	1226	3.41%	5,46
Terminator UGA for E1a 9s mRNA 6K polypeptide	1313	3.65%	-
*Xba*I	1336	3.71%	-
Terminator UAA for E1a polypeptides	1540	4.28%	-
*Hpa*I	1569	4.36%	-
AAUAAA for E1a mRNAs	1608	4.47%	-
Poly-A addition site for E1a mRNAs	1630	4.53%	5
Cap site for E1b mRNAs	1699	4.72%	9,44
Initiator AUG for E1b 15K polypeptide	1711	4.76%	47
1st AUG in 18.1K URF	1712	4.76%	-
*Sac*I	1767	4.91%	-
Initiator AUG for E1b 57K polypeptide	2016	5.60%	48
*Kpn*I	2045	5.69%	-
*Bal*I	2063	5.74%	-
Terminator UGA for E1b 15K polypeptide	2236	6.22%	-
DONOR splice site for E1b 13s, 1.26 & 1.31kb mRNAs	2249	6.25%	49
*Bal*I	2767	7.69%	-
*Hind*III	2798	7.78%	-
*Bal*I	3156	8.78%	-
ACCEPTOR splice site for E1b 1.31kb mRNA	3212	8.93%	50
Terminator UAA for E1b 1.31kb mRNA	3254	9.05%	-
ACCEPTOR splice site for E1b 1.26kb mRNA	3270	9.09%	50
*Bgl*II	3322	9.24%	-
Terminator UGA for E1b 57K polypeptide	3501	9.74%	-

DONOR splice site for E1b 22s, 1.26 & 1.31kb mRNAs	3504	9.75%	49,50
Cap site for polypeptide IX mRNA	3576	9.95%	7,9
ACCEPTOR splice site for all E1b mRNAs	3589	9.98%	49
Initiator AUG for polypeptide IX	3600	10.01%	47
*Sac*I	3632	10.10%	-
*Xma*I	3931	10.93%	-
Terminator UAA for polypeptide IX	4020	11.18%	-
AAUAAA for E1b and polypeptide IX mRNAs	4029	11.21%	-
Poly-A addition site for IVa2 mRNA	4050	11.26%	7
*Bcl*I	4056	11.28%	-
Poly-A addition site for E1b & polypeptide IX mRNA	4061	11.30%	7
Terminator UAA for IVa2	4083	11.36%	-
AAUAAA for IVa2 mRNA	4085	11.36%	-
*Xma*I	4110	11.43%	-
Terminator UAG for DNA polymerase	5189	14.43%	-
Terminator UGA for 12.2K URF	5329	14.82%	-
*Bal*I	5360	14.91%	-
ACCEPTOR splice site for IVa2 mRNA	5417	15.07%	51 (Ad5)
*Sac*I	5634	15.67%	-
1st AUG in 12.2K URF	5674	15.78%	-
DONOR splice site for IVa2 mRNA	5696	15.84%	51 (Ad5)
1st AUG in IVa2 reading frame	5708	15.88%	-
*Xho*I	5778	16.07%	-
Cap site for IVa2 mRNA	5826	16.21%	9
Cap site for major late transcript	6039	16.80%	9
DONOR splice site for 1st late leader	6079	16.91%	13,11,15,14
*Hind*III	6231	17.33%	-
1st AUG in 13.5K polypeptide reading frame	6280	17.47%	-
Terminator UGA for 12.8K URF	6444	17.93%	-
*Xma*I	6573	18.29%	-
Terminator UAG for 13.5K polypeptide	6598	18.35%	-
1st AUG in 12.8K URF	6780	18.86%	-
ACCEPTOR splice site for 2nd late leader	7101	19.75%	13,11,15,14
DONOR splice site for 2nd late leader	7172	19.95%	13,11,15,14
*Sac*I	7611	21.17%	-

*Sac*I	7684	21.38%	-
*Bal*I	7751	21.56%	-
ACCEPTOR splice site for 'i' leader	7942	22.09%	17,16
Initiator AUG for 13.6K (16K) polypeptide	7968	22.17%	52
*Xho*I	8244	22.94%	-
1st AUG in DNA polymerase reading frame	8357	23.25%	-
*Sac*I	8364	23.27%	-
DONOR splice site for 'i' leader	8381	23.32%	17,53,16
Terminator UGA for 22.8K URF	8385	23.33%	-
*Sac*I	8391	23.34%	-
*Kpn*I	8523	23.71%	-
Terminator UAG for Terminal protein	8575	23.86%	-
*Bgl*II	8904	24.77%	-
1st AUG in 22.8K URF	9030	25.12%	-
*Sac*I	9062	25.21%	-
*Hpa*I	9270	25.79%	-
1st AUG in 17.7K URF	9294	25.86%	-
*Sac*I	9296	25.86%	-
*Sal*I	9452	26.30%	-
ACCEPTOR splice site for 3rd late leader	9634	26.80%	13,11,15,14
Terminator UGA for 13.6K polypeptide	9655	26.87%	-
*Xho*I	9689	26.96%	-
DONOR splice site for 3rd late leader	9723	27.05%	13,11,15,14
Terminator UAA for 17.7K URF	9798	27.26%	-
*Sal*I	9831	27.35%	-
*Bal*I	10210	28.41%	-
*Hpa*I	10226	28.45%	-
*Sac*I	10253	28.53%	-
1st AUG in 14.4K URF	10421	28.99%	-
1st AUG in Terminal protein reading frame	10534	29.31%	-
*Xba*I	10579	29.43%	-
5' end of VA I RNA	10610	29.52%	31,54,32,55
*Bam*HI	10680	29.71%	-
Terminator UGA for 12.76K URF	10746	29.90%	-
3' end of VA I RNA	10766	29.95%	31,54,32
*Bal*I	10810	30.08%	-
Terminator UAA for 14.4K URF	10832	30.14%	-

5′ end of VA II RNA	10866	30.23%	31
3′ end of VA II RNA	11023	30.67%	31
1st AUG in 52,55K polypeptide reading frame	11040	30.72%	-
ACCEPTOR splice site for 52,55K mRNA	11040	30.72%	56
1st AUG in 12.76K URF	11109	30.91%	-
Terminator UAG for 13.6K URF	11252	31.31%	-
*Hin*dIII	11555	32.15%	-
1st AUG in 13.6K URF	11618	32.32%	-
*Sac*I	12039	33.50%	-
Terminator UAA for 52,55K polypeptide	12285	34.18%	-
*Bcl*I	12301	34.22%	-
1st AUG in polypeptide IIIa reading frame	12308	34.24%	-
*Xma*I	13053	36.32%	-
*Hin*dIII	13636	37.94%	-
*Hin*dIII	13711	38.15%	-
Terminator UAA for 13.54K URF	13730	38.20%	-
Terminator UAA for polypeptide IIIa	14063	39.13%	-
AAUAAA for L1 mRNAs	14092	39.21%	-
Poly-A addition site for L1 mRNAs	14113	39.27%	34
1st AUG in 13.54K URF	14114	39.27%	-
ACCEPTOR splice site for penton mRNA	14150	39.37%	34
1st AUG in penton reading frame	14151	39.37%	-
*Kpn*I	14281	39.73%	-
Terminator UAG for 16.4K URF	14426	40.14%	-
*Xma*I	14456	40.22%	-
1st AUG in 16.4K URF	14861	41.35%	-
*Hin*dIII	15032	41.82%	-
*Bcl*I	15283	42.52%	-
*Kpn*I	15367	42.76%	-
*Bam*HI	15403	42.86%	-
Terminator UGA for penton	15864	44.14%	-
Initiator AUG for Pro-VII	15873	44.16%	36
Gly/Ala cleavage site for Pro-VII to VII	15945	44.36%	36
Terminator UAG for Pro-VII	16467	45.82%	-
ACCEPTOR splice site for pV mRNA	16516	45.95%	35
1st AUG in pV reading frame	16539	46.02%	-
*Bgl*II	16588	46.15%	-

*Sal*I	16744	46.59%	-
*Kpn*I	17066	47.48%	-
1st AUG in 17.4K URF	17284	48.09%	-
Terminator UAA for pV	17646	49.10%	-
Terminator UGA for 17.4K URF	17761	49.42%	-
AAUAAA for L2 mRNAs	17949	49.94%	-
Poly-A addition site for L2 mRNAs	17969	50.00%	37,34
ACCEPTOR splice site for pVI mRNA	18000	50.08%	37
Initiator AUG for pVI	18001	50.09%	38
Terminator UAG for 19.0K URF	18161	50.53%	-
*Bal*I	18281	50.86%	-
*Hind*III	18316	50.96%	-
1st AUG in 19.0K URF	18707	52.05%	-
Terminator UAA for pVI	18751	52.17%	-
ACCEPTOR splice site for hexon mRNA	18802	52.31%	19,13
Initiator AUG for hexon	18838	52.41%	57
*Xma*I	18913	52.62%	-
*Xho*I	19331	53.79%	-
*Xma*I	20416	56.81%	-
1st AUG in 15.7K URF	20504	57.05%	-
*Hpa*I	20650	57.46%	-
*Sac*I	20667	57.50%	-
Terminator UGA for 15.7K URF	20933	58.24%	-
*Bal*I	21047	58.56%	-
*Bal*I	21197	58.98%	-
*Eco*RI	21338	59.37%	-
*Bam*HI	21606	60.12%	-
ACCEPTOR splice site for 23K mRNA	21650	60.24%	23
Terminator UAA for Hexon	21742	60.50%	-
1st AUG in 23K polypeptide reading frame	21778	60.60%	-
*Bgl*II	21816	60.70%	-
*Kpn*I	22233	61.86%	-
Terminator UAA for 23K polypeptide	22390	62.30%	-
AAUAAA for L3 mRNAs	22418	62.38%	-
Poly-A addition site for E2a mRNAs	22420	62.38%	23,34
AAUAAA for E2a mRNAs	22439	62.43%	-
Poly-A addition site for L3 mRNAs	22443	62.45%	23,34

Terminator UAA for 72K protein	22492	62.58%	-
*Bal*I	22765	63.34%	-
*Bgl*II	23086	64.24%	-
1st AUG in 100K reading frame	23526	65.46%	-
1st AUG in 13.0K URF	23782	66.17%	-
*Bal*I	23836	66.32%	-
*Xho*I	23924	66.57%	-
*Bal*I	24073	66.98%	-
1st AUG in 72K protein reading frame	24079	67.00%	-
ACCEPTOR splice site for main body of 72K mRNA	24088	67.02%	26,58
ACCEPTOR splice site for 100K mRNA	24095	67.04%	59
1st AUG in 100K polypeptide mRNA	24108	67.08%	-
Terminator UAA for 13.0K URF	24136	67.16%	-
1st AUG in 14.2K URF	24481	68.12%	-
DONOR splice site for 2nd leader in 72K RNA	24715	68.77%	26,58
ACCEPTOR splice site for 2nd leader in 72K mRNA	24791	68.98%	26,58
Terminator UAG for 14.2K URF	24865	69.19%	-
*Sac*I	25251	70.26%	-
*Eco*RI	25633	71.32%	-
*Kpn*I	25881	72.01%	-
DONOR splice site for 1st leader of 72K mRNA late	25886	72.03%	59
Cap site for 72K mRNA late	25954	72.22%	9
Alternative cap site for 72K mRNA late	25956	72.22%	9
*Bal*I	26087	72.59%	-
*Sac*I	26175	72.83%	-
1st AUG in 33K polypeptide reading frame	26239	73.01%	-
*Hin*dIII	26369	73.37%	-
Terminator UAG for 100K polypeptide	26523	73.80%	-
Possible DONOR splice site for 33K mRNA	26551	73.88%	60
Possible ACCEPTOR splice site for 33K mRNA	26754	74.44%	60
DONOR splice site for 1st leader of 72K mRNA	27025	75.20%	29,26
Cap site for 72K mRNA early	27091	75.38%	9
Cap site for 72K mRNA early	27092	75.38%	9
Terminator UAG for 33K polypeptide	27125	75.47%	-
1st AUG in pVIII reading frame	27215	75.72%	-
*Xma*I	27339	76.07%	-

*Eco*RI	27372	76.16%	-
*Xma*I	27569	76.71%	-
Cap site for E3 mRNAs	27609	76.82%	9
*Sac*I	27634	76.89%	-
Terminator UGA for pVIII	27896	77.62%	-
1st AUG in 12.4K URF	27899	77.63%	-
DONOR splice site for 'x' leader	27980	77.85%	61
*Bgl*II	28174	78.39%	-
AAUAAA for L4 mRNAs	28205	78.48%	-
Terminator UAG for 12.4K URF	28220	78.52%	-
Poly-A addition site for L4 mRNAs	28223	78.53%	34
ACCEPTOR splice site for 'y' leader	28376	78.96%	61,53,15
DONOR splice site for 'y' leader	28559	79.46%	61,53,15
*Sac*I	28562	79.47%	-
*Hin*dIII	28653	79.73%	-
Initiator AUG for E3 19K glycoprotein	28812	80.17%	62
*Hin*dIII	28962	80.59%	-
Terminator UGA for E3 19K glycoprotein	29289	81.50%	-
*Bcl*I	29433	81.90%	-
1st AUG in E3 11.6K reading frame	29468	81.99%	63
Terminator UAA for E3 11.6K polypeptide	29771	82.84%	-
*Xho*I	29788	82.88%	-
Poly-A addition site for E3-1 mRNAs	29792	82.90%	40
Poly-A addition site for E3-1 mRNAs	29799	82.92%	40,61
Poly-A addition site for E3-1 mRNAs	29801	82.92%	40
Poly-A addition site for E3-1 mRNAs	29804	82.93%	40
*Eco*RI	30046	83.60%	-
1st AUG in 14.5K URF	30059	83.64%	-
ACCEPTOR splice site for 'z' leader	30438	84.69%	61,53
1st AUG in 14.7K URF	30444	84.71%	-
Terminator UGA for 14.5K URF	30449	84.72%	-
*Xba*I	30455	84.74%	-
*Bgl*II	30458	84.75%	-
*Xba*I	30461	84.76%	-
*Hpa*I	30558	85.03%	-
DONOR splice site for 'z' leader	30582	85.09%	61,53
*Bgl*II	30809	85.73%	-

Terminator UAA for 14.7K URF	30828	85.78%	-
AAUAAA for E3 mRNAs	30842	85.82%	-
Terminator UAA for 13.5K URF	30854	85.85%	-
Poly-A addition site for E3-2 mRNAs	30864	85.88%	61
ACCEPTOR splice site for fiber mRNA	31030	86.34%	53,15,64,14
Initiator AUG for fiber	31030	86.34%	47
1st AUG in 13.5K URF	31211	86.84%	-
Terminator UGA for 10.7K URF	31292	87.07%	-
1st AUG in 10.7K URF	31604	87.94%	-
*Xba*I	31830	88.57%	-
*Hin*dIII	32172	89.52%	-
*Eco*RI	32264	89.77%	-
*Bgl*II	32358	90.04%	-
AAUAAA for L5 mRNAs	32774	91.19%	-
Terminator UAA for fiber	32776	91.20%	-
Poly-A addition site for L5 mRNAs	32798	91.26%	34
Poly-A addition site for E4 mRNAs	32802	91.27%	65,66
AAUAAA for E4 mRNAs	32821	91.32%	-
*Bcl*I	32891	91.52%	-
Final terminator UGA in E4	32916	91.59%	-
*Xma*I	33091	92.08%	-
Main ACCEPTOR splice site for large E4 intron	33192	92.36%	65,67,66
Terminator UAG for 34.1K URF	33195	92.36%	-
Secondary ACCEPTOR splice site for large E4 intron	33283	92.61%	65
ACCEPTOR splice site in E4	33376	92.87%	67
ACCEPTOR splice site in E4	33403	92.94%	67
ACCEPTOR splice site in E4	33451	93.08%	67
*Bal*I	33491	93.19%	-
*Kpn*I	33594	93.48%	-
ACCEPTOR splice site in E4	33609	93.52%	67
*Bal*I	33651	93.63%	-
ACCEPTOR splice site in E4	33678	93.71%	67
ACCEPTOR splice site in E4	33874	94.25%	67
DONOR splice site for large intron in E4	33904	94.34%	65,67,66
1st AUG in 18.8K URF	33956	94.48%	-
Terminator UAG in 13.3K URF	34000	94.61%	-
1st AUG in 34.1K URF	34077	94.82%	-

ACCEPTOR splice site in E4	34082	94.83%	65
*Bgl*II	34115	94.93%	-
ACCEPTOR splice site in E4	34241	95.28%	65
DONOR splice site in E4	34288	95.41%	66
ACCEPTOR splice site in E4	34329	95.52%	65
1st AUG in 13.3K URF	34342	95.56%	-
Variable run of A's (13-16) in E4	34357	95.60%	1
Terminator UAA for E4 11K protein	34358	95.60%	-
ACCEPTOR splice site in E4	34379	95.66%	67
*Bgl*II	34390	95.69%	-
ACCEPTOR splice site in E4	34435	95.82%	65
Terminator UAA for 18.8K URF	34454	95.87%	-
1st AUG in 12.7K URF	34470	95.91%	-
DONOR splice site in E4	34606	96.29%	65
Terminator UGA for 15.3K URF	34705	96.57%	-
Initiator AUG for E4 11K protein	34706	96.57%	68 (Ad5)
ACCEPTOR splice site in E4	34735	96.65%	65
Terminator UGA for 12.7K URF	34806	96.85%	-
*Hind*III	34933	97.20%	-
2nd AUG in 15.3K URF	35095	97.65%	-
ACCEPTOR splice site in E4	35107	97.69%	66
Terminator UAA for 14.3K URF	35148	97.80%	-
*Hpa*I	35323	98.29%	-
*Xma*I	35360	98.39%	-
1st AUG in 14.3K URF	35532	98.87%	-
DONOR splice site for 1st leader of E4 mRNAs	35548	98.91%	65,66
Cap sites for E4 mRNAs	35609 - 35614	99.10%9,69	
End of inverted terminal repetition	35836	99.71%	3,2

REFERENCES

1. Gingeras, T.R., Sciaky, D., Gelinas, R.E., Bing-Dong, J., Yen, C.E., Kelly, M.M., Bullock, P.A., Parsons, B.L., O'Neill, K.E. and Roberts, R.J. (1982). *J. Biol. Chem.* **257**, 13475-13491.
2. Shinagawa, M. and Padmanabhan, R. (1979). *Biochem. Biophys. Res. Com.* **87**, 671-678.
3. Arrand, J.R. and Roberts, R.J. (1979). *J. Mol. Biol.* **128**, 577-594.
4. Osborne, T.F., Schell, R.E., Burch-Jaffe, E., Berget, S.J. and Berk, A.J. (1981). *Proc. Natl. Acad. Sci. USA* **78**, 1381-1385.

5. Perricaudet, M., Akusjarvi, G., Virtanen, A. and Pettersson, U. (1979). *Nature* **281**, 694-696.
6. Fraser, N.W., Baker, C.C., Moore, M.A. and Ziff, E.B. (1982). *J. Mol. Biol.* **155**, 207-233.
7. Alestrom, P., Akusjarvi, G., Perricaudet, M., Mathews, M.B., Klessig, D.F. and Pettersson, U. (1980). *Cell* **19**, 671-681.
8. Alestrom, P., Akusjarvi, G., Pettersson, M. and Pettersson, U. (1982). *J. Biol. Chem.* **257**, 13492-13498.
9. Baker, C.C. and Ziff, E.B. (1981). *J. Mol. Biol.* **149**, 189-221.
10. Baker, C.C. and Ziff, E.B. (1979). *Cold Spring Harb. Symp. Quant. Biol.* **44**, 415-428.
11. Akusjarvi, G. and Pettersson, U. (1979). *J. Mol. Biol.* **134**, 143-158.
12. Ziff, E.B. and Evans, R.M. (1978). *Cell* **15**, 1463-1475.
13. Akusjarvi, G. and Pettersson, U. (1979). *Cell* **16**, 841-850.
14. Zain, S., Sambrook, J., Roberts, R.J., Keller, W., Fried, M. and Dunn, A.R. (1979). *Cell* **16**, 851-861.
15. Zain, B.S., Gingeras, T.R., Bullock, P., Wong, G. and Gelinas, R.E. (1979). *J. Mol. Biol.* **135**, 413-433.
16. Virtanen, A., Alestrom, P., Persson, H., Katze, M.G. and Pettersson, U. (1982). *Nucleic Acids Res.* **10**, 2539-2548.
17. Falvey, E. and Ziff, E. (1983). *J. Virol.* **45**, 185-191.
18. Mautner, V. and Boursnell, M.E.G. (1983). *Virology* **131**, 1-10.
19. Akusjarvi, G. and Pettersson, U. (1978). *Proc. Natl. Acad. Sci. USA* **75**, 5822-5826.
20. Jornvall, H., Alestrom, P., Akusjarvi, G., Von Bahr-Lindstrom, H., Philipson, L. and Pettersson, U. (1981). *J. Biol. Chem.* **256**, 6204-6212.
21. Akusjarvi, G., Alestrom, P., Pettersson, M., Lager, M., Jornvall, H. and Pettersson, U. *J. Biol. Chem.*, submitted.
22. Akusjarvi, G. and Pettersson, U. (1978). *Virology* **91**, 477-480.
23. Akusjarvi, G., Zabielski, J., Perricaudet, M. and Pettersson, U. (1981). *Nucleic Acids Res.* **9**, 1-17.
24. Buttner, W. and Veres-Molnar, Z. (1980). *FEBS Lett.* **122**, 317-321.
25. Kruijer, W., van Schaik, F.M.A. and Sussenbach, J.S. (1982). *Nucleic Acids Res.* **10**, 4493-4500.
26. Goldenberg, C.J. and Hauser, S.D. (1983). *Nucleic Acids Res.* **11**, 1337-1348.
27. Roberts, R.J., O'Neill, K.E. and Yen, C.E. *J. Biol. Chem.*, submitted.
28. Galibert, F., Herisse, J. and Courtois, G. (1979). *Gene* **6**, 1-22.
29. Baker, C.C., Herisse, J., Courtois, G., Galibert, F. and Ziff, E. (1979). *Cell* **18**, 569-580.
30. Pan, J., Celma, M.L. and Weissman, S.M. (1977). *J. Biol. Chem.* **252**, 9047-9054.
31. Akusjarvi, G., Mathews, M.B., Andersson, P., Vennstrom, B. and Pettersson, U. (1980). *Proc. Natl. Acad. Sci. USA* **77**, 2424-2428.
32. Ohe, K. and Weissman, S.M. (1971). *J. Biol. Chem.* **246**, 6991-7009.
33. Celma, M.L., Pan, J. and Weissman, S.M. (1977). *J. Biol. Chem.* **252**, 9032-9042.
34. LeMoullec, J.M., Akusjarvi, G., Stalhandske, P., Pettersson, U., Chambraud, B., Gilardi, P., Nasri, M. and Perricaudet, M. (1983). *J. Virol.* **48**, 127-134.
35. Alestrom, P., Akusjarvi, G., Lager, M., Yeh-kai, L. and Pettersson, U. *J. Biol. Chem.*, submitted.
36. Sung, M.T., Cao, T.M., Coleman, R.T. and Budelier, K.A. (1983). *Proc. Natl. Acad. Sci. USA* **80**, 2902-2906.
37. Akusjarvi, G. and Persson, H. (1981). *J. Virol.* **38**, 469-482.

38. Sung, M.T., Cao, T.M., Lischwe, M.A. and Coleman, R.T. (1983). *J. Biol. Chem.* **258**, 8266-8272.
39. Herisse, J., Courtois, G. and Galibert, F. (1980). *Nucleic Acids Res.* **8**, 2173-2192.
40. Ahmed, C.M.I., Chanda, R.S., Stow, N.D. and Zain, B.S. (1982). *Gene* **20**, 339-346.
41. Herisse, J. and Galibert, F. (1981). *Nucleic Acids Res.* **9**, 1229-1240.
42. Herisse, J., Rigolet, M., Dupont de Dinechin, S. and Galibert, F. (1981). *Nucleic Acids Res.* **9**, 4023-4042.
43. Shinagawa, M., Padmanabhan, R.V. and Padmanabhan, R. (1980). *Gene* **9**, 99-114.
44. Hashimoto, S., Wold, W.S.M., Brackmann, K.H. and Green, M. (1981). *Biochemistry* **20**, 6640-6647.
45. Cladaras, and Wold, W. (1984). in press.
46. Virtanen, A. and Pettersson, U. (1983). *J. Mol. Biol.* **165**, 496-499.
47. Anderson, C.W. and Lewis, J.B. (1980). *Virology* **104**, 27-41.
48. Anderson, C.W., Schmitt, R.C., Smart, J.E. and Lewis, J.B. (1984). *J. Virol.*, in press.
49. Perricaudet, M., Michel le Moullec, J. and Pettersson, U. (1980). *Proc. Natl. Acad. Sci. USA* **77**, 3778-3782.
50. Virtanen, A. and Pettersson, U., submitted for publication.
51. van Beveren, C.P., Maat, J., Dekker, B.M.M. and van Ormondt, H. (1981). *Gene* **16**, 179-189.
52. Lewis, J.B. and Anderson, C.W. (1983). *Virology* **127**, 112-123.
53. Uhlen, M., Svensson, C., Josephson, S., Alestrom, P., Chattapadhyaya, J.B., Pettersson, U. and Philipson, L. (1982). *EMBO J.* **1**, 249-254.
54. Celma, M.L., Pan, J. and Weissman, S.M. (1977). *J. Biol. Chem.* **252**, 9043-9046.
55. Vennstrom, B., Pettersson, U. and Philipson, L. (1978). *Nucleic Acids Res.* **5**, 195-204.
56. Akusjarvi, G. and Persson, H. (1981). *Nature* **292**, 420-426.
57. Jornvall, H., Ohlsson, H. and Philipson, L. (1974). *Biochem. Biophys. Res. Comm.* **56**, 304-310.
58. Kruijer, W., van Schaik, F.M.A. and Sussenbach, J.S. (1981). *Nucleic Acids Res.* **9**, 4439-4457.
59. Kruijer, W., van Schaik, F.M.A., Speijer, J.G. and Sussenbach, J.S. (1983). *Virology* **128**, 140-153.
60. Oosterom-Dragon, E.A. and Anderson, C.W. (1983). *J. Virol.* **45**, 251-263.
61. Stalhandske, P., Persson, H., Perricaudet, M., Philipson, L. and Pettersson, U. (1983). *Gene* **22**, 157-165.
62. Persson, H., Jornvall, H. and Zabielski, J. (1980). *Proc. Natl. Acad. Sci. USA* **77**, 6349-6353.
63. Downey, J.F., Evelegh, C.M., Branton, P.E. and Bayley, S.T. (1984). *J. Virol.* **50**, 30-37.
64. Zain, B.S. and Roberts, R.J. (1979). *J. Mol. Biol.* **131**, 341-352.
65. Freyer, G.A., Katoh, Y. and Roberts, R.J. (1984). *Nucleic Acids Res.* **12**, 3503-3519.
66. Virtanen, A., Gilardi, P., Naslund, A., LeMoullec, M., Pettersson, U. and Perricaudet, M., *J. Virol.*, in press.
67. Tigges, M.A. and Raskas, H.J. (1984). *J. Virol.* **50**, 106-117.
68. Downey, J.F., Rowe, D.T., Bacchetti, S., Graham, F.L. and Bayley, S.T. (1983). *J. Virol.* **45**, 514-523.
69. Hashimoto, S., Pursley, M.H. and Green, M. (1981). *Nucleic Acids Res.* **9**, 1675-1689.

2

STRUCTURE AND FUNCTION OF THE ADENOVIRUS-2 GENOME

GÖRAN AKUSJÄRVI, ULF PETTERSSON and RICHARD J. ROBERTS*
Department of Medical Genetics, Biomedical Center, Box 589, S-751 23
Uppsala, Sweden
*Cold Spring Harbor Laboratory, P.O. Box 100, New York 11724, USA

SUMMARY

As a result of work in several laboratories the complete DNA sequence of the 35,937 base pairs long Ad2 genome has been established. In this paper we provide an overview of the biological implications of the sequence.

INTRODUCTION

Adenovirus was first discovered in 1953 as an agent causing upper respiratory tract infections in man (1,2). Subsequent studies have shown that adenoviruses are widespread in nature and around 100 different serotypes have been isolated (for reviews see refs. 3 and 4). The human adenoviruses comprise 41 distinct serotypes which cause a variety of ailments such as acute respiratory, ocular, gastrointestinal and urinary tract diseases (see ref.3). Some serotypes cause latent infections without any apparent symptoms.

The discovery in 1962 (5,6) that certain adenovirus serotypes have an oncogenic potential and are able to induce tumors in rodents caused a tremendous surge of interest in the study of the molecular biology of human adenoviruses. This work has led to many important discoveries which have had implications reaching far beyond the adenovirus field. Most of our current knowledge stems from studies of the human serotypes 2 and 5, which belong to the non-oncogenic subgroup C (7). They serve as the prototypes for the human adenovirus family.

Adenoviruses are non-enveloped, icosahedral viruses, that replicate and assemble their virions in the nucleus of the infected cell. The capsid consists of 252 capsomers; 240 hexons form the facets of the icosahedron, and 12 pentons are located at the corners of the virus particle. The pentons consist of the penton base or the vertex capsomer

W. Doerfler (ed.), ADENOVIRUS DNA.

and a noncovalently attached protein, known as the fiber (for a review see ref. 8). The fiber serves as the attachment organ when the virion binds to its receptor on the surface of the host cell. The HeLa cell receptor appears to be a 40-42 K glycoprotein which exists in approximately 10^5 copies per cell (9). After attachment of the virion to the cell, the viral DNA is transported to the nucleus where early transcription takes place. The incoming viral DNA is associated with basic histone-like proteins (polypeptides V and VII) which are encoded by the viral genome. Early after infection the adenovirus chromatin changes to a nucleosomal structure, probably by replacement of the viral core proteins with cellular histones (10,11). At late times after infection, when viral DNA replication provides an excess of DNA for virus assembly, the DNA again becomes associated with the viral core proteins and is encapsidated in a preformed head, aided by a specific sequence located at the left end of the genome (12) (see ref. 13 for a review of adenovirus assembly).

The genomes of adenoviruses are linear, double-stranded DNA molecules with molecular weights of $20\text{-}25 \times 10^6$ (14,15). In the case of Ad2 the genome has a length of $35{,}937 \pm 9$ base pairs (16). The ambiguity in length is the result of two regions of heterogeneity that are found both within individual Ad2 stocks and between strains held in different laboratories. Both ends of the viral genome function as origins for DNA replication (17,18,19) and contain a protein, the terminal protein (TP), which is covalently linked to the 5'-end of each DNA strand (20,21) via a serine residue (22). The ends of the viral DNA contain an inverted terminal repetition (23,24) with a length of 102 base pairs (25).

Adenovirus gene expression is strictly regulated at multiple levels during a productive infection. The lytic cycle is traditionally divided into two distinct phases; an early phase precedes viral DNA replication and is triggered by gene products from early region 1A, and a late phase which follows DNA replication and is characterized by the expression of the structural proteins (Fig. 1).

This article outlines the key functions encoded by the viral genome which lead to a successful virus multiplication. It is organized so as to reflect the usual pathway occurring during a lytic infection by the virus.

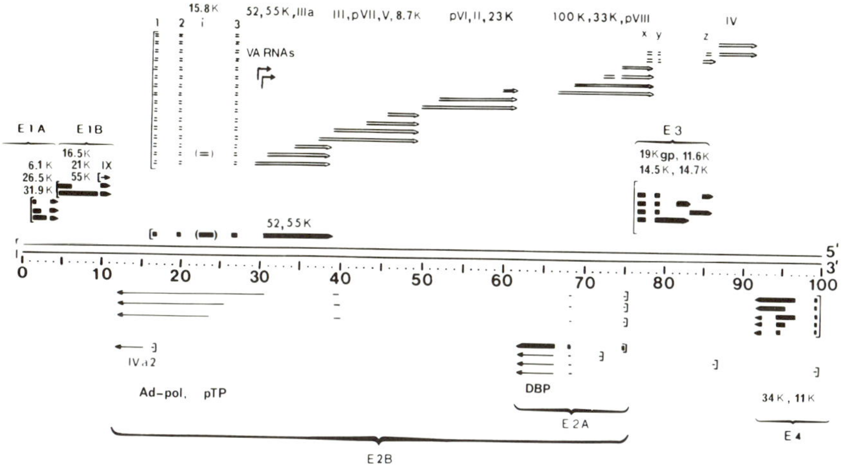

Fig. 1: Organization of the Ad2 genome. Arrows show the location of the major RNA species. Thick lines illustrate mRNAs expressed early after infection and thin lines mRNAs expressed at intermediate times after infection. Open arrows show sequences present in late mRNA. Polypeptides that have been assigned to the different regions are indicated (Modified from ref. 32).

EARLY REGION 1

Region E1 which is located at the left hand end of the viral genome (coordinate 1.4-11.3) contains the genes which are involved in the regulation of early viral transcription (26,27,28) and are responsible for cell transformation (29,30). Mapping of promoter sites and mRNAs within this region (31,31,33,34) has shown the presence of three separate transcription units, designated E1A, E1B and pIX (Fig. 1). The E1A region (coordinates 1.4 to 4.5) and the E1B region (coordinates 4.7 to 11.3) together specify the genes required for cell transformation. The pIX transcription unit overlaps completely with region E1B (coordinates 9.9 to 11.3) and encodes the structural virion polypeptide IX.

The existence of a fourth transcription unit encoded by the l-strand of region E1 has been reported (35). A 22S mRNA, which spans the border between E1A and E1B and encodes an 11K polypeptide, has been identified by in vitro translation of hybridization-selected RNA. However, the location of the promoter and the exact molecular structure of this transcription unit remain to be established.

EARLY REGION 1A (1.4-4.5)

The transcription unit

The structures of the mRNAs transcribed from region E1A have been studied by a variety of methods (31,32,36,37,38). The conclusion drawn from these studies is that three mRNAs of sizes 13S, 12S and 9S are generated by differential splicing of a common nuclear precursor RNA. The mRNAs have common 5'- and 3'-ends but differ from each other by the size of the intron which is removed during mRNA maturation. The E1A mRNAs have one major cap site located at position 499 (34) and multiple minor 5'-ends mapping up to 375 nucleotides upstream from the major cap site (Fig. 1) (39,40,41). These minor initiation sites are predominantly used after the onset of viral DNA replication (40). The cap site is preceded by a TATA motif (42,43) although this is not required for efficient transcription in vivo. Deletion of the TATA box results in only slightly decreased mRNA levels and the mRNAs initiate heterogenously at multiple sites upstream of the major cap site (44,45). The use of upstream transcriptional start sites does not change the coding capacity of the E1A mRNAs (46).

A DNA segment which is located between positions -141 and -305 relative to the E1A cap site has been shown to increase E1A transcription approximately 20-fold early after infection (47).This segment has the properties of an enhancer element and contains two repeated regions which show extensive homology with other eucaryotic enhancers (reviewed in ref. 48). A cis-acting packaging sequence has also been shown to overlap with the enhancer region (12,47). Using transient expression assays additional enhancer elements have been mapped to the E1A region (49,50). A cis-acting transcriptional control signal is present within the protein coding sequence of E1A (51).

The poly(A) addition site for all E1A mRNAs is located at nucleotide 1630 (37). It is preceded by the hexanucleotide sequence AAUAAA, which is specifically required for the cleavage of the precursor RNA prior to its polyadenylation (52).

The E1A mRNAs accumulate with different kinetics during the infectious cycle. The 13S and 12S mRNAs are the most abundant RNA species early after infection, while the 9S mRNA represents less than 5% of the total E1A mRNA. In contrast, at late times, a shift in the steady-state levels of the E1A mRNAs occurs and the 9S mRNA becomes the most abundant species (32,33,53).

The polypeptides which are specified by the 12S and 13S mRNAs are predicted to be 243 and 289 amino acid residues long, with molecular

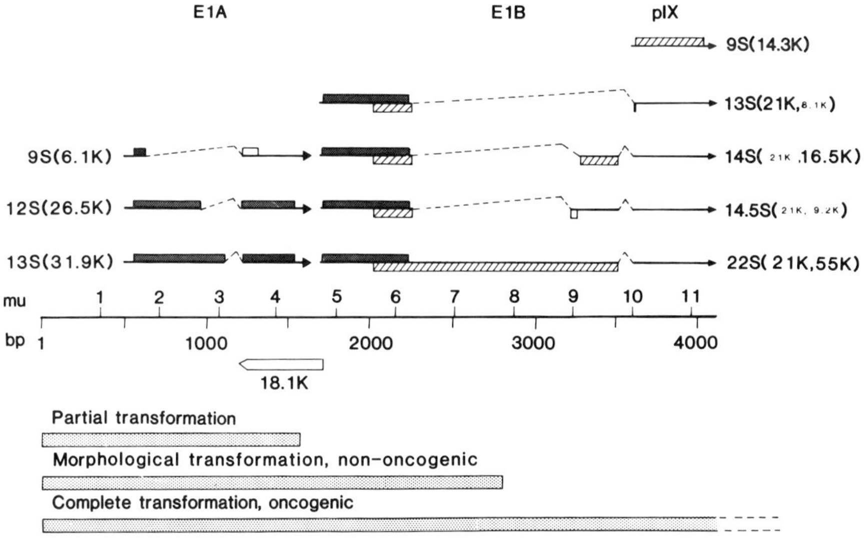

Fig.2: Schematic drawing illustrating the structures of the mRNAs encoded by region E1. Different parts of the mRNAs are translated from different reading frames as indicated by the stipled, hatched and open boxes. The molecular weights of polypeptides encoded by the different mRNAs are shown in bold letters within parenthesis. Small letters indicate sizes of those polypeptides which are likely to be encoded by the E1B mRNAs due to their polycistronic nature (87). The lower part of the figure shows the location of DNA fragments that induce different transformed phenotypes. mu, map units; bp, base pairs from the left hand end of the genome.

weights of 26.5K and 31.9K, respectively (37). They are closely related to each other and contain identical N-terminal and C-terminal ends.

They differ by the deletion of 46 internal amino acids in the shorter polypeptide (Fig. 2). The proteins observed in vivo or by in vitro translation are heterogeneous and have molecular weights much larger than predicted (54,55,56,57,58). This probably reflects a combination of post-translational modification (phosphorylation, etc.) and an unusual amino-acid composition. The C-termini are especially rich in glutamic acid and proline residues.

The polypeptide specified by the 9S mRNA is predicted to be 55 amino acids long. It starts at the same AUG triplet as is used for initiation in the 12S and 13S mRNAs, but a different reading frame is used beyond the splice junction (38). Thus the 26 N-terminal amino acids of the 9S polypeptide are common to the 26.5K and 31.9K polypeptides whereas the C-terminus contains 29 unique amino acids. The predicted molecular weight of this polypeptide is 6.1 K, which is far less than the value of 28K that has been reported for the protein product translated in vitro from this mRNA (59,60).

E1A and regulation of adenovirus transcription

During the infectious cycle the five early transcription units are expressed with different kinetics (61). Region E1A is the first to be expressed and transcripts can be detected within 45 min. after infection. Studies using viral mutants have shown that a gene product from E1A stimulates transcription from the other viral promoters approximately 50-fold early after infection (26,27,28). In contrast, transcription of E1A itself is stimulated only 5-fold by this same E1A regulatory protein (27,51).

Characterization of viral mutants (62,63,64), results from transient expression experiments (65,66,67), microinjection of plasmids (68) and microinjection of purified E1A protein made in E.coli (69) have shown that the 289 amino acid residues long (289R) protein encoded by the E1A 13S mRNA is responsible for the stimulation of early viral transcription. As expected for a protein whose primary function seems to be in the regulation of transcription, the E1A 289R protein has been shown to be a nuclear antigen (57,58). Although the 243 amino acid residues long (243R) product encoded by the E1A 12S mRNA is inefficient in stimulating viral transcription (70,71), it has been shown that in a transient expression assay both the 289R and 243R proteins are capable of stimulating transcription from the E3 and the E2A early promoters (72).

The 243R protein is not required for viral replication under standard cell culture conditions (64). However, it is needed for efficient viral replication in growth-arrested cells (71). It may therefore be responsible for inducing cells to progress from G1 to S phase during the cell cycle (73).

Adenovirus gene expression can also be activated by the pseudorabies immediate early gene product supplied either by an infecting virus (73a) or by DNA transfection (49). In fact, the herpes mediated activation of early viral transcription is more efficient than the E1A mediated activation. The existence of a cellular E1A activity in some cell lines which is responsible for the induction of viral transcription in the absence of E1A has been suggested (73b).

In addition to stimulating transcription of viral genes the E1A regulatory protein has been shown to induce the synthesis of the cellular heat shock proteins (74,75) and enhance the transcription of the cellular ß-tubulin genes (75a) during the early phase of the infectious cycle. The 289R protein can also stimulate expression of non-viral genes such as rabbit and human ß-globin genes (66,76) and rat preproinsulin I gene (77) when introduced into cells by transfection or infection. However, the corresponding resident genes present in the cellular chromosome remains quiescent (76,77). The stimulation of cellular gene transcription is much less than that observed for adenovirus genes (66).

In contrast to these results the E1A regulatory protein induces expression from adenoviral promoters both when they are present as episomes and when they are integrated into the chromosomal DNA (65,78,79,80). It appears that several pathways must exist for the stimulation of transcription by E1A. At least one pathway involves activation by a change in the state of the viral template (81). A detailed understanding of these mechanisms is an important goal for the future.

EARLY REGION 1B (4.7 - 11.3)

The transcription unit

Region E1B is located immediately adjacent to region E1A and is also transcribed in the rightward direction (Fig. 2). Two major mRNAs (22S and 13S) and two minor mRNAs (14.5S and 14S) are generated by splicing of a common precursor RNA (31,32,36,82). The E1B mRNAs differ from each other by the sizes of the introns that are removed (Fig. 2). The 22S and 13S mRNAs are the predominent species whereas the 14.5S and 14S mRNAs represent less than 5% of the total steady-state level of E1B mRNAs (82).

The accumulation of E1B mRNAs is subjected to post-transcriptional regulation (32,53,83). Early after infection the E1B 22S and 13S mRNAs are present in equal amounts. However, by 20 hours postinfection (hpi) approximately 20-fold more 13S mRNA than 22S mRNA has accumulated (53). This change in abundance of cytoplasmic E1B mRNAs is in part due to a 5- to 10-fold increase in the half-life of the 13S mRNA as compared to the 22S mRNA (83). The stability of the E1B mRNAs is influenced by the E2A-72K DNA binding protein (84) and at late times, probably by a change in the specificity of the RNA splicing machinery (85).

Region E1B specifies three long open translational reading frames (ORFs) which encode polypeptides with predicted molecular weights of 21K, 55K and 14.3K (Fig. 2). The 14.3K polypeptide corresponds to virion polypeptide IX (pIX) (see below).

The 21K and the 55K ORFs overlap each other for 220 base pairs (bp). The intron in the 13S mRNA removes most of the 55K ORF and translation of the 13S mRNA gives rise to the 21K polypeptide (60,86). Both ORFs are present in their entirety in the 22S mRNA and experimental evidence has shown that this mRNA is functionally polycistronic. It can be translated to give both the 21K and 55K polypeptides (87).

The 14.5S and 14S mRNAs have the capacity to encode both the intact 21K polypeptide and truncated versions of the 55K polypeptide (82). The novel translational product predicted to be synthesized from the 14S mRNA has a molecular weight of 16.5K with N- and C-terminal sequences in common with the 55K polypeptide (82,88). This protein has been found both in vivo and after in vitro translation of size fractionated adenovirus mRNA (60,88,89). The polypeptide, predicted to be translated from the 14.5S mRNA would have a molecular weight of 9.2K. The N-terminus is expected to be identical with that of the 55K polypeptide but, as a result of splicing, it would be connected to a novel C-terminus.

Function of E1B proteins during lytic infection

The E1B-21K polypeptide is associated with both the nuclear and plasma membrane fractions of infected and transformed cells (90,91). It is neither phosphorylated nor glycosylated. Its localization on the cell surface makes it a likely candidate for the tumor specific transplantation antigen (TSTA) expressed by adenovirus-transformed cells.

Analysis of a number of mutants with specific lesions in the 21K ORF has shown that in its absence an enhanced cytopathic effect (cyt phenotype) is observed and both viral and host cell DNA are degraded (92,93,94,95). Since the cyt phenotype, which is characterized by extensive cellular destruction, can be separated from the DNA degradation phenotype, it appears that DNA degradation is not simply a result of the altered cell morphology. Rather it would seem that one specific function of the 21K polypeptide is to serve as a nuclease inhibitor.

Mutants in the 55K protein are reduced approximately 100-fold in virus yield and their phenotype suggests that this protein is involved in the translational regulation seen late after infection (96,97). In mutant-infected cells the inhibition of host cell protein synthesis is not complete and the translation of late proteins and the synthesis of late cytoplasmic mRNAs are drastically reduced. It appears that the 55K protein is required for an efficient transition from the early to the late mode of gene expression.

In adenovirus-transformed cells the E1B-55K polypeptide is found complexed with a cellular protein designated p53 (98) . However, during lytic infection the same 55K protein is found in a physical complex with the adenovirus encoded E4-34K polypeptide (99). Mutants with an altered expression of the E4-34K polypeptide are defective in their inhibition of host cell protein synthesis supporting the idea that an E1B-55K/E4-34K protein complex is involved in the shut-off of cellular gene expression late during the infectious cycle.

Analysis of the cellular localization of the E1B-55K polypeptide has shown that the protein is evenly distributed between the nucleus and cytoplasm at early times whereas it is localized almost exclusively in the cell nucleus at late times (91). The protein is phosphorylated (100,101) and indirect evidence suggests that it may be a protein kinase or at least may be in close association with a protein kinase activity (102).

Role of regions E1A and E1B in morphological transformation and oncogenicity

Human adenoviruses have been divided into three subgroups based on their oncogenic potential (7). However, all human adenoviruses are capable of transforming cultured cells in vitro. In fact, the nononcogenic Ad5 transforms primary rat cells equally well (103) or even better (104) than the highly oncogenic Ad12.

Studies of viral DNA sequences present in adenovirus transformed cells have shown that a DNA fragment containing the E1 region is sufficient to cause morphological transformation (105,106). It is now apparent that expression of the E1A polypeptides together with the E1B-21K polypeptide and the amino terminal end of the E1B-55K polypeptide is sufficient for morphological transformation, whereas expression of the complete E1B-55K polypeptide appears to be required to obtain tumor induction in nude mice (Fig. 2) (107,108).

Cells transformed by a DNA fragment comprising region E1A alone (Ad5 HpaI-E fragment; 0-4.4%) become immortalized but lack many of the characteristics typical for completely transformed cells (109).

Studies of cellular oncogenes have shown that the expression of two separate genes is often required to transform primary cells (110,111). It has been shown that the Ad2 E1A region in conjunction with T24 Harvey ras1 could transform primary cells (110,111). Since Ad2 E1A could substitute for myc in this assay it seems that the E1A and myc proteins might be functionally related. Indeed their primary sequences show certain homologies (112). Both proteins are nuclear antigens (57,113)and recent experiments have shown that they both possess transcriptional activation properties and can induce the heat shock promoter (114).

Both the 289R and 243R proteins are required to obtain complete oncogenic transformation (62,63,70,71,115,116,117). The 243R protein encoded by the E1A 12S mRNA was shown to be particularly important for the anchorage independent growth of the resulting transformants (71).

The immune system of the host is important in determining oncogenic potential. For Ad12 it has been shown that cells expressing the E1A region are oncogenic in immunocompetent rats because they lack class I antigens on the cell surface and thus are not recognized by the cellular immune defense system (118,119). The Ad12 E1A 13S mRNA has been shown to encode the product that causes suppression of class I antigen synthesis (119).

Genes located outside of region E1 are also of importance in determining the transforming and oncogenic properties of adenoviruses. For example the E1 region of Ad12 encodes all the functions which are required for oncogenic transformation as studied in transfection experiments. However, a nondefective hybrid virus containing the Ad12 E1 region linked to Ad5 sequences is non-oncogenic in newborn hamsters (120), emphasizing that oncogenic transformation by adenovirus is a complicated process which involves sequences located outside the normally recognized transforming region (for more elaborate reviews on adenovirus transformation see other chapters in this book).

EARLY REGION 2

The transcription unit (coordinates 11.3-75.4).

Region E2 which is transcribed from the viral l-strand, differs from other adenovirus transcription units by using alternative promoter sites for the initiation of transcription (Fig. 1) (32). At early times after infection a major promoter, located at coordinate 75.4 (E2-E promoter) is activated. mRNAs from this region appear in the cytoplasm approximately 2.5-3 hpi (61). Following the switch from the early to the late phase, a promoter shift occurs, and the E2 mRNAs are preferentially transcribed from a promoter located at coordinate 72.2 (E2-L promoter). Additional minor start sites for E2 transcription have also been reported (32,121).

The two major E2 promoters are regulated differently; the E2-E promoter appears to be stimulated by the E1A gene products, whereas expression from the E2-L promoter seems to be inhibited by E1A (122). The sequence requirements for both transcription and E1A transactivation of the E2-E promoter have been characterized in great detail (80,123,124). A critical sequence required for both transcription and induction by E1A is located within the first 79 nucleotides preceding the E2-E cap site. This upstream sequence resembles an E1A regulated enhancer element and can function independently of its orientation and position (124).

Two major classes of transcripts are generated from region E2. The first mRNA class, designated E2A, extends from the transcriptional initiation site to a polyadenylation site at coordinate 62.4 (Fig. 1). One principal mRNA is generated by the stepwise removal of two intron sequences from the RNA precursor (125,126). The second set of transcripts bypass the polyadenylation signal used by the E2A mRNAs and extends to a second polyadenylation site located at coordinate 11.3. (127). This site appears to coincide with the polyadenylation site used to generate IVa2 mRNA (see below). Because these mRNAs are controlled by the same promoter as the mRNAs from region E2A, this region has been designated E2B.

The 3'-ends of the mature E2A mRNAs are formed by endonucleolytic cleavage of a larger precursor RNA followed by polyadenylation (128). It has been shown that an AT-rich sequence which is located between 20 and 35 nucleotides downstream of the poly(A) site is essential for the formation of the mature polyadenylated E2A mRNAs (129). Similar sequences have been found distal to most other polyadenylation sites in the adenovirus genome (130).

Three differentially spliced mRNAs have been identified from region E2B (127). They accumulate to about 10% of the level of the E2A mRNAs. All three mRNAs have a common set of leader exons from coordinates 76, 68.5 and 39 joined to main bodies which begin at coordinate 30, 26 and 23 respectively (Fig 3).

Since the exact splice points of the E2B mRNAs have not been defined at the nucleotide sequence level the coding potential of the E2B mRNAs cannot be predicted accurately. It is likely that an upstream leader

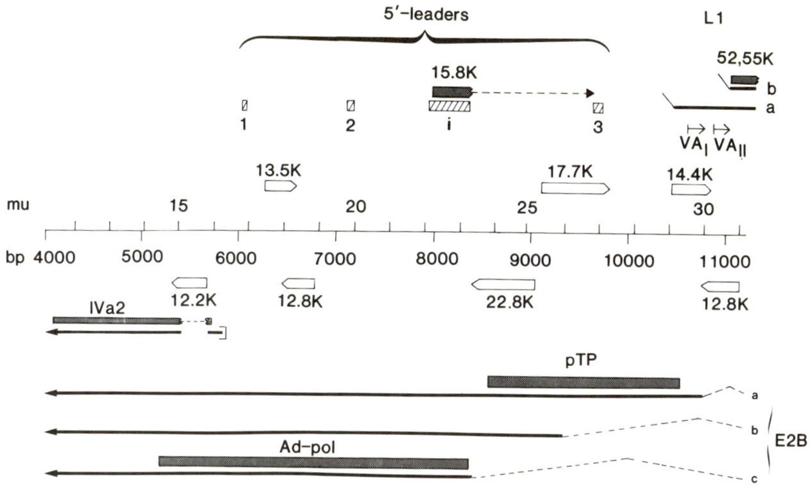

Fig. 3: Position of open reading frames and mRNAs in the r- and l-strand sequence between coordinates 11 and 31. The open blocks signify ORF's with a theoretical capacity to encode polypeptides larger than 10K. Stipled boxes indicate position of ORF's encoding well characterized adenovirus proteins. mu, map units; bp, distance in base pairs from the left hand end of the genome.

exon contains the initiation codon for translation (43,131). By aligning the ORFs present in the l-strand sequence between coordinates 30 and 11 (43,131) with the approximate positions of the E2B mRNAs an estimate of the coding potentials for the different mRNAs can be obtained. The longest of the E2B mRNAs (species a, Fig. 3) encompasses the ORF located between coordinates 29.3 and 23.9. This ORF encodes the 87K terminal protein precursor (pTP) (132) which serves as the primer for DNA replication and which is later cleaved to its mature 55K form by a viral-encoded protease (127). The shortest of the three E2B mRNAs (species c, Fig. 3) encodes a 140K polypeptide which is the DNA polymerase that is required to replicate the viral DNA (133). No protein product has so far been assigned to the intermediate-sized E2B mRNA (species b, Fig. 3). However, an ORF is present between coordinates 25.1 and 23.3 with the capacity to encode a 22.8K polypeptide which is likely to be the protein product specified by this mRNA.

Function of the E2 proteins.

The E2 transcription unit is reminiscent of a bacterial operon in its organization in that the genes known to be required for adenovirus DNA replication are coordinately expressed from the E2 promoter.

Three viral proteins are required for replication of the viral DNA: the 72K DNA binding protein (DBP), the terminal protein precursor (pTP), and the DNA polymerase (Ad-pol). In addition two cellular proteins, nuclear factor I and nuclear factor II, which have been purified from uninfected cells are required for efficient replication of adenovirus DNA in vitro (134,135).

Mutational analysis has shown that a domain encompassing the first 18 bp of the inverted terminal repeat (ITR) is sufficient for initiation of replication (136,137). This sequence contains the core element (positions 9 to 18) which is preserved between all adenovirus serotypes (138,139) and has been suggested to be a binding site for the pTP (140). The sequence between positions 19 and 67 has been shown to enhance the replication efficiency considerably (141,142). Foot-printing experiments have shown that nuclear factor I specifically binds to a domain lying between nucleotides 17 and 48 of the ITR (135). It has been suggested that by binding to the ITR, nuclear factor I unwinds a single stranded region near the termini which then facilitates the binding of the Ad-pol/pTP-dCMP complex. DBP, which is a single strand specific DNA binding protein, interacts with the displaced DNA strand that is generated during replication.

The DBP is a multifunctional protein, which is also involved in the regulation of viral gene expression. In the absence of a functional DBP the normal turn off of early mRNA expression does not occur (143,144). The negative regulation of early mRNA synthesis appears to act at the level of initiation of transcription for region E4 (145, 146) and at the level of mRNA stability for region E1 (84). DBP also influences the host range of the virus; mutants in DBP have been isolated which allow adenovirus to replicate efficiently in monkey cells (147). During an infection of monkey cells by wild type virus the L5-fiber mRNA is abberantly spliced and this block to a successful infection is overcome in these mutants (148).

EARLY REGION 3

The transcription unit (coordinates 76.8-85.9).

Region E3 is transcribed from the viral r-strand and is located completely within the major late transcription unit (31,32,36). The E3 mRNAs are generated by differential processing of two major RNA precursors (Fig. 4). These have a common cap site at coordinate 76.8

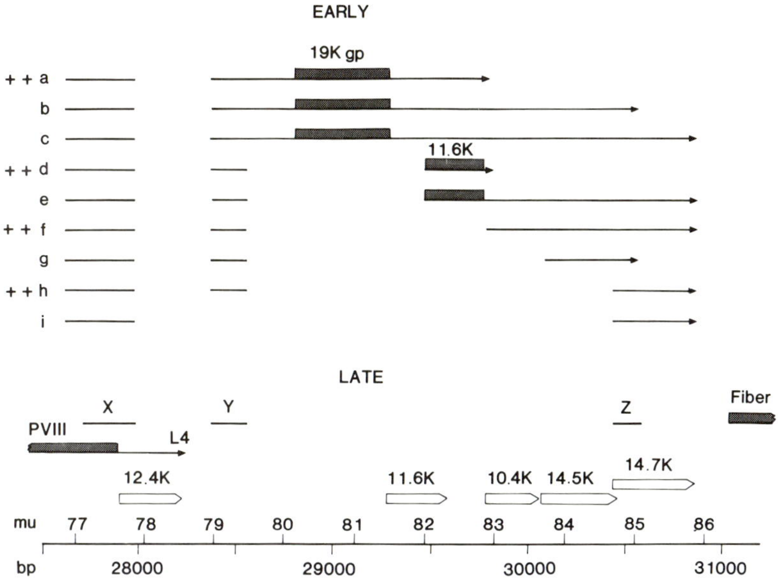

Fig. 4: Position of open reading frames and r-strand specific mRNAs in region E3. Early, mRNAs expressed from the E3 transcription unit. Late, mRNAs expressed from the major late transcription unit. ++, denotes the most abundant E3 mRNAs. Other conventions follow those in Fig. 3.

(34) and differ from each other by the position of their 3'-termini. A major E3 polyadenylation site is located at coordinate 85.9 (149) and is preceded by the hexanucleotide sequence AAUAAA. The second poly(A) addition site which is located at coordinate 82.9 is less frequently used. At this site, the usual polyadenylation sequence AAUAAA is replaced by the sequence AUUAAA and it is noteworthy that the 3'-ends which have been mapped to this position are slightly heterogeneous in length (150). A third minor poly(A) addition site has been mapped to

coordinate 85 by electron microscopy (32).

At least nine E3 mRNAs are generated by differential splicing (Fig. 4) and most of the splice junctions have been characterized by sequence analysis (149,150). It is noteworthy that many of the splice sites used by the supplementary leaders x, y and z, present on some fiber mRNAs, coincide with these splice junctions (151,152). Eight ORFs which could encode proteins larger than 100 amino acids are present (153,154), but only a 19K glycoprotein (155,156) and an 11.6K polypeptide (157) have been unequivocally assigned to this region (Fig. 4). A 14K polypeptide has been purified and mapped to E3 (158), but its precise coding location is not known. Both the 19K glycoprotein and the 11.6K polypeptide appear to be translated from multiple E3 mRNAs (159) (Fig. 4). Since most of the reading frames are conserved between adenovirus serotypes 2, 3 and 5 (160,161) it seems likely that additional polypeptides will be assigned to this region in the future.

Functions of the E3 proteins during lytic infection.

Studies of recombinant adenoviruses lacking all or part of the E3 region have shown that E3 is dispensible for viral growth in tissue culture cells (162,163,164). Since the overall organisation of the E3 region is well conserved between distantly related adenovirus serotypes it seems likely that E3 serves an important function when the virus infects its natural host. The E3-19K glycoprotein forms a complex with the heavy chain of the class I transplantation antigens both in Ad2-transformed rat and hamster cells and also in productively infected HeLa cells (155,165,166). The E3-19K glycoprotein competes efficiently with β_2-microglobulin for binding to the class I antigens (167) and exhibits a domain structure and limited sequence homology with immune system proteins such as β_2-microglobulin (168).

Since the E3-19K class I complex found on transformed cells appears to elicit synthesis of cytotoxic T-cells in syngeneic rats (165) it might seem that expression of this glycoprotein would be detrimental to the virus. However, recent experiments indicate that the function of the E3-19K protein may be quite the opposite during lytic infection. Complex formation appears to result in a reduction of class I expression on the cell surface, thus protecting the virus-infected cell against T-cell recognition (169,170). The E3-19K protein may therefore serve a function during lytic infection that is similar to that of the Ad12 E1A region during oncogenic transformation of primary rat cells (118,119).

The amino acid sequences of many of the predicted E3 proteins suggest that they may be membrane-associated (160,161) in which case the E3 region may provide another example of a specialized transcription unit.

EARLY REGION 4

The transcription unit (coordinates 91.3-99.1).

Region E4 is transcribed from the l-strand and is located at the extreme right end of the genome (Fig. 5)(31,32,36). The promoter, which is located at coordinate 99.1, initiates transcription heterogeneously within a region of seven nucleotides (34,171). Mutational analysis has identified the sequences required for both E4 transcription in vitro and E1A transactivation in vivo (67,172). Transcription from the E4 promoter is subjected to both positive and negative regulation during the infectious cycle. The E1A 289R protein induces E4 transcription early after infection (see the E1A section) whereas the E2A 72K-DBP has been shown to down regulate E4 transcription at intermediate to late times (145). This latter inhibition of E4 transcription has been demonstrated in vitro (173).

The primary transcript from E4, which extends from coordinate 99.1 to coordinate 91.3, is spliced into a complicated set of cytoplasmic mRNAs (32). Fig. 5 shows the structure of twelve E4 mRNAs which have been characterized by sequence analysis of cDNA clones (174,175). Additional minor variants have been detected by S1 nuclease analysis (176,177).

The DNA sequence of E4 predicts the existence of six ORFs in the l-strand sequence which can encode polypeptides of at least 10K molecular weight (43,178). An alignment of the mRNAs with the ORFs

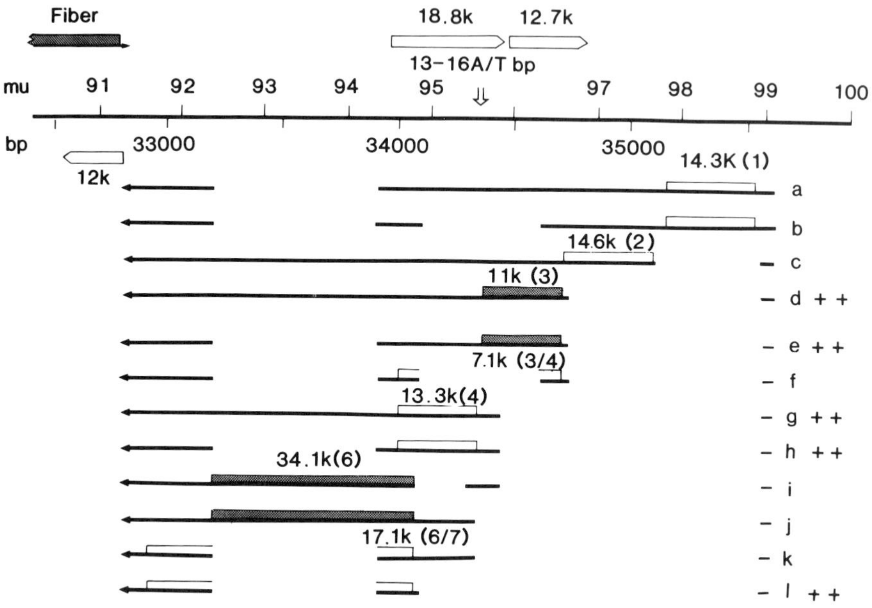

Fig. 5: Position of ORF's and mRNAs in region E4. The position of ORF's (numbered according to ref. 178) with their calculated coding capacities are indicated. Stipled boxes denote well characterized polypeptides. The position of an AT-tract which varies in length both within individual Ad2 stocks and strains held in various laboratories is indicated (Modified from ref. 175). Other conventions follow those in Figs. 3 and 4.

shows that all major reading frames with the exception of the 34.1K reading frame (ORF-6) are readily accessible in a unique mRNA. Partial amino acid sequence analysis of an 11K polypeptide (179,180) and a 34K polypeptide (99) has demonstrated that they are encoded by ORF-3 and ORF-6 respectively (Fig. 5).

It could be noted that a 15 bp long oligo-AT tract located between ORF 2 and 3 in region E4 varies in length between different Ad2 stocks (43,178).

Function of E4 proteins during lytic infection.

Certain mutants that contain deletions in region E4 are defective in the synthesis of late mRNAs and proteins (181,182). Viral DNA replication appears normal but the shut off of host cell macromolecule synthesis seems to be incomplete. The mutation causing these effects maps to the E4-34K reading frame (ORF-6) (97;Shenk personal communication). Since mutants in the E1B-55K polypeptide are also defective in late mRNA synthesis and in the control of cellular gene expression (96,97) it seems likely that the physical complex formed between the E1B-55K and the E4-34K polypeptides (99) is of major significance for the switch from early to late adenovirus gene expression. A product from region E4 is also required for adeno-associated virus (AAV) to replicate in human cells (183).

Nothing is known about the functions of the other E4 proteins. Several mutants with alterations in different parts of region E4 have been constructed but they do not seem to impede viral growth in tissue culture cells. It is possible that these other E4 proteins, like the E3 proteins, are specifically required for the growth of adenoviruses in their natural hosts.

THE VIRUS-ASSOCIATED RNAs AND REGULATION OF LATE mRNA TRANSLATION

The adenovirus genome encodes two low molecular weight RNAs designated virus associated RNA (VA RNA) I and II (184,185,186). Both RNAs are approximately 160 nucleotides in length and are transcribed in a rightward direction from two separate promoters around coordinate 30 (187,188). The exact length of the VA RNAs varies slightly due to small heterogeneities at both the 5'- and 3'-ends (189,190). The VA RNAs exhibit scattered regions of primary sequence homology and are capable of forming extensive intramolecular base paired regions (188,191). They are both transcribed by RNA polymerase III (192) and accumulate in massive amounts late during infection (10^5-10^6 copies per cell). As with other genes transcribed by RNA polymerase III, intragenic control regions have been identified (193,194,195). In VA RNA_I two sequence blocks, located at position +10 to +18 (block A) and +54 to +69 (block B), appear to control transcription (195).

Since the VA RNA genes compete for the same transcription factors both species are made in approximately equal amounts early during infection (193,194,196). However, the VA RNA_I gene binds the initiation factors more efficiently (197), so that VA RNA_I continues to accumulate and outnumbers VA RNA_{II} by a factor of 40 at late times (186,196). A small proportion of the VA RNAs is found as a ribonucleoprotein particle in association with the cellular lupus antigen, La, in infected cells (198,199). The La antigen, which is a 46K phosphoprotein, binds specifically to the U residues found at the 3'-termini of VA RNA (200,201).

Although the VA RNAs were discovered in 1966 (184) only recently has a specific function been assigned to them. Studies of deletion mutants have shown that a virus that lacks the VA RNA_I species synthesizes normal amounts of late mRNA but is defective in the translation of the late proteins (163). A double mutant lacking both VA RNAs has been shown to grow very poorly, producing 5- to 6-fold less virus compared to a VA RNA_I negative mutant (196). Since a virus lacking the VA RNA_{II} species (VA RNA_I^+/VA RNA_{II}^-) grows like wild type, it is likely that VA RNA_{II} serves a similar function to VA RNA_I and probably can partially substitute for it during lytic growth.

Using cell free translation systems prepared from VA RNA mutant infected cells, the site of VA RNA_I action has recently been investigated (202,203). The conclusion drawn from these studies is that VA RNA_I functions at a very early stage during the initiation of protein synthesis, probably by maintaining the catalytic activity of initiation factor eIF-2 (203).

The Epstein-Barr virus genome also encodes two low molecular weight RNAs (204), which are similar in size to the VA RNAs and can, with a very low efficiency, substitute for VA RNA_I during a lytic infection (205).

At late times during infection a dramatic inhibition of host cell protein synthesis occurs and results in the almost exclusive synthesis of adenovirus specific polypeptides (206,207). Synthesis of cellular hnRNA appears normal but the transport of cellular mRNA sequences from the nucleus to the cytoplasm is inhibited (208,209). In addition, those cellular mRNAs which survive in the cytoplasm are not efficiently translated (210). Since VA RNA_I functions as a general stimulator of translation, enhancing the efficiency with which both viral and non-viral mRNAs are translated (211,212), some additional mechanism must exist to restrict cellular mRNA translation.

GENES EXPRESSED AT INTERMEDIATE TIMES

At intermediate times after infection (5-10 hpi) several regions of the viral genome are transcriptionally active. In addition to the early genes two transcription units, encoding polypeptides IX (pIX) and IVa2, appear to be selectively activated. Both genes are commonly referred to as intermediate genes, since small amounts of their mRNAs can be detected prior to the onset of viral DNA replication (213,214). However, an efficient synthesis of both mRNAs has recently been shown to require viral DNA replication (215). Superinfection experiments have demonstrated that the viral and cellular factors present in a late infected cell are not sufficient to support pIX transcription and the activation of the pIX promoter is probably dependent on a physical change in the template structure (215).

The transcription unit for polypeptide IX

The synthesis of pIX is controlled by a r-strand specific transcription unit which is located within region E1B (Fig. 1,2) (32,33,216,217). The promoter has been positioned to coordinate 9.9 (34,217) and lies within the intron common to all E1B mRNAs. The polyadenylation site is identical with that used by the E1B mRNAs. pIX mRNA is unique among adenovirus mRNAs in being the only one which is not spliced (217). It is also the mRNA that is translated with the highest efficiency late during infection (218).

pIX is a minor structural component of the virion believed to be located between the hexons in the adenovirus icosahedron (219). Mutants lacking this gene are viable, although the virions are less heat stable than wild type virions (220), suggesting that the protein plays a role in stabilizing the capsid structure.

Since transcription of pIX mRNA is under the control of its own promoter and the gene is physically separated from those of the other late proteins it is possible that pIX has some additional function.

The transcription unit for polypeptide IVa2

The mRNA for polypeptide IVa2 is transcribed from the viral l-strand and is controlled by a promoter located at coordinate 16.2 (Fig. 1,3). The polyadenylation site is located at coordinate 11.3. The adenovirus major late promoter which operates in the rightward direction (see below) lies adjacent to the IVa2 promoter and the initiation sites for transcription are separated by only 210 base pairs (34). The IVa2 promoter lacks a TATA box, although in vitro transcription experiments have shown that a pyrimidine-rich sequence located between positions -38 and -49 relative to the cap site may serve a similar function (221). Deletion of the cap site and TATA box in the major late transcription unit increases IVa2 transcription approximately 15-fold suggesting that the two promoters compete for the same transcriptional factors, at least in vitro (221).

The N-terminal five amino acids of IVa2 are encoded in the same ORF from which the E2B DNA polymerase is translated. However, as a result of the splice which takes place when the IVa2 mRNA is generated, the remainder of the polypeptide is translated from a different, but overlapping, reading frame and has a predicted molecular weight of 51K. The two reading frames overlap for 228 nucleotides. The IVa2 polypeptide is present in adenovirus assembly intermediates and has been suggested to function as a maturation protein during adenovirus morphogenesis (222). It appears to be a DNA binding protein (223) and may therefore interact with the viral DNA as it enters the preformed capsid.

THE MAJOR LATE TRANSCRIPTION UNIT AND THE SWITCH FROM EARLY TO LATE GENE EXPRESSION

Late after infection, transcription initiates predominantly at the major late (ML) promoter located at coordinate 16.8 (Fig. 1,6)(224). Transcription from the ML promoter accounts for approximately 30% of the total RNA synthesis at late times (225). This does not necessarily mean that the ML promoter is a strong promoter, since efficient transcription from the promoter requires viral DNA synthesis (226) and an infected cell may contain up to 100,000 copies of viral DNA (227). In fact, electron microscopy has shown that only 5% of the viral DNA is being actively transcribed at late times and that each transcribing template has, on an average, only four nascent RNA chains distributed over the whole transcription unit (228,229).

The ML promoter was the first eukaryotic RNA polymerase II promoter to be studied in great detail (224,230). The sequences required for both in vitro and in vivo transcription have been characterized by mutational analysis (231,232,233). The results show that the TATA motif is important for efficient and selective transcription, presumably because initiation factors present in the cell bind to the TATA box and form a tight pretranscriptional complex (234). An upstream sequence located between -97 and -34, relative to the cap site, is also required for efficient transcription (233).

The primary transcript which is initiated from this promoter extends in the rightward direction for about 28,000 nucleotides and appears to terminate at a position close to coordinate 99 on the viral genome (235). With a transcription rate of approximately 50 nucleotides per second (236) it takes the RNA polymerase almost 10 min to complete a single primary transcript. Approximately 80% of the RNA polymerases

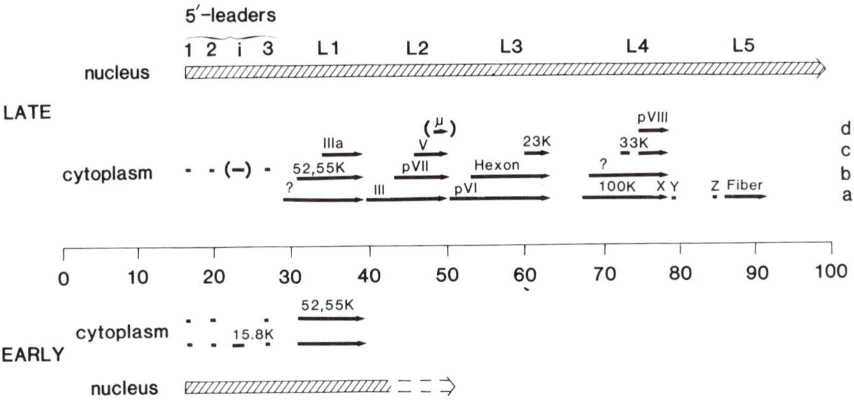

Fig. 6: Primary transcripts and mRNAs expressed from the major late transcription unit at early and late times after infection.

which initiate transcription at the ML promoter appear to terminate prematurely within the first 800 nucleotides (237,238). These incomplete RNA chains seem to consist of a mixture of RNAs which have specifically terminated at positions +120 and +175 relative to the cap site, together with RNA chains which have paused during elongation (239). The primary transcript becomes polyadenylated at one of five locations along the genome, generating five families of RNAs with coterminal 3'-ends (L1 to L5, Fig.6) (240,241,242). The 3'-ends are generated by endonucleolytic cleavage of the nuclear RNA precursor which occurs before the RNA polymerase has completed the synthesis of the primary transcript (243,244,245). Although poly(A) addition frequently precedes RNA splicing (243) it is not an absolute requirement. Non-polyadenylated RNAs are both spliced and transported, but they are rapidly degraded in the cytoplasm of the infected cell presumably because of their reduced stability (246).

Following the selection of a 3'-end the primary transcript is processed by splicing in such a way that each mature mRNA gains a common set of three short 5'-leader segments, the tripartite leader, derived from coordinates 16.8, 19.8 and 26.9 (247,248). This tripartite leader, which is 201 nucleotides long can be joined to at least 15 different downstream acceptor splice sites (Fig. 6). The tripartite leader sequence does not contain an AUG (249,250) and all mRNAs generated within the ML transcription unit use the first AUG present in the mRNA main body for initiation of their translation.

The first step in assembly of the tripartite leader appears to be the removal of the intron located between leader 1 and leader 2 (251, 252) a reaction which is fast and probably precedes the selection of the poly(A) addition site (253). With the recent development of extracts, which are capable of accurately splicing RNA precursors in vitro (254,255,256) a model for the excision of the intron between leader 1 and leader 2 has been proposed (257). This involves an intermediate lariat structure in which the 5'-end of the intron has been joined via an unusual 2'-5' phosphodiester bond (258) to an A residue located approximately 30 nucleotides from the 3' splice acceptor site.

A polypeptide, with a molecular weight of 13.5K, has been shown to be encoded by a r-strand specific mRNA located within the intron between leader 1 and leader 2 (259). The structure of this mRNA, which is preferentially transcribed under conditions of stringent protein synthesis inhibition, has not been determined.

The major event governing the transition from the early to late mode of gene expression appears to be DNA replication, since unreplicated DNA, when introduced into adenovirus infected cells which are in the late phase, still fails to express the late viral genes (226). One major difference between early and late times is found in the ML transcription unit (Fig. 6). At early times, it is active at a level comparable to the other early transcription units (32,36,260). The start site is the same both early and late but the length of the primary transcript and the structures of the resulting mRNAs differ considerably. At early times transcription from this promoter terminates near the middle of the genome, whereas at late times it continues to a position close to the right terminal end (260,261,262). Thus, one mode of regulation of transcription from the ML promoter involves the termination step.

A second level of regulation involves virus-induced changes in the splicing machinery and is most apparent in L1 mRNA synthesis (see below) (226,261,262).

Late region 1

The L1 mRNAs are characterized by having a common poly(A) addition site at coordinate 39.3 (Fig. 6). It is noteworthy that a stretch of 21 AT base pairs is located just upstream of the poly(A) site (130,263). This oligo-AT tract appears to be heterogenous in length both within individual Ad2 stocks and between strains held in different laboratories (263). Three major mRNAs, with splice acceptor sites located at coordinate 29.0 (species $L1_a$), 30.7 (species $L1_b$) and 34 (coordinate $L1_c$), have been identified from region L1 (32,261,264).

The splice site of species $L1_a$ has been mapped to coordinate 29.0 on the Ad2 genome (264). No protein product has so far been assigned to this mRNA, but since the predicted polypeptide has a molecular weight of only 8.3K it may have escaped detection because of its small size. Both the VA RNA_I and VA RNA_{II} sequences are present as part of the 3' non coding region of the $L1_a$ mRNA.

The L1 nuclear RNA precursor is the same both early and late after infection. However, at early times one L1 mRNA, $L1_b$, accumulates exclusively (Fig.6)(32,261). This shift in the spliced structure of the L1 mRNAs is due to the regulation of splicing rather than resulting from changes in mRNA stability or transport (262). About half of the early L1 mRNA population contains the normal tripartite leader sequence and encodes two structurally related polypeptides of molecular weights 52K and 55K (259,261,265). The splice acceptor site used to generate the 52,55K mRNA maps very close to the VA RNA genes. In fact, the T-cluster which terminates VA RNA_{II} transcription is separated by only 11 nucleotides from the acceptor site for species $L1_b$ (261). The function of the 52,55K polypeptide is at present unknown and the protein has only been observed by in vitro translation of hybridization selected RNAs (259,261,265).

The other half of the early L1 mRNA population contains an extra 440 nucleotide long leader, derived from coordinate 22.7 (32,261,266). This leader, which has been designated the "i" leader (intermediate leader), contains an AUG triplet followed by an open translational reading frame (Fig.3) (266,267). Its presence in the L1 mRNA results in the synthesis of a 15.8K polypeptide (literature values 13.6K, 14K and 16K) both in vitro and in vivo (261,268). The first 120 amino acids of the 15.8K polypeptide are encoded by the "i" leader segment at coordinate 22.7 while the five C-terminal amino acids are encoded by the third leader segment at coordinate 26.9 (266). The "i" leader is also present in L2, L3, L4, and L5 mRNAs at late times after infection, albeit at a very low frequency (152,261).

Species $L1_c$ which becomes the most abundant L1 mRNA late after infection encodes virion polypeptide IIIa (261,265,269). This polypeptide which exists in only 60 copies per virion (270) is a minor component of the virus particle. It appears to be located in the vertex region of the virion probably acting as a bridge between the pentons and peripentonal hexons (270).

Late region 2

The L2 mRNAs are characterized by a common poly(A) addition site at coordinate 50.0 (Fig.6). Three major mRNAs with acceptor splice sites at coordinates 39.4 (species $L2_a$), 44.1 (species $L2_b$) and 45.9 (species $L2_c$) have been identified (213,271).

Three polypeptides, designated III, V and pVII, are coded in region L2 based upon in vitro translation of hybridization-selected RNA (265,269,272).

Polypeptide III which is encoded by $L2_a$ mRNA has a predicted molecular weight of 63.3K. This value is significantly smaller than the experimentally observed value of 82K, but the reason for the large discrepancy is not known. Polypeptide III, which is the building block of the penton base, associates with the fiber to form the penton, i.e. the capsomers which are located at the twelve corners of the adenovirus icosahedron.

Polypeptides V and pVII are encoded by mRNA species $L2_b$ and $L2_c$ and together constitute the major proteins found in adenovirus chromatin (273,274). The viral core appears to assemble into a regular beads-on-a-string pattern although the repeating unit seems to be smaller than that of the cellular nucleosomes (275,276,277).

Polypeptide VII which is tightly bound to the viral DNA is synthesized as a 21.8K precursor (pVII). During the final stages of virion maturation pVII is cleaved by the virus encoded protease to its final size of 19.4K. Polypeptide VII is very basic in its amino acid composition (24% arginine) and has a distribution of basic residues and potential helical segments suggesting that it may function as a protamine (278).

Polypeptide V which is moderately basic (13% arginine, 7% lysine) has a molecular weight of 41.6K and is less tightly associated with the viral DNA than polypeptide VII. It has been proposed that it forms a protein shell around a nucleoprotein particle consisting of polypeptide VII and the viral DNA (279).

A fourth ORF is present in L2 immediately following the terminator for polypeptide V. It has a length of 85 amino-acids (85R) and a corresponding protein product has recently been detected (Anderson and Lewis, unpublished). This polypeptide may be the precursor of the μ polypeptide which is a highly basic, low molecular weight polypeptide that has been found in viral chromatin preparations (277,280). The amino acid composition of the μ polypeptide agrees well with a portion of this 85R protein mapping at the extreme 3'-end of region L2 (271).

Late region 3

The L3 mRNAs are characterized by a common poly(A) addition site at coordinate 62.4 (Fig.6). Their 3'-ends overlap by 22 nucleotides with the 3'-ends of the mRNAs transcribed from region E2A (281). Three major mRNAs have been mapped to this region and they encode three well-characterized polypeptides designated pVI, hexon, and 23K (265,269,281).

Polypeptide pVI which is translated from mRNA species $L3_a$ (coordinates 50.1-62.4) consists of 249 amino acids and has a predicted molecular weight of 27.0K (272). It is processed by proteolytic cleavage to a 24K polypeptide during virion maturation. The location in the capsid of polypeptide VI has not been firmly established. It is believed to be associated with the hexon capsomer on the inside of the virus particle, perhaps forming a bridge between the viral core and the capsid structure. Since it has DNA binding properties (223), it may turn out to be a core protein that is only loosely associated with the viral chromatin.

The hexon which is encoded by mRNA species $L3_b$ (coordinates 52.3-62.4) is the major structural component of the virion. Each adenovirus particle contains 240 hexon capsomers which form the facets of the icosahedron. The hexon capsomer is, in turn, composed of three identical polypeptide chains, each consisting of 967 amino acids and having a molecular weight of 109,000 (282,283).

Many Ad2 polypeptides undergo proteolytic cleavage during their maturation (207). The protease responsible for this cleavage has been shown to be the 23K polypeptide encoded by mRNA species $L3_c$ (coordinates 60.2-62.4) (284,285). This protease is responsible for the cleavage of four precursor polypeptides to their mature forms, namely the L2-pVII, L3-pVI, L4-pVIII and E2B-pTP polypeptides. It has also been proposed that polypeptide IIIa matures from a slightly larger precursor (286). From the protease cleavage sites so far identified a tentative consensus sequence that is recognized by the 23K endopeptidase can be constructed. The cleavage appears to occur between the Gly-Ala residues in the sequence Met-X-Gly-Gly-Ala/Val.

Late region 4

The L4 mRNAs are characterized by a common poly(A) addition site at coordinate 78.5 (Fig.6). The 3'-end of the L4 mRNAs is thus located within an intron that is common to all E3 mRNAs. Region L4 encodes three well characterized polypeptides. Two of these, the 100K and the 33K polypeptides are non-structural whereas the third, polypeptide pVIII, constitutes a minor component of the virus particle. Four major mRNAs have been mapped to this region by electron microscopy (151).

The 100K polypeptide is most likely encoded by mRNA species $L4_a$. The acceptor splice site that is used for generation of the 100K mRNA has been mapped for the closely related Ad5 (287) and by analogy with this it has been suggested that the Ad2 100K mRNA body is located between coordinates 67.0 to 78.5 (263). A surprising feature of the 100K reading frame is that it continues upstream for a considerable distance and could theoretically encode a polypeptide which would be approximately 20K larger at the N-terminus. However, no mRNA which corresponds to this extended 100K polypeptide has yet been identified. The 100K polypeptide is made in large quantities late after infection although the protein is absent from purified virus particles. It is believed to function during the assembly of the hexon capsomers and may be responsible for their transport from the cytoplasm into the nucleus (288,289,290).

The exact location of the gene for the L4-33K polypeptide has been mapped on the Ad2 genome, by aligning a partial protein sequence with the DNA sequence (291). From the results it was proposed that mRNA species $L4_c$ (coordinates 73-78.5) encodes the non-structural 33K polypeptide. The $L4_c$ mRNA is unique among late mRNAs in that it contains a splice within the coding sequence (291). No function has so far been assigned to the 33K polypeptide although studies of its intracellular location suggest that it is a nuclear protein (290).

Polypeptide VIII has a molecular weight of 13K as estimated by SDS polyacrylamide gel electrophoresis (207,273). It is synthesized as a 27K precursor which is cleaved by the L3-23K endopeptidase. Polypeptide pVIII is most likely translated from mRNA species $L4_d$ (coordinates 75-78.5). It is not known where the proteolytic cleavage occurs which is responsible for the maturation of polypeptide pVIII. However, the pVIII sequence contains a hypothetical cleavage site for the L3-23K endopeptidase (see section about L3). If this sequence indeed is recognized it would leave a 13K N-terminal fragment which might correspond to virion polypeptide VIII.

Late region 5

The L5 mRNAs are characterized by a poly(A) addition site at coordinate 91.3 (Fig.6). In contrast to the other late regions, a single polypeptide is encoded within region L5 (265,269). This protein is the fiber protein which decorates the vertices of virions and functions as the attachment organ when adenovirus binds to the cellular receptor. The native fiber protein is a dimeric structure with a characteristic rod and knob appearance. The shaft portion of the fiber has been shown to be composed of two parallel ß sheet structures (292,293).

The fiber mRNA (coordinates 86.3-91.3) differs from other late mRNAs in that one mRNA body can be connected to a number of different 5'-leader sequences. In addition to the normal tripartite leader a substantial fraction of the fiber mRNA population (approximately 30%; 152) has been shown to be connected to the ancillary x,y and z leaders or the "i" leader in various combinations (151,152,250). Many of the splice sites of the ancillary leaders coincide precisely with splice junctions used for maturation of the E3 mRNAs (149,150,152). Addition of the y leader to the fiber mRNA does not alter the coding capacity, or the translational efficiency of the fiber polypeptide in vitro (294). However, recent experiments indicate that the x and y leaders may be of importance for an efficient translation of the fiber mRNA during in vivo conditions (295).

ORGANISATION OF THE ADENOVIRUS GENOME

Several interesting findings have emerged now that the complete nucleotide sequence of the Ad2 genome is available. The information content of the genome is extremely high, with few if any surplus nucleotides. Even within those regions of the genome that do not code for proteins, there are often multiple signals that control the viral life-cycle. For example, within the 498 nucleotides at the left end, which precede the cap site for the E1A transcription unit, there are signals for the initiation of DNA replication, signals for packaging the DNA within the virion, several regions that serve as enhancer elements as well as a traditional RNA-polymerase recognition site. Typical intergenic distances are short and often contain signals that direct transcription or RNA processing.

Because of the extensive RNA splicing that takes place, there are numerous examples of splice signals overlayed on coding sequences. The most crowded region of the genome lies between co-ordinates 13 and 30. Here the 22.8K ORF overlaps both the C-terminus of the pTP gene and the N-terminus end of the DNA polymerase gene. The C-terminus of the DNA polymerase gene overlaps the N-terminus of the IVa2 gene. Moreover, the 13.6K and 15.8K polypeptides whose mRNAs are transcribed from the viral r-strand wholly overlap the DNA polymerase gene on the complementary strand. Superimposed upon these coding sequence overlaps are the promoters for the IVa2 gene and the major late transcription unit together with the three leader segments of the late mRNAs and the signals necessary to excise them.

One intriguing feature of the genome organization concerns the regulation and spatial locations of related sets of genes. E2, which specifies three gene products vital to DNA replication, encompasses coding sequences scattered over half of the genome. Control is exercised from a single promoter at early times and a different promoter at late times, while polyadenylation and RNA splicing direct the fine details of mRNA production. The organization is highly reminiscent of that found in bacterial operons. Perhaps this is the equivalent eukaryotic scheme to provide synchronous control of a group of related genes. Similar strategies prevail for the L2 family of chromatin genes and the E3 membrane-associated proteins. One feature of transcription that is especially pronounced is the interplay of polyadenylation and RNA splicing. This is of particular importance in the major late transcription unit, where five families of mRNAs are generated firstly by controlling polyadenylation at five distinct sites and then by differential splicing of each of these precursors to generate a set of 3 or 4 mRNAs within each family.

A curious feature of the genetic organization of Ad2 is that many early genes contain introns located within their coding sequences, whereas most of the late genes contain introns only within untranslated sequences. Why is this so? Consider the E1A gene block. Here a key protein is encoded, the 289R polypeptide, that is responsible for the activation of all other early genes. It is reasonable to think of the 289R polypeptide as a two domain protein. In the absence of RNA splicing the domains remain separated and conceivably are without function. Only when the mRNA is spliced correctly can a functional protein be made. Thus the absolute levels of the 289R mRNA can be controlled at two stages. Firstly by the strength of the promoter, and secondly by the specificity and efficiency of splicing. Most importantly, abberant transcripts from this region would not lead to expression of the 289R protein. For even if they were able to reach the translational machinery in the absence of splicing no functional information would be present in a linear form on any RNA molecule. From this perspective, the presence of introns serves to protect the virus against the possible deleterious or lethal effects resulting from overexpression of key regulatory proteins. In the case of the late structural proteins such overexpression might lead to inefficiency but would probably never prove lethal.

A comparison between the organization of the adenovirus genome and that of another well-characterized DNA animal virus, SV40, reveals similarities as well as clear-cut differences. The adenovirus genome is approximately 7 times larger than the SV40 genome and has a much more sophisticated organization. In both cases the genes for the structural proteins consume approximately 50% of the genetic information. Components which are unique to adenovirus include the histone-like proteins (V and VII), the virion maturation protein (IVa2), the processing endopeptidase and the 100K protein required for hexon assembly.

The increased complexity of the adenovirus genome is most apparent in the early genes. The SV40 genome encodes only two early polypeptides, the small t- and the large T-antigens, whose mRNAs are generated by differential splicing of a single primary transcript. In contrast, Ad2 uses four early regions which include five different transcription units and a plethora of spliced mRNAs. These encode more than 20 early polypeptides, including at least three functions which allow viral replication to take place by a unique mechanism that is independent of that of the host. Another difference from SV40 is found in the regulatory cassette that is embedded in the E1A region. This provides the initial interaction with the host and, once activated, leads to a cascade of viral functions. While the more complex design of the adenovirus particle may be a necessity to accomodate the larger genome, it probably also provides the virus with additional ways to interact with its host, including an increase in its range. The difference in complexity of the early functions highlights the fact that SV40 relies upon its host for many functions, whereas Ad2 is much more an adolescent among viruses.

Students of viruses frequently argue that lessons learnt from the virus will find parallels in the host. Certainly this view was fulfilled in a most spectacular manner with the discovery of RNA-splicing in the Ad2 system. However other phenomena have also been pioneered in viral systems - oncogenes and enhancers being recent examples. With the extensive knowledge already accumulated about Ad2 and the availability of its complete nucleotide sequence many experiments can now be planned with a precision hitherto impossible. We anticipate that these will lead to dramatic new insights about both the virus and its host.

ACKNOWLEDGEMENT

We are most grateful to Jeanette Backman and Gunvor Lindman for excellent secretarial aid. We also thank Peter Aleström for discussions.

REFERENCES

1. Rowe, W.P., Huebner, R.J., Gilmore, L.K., Parrott, R.H. and Ward, T.G. Proc. Soc. Exp. Biol. Med. 84:570-573, 1953.
2. Hilleman, M.R. and Werner, J.H. Proc. Soc. Exp. Biol. Med. 85:183-188, 1954.
3. Straus, S.E. In: The Adenoviruses (Ed. Ginsberg), H.S. Plenum Press, New York, 1984, pp. 451-496.
4. Ishibashi, M. and Yasue, H. In: The Adenoviruses. (Ed. Ginsberg, H.S.), Plenum Press, New York, 1984, pp. 497-562.
5. Trentin, J.J., Yabe, Y. and Taylor, G. Science. 137:835-841, 1962.
6. Huebner, R.J., Rowe, W.P and Lane, W.T. Proc. Natl. Acad. Sci. USA, 48:2051-2058, 1962.
7. Huebner, R.J. In: Perspect in Virol. V.(Ed. M. Pollard). Academic Press, New York, 1967, pp 147-167.
8. Pettersson, U. In: The Adenoviruses. (Ed. Ginsberg, H.S.) Plenum Press, New York, 1984, pp. 205-270.
9. Svensson, U., Persson, R. and Everitt, E. J. Virol. 38:70-81, 1981.
10. Tate, V. and Philipson, L. Nucl. Acids. Res. 6:2769-2785, 1979.
11. Sergeant, A., Tigges, M.A. and Raskas, H.J. J. Virol. 29:888-898, 1979.
12. Hammarsköld, M-L., Winberg, G. Cell 20:787-795, 1980.
13. Philipson, L. In: The Adenoviruses. (Ed. Ginsberg, H.S.), Plenum Press, New York, 1984, pp. 309-338.
14. Green, M., Pina, M., Kimes, R.C., Wensink, P.C., MacHattie, L.A. and Thomas, C.A. Jr. Proc. Natl. Acad. Sci. USA. 57:1302-1309, 1967.
15. van der Eb, A.J., van Kesteren, L.W. and van Bruggen, E.F.J. Biochem. Biophys. Acta. 182:530-541, 1969.
16, Roberts, R.J., Akusjärvi, G., Aleström, P., Gelinas, R.E., Gingeras, T.R., Sciaky, D. and Pettersson, U. This volume, 1985.
17. Tolun, A. and Pettersson, U. J. Virol. 16:759-766, 1975.
18. Weingärtner, B., Winnacker, E.L., Tolun, A. and Pettersson, U. Cell 7,557-566, 1976.
19. Lechner, R.L. and Kelly, T.J., Jr. Cell 12:1007-1020, 1977.
20. Robinson, A.J., Younghusband, H.B. and Bellet, A.J.D. Virology 56:54-69, 1973.
21. Rekosh, D.M.K., Russel, W.C., Bellett, A.J.D. and Robinson, A.J. Cell 11:283-295, 1977.
22. Desiderio, S.V. and Kelly, T.J., Jr. J. Mol. Biol. 145:319-337, 1981.
23. Garon, C.F., Berry, K.W. and Rose, J.A. Proc. Natl. Acad. Sci. USA 69:2391-2395, 1972.
24. Wolfson, J. and Dressler, D. Proc. Natl. Acad. Sci. USA 69:3054-3057, 1972.
25. Arrand, J.R. and Roberts, R.J. J. Mol. Biol.128:577-594, 1979.
26. Jones, N. and Shenk, T. Proc. Natl. Acad. Sci. USA 76:3665-3669, 1979.
27. Berk, A.J., Lee, F., Harrison, T., Williams, J. and Sharp, P.A. Cell 17:935-944, 1979.
28. Nevins, J.R. Cell 26:213-220, 1981.
29. Graham, F.L., Abrahamas, P.J., Mulder, C., Heijneker, H.L., Warnaar, S.O., de Vries, F.A.J., Fiers, W. and van der Eb, A.J. Cold Spring Harbor Symp. Quant. Biol. 39,637-650, 1974.

30. van der Eb, A.J., Mulder, C., Graham, F.L. and Houwelling, A. Gene 2, 115-132, 1977.
31. Berk, A.J. and Sharp, P.A. Cell 14:695-711, 1978.
32. Chow, L.T., Broker, T.R. and Lewis, J.B. J. Mol. Biol. 134:265-303, 1979.
33. Wilson, M., Fraser, N. and Darnell, J. Virology 94:175-194, 1979.
34. Baker, C.C. and Ziff, E.B. J. Mol. Biol. 149:189-221, 1981.
35. Katze, M.G., Persson, H. and Philipson, L. EMBO J. 1:783-789, 1982.
36. Kitchingman, G.R. and Westphal, H. J. Mol. Biol. 137:23-48, 1980.
37. Perricaudet, M., Akusjärvi, G., Virtanen, A. and Pettersson, U. Nature 281:694-696, 1979.
38. Virtanen, A. and Pettersson, U. J. Mol. Biol. 165:496-499, 1983.
39. Sambrook, J., Greeen, R., Stringer, J., Mitchison, T., Hu, S-L. and Botchan, M. Cold Spring Harbor Symp. Quant. Biol. 44:569-584, 1979.
40. Osborne, T.F. and Berk, A.J. J. Virol. 45:594-599, 1983.
41. Svensson, C., Pettersson, U. and Akusjärvi, G. J. Mol. Biol.165:475-499, 1983.
42. van Ormondt, H., Maat, J. and van Beveren, C.P. Gene 11:299-309, 1980.
43. Gingeras, T.R., Sciaky, D., Gelinas, R.E., Bing-Dang,J., Yen, C., Kelly, M., Bullock, P., Parsons, B., O'Neill, K. and Roberts, R.J. J. Biol. Chem. 257:13475-13491, 1982.
44. Osborne, T.F., Gaynor, R.B. and Berk, A.J. Cell 29:139-148, 1982.
45. Hearing, P. and Shenk, T. J. Mol.Biol. 167:809-822,1983.
46. Spindler, K.R. and Berk, A.J. J. Virol. 51:884-887, 1984.
47. Hearing, P. and Shenk, T. Cell, 33:695-703,1983.
48. Khoury, G. and Gruss, P. Cell 33:313-314, 1983.
49. Impereale, M.J., Feldman, L.T. and Nevins, J.R. Cell 35:127-136, 1983.
50. Hen, R., Borelli, E., Sassone-Corsi, P. and Chambon, P. Nucleic Acids Res. 11:8747-8760, 1983.
51. Osborne, T.F., Arvidson, D.N., Tyan, E.S. Dunsworth-Browne, M. and Berk, A.J. Mol. Cell. Biol. 4, 1293-1305, 1984.
52. Montell, C., Fisher, E.F. Caruthers, M.H. and Berk, A.J. Nature 305:600-605, 1983.
53. Spector, D.J., McGrogan, M. and Raskas, H.J. J. Mol. Biol. 126:395-414, 1978.
54. Harter, M.L. and Lewis, J.B. J. Virol. 26:736-749, 1978.
55. Smart, J.E., Lewis, J.B., Mathews, M.B., Harter, M.L. and Anderson, C.W. Virology 112:703-713, 1981.
56. Lupker, J.H., Davis, A., Jochemsen, H. and van der Eb, A.J. J. Virol. 37:524-529, 1981.
57. Feldman, L.T. and Nevins, J.R. Mol. Cell. Biol. 3:829-838, 1983.
58. Spindler, K.R., Rosser, D.S.E. and Berk, A.J. J. Virol. 49:132-141, 1983.
59. Spector, D.J., Crossland, L.D., Halbert, D.N. and Raskas, H.J. Virology 102:218-221, 1980.
60. Esche, H., Mathews, M.B. and Lewis, J.B. J. Mol. Biol.142:399-417, 1980.
61. Nevins, J.R., Ginsberg, H.S., Blanchard, J.M., Wilson, M.C. and Darnell, J.E. J. Virol. 32:727-733, 1979.
62. Riccardi, R.P., Jones, R.L., Cepko, C.L., Sharp, P.A. and Roberts, B.E. Proc. Natl. Acad. Sci. USA 78:6121-6125, 1981.

63. Carlock, L. and Jones, N.C. Nature 294:572-574, 1981.
64. Montell, C., Fisher, E.F., Caruthers, M.H. and Berk, A.J. Nature 295:380-384, 1982.
65. Weeks, D.L. and Jones, N.C. Mol. Cell. Biol. 3:1222-1234, 1983.
66. Svensson, C. and Akusjärvi, G. EMBO J. 3:789-794, 1984.
67. Gilardi, P. and Perricaudet, M. Nucleic Acids Res. 12:7877-7888, 1984.
68. Jones, N.C., Richter, J.D., Weeks, D.L. and Smith, L.D. Mol. Cell Biol. 3:2131-2142, 1983.
69. Ferguson, B., Jones, N., Richter, J. and Rosenberg, M. Science 224:1343-1346, 1984.
70. Winberg, C. and Shenk, T. EMBO J. ?:1907-1912.
71. Montell, C., Geneviéve, C., Eng, C. and Berk, A. Cell 36:951-961, 1984.
72. Leff, T., Elkaim, R., Goding, C.R., Jalinot, P., Sassone-Corsi, P., Perricaudet, M., Kedinger, C. and Chambon, P. Proc. Natl. Acad. Sci. 81:4381-4385, 1984.
73. Braithwaite, A.W., Cheetham, B.F., Li, P., Parish, C.R., Waldron-Stevens, L.K. and Bellet, A.J.D. J. Virol. 45:192-199, 1983.
73a. Feldman, L.J., Impereale, M.J. and Nevins, J.R. Proc.Natl.Acad Sci USA 79:4952-4956, 1982.
73b. Impereale, M.J., Kao, H.-T., Feldman, L.T., Nevins, J.R. and Strickland, S. Mol.Cell.Biol. 4:867-874, 1984.
74. Nevins, J.R. Cell 29:913-919, 1982.
75. Kao, H.-T. and Nevins, J.R. Mol. Cell. Biol. 3:2058-2065, 1983.
75a. Stein, R. and Ziff, E.B. Mol. Cell. Biol. 4:2792-2801, 1984.
76. Green, M.R., Treisman, R. and Maniatis, T. Cell 35:137-148, 1983.
77. Gaynor, R.B., Hillman, D. and Berk, A.J. Proc.Natl.Acad.Sci. 81:1193-1197, 1984.
78. Bos, J.L. and ten Wolde-Kraamwinkel, H.C. EMBO J. 2:73-76, 1983.
79. Curtois, G. and Berk, A. EMBO J. 3, 1145-1149, 1984.
80. Kingston, R.E., Kaufman, R.J. and Sharp, P.A. Mol. Cell Biol. 4:1970-1977, 1984.
81. Gaynor, R.B. and Berk, A.J. Cell 33:683-693, 1983.
82. Virtanen, A. and Pettersson, U. J. Virol. in press, 1985.
83. Wilson, M.C. and Darnell, J.E. J. Mol. Biol. 148:231-251, 1981.
84. Babich, A. and Nevins, J.R. Cell 26:371-379, 1981.
85. Montell, C., Fisher, E.F., Caruthers, M.H. and Berk, A.J. Mol. Cell. Biol. 4:966-972, 1984.
86. Andersson, C.W. and Lewis, J.B. Virology 104:27-41, 1980.
87. Bos, J.L., Polder, L.J., Bernards, R., Schrier, P.I., van den Elsen, P.J., van der Eb, A.J. and van Ormondt, H. Cell 27:121-131, 1981.
88. Andersson, C.W., Schmitt, R.C., Smart, J.E. and Lewis, J.B. J. Virol. 50:387-396, 1984.
89. Lucher, L.A. Brackman, K.M., Symington, J.S. and Green, M. Virology 132:217-221, 1984.
90. Persson, H., Katze, M.G. and Philipson, L. J. Virol. 42:905-917, 1982.
91. Rowe, D.T., Graham, F.L. and Branton, P.E. Virology 129:456-468, 1983.
92. Subramanian, T., Kuppuswamy, M, Gysbers, J., Mak, S. and Chinnadurai, G. J. Biol. Chem. 259:11777-11783, 1984.
93. Pilder, S., Logan, J. and Shenk, T. J. Virol. 52:664-671, 1984.

94. White, E., Grodzicker, T. and Stillman, B. J. Virol. 52:410-419, 1984.
95. Takemori, N., Cladaras, C., Bhat, B., Conley, A.J. and Wold, W.S.M. J. Virol. 52:793-805, 1984.
96. Babiss, L.E. and Ginsberg, H.S. J. Virol. 50:202-212, 1984.
97. Pilder, S., Logan, J. and Shenk, T. J. Virol. In press, 1985.
98. Sarnow, P., Ho, Y.S., Williams, J. and Levine, A.J. Cell 28:387-394, 1982.
99. Sarnow, P., Hearing, P., Anderson, C.W., Halbert, D.N., Shenk, T. and Levine, A.J. J. Virol. 49:692-700, 1984.
100. Ross, S.R., Levine, A.J., Galos, R.S., Williams, J. and Shenk, T. Virology, 103:475-492, 1980.
101. Mallete, P., Yee, S.-P. and Branton, P.E. J. Gen. Virol. 64:1069-1078, 1983.
102. Branton, P.E., Lassam, N.J., Downey, J.F., Yee, S.-P., Graham, F.L., Mak, S. and Bayley, S.T. J. Virol. 37:601-608, 1981.
103. Gallimore, P.H. and Paraskeva, C. Cold Spring Harbor Symp. Quant. Biol. 44:803-813, 1979.
104. Bernards, R., Houweling, A., Schrier, P.I., Bos, J.L. and van der Eb, A.J. Virology 120:422-432, 1982.
105. Gallimore, P.H., Sharp, P.A. and Sambrook, J. J. Mol. Biol. 89:49-72, 1984.
106. Graham, F.L., van der Eb, A.J. and Heijneker, M.L. Nature 251:687-691, 1974.
107. van der Eb, A.J., van Ormondt, H., Schrier, P.I., Lypker, J.H., Jochemsen, H., van der Elsen, P.J. de Leys, R.J., Maat, J., van Beveren, C.P., Dijkema, R. and de Waard, A. Cold Spring Harbor Symp. Quant. Biol. 44:383-399, 1979.
108. Jochemsen, H., Daniels, G.S.G., Hertoghs, J.J.L., Schrier, P.I., van der Elsen, P.J. and van der Eb, A.J. Virology 121:15-28, 1982.
109. Houwelling, A., van den Elsen, P.J., and van der Eb, A.J. Virology 105:537-550, 1980.
110. Land, H., Parada, L.F. and Weinberg, R.A. Nature 304:596-602, 1983.
111. Ruley, H.E. Nature 304:602-605, 1983.
112. Ralston, R. and Bishop, J.M. Nature 306:803-806, 1983.
113. Donner, P., Greiger-Wilkie, I. and Moelling, K. Nature 296:262-266, 1982.
114. Kingston, R.E., Baldwin, A.S. and Sharp, P.A. Nature 312:280-281, 1984.
115. Graham, F.L., Harrison, T. and Williams, J. J. Virology 86:10-21, 1978.
116. Haley, K.P., Overhauser, J., Babiss, L.E., Ginsberg, H.S. and Jones, N.C. Proc. Natl. Acad. Sci. 81:5734-5738, 1984.
117. Huruwitz, D.R. and Chinnadurai, G. Proc. Natl. Acad. Sci. USA. In press, 1985.
118. Schrier, P.I., Bernards, R., Vaessen, R.T.M.J., Houweling, A. and van der Eb, A.J. Nature 305:771-775, 1983.
119. Bernards, R., Schrier, P.I. Houweling, A., Bos, J.L., van der Eb, A.J., Zijlstra, M. and Melief, C.J.M. Nature 305:7i76-779, 1983.
120. Bernards, R., de Leeuw, M.G.W., Vaessen, M-J., Houweling, A. and van der Eb, A.J. J. Virol. 50:847-853, 1984.
121. Mathis, D.J., Elkaim, R., Kedinger, C., Sassone-Corsi, P. and Chambon, P. Proc. Natl. Acad. Sci. USA 78:7383-7387, 1981.
122. Rossini, M. Virol. 131:49-58, 1983.

123. Elkaim, R., Goding, C. and Kédinger, C. Nucl. Acids.Res. 11:7105-7117, 1983.
124. Imperiale, M.J. and Nevins, J.R. Mol. Cell. Biol. 4:875-882, 1984.
125. Goldenberg, C.J. and Raskas, H.J. Cell 16:131-138, 1979.
126. Weber, J., Blanchard, J.M., Ginsberg, H.S. and Darnell, J.E. J. Virol. 33:286-296, 1980.

127. Stillman, B.W., Lewis, J.B., Chow, L.T., Mathews, M.B. and Smart, J.E. Cell 23:497-508, 1981.
128. Nevins, J.R., Blanchard, J.-M. and Darnell, J.E. J. Mol. Biol. 144:377-386, 1980.
129. McDevitt, M.A., Imperiale, M.J., Ali, H. and Nevins, J.R. Cell 37:993-999, 1984.
130. LeMoullec, J.M., Akusjärvi, G., Stålhandske, P., Pettersson, U., Chambraud, B., Gilardi, P., Nasri, M. and Perricaudet, M. J. Virol. 48:127-134, 1983.
131. Alestrõm, P., Akusjärvi, G., Pettersson, M. and Pettersson, U. J. Biol. Chem. 257:13492-13498, 1982.
132. Smart, J.E. and Stillman, B.W. J. Biol. Chem. 257:13499-13505, 1982.
133. Stillman, B., Tamanoi, F. and Mathews, M.B. Cell 31:613-620, 1982.
134. Nagata, K. Guggenheimer, R.A., Enomoto, T., Lichy, J.M. and Hurwitz, J. Proc. Natl. Acad. Sci. USA 79:6438-6442, 1982.
135. Nagata, K., Guggenheimer, R.A. and Hurwitz, J. Proc. Natl. Acad. Sci. USA 80:6177-6181, 1983.
136. Tamanoi, F. and Stillman, B.W. Proc. Natl. Acad. Sci. USA 80:6446-6450, 1983.
137. Challberg, M.D. and Rawlins, D.R. Proc. Natl. Acad. Sci. USA 81:100-104, 1984.
138. Tolun, A., Alestrõm, P. and Pettersson, U. Cell 17:705-713, 1979.
139. Stillman, B.W., Topp, W.C. and Engler, J.A. J. Virol. 44:530-537, 1982.
140. Rijinders, A.W.M., van Bergen, B.G.M., van der Vliet, P.C. and Sussenbach, J.S. Nucleic Acids Res. 11:8777-8789, 1983.
141. Rawlings, D.R., Rosenfeld, P.J., Widers, R.J., Challberg, M.D. and Kelly, T.J. Cell 37:309-319, 1984.
142. Guggenheimer, R.A., Stillman, B.W., Nagata, K., Tamanoi, F. and Hurwitz, J. Proc. Natl. Acad. Sci. USA 81:3069-3073, 1984.
143. Carter, T.H and Blanton, R.A. J. Virol. 25:664-674, 1978.
144. Carter, T.H. and Blanton, R.A. J. Virol. 28:450-456, 1978.
145. Nevins, J.R. and Winkler, J.J. Proc. Natl. Acad. Sci. USA 77:1893-1897, 1980.
146. Handa, M., Kingston, R.E. and Sharp, P.A. Nature 302:545-547, 1983.
147. Klessig, D.F., J. Virol. 21:1243-1246, 1977.
148. Klessig, D.F. and Chow, L.T. J. Mol.Biol. 139:221-242, 1980.
149. Stålhandske, P., Persson, H., Perricaudet, M., Philipson, L. and Pettersson, U. Gene 22:157-165, 1983.
150. Ahmed, C.M.I., Chanda, R.S., Stow, N.D. and Zain, B.S. Gene 20:339-346, 1982.
151. Chow, L.T. and Broker, T.R. Cell 15:497-510, 1978.
152. Uhlén, M., Svensson, M., Josephson, S., Alestrõm, P., Chattopadhyaya, J.B., Pettersson, U. and Philipson, L. EMBO J. 1:249-254, 1982.
153. Hérissé, J., Courtois, G. and Galibert, F. Nucl. Acids. Res. 8:2173-2192, 1980.

154. Herissé, J. and Galibert, F. Nucleic Acids Res. 9:1229-1249, 1981.
155. Persson, H., Signäs, C. and Philipson, L. J. Virol. 29, 938-948, 1979.
156. Persson, H. Jansson, M. and Philipson, L. J. Mol. Biol. 136:375-394, 1980.
157. Wold, W.S.M., Cladaras, C., Magie, S.C. and Yacoub, N. J. Virol. 52:307-313, 1984.
158. Persson, H., Öberg, B. and Philipson. L. J. Virol. 28:119-139, 1978.
159. Persson, H., Jörnvall, H. and Zabielski, J. Proc. Natl. Acad. Sci. USA 77:6349-6353, 1980.
160. Cladaras, C. and Wold, W.S.M. Submitted. 1985.
161. Signäs, C., Akusjärvi, G. and Petterson, U. Submitted. 1985.
162. Lewis, A.M., Rabson, A.S. and Levine, A.S. J. Virol. 13:1291-1301, 1974.
163. Thimmapaya, B., Weinberger, C., Schneider, R.J. and Schenk, T. Cell 31:543-551, 1982.
164. Berkner, K.L. and Sharp, P.A. Nucleic Acids Res. 11:6003-6020, 1983.
165. Kvist, S., Östberg, L., Persson, H., Philipson, L. and Pettersson, P.A. Proc. Natl. Acad. Sci. USA 75:5674-5678, 1978.
166. Signäs, C., Katze, M.G., Persson, H. and Philipson, L. Nature 299:175-178, 1982.
167. Kämpe, O., Bellgram, D., Hammerling, U., Lind, P., Pääbo, S., Severinsson, L. and Peterson, P.A. J. Biol. Chem. 258:10594-10598, 1983.
168. Chatterjee, D. and Maizel, J.V. Jr. Proc. Natl. Acad. Sci. 81:6039-6043, 1984.
169. Severinsson, L. and Peterson, P. J. Cell Biol. Submitted. 1985.
170. Pääbo, S., Kämpe, O., Severinson, L., Andersson, M., Fernandez, C., and Peterson, P.A. Prog. Allergy, Karger Basel, 36:114-136, 1985.
171. Hashimoto, S., Pursley, M.H. and Green, M. Nucleic Acids Res. 9:1675-1689, 1981.
172. Handa, H. and Sharp, P.A. Mol. Cell Biol. 4:791-798, 1984.
173. Handa, H., Kingston,R.E. and Sharp,P.A. Nature 302:545-547,1983.
174. Freyer, G.A., Kateh, Y. and Roberts, R.J. Nucleic Acids Res. 12:3503-3518, 1984.
175. Virtanen, A., Gilardi, P., Näslund, A., Le Moullec, J.M., Pettersson, U. and Perricaudet, M. J. Virol. 51:822-831, 1984.
176. Tigges, M.A. and Raskas, H.J. J. Virol. 50:106-117, 1984.
177. Rigolet, M. and Galibert, F. Nucleic Acids Res. 12:7649-7661, 1984.
178. Herissé, J., Rigolet, M., Dupont de Dinechin, S. and Galibert, F. Nucleic Acids Res. 9:4023-4042, 1981.
179. Downey, J.F., Rowe, D.T., Bacchetti, S., Graham, F.L. and Bayley, S.T. J. Virol. 45:514-523, 1983.
180. Sarnow, P., Hearing, P., Andersson, C.W., Reich, N. and Levine, A.J. J. Mol. Biol. 162:565-583, 1982.
181. Challberg, S.S. and Ketner, G. Virology 114:196-209, 1981.
182. Weinberg, D.H. and Ketner, G. Proc. Natl. Acad. Sci. USA 80:5383-5386, 1983.
183. Richardson, W.D. and Westphal, H. Cell 27:141-144, 1981.
184. Reich, P.R., Forget, B.G., Weissman, S.H. and Rose, J.A. J. Mol. Biol. 17:428-439, 1966.
185. Mathews, M.B. Cell 6:223-229, 1975.

186. Söderlund, H., Pettersson, U., Vennström, B., Philipson, L. and Mathews, M.B. Cell 7:585-593, 1976.
187. Mathews, M.B. and Pettersson, U. J. Mol. Biol. 119:293-328, 1978.
188. Akusjärvi, G., Mathews, M.B., Andersson, P., Vennström, B. and Pettersson, U. Proc. Natl. Acad. Sci. USA 77:2424-2428, 1980.
189. Celma, M.L., Pan, J. and Weissman, S.M. J. Biol. Chem. 252:9032-9042, 1977.
190. Celma, M.L., Pan, J. and Weissman, S.M. J. Biol. Chem. 252:9043-9046, 1977.
191. Monstein, H.-J. and Philipson, L. Nucleic Acids Res. 9:4239-4250, 1981.
192. Weinmann, R., Raskas, H.J. and Roeder, R.G. Cell 7:557-566, 1976.
193. Fowlkes, D.M. and Shenk, T. Cell 22:405-413, 1980.
194. Guilfoyle, R. and Weinmann, R. Proc. Natl. Acad. Sci. USA 78:3378-3382, 1981.
195. Bhat, R.A., Metz, B. and Thimmapaya, B. Mol. Cell Biol. 3:1996-2005, 1983.
196. Bhat, R.A. and Thimmappaya, B. Nucleic Acids Res. 12:7377-7388, 1984.
197. Lassar, A.B., Martin, P.L. and Roeder, R.G. Science 222:740-748, 1983.
198. Lerner, M.R., Boyle, J.A., Hardin, J.A. and Steitz, J.A. Science 211:400-402, 1981.
199. Habets, W.J., den Brok, J.H., Boerbooms, A.M., van de Putte, L.B.A. and van Venrooij, W.J. EMBO J. 2:1625-1631, 1983.
200. Francoeur, A.M. and Mathews, M.B. Proc. Natl. Acad, Sci. 79:6772-6776, 1982.
201. Mathews, M.B. and Francoeur, A.M. Mol. Cell. Biol. 4:1134-1140, 1984.
202. Schneider, R.J., Weinberger, C. and Shenk, T. Cell 37:291-298, 1984.
203. Reichel, P.A., Merrick, W.C., Siekierka, J. and Mathews, M.B. Nature 313:196-200, 1985.
204. Lerner, M.R., Andrews, N.C., Miller, G. and Steitz, J.A. Proc. Natl. Acad. Sci. USA 78:805-809, 1981.
205. Bhat, R.A. and Thimmappaya, B. Proc. Natl. Acad. Sci. USA 80:4789-4793, 1983.
206. Bello, L.J. and Ginsberg, H.S. J. Virol. 1:843-850, 1967.
207. Andersson, C.W., Baum, P.R. and Gesteland, R.F. J. Virol. 12:241-255, 1973.
208. Beltz, G.A. and Flint, S.J. J. Mol. Biol. 131:353-373, 1979.
209. Babich, A., Feldman, L.T., Nevins, J.R., Darnell, J. E. and Weinberger, C. Mol. Cell Biol. 3:1212-1221, 1983.
210. Khalili, K. and Weinmann, R. J. Mol. Biol. 175:453-468, 1984.
211. Svensson, C. and Akusjärvi, G. Mol. Cell Biol. 4:736-742, 1984.
212. Svensson, C. and Akusjärvi, G. EMBO J. In press, 1985.
213. Chow, L.T., Lewis, J.B. and Broker, T.R. Cold Spring Harbor Symp. Quant. Biol. 44:401-414, 1980.
214. Persson, H., Pettersson, U. and Mathews, M.B. Virology 90:67-79, 1978.
215. Crossland, L.D. and Raskas, H.J. J. Virol. 46:737-748, 1983.
216. Pettersson, U. and Mathews, M.B. Cell 12:741-750, 1977.
217. Aleström, P., Akusjärvi, G., Perricaudet, M., Mathews, M.B. Klessig, D.F. and Pettersson, U. Cell 19:671-681, 1980.
218. Lawrence, C.B. and Jackson, K.J. J. Mol. Biol. 162:317-334, 1982.

219. Everitt, E., Sundquist, B., Pettersson, U. and Philipson, L. Virology 52:130-147, 1973.
220. Colby, W.W. and Shenk, T. J. Virol. 39:977-980, 1981.
221. Natarajan, V., Madden, M.J. and Salzman, N.P. Proc. Natl. Acad. Sci. 81:6290-6294, 1984.
222. Persson, H., Mathisen, B., Philipson, L. and Pettersson, U. Virology 93:198-208, 1979.
223. Russell, W.C. and Precious, B. J. Gen. Virol. 56:393-401, 1982.
224. Ziff, E.B. and Evans, R.M. Cell 15:1463-1475, 1978.
225. Lucas, J.J. and Ginsberg, H.S. J. Virol. 8:203-218, 1972.
226. Thomas, G.G. and Mathews, M.B. Cell 22:523-532, 1981.
227. Simmons, T., Heywood, P. and Hodge, L.D. J. Mol. Biol. 89:423-433, 1974.
228. Wolgemuth, D.J. and Hsu, M.-T. J. Mol. Biol. 147:247-268, 1981.
229. Beyer, A.L., Bouton, A.H., Hodge, L.D. and Miller, O.L. J. Mol. Biol. 147:269-295, 1981.
230. Manley, J.L., Fire, A., Cano, A., Sharp, P.A. and Gefter, M.L. Proc. Natl. Acad. Sci USA 77:3855-3859, 1980.
231. Corden, J., Wasylyk, B., Buchwalder, A., Sassone-Corsi, P., Kedinger, C. and Chambon, P. Science 209:1406-1414, 1980.
232. Hu, S. and Manley, J.L. Proc. Natl. Acad. Sci. USA 78:820-824, 1981.
233. Hen, R., Sassone-Corsi, P. Corden, J., Gaub, M.P. and Chambon, P. Proc. Natl. Acad. Sci 79:7132-7136, 1982.
234. Davidson, B.L., Egly, J.M., Mulvihill, E.R. and Chambon, P. Nature 301:680-686, 1983.
235. Fraser, N.W., Nevins, J.R., Ziff, E. and Darnell, J.E. J. Mol. Biol. 129:643-656, 1979.
236. Sehgal, P., Derman, E., Molloy, G.R., Tamm, I and Darnell, J.E. Science 194:431-433, 1976.
237. Fraser, N., Sehgal, P. and Darnell, J. Proc.Natl.Acad.Sci. USA 76:2571-2575,1979.
238. Evans, R., Weber, J., and Darnell, J.E. Nature 278:367-370, 1979.
239. Materious, A. and Chen-Kiang, S. Proc. Natl. Acad. Sci 81:5931-5935, 1984.
240. Nevins, J.R. and Darnell, J.E. J. Virol. 25:811-823, 1978.
241. Ziff, E. and Fraser, N. J. Virol. 25:897-906, 1978.
242. McGrogan, M. and Raskas, H.J. Proc. Natl. Acad. Sci. USA 75:625-629, 1978.
243. Nevins, J.R. and Darnell, J.E. Cell 15:1477-1493, 1978.
244. Chen-Kiang, S., Wolgemuth, D.J., Hsu, M.-T. and Darnell, J.E. Cell 28:575-582, 1982.
245. Manley, J.L., Sharp, P.A. and Gefter, M.L. J. Mol. Biol. 159:582-595, 1982.
246. Zeevi, M., Nevins, J.r. and Darnell, J.E. Cell 26:39-46, 1981.
247. Berget, S.M., Moore, C. and Sharp, P.A. Proc. Natl. Acad. Sci USA 74:L3171-3175, 1977.
248. Chow, L.T., Gelinas, R.E., Broker, T.R. and Roberts, R.J. Cell 12:1-8, 1977.
249. Akusjärvi, G. and Pettersson, U. Cell 16:841-850, 1979.
250. Zain, S., Sambrook, J., Roberts, R.J., Keller, W., Fried, M. and Dunn, A.R. Cell 16:851-861, 1979.
251. Berget, S.M. and Sharp, P.A. J. Mol. Biol. 129:547-565, 1979.
252. Mariman, E.C.M., van Beek-Reindevs, R.J. and van Venrooij. J. Mol. Biol. 163:239-256, 1983.

253. Keohavong, P., Gattoni, R., LeMoullec, J.M., Jacob, M and Stévenin, J. Nucl. Acids Res. 10:1215-1229, 1982.
254. Green, M.R., Maniatis, T. and Melton, D. Cell 32:681-694, 1983.
255. Padgett, R.A., Hardy, S.F. and Sharp, P.A. Proc. Natl. Acad. Sci USA 80:5230-5234, 1983.
256. Hernandez, N. and Keller, W. Cell 35:89-99, 1983.
257. Padgett, R.A., Konarska, H.M., Grabowski, P.J., Hardy, S.F. and Sharp, P.A. Science 225:898-903, 1984.
258. Wallace, J.C. and Edmonds, M. Proc. Natl. Acad. Sci. USA 80:950-954, 1983.
259. Lewis, J.B. and Mathews, M.B. Cell 21:303-313, 1980.
260. Shaw, A.R. and Ziff, E.B. Cell 22:905-916, 1980.
261. Akusjärvi, G. and Persson, H. Nature 292:420-426, 1981.
262. Nevins, J.R. and Wilson, M.C. Nature 290:113-118, 1981.
263. Roberts, R.J., O'Neill, K.E. and Yen, C.E. J. Biol. Chem. 259:13968-13975, 1984.
264. Akusjärvi, G. Submitted. 1985.
265. Miller, J.S., Ricciardi, R., Roberts, B.E., Paterson, B. and Mathews, M.B. J. Mol. Biol. 142:455-488, 1980.
266. Virtanen, A., Aleström, P., Persson, H., Katze, M. and Pettersson, U. Nucl. Acids Res. 10:2539-2548,1982.
267. Falvey, E. and Ziff, E. J. Virol. 45:185-191, 1983.
268. Lewis, J.B. and Andersson, C.W. Virology 127:112-123, 1983.
269. Lewis, J.B., Anderson, C.W. and Atkins, J.F. Cell 12:37-44, 1977.
270. Everitt, E., Lutter, L. and Philipson, L. Virology 67:197-208, 1975.
271. Aleström, P., Akusjärvi, G., Lager, M., Yeh-kai, L. and Pettersson, U. J. Biol. Chem. 259:13980-13985, 1984.
272. Akusjärvi, G. and Persson, H. J. Virol. 38:469-482, 1981.
273. Maizel, J.V., White, D.O. and Scharff, M.D. Virology 36:126-136, 1968.
274. Laver, W.G., Pereira, H.G., Russell, W.C. and Valentine, R. J. Mol. Biol. 37:379-389, 1968.
275. Corden, J., Engelking, H.M. and Pearson, G.D. Proc. Natl. Acad. Sci. USA 73:401-404, 1976.
276. Mirza, M.A. and Weber, J. Biochim. Biophys. Acta 696:76-89, 1982.
277. Vayda, M.E., Rogers, A.E. and Flint, S.J. Nucl. Acids Res. 11:441-460, 1983.
278. Sung, M., Cao, T.M., Coleman, R.T. and Budelier, K.A. Proc. Natl. Acad. Sci. USA 80:2902-2906, 1983.
279. Nermut, M.V. Arch. Virol. 62:101-112, 1979.
280. Hosokawa, K. and Sung, M.T. J. Virol. 17:924-934, 1976.
281. Akusjärvi, G., Zabielski, J. Perricaudet, M. and Pettersson, U. Nucl. Acids. Res. 9:1-17, 1981.
282. Jörnvall, H., Akusjärvi, G., Aleström, P., von Bahr-Lindström, H., Pettersson, U., Apella, E., Fowler, A.V. and Philipson, L. J. Biol. Chem. 256:6181-6186, 1981.
283. Akusjärvi, G., Aleström, P., Pettersson, M., Lager, M., Jörnvall, M. and Pettersson, U. J. Biol. Chem. 259:13976-13979, 1984.
284. Bhatti, A.R. and Weber, J. J. Biol. Chem. 254:12265-12268, 1979.
285. Yeh-Kai, L., Akusjärvi, G., Aleström, P., Pettersson, U., Tremblay, M. and Weber, J. J. Mol. Biol. 167:217-222, 1983.
286. Boudin, M.-L., D'Halluin, J.C., Cousin, C. and Boulanger, P. Virology 101:144-152, 1980.

287. Kruijer, W., van Schaik, F.M.A. and Sussenbach, J.S. Nucleic Acids Res. 10:4493-4500, 1982.
288. Oosterom-Dragon, E.A. and Ginsberg, H.S. J. Virol. 40:491-500, 1981.
289. Cepko, C.L. and Sharp, P.A. Cell 31:407-415, 1982.
290. Gambke, D. and Deppert, W. Virology 124:1-12, 1984.
291. Oosterom-Dragon, E.A. and Andersson, C.W. J. Virol. 45:251-263, 1983.
292. Green, N.M., Wrigley, N.G., Russell, W.C., Martin, S.R. and McLachlan, A.D. EMBO J. 2:1357-1365, 1983.
293. Signäs, C., Akusjärvi, G. and Pettersson, U. J. Virol. In press, 1985.
294. Dunn, A.R., Mathews, M.B., Chow, L.T., Sambrook, J. and Keller, W. Cell 15:511-526, 1978.
295. Andersson, K.P. and Klessig, D.F. Proc. Natl. Acad. Sci 81:4023-4027, 1984.

3

ON THE MECHANISM OF ADENOVIRUS DNA REPLICATION

FUYUHIKO TAMANOI
Dept. of Biochemistry and Molecular Biology, University of Chicago, 920 East 58th Street, Chicago, IL 60637

1. INTRODUCTION

Adenovirus DNA is a linear double stranded DNA of approximately 35,000 base pairs (1-3). A protein, termed terminal protein, is covalently linked to the 5′ terminus of each strand (4-6). Overall view of adenovirus DNA replication is shown in Fig. 1. Initiation takes place at either end of the DNA molecule and proceeds by strand displacement mechanism (7-13). During the initiation, a precursor to the terminal protein (terminal protein precursor) comes to the end of the DNA molecule and becomes covalently attached to the nascent DNA (14-18). Replication continues to the opposite end of the DNA molecule thus producing displaced single-stranded DNA. This process of replication which takes place on a double-stranded DNA is called type I replication (19). Another type of replication, type II replication, takes place on the displaced single-stranded DNA. Late in infection, after progeny DNA is packaged into virion particles, the 87K (or 80K) terminal protein precursor is processed to a mature 55K terminal protein by a protease coded by adenovirus DNA (20-21).

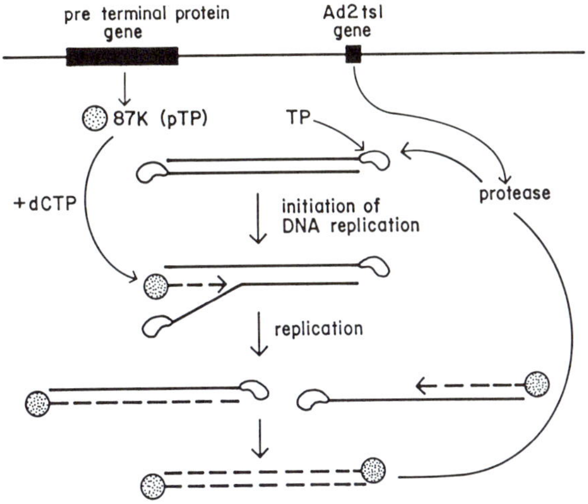

Fig. 1. Model for adenovirus DNA replication (see text for detail). From Stillman et al. (21).

One of the remarkable features of this replication process is its use of a protein to prime DNA synthesis. Since all known DNA polymerases are unable to initiate DNA synthesis de novo, a primer is needed (22,23). In the case of adenovirus DNA, the virus itself produces a primer protein (21,24-26). In many other cases short oligoribonucleotides are used as primers. SS -- RF DNA synthesis of small phages M13 or ϕX174 initiates by utilizing such RNA primers (23,27). Initiation of plasmid DNA replication such as ColE1 also utilizes a RNA primer (28) and impressive amounts of evidence are accumulated which show that Okazaki fragments contain RNA primers attached at their 5′ end (29). In yet another case, the 3′ end of the DNA chain is used as a primer and the 3′ end is supplied by forming a hairpin structure (adeno-associated virus; 30) or by nicking double-stranded DNA (RF -- RF replication of ϕX174 or M13 DNA; 23).

Besides adenovirus, many other systems appear to utilize a protein-priming mechanism for the initiation of DNA replication. The presence of a covalently bound protein on a genome is widely observed. DNAs of some Bacillus subtilis phages such as Ø29, Ø15, Nf, M2Y and GA-1 contain covalently bound proteins (31-36). Ø29 is the most extensively studied case among these and it has been demonstrated that the mode of DNA replication is very similar to that of adenovirus (see ref. 37 for review). The protein with a molecular weight of 28,000 is attached to the Ø29 DNA via a phosphodiester bond between the β-OH of a serine residue and the 5′ deoxyadenosine residue (38). Some eukaryotic virus DNAs besides adenovirus have also been shown to have protein associated with them. These include parvovirus H1, minute virus of mice and hepatitis B virus (39,40). RNA viruses such as poliovirus, encephalo-myocarditis virus (41) and some plant viruses (42) have been reported to have proteins on their genomes. Recent additions to this list are linear double-stranded plasmid DNAs found in yeast, Kluyveromyces lactis (43). Two kinds of plasmids, pGK1 and pGK2, have been found and DNAs from both plasmids are reported to contain small proteins at their termini (44).

The protein priming mechanism also provides a unique way to preserve the ends of linear DNA. As pointed out by Watson (45), linear DNAs pose a special problem of preserving the end of DNA. Especially with systems utilizing RNA primers, one is left with a problem on how to replicate the DNA after the RNA primer is removed. Phages T4 and T7 form concatemers to get around this problem (23,27). Phage lambda circularizes its DNA by the use of sticky ends to overcome this problem (23,27). With protein priming, synthesis could begin from the very end, thus bypassing any need for the formation of special structures.

Recent progress in understanding adenovirus DNA replication processes in molecular terms has become possible by the development of an *in vitro* system which accurately carries out replication (46,47). The typical system utilizes nuclear extracts from adenovirus-infected cells and replicates exogenously added DNA. The success in developing this system is due to the use of hydroxyurea to block viral DNA replication during the infection and at the same time to accumulate early proteins required for DNA replication. This system mimicked *in vivo* replication (46-49). Specific initiation from the terminus was obtained and full-length adenovirus DNA was produced. About 5% of the molecules were type I replicative intermediates which was a double-stranded DNA with a displaced single-stranded DNA. However, this system did not carry out type II replication to a detectable extent (47). Use of temperature-sensitive mutants of adenovirus and fractionation of this system allowed the identification and purification of factors required for adenovirus DNA replication. Furthermore, coupled with the development of an assay to detect initiation, use of this system led to the defining of the origin of DNA replication. In this review, I would like to describe substrates and proteins required for replication and discuss the replication process.

2. ADENOVIRUS DNA

2.1. Structure of viral DNA

Two unique features of adenovirus DNA are relevant for DNA replication. One is the presence of the terminal protein covalently attached to each end of the DNA. This protein will be discussed later in chapter 3.1.1. The other feature is the presence of inverted terminal repeats at the ends of the DNA molecule, which ranges from 63 bp (CELO virus) to 165 bp (Ad12 and Ad18). DNA sequences of the inverted terminal repeat from human adenovirus, simian adenovirus SA7, mouse adenovirus AdFl, Tupaja adenovirus (TAV) and CELO virus have been determined (50-62, also see 63,64 for review). The first nucleotide on the 5′ end is always dC with the exception of CELO virus which has the terminal dG.

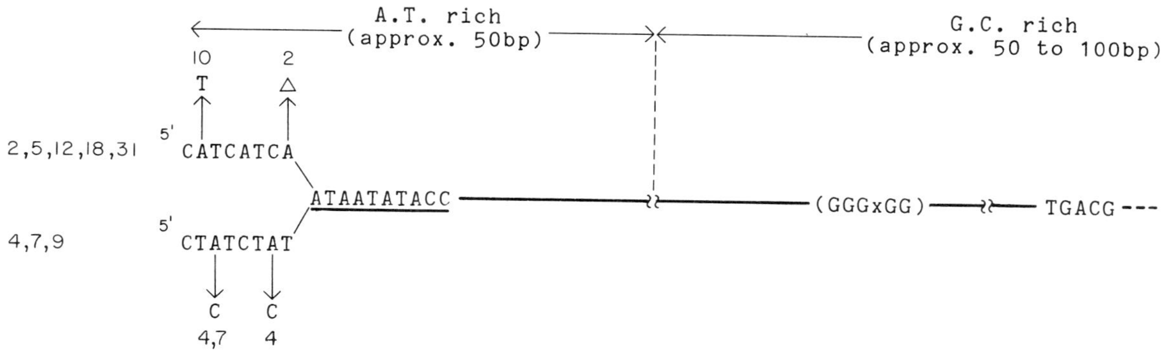

Fig. 2. Structure of the inverted terminal repetition in human adenovirus DNA. 5'-strand sequence of the inverted terminal repetition from all known human adenoviruses can be summarized as shown here. Two prototype sequences at the end of adenovirus DNAs are shown and the perfectly conserved 10-bp region is underlined. The numbers refer to the adenovirus serotype and the arrows point to single base changes in some serotype DNAs. Conserved blocks within the G-C rich region are also shown. From Tamanoi and Stillman (63).

The inverted terminal repeat can be divided into two regions; the terminal A-T rich region and the internal G-C rich region. In the case of human adenovirus DNA, the A-T rich region constitutes 50-52 base pairs at the terminus and is adjacent to the G-C rich region (50-110 base pairs) which occupies the rest of the inverted terminal repeat. DNA sequences within the A-T rich region are highly conserved among all human adenoviruses and also partially conserved in the simian, murine, canine, equine and avian adenoviruses. Particularly notable is a 10 base-pair sequence ATAATATACC which is perfectly conserved in all human adenovirus DNAs. This sequence plays an important role in the initiation of DNA replication as described below.

The G-C rich region, on the other hand, is not conserved and is also variable in size. However, some short sequences of conserved block can be found within this G-C rich region. One such sequence is GGGCGG which appears as multiple copies in the G-C rich region of all human adenoviruses and also in the G-C rich region of the simian and equine adenoviruses. Interestingly this sequence is similar to the sequence GGGXGGAG which is present in multiple copies at the origin of the BK, SV40 and polyoma viruses (65). The other sequence is TGACG that is present at or near the end of inverted terminal repeats of all human adenoviruses. A similar sequence is found in murine (TCACG, right end; TGACG, left end) and in avian adenovirus (TGTCG). The function of these conserved sequences, however, is not known.

In addition to the above features other aspects of adenovirus DNA may be relevant in understanding the DNA replication process. One aspect concerns the structure of adenovirus chromatin. The viral DNA within the virion is complexed with viral-coded proteins (66,67) and the major core protein, pVII is bound tightly to viral DNA (68). The pVII is a small arginine-rich protein and the N terminal domain is similar to that of histones (68,69). Therefore, adenovirus DNA is not arranged in the nucleosomes that are observed with cellular DNA. In fact, micrococcal nuclease digestion of viral cores and of intracellular adenovirus DNA from cells late in infection does not show a repeat pattern typical to cellular or papovavirus DNAs (70,71). However, similar treatment of parental virus DNA extracted early in infection exhibits a pattern similar to the repeat obtained with cellular nucleosomes (72,73). The implications of these findings are that the incoming viral DNA is covered with cellular histones which have replaced pVII early after infection. In the late stage of infection when histone synthesis has ceased, newly synthesized pVII will associate with newly synthesized DNA.

Recently covalently closed circles of adenovirus 5 DNA were isolated (74). These structures arise by head to tail joining of adenovirus DNA and amount to 10-15% of the intracellular viral DNA in BRK cells infected with the hr-1 mutant of Ad5. The circular DNA was detected as early as 3-5 hr postinfection. In wild-type Ad5-infected BRK cells, the circular DNAs were detected, however, only at later times. The Ad5 circular DNAs have been cloned into E. coli. These plasmid DNAs generate infectious virus with an efficiency comparable to virion DNA following transfection into human cells (75). This indicates that linear progeny DNA can be generated from the circles thus suggesting the function of circular DNAs in DNA replication. However, circumstantial evidence points more to their role as intermediates in the integration of viral DNA. In fact, there are examples of integrated viral DNA which has the structure of left and right termini linked together (76-80). In addition, a high percentage of circular DNAs were observed in hr-1 infected rat cells. This is the condition which results in highly efficient transformation of primary baby rat kidney cells. Further work is needed to establish the role, if any, of the circular adenovirus DNA in DNA replication.

2.2 Plasmid DNA

Plasmid DNAs containing adenovirus terminal sequences have been useful in elucidating the origin of DNA replication. Two such DNAs are shown in Fig. 3. pLA1 DNA contains 3.3 kb adenovirus type 5 terminal fragment. The cloning was done by removing the terminal protein from BglII E fragments with piperidine treatment and by using EcoRI linker to clone into plasmid pAT (81). XD7 contains XbaI terminal fragment of adenovirus type 2 (82). These plasmid DNAs can be replicated in vitro when the DNA was linearized such that the adenovirus terminal sequence was located at the end of the linear DNA. In vitro mutagenesis of these plasmid DNAs were utilized to define the origin of DNA replication (see chapter 4.1.2.). Cloning of the entire adenovirus genome has been accomplished by Hanahan and Gluzman (83). A variant of adenovirus type 5 genome which lacked EcoRI site was cloned in a plasmid after the addition of EcoRI linker to its ends. Recombinant plasmid pXAd thus constructed produced infectious virus upon transfection into 293 cells provided that the plasmid DNA was digested with EcoRI. Similar experiments are reported by Berkner and Sharp (84).

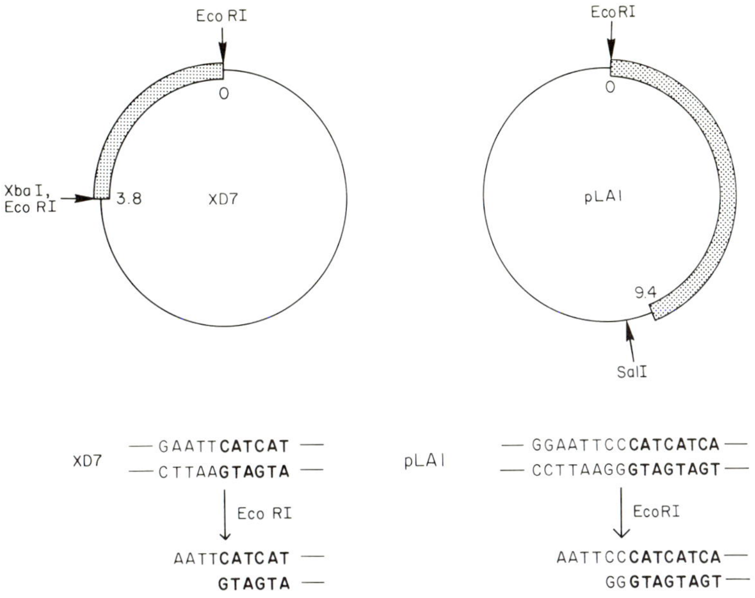

Fig. 3. Plasmid DNAs containing adenovirus terminal fragments. Adenovirus DNA sequence is shown as the shaded area. The numbers refer to the map units (in percent) on the adenovirus genome. See text for detail. Bottom: the DNA sequences of the linearized plasmid DNAs obtained by digestion with EcoRI. Bold letters indicate adenovirus DNA sequences.

Replication of plasmid DNAs *in vivo* was demonstrated by Hay et al. (85). Two kinds of plasmid DNAs used for this study are shown in Fig. 4. Plasmid 3.1 contains two inverted terminal repeats from the adenovirus left terminus (one 0-2.9 map units and the other 0-4.5 map units) and the PstI site is present at the terminal sequence. Plasmid pARKR contains two copies of the adenovirus right terminus (98.4 to 100 map units) separated by the bacterial kanamycin resistance gene. In this case, EcoRI site is located close to the adenovirus terminal sequence. When PstI cleaved 3-1 or EcoRI cleaved pARKR was introduced into 293 cells with an intact Ad2 DNA as helper, replication of fragments containing adenovirus terminal sequence was detected by Southern blotting. The replicated fragments contained covalently bound proteins, which suggests that the plasmid replication are initiated by the same mechanism as the Ad2 genome. This is in line with the *in vitro*

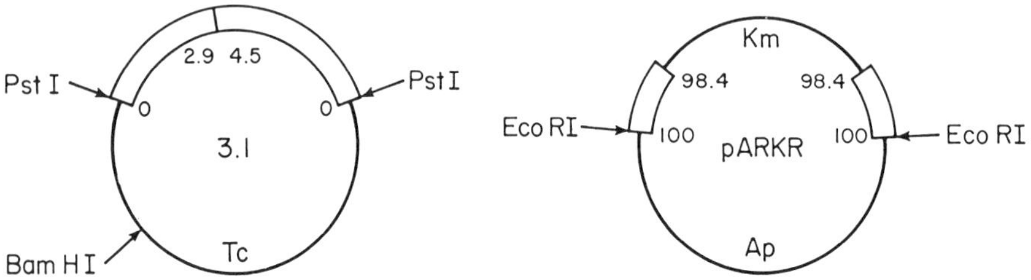

Fig. 4. Plasmid DNAs used for the study of replication *in vivo*. Adenovirus type 2 DNA sequences are shown as boxed areas. Numbers adjacent to the boxed areas represent map units on the adenovirus genome. From Hay et al. (85).

experiments which show that a linear DNA containing adenovirus terminal sequence is capable of replication. In addition to these results the *in vivo* experiment revealed a surprising feature of adenovirus DNA replication. When uncut plasmid DNA was used instead of linearized DNA, replication of the same fragment was still detected, although the efficiency was lower than that observed with the linearized DNA. This suggests that a mechanism exists *in vivo* to replicate adenovirus origin embedded in circular or long linear molecule. Hay et al. proposed a model which involves the formation of a panhandle structure (85).

3. REPLICATION PROTEINS

Proteins required for adenovirus DNA replication are listed in Table 1. Five proteins, three viral and two cellular proteins, are needed for the synthesis of full-length adenovirus DNA. The viral proteins are required for initiation of DNA replication as well as for elongation. Nuclear factor I is required for the initiation as well as for partial replication whereas the factor II is only required for the elongation.

Table 1. Proteins Required for Adenovirus DNA Replication

	Apparent MW	Function
Viral proteins		
terminal protein precursor	80K	- serves as a primer protein
adenovirus DNA polymerase	140K	- polymerization
adenovirus DNA binding protein	72K	- bind single-stranded DNA - interacts with adenovirus DNA polymerase
Cellular proteins		
nuclear factor I	47K	- specific binding to origin
nuclear factor II	<15K	- topoisomerase I activity

From reconstitution experiments it appears that these five proteins are sufficient for the synthesis of full-length adenovirus DNA (86). Two other factors, however, may also affect the replication. One is a cellular protein which could substitute adenovirus DNA-binding protein. It has been demonstrated that crude extracts prepared from cells infected with H5*ts*125 support the initiation even at non-permissive temperature (88), which suggests that the DNA-binding protein is not required for the initiation. However, when purified proteins are used to reconstitute the initiation, the DNA-binding protein is clearly required (see below, chapter 3.2.1.). These results raise the possibility that there is a factor (factors) in uninfected nuclear extracts which substitutes for the adenovirus DNA-binding protein. It has also been reported that RNA from uninfected cells stimulates adenovirus DNA replication carried out in crude extracts (89). The stimulatory RNA appears to be rather non-specific.

3.1. Viral proteins required for adenovirus DNA replication

Three viral proteins required for adenovirus DNA replication are all encoded in the early transcribing region E2. As shown in Fig. 5, the region E2A codes single-stranded DNA-binding protein (90,91) and the region E2B codes the terminal protein precursor and DNA polymerase (21,26,92). The latter two proteins are produced from two different mRNAs resulting from splicing of E2B mRNA (92). Several DNA negative ts mutants of adenovirus have been obtained and they have been classified into two complementation groups. Mutations in the first group (HDts125 and H5ts107) map in the region E2A and affect the activity of DNA-binding protein (93-95). Mutations in the second group (N complementation group H5ts149 and H5ts36) map in the region E2B between coordinates 18 and 22 from the left end of the adenovirus genome and affect the DNA polymerase activity (92, 96-101).

3.1.1. Terminal protein precursor. As described before the terminal protein is covalently linked to the 5′ terminus of each strand. The linkage occurs via the phosphodiester bond between a β-hydroxyl group of a serine residue in the protein and the 5′ phosphate group of the terminal deoxycytidine residue (102). The protein has a molecular weight of 55,000 and is processed from a precursor form which has a molecular weight of 80,000 (terminal protein precursor) (20,21). The processing might involve an intermediate form which has a molecular weight of 62,000 (21,26) and occurs late in infection, after the DNA is packaged into the virion, by a virus-coded protease (20,21). Thus the 80K protein is found at the 5′ end of newly replicated DNA and at the 5′ end of the DNA of adenovirus mutant Ad2ts1 which produces a defective protease (20,21). Viral DNA having the 80K protein replicates as well as the DNA having the 55K protein (21). The virus containing DNA with covalently bound 80K protein can infect cells, however, no progeny viruses are produced (103,104).

The terminal protein precursor is encoded in the region E2B between 23.5 to 28 map units (21,24-26). Cell-free translation of mRNA, hybrid selected against the viral l-strand between coordinates 11.2 and 31.5, produced an 80K protein (21,105). The 80K protein is shown to be structurally related to the 55K protein by tryptic peptide analysis (21). Region E2B codes transcripts which originate from a promoter at 75 map units. The transcripts have leader sequences at 75, 68.5 and 39 map units and splice to three main

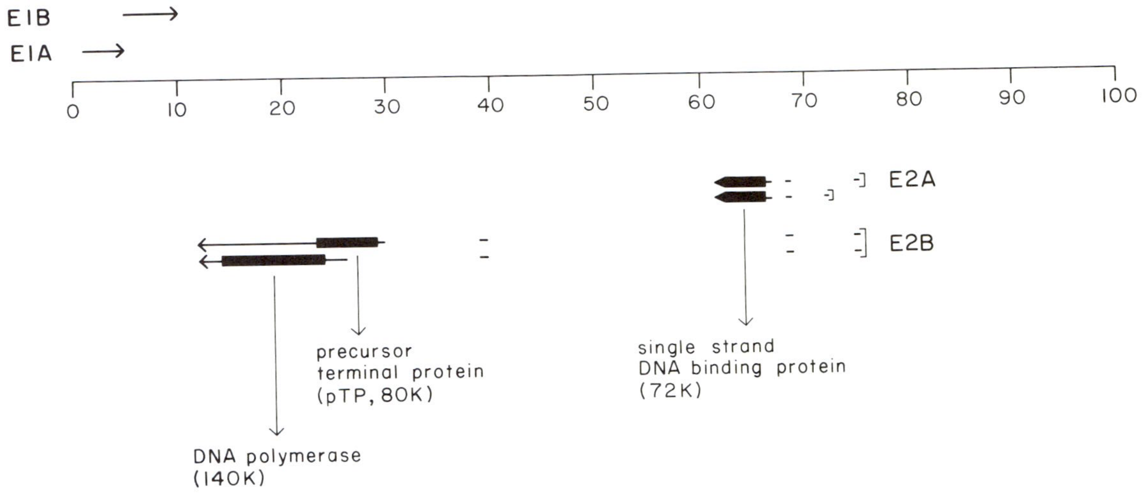

Fig. 5. Map locations of regions coding for three viral proteins required for adenovirus DNA replications. Numbers at the top show map units (in percentage) on the adenovirus genome. mRNAs are shown in arrows and coding regions are shown in thick lines. The multiple mRNA species transcribed in regions E1A and E1B are not shown. See text for detail.

bodies at positions 30.3, 26.0 or 23.1. All these transcripts terminate at around 11.1 map units. One of the long reading frames begins at 28.97 map units and ends at 23.48 map units and encodes a polypeptide of 74.6K dalton (21,24-26). Definitive proof that the terminal protein precursor is encoded by adenovirus was obtained by matching the methionine-containing tryptic peptides of the terminal protein precursor to the peptides predicted from the DNA sequence in this open reading frame (26). A site on the terminal protein precursor where DNA is bound was determined. The site includes serine residues corresponding to position 24.2 on the genome (26). Tryptic mapping also indicated that the 55K terminal protein is derived from carboxy terminal end of the 80K terminal protein precursor (26). Exact site of cleavage is not known, but a sequence of the terminal protein precursor is asp-met-thr-gly-gly-val-phe, which has similarity to the cleavage site for the maturation of pVI to virion protein VI (asn-met-ser-gly-gly-ala-phe).

Purification of the terminal protein precursor in a functional form was accomplished by Enomoto et al. (106). This was possible by the development of an assay to look for an initiation event (107). The assay detects the formation of a covalent complex between the terminal protein precursor and dCMP (pTP-dCMP complex). Using this assay the terminal protein precursor was purified at least 1000-fold over the crude extract. Interestingly, the purified fraction contained a protein of 140K dalton in addition to the 80K dalton terminal protein precursor. Separation of these proteins by glycerol gradient in the presence of 2.8M urea showed that the 140K protein exhibited a DNA polymerase activity (92,99,108). However, both proteins are required for the formation of the pTP-dCMP complex (92,99,108).

3.1.2. Adenovirus DNA polymerase. During the purification of the terminal protein it was realized that a DNA polymerase activity co-purified through several chromatographic steps (105). From chromatographic behavior and from sensitivity to drugs or salt it was suggested that the polymerase was distinct from known cellular DNA polymerases α, β or γ (105). Soon it was realized that it was a novel DNA polymerase coded by adenovirus. This came from the use of temperature-sensitive mutants of adenovirus, Ad5*ts*149 or Ad5*ts*36 (92,99-101,109). Crude extracts prepared from cells infected with these mutants are negative for adenovirus specific DNA replication. It was also shown that the initiation was defective. Complementation with column fractions from wild-type adenovirus infected cells provided an assay to purify the protein and thus led to the purification of the adenovirus DNA polymerase (Ad DNA polymerase). It has not been possible to show temperature-sensitive phenotype of Ad5ts149 extracts made from cells grown at the permissive temperature suggesting that the temperature sensitivity is in the association of the Ad DNA polymerase with the terminal protein precursor (see 92 for discussion).

Proof that the polymerase is coded by adenovirus again came from cell-free translation experiments (92). mRNA selected against E2B region between 20.60 and 24.36 map units produced proteins of 140K and 80K upon translation with reticulocyte lysate whereas mRNA selected against the region between 24.86 and 28.91 map units produced only a protein of 80K. By using this kind of approach, the rightmost limit for the main body of the 140K mRNA was mapped to be between 23.88 and 24.86 map units (92). A long reading frame extends from 24 to 14.2 map units (24,25).

The adenovirus DNA polymerase is sensitive to N-ethylmaleimide, NaCl and cytosine β-D-arabino-furanoside-5'-triphosphate (Ara-CTP) (106). This resembles the property of DNA polymerase α (106). However, the Ad DNA polymerase is sensitive to ddTTP in contrast to DNA polymerase α. Furthermore, the Ad DNA polymerase is resistant to aphidicolin which is a potent inhibitor of DNA polymerase α. Aphidicolin at 100 μM inhibited DNA polymerase α activity about 90%, while the Ad DNA polymerase activity was unaffected (106). This is interesting since adenovirus DNA replication in vivo, as well as in vitro, is sensitive to the aphidicolin (110-113). It has been thought that this indicates involvement of cellular DNA polymerase β or γ in the adenovirus DNA replication (111, 114-116). More recent results, however, indicate that the aphidicolin sensitivity is conferred when the Ad DNA polymerase is functioning together with the terminal protein precursor, DNA-binding protein and host factors I and II (86, also see chapter 4.2).

Properties of adenovirus DNA polymerase have further been investigated using a complex of the Ad DNA polymerase and the terminal protein precursor (117). The complex (pTP-Ad pol) utilizes a variety of homopolymer template-primer combinations including poly(dC)-oligo(dG), poly(dA)-oligo(dT), poly(dT)-oligo(dA) and poly(dT)-oligo(rA). The pTP-Ad pol complex is at least 15 times more active than DNA polymerase α on poly(dA)-oligo(dT). On the other hand, DNA polymerase α was at least 10 times more active than the pTP-Ad pol with poly(dT)-oligo(rA), poly(dC)-oligo(rI) or RNA primed fd single-stranded circular DNA as template-primers. Thus, the pTP-Ad pol utilizes RNA primers much less efficiently than DNA polymerase α. Nicked salmon sperm DNA can also serve as a template for the pTP-Ad pol but poly(rA)-oligo(dT) does not work as a template. The pTP-Ad pol does not have the ability to utilize unprimed poly(dT) template.

Adenovirus DNA-binding protein appears to exhibit a profound effect on the action of the pTP-Ad pol (117). With poly(dT) as the template and oligo(rA) or oligo(dA) as the primer, 10 to 200-fold stimulation of DNA synthesis was observed by the addition of the DNA binding protein. ATP also stimulates the activity of the pTP-Ad pol (117). However, the stimulation was seen only under the condition of low amount of DNA binding protein relative to the template. Apparently the ratio of DNA binding protein to template primer determines whether ATP will stimulate DNA synthesis by pTP-Ad pol. Under the optimal condition the adenovirus DNA polymerase can be a highly processive enzyme, able to synthesize a DNA of 30,000 nucleotides. The long DNA chain was synthesized when poly(dT) of 50 nucleotides and oligo(dA) of 15 nucleotides were used, which indicates that 'slippage' of poly(dA) synthesis takes place.

The adenovirus DNA polymerase appears to contain an intrinsic exonuclease activity (117). The exonuclease hydrolyzes single-stranded DNA in a 3′ → 5′ direction and is at least 10-fold more active on single-stranded DNA than on duplex DNA. The products of the hydrolysis are 5′ deoxynucleoside monophosphates. The hydrolysis is inhibited by the addition of adenovirus DNA-binding protein. Similar nuclease activities have been shown to be intrinsic to several prokaryotic and viral DNA polymerases (23,27), such as E. coli DNA polymerases, phage DNA polymerases and some eukaryotic DNA polymerases including vaccinia virus DNA polymerases. It is thought that an associated nuclease removes misincorporated and mispaired nucleotides thus increasing the fidelity of DNA synthesis (23,27).

3.1.3. Adenovirus DNA binding protein. Adenovirus (Ad) DNA-binding protein represents one of the best-characterized proteins involved in eukaryotic DNA replication. The protein is encoded in region E2A between map units 61.6 and 66.5 on the viral l-strand (90,91). It has been determined from the nucleotide sequence that the protein contains 529 amino acid residues which gives the molecular weight of 59K (90,91). The apparent molecular weight from its mobility in SDS-polyacrylamide gels, however, is 72K. The protein binds efficiently to single-stranded DNA and less efficiently to double-stranded

DNA (118-121). It also binds to DNA termini (119,121). Single-stranded DNA complexed with the Ad DNA-binding protein has been detected by electron microscopy in adenovirus-infected cells (122,123).

Mutants of adenovirus H5*ts*125 and H5*ts*107 produce thermolabile protein and are defective for viral DNA replication at non-permissive temperature (93-95). Both mutants carry a proline -- serine substitution at position 413 in the C-terminal region of the protein (124). When revertants of the H5*ts*125 or H5*ts*107 mutants were isolated, they were shown to carry second-site mutations, in addition to the original mutation, all located at the C-terminal portion (124). These revertants support DNA replication at high temperature. Thus, it appears that the region of the DNA-binding protein responsible for DNA replication lies at the C terminus. This conclusion was further substantiated by the use of proteolytic fragments of the DNA binding protein. Extracts prepared from H5*ts*125 infected cells were inactive at high temperatures (49,100). The activity could be restored by the addition of wild-type Ad DNA binding protein. Ariga et al. (125) showed that a 44K C-terminal fragment produced by chymotryptic digestion of the DNA binding protein could complement the *ts*125 defect. Furthermore Friefeld et al. (88) showed that a 34K fragment derived from the 44K fragment was almost as active as intact 72K DNA binding protein in supporting DNA replication of *ts*125 infected extracts. Therefore the C terminal portion of the protein is the region responsible for DNA replication.

The amino terminal region of the DNA-binding protein, on the other hand, appears to be responsible for another function. Wild-type adenovirus grows poorly in monkey cells because of its inability to express late genes. Expression of early genes and viral DNA replication are normal (126,127). Mutants of adenovirus have been isolated which alter this host range. The host range mutants, such as H5*hr*404, make normal amounts of late proteins and thus are able to grow in monkey cells (128). The mutations map in the amino terminal portion of the DNA binding protein (90,124,128). The amino terminal portion is also a site for phosphorylation of the protein. Protease digestion of the DNA binding protein showed that the N-terminal 26000-dalton fragment carried most of the phosphate groups whereas the C-terminal 34000-dalton fragment was essentially free of phosphate residues (129,130). Therefore, the phosphorylation is not absolutely required for DNA replication nor for binding of the protein to DNA.

Comparison of amino acid sequences of the DNA binding protein between

Ad5 and Ad12 have been carried out by Kruijer et al. (131). Ad5 and Ad12 DNA binding proteins contain 529 and 484 amino acids, respectively. The C-terminal regions exhibit high degree of homology that ranges to 80%. On the other hand, the N-terminal region exhibits only 45% homology and alterations such as deletions or insertions are found. This is in line with the notion that the C-terminal portion is responsible for a function common to all adenoviruses such as DNA replication.

3.2. Cellular proteins

Nuclear extracts prepared from uninfected HeLa cells support the replication of adenovirus DNA when mixed with the terminal protein precursor, adenovirus DNA polymerase and adenovirus DNA binding protein. Fractionation of the nuclear extracts resulted in the detection of two factors termed nuclear factor I and nuclear factor II.

3.2.1. Nuclear factor I. Nuclear factor I has been purified from nuclear extracts of uninfected HeLa cells and exhibits the following properties (132). It has a molecular weight of 47,000 and sediments as a monomer in glycerol gradients. It is distinct from DNA polymerases α, β or γ and contains no detectable nuclease, topoisomerase I, ATPase or RNA polymerase activities. Using viral DNA (adenovirus DNA-protein complex) as a template, low levels of the dCMP complex can be formed by the action of two viral proteins; terminal protein precursor and adenovirus DNA polymerase. The addition of nuclear factor I or ATP stimulates this reaction. When the third viral protein, adenovirus DNA-binding protein is added the pTP-dCMP complex formation is extremely low in the absence of the nuclear factor I. Nuclear factor I is definitely required under this condition. Further stimulation is obtained by the addition of ATP. The stimulatory activity of the nuclear factor I is sensitive to N-ethylmaleimide treatment. It is not affected by heat treatment at $55^{0}C$ for 15 min but is totally inactivated by the treatment at $90^{0}C$ for 2 min.

Nuclear factor I binds to a specific DNA sequence within the origin of DNA replication. This was first shown by using plasmid pLA1 DNA which contained a 3,290 base-pair fragment derived from the left-hand terminus of Ad5 (133). A 451 bp restriction fragment of the pLA1 DNA was bound to the nitrocellulose filter when the fragment was incubated with nuclear factor I.

The nuclear factor I bound preferentially to double-stranded DNA since heat denaturation greatly reduced the binding. Retention of the DNA on filters by nuclear factor I did not require Mg^{2+} or ATP and was unaffected by the length of incubation or by temperature. E. coli tRNA did not influence the binding of the 451 bp fragment by nuclear factor I.

The DNase I protection experiment further narrowed down the region which nuclear factor I binds (133). It binds to a nucleotide sequence located at 17-48 nucleotides, close to the terminus of adenovirus DNA. The DNaseI protection experiments carried out in the presence of replication proteins (terminal protein precursor, adenovirus DNA polymerase and adenovirus DNA binding protein), showed essentially the same protection pattern. Assuming that the nuclear factor I is a globular protein, its size is enough to cover approximately 27A. Since the length of a 32 bp fragment is about 110A it is suggested that two molecules of nuclear factor I are needed to protect the entire sequence (133). Further study on the DNA binding property of the nuclear factor I is described in the later chapter 4.2.3.

3.2.2. Nuclear factor II. Purification of nuclear factor II utilized its ability to complement the replication of adenovirus DNA with terminal protein precursor, adenovirus DNA polymerase and adenovirus DNA binding protein in a nuclear factor I dependent fashion (86). The activity is separated from nuclear factor I independent activity or from DNA polymerases α, β or γ. On gel filtration, nuclear factor II behaves as a protein with a native molecular mass between 25 and 45,000 daltons, possessing a Stokes radius of 26A. SDS polyacrylamide gel electrophoresis revealed two major bands of 14.5 and 15.5 kilodalton and one minor band of 30.5 kilodalton (86).

In the presence of the three viral proteins, (terminal protein precursor, adenovirus DNA polymerase and DNA binding protein) synthesis on the viral DNA was stimulated by the addition of nuclear factor I, but not by the addition of factor II. If both factors were added simultaneously, DNA synthesis was stimulated 15-fold (86). This is more than an additive effect. Thus, one of the interesting features of the factor II is its complete dependence on the presence of factor I. Nuclear factor II has no effect on the formation of pTP-dCMP complex or the replication of terminal XbaI fragments of adenovirus DNA. Synthesis of full-length adenovirus DNA was obtained with these three viral proteins and two cellular proteins. The replication required ATP and was sensitive to aphidicolin to the same extent observed with crude extracts. This point will be further discussed later.

The nuclear factor II exhibits topoisomerase I activity (86). The activity changed the linking number of a plasmid pNT7 by steps of one, which is one of the characteristics of topoisomerase I. It does not require Mg^{2+} or ATP for the activity. Heat treatment at 45°C for 15 min and at 90°C for 2 min resulted in the 30% and 60% inactivation, respectively. Type I topoisomerase purified from HeLa cells or calf thymus substituted for nuclear factor II for the adenovirus DNA replication. E. coli topoisomerase I, on the other hand, did not substitute. Whether the nuclear factor II is identical to the HeLa cell topoisomerase I is unclear since the molecular weight of these proteins are different. The requirement of topoisomerase I activity for adenovirus DNA replication suggests that there is a requirement to relieve restraint ahead or behind a replication fork. Whether this restraint arises by circular structures formed by the interaction of the two terminal proteins or by the size of the linear DNA remains to be elucidated.

4. REPLICATION PROCESS

4.1. Initiation

4.1.1. Assays for the initiation. A breakthrough in the study of initiation came when Lichy et al. (107) found out that a covalent complex between the terminal protein precursor and the first nucleotide dCMP could be detected on SDS-polyacrylamide gel. Briefly, nuclear extracts prepared from adenovirus-infected cells are incubated with α-^{32}P-dCTP in the presence of viral DNA and ATP and then run on a SDS-polyacrylamide gel. The covalent complex between the terminal protein precursor and dCMP is detected as a radioactive band migrating at ~ 80,000 dalton protein position (81,87,107). No other triphosphates including rCMP formed the complex. When crude extracts are used it is necessary to add dideoxy ATP (ddATP) to block elongation. Alternatively endogenous triphosphates could be removed by passing over Sephadex G-25 (87).

When dATP, dTTP and ddGTP are added to α-^{32}P-dCTP, DNA chain elongation continues to the 26th nucleotide. A complex between the terminal protein precursor and a DNA of 26 nucleotides is detected as an 88K band (107). Thus, this assay can also be used to detect early elongation.

The initiation can also be assayed by using restriction fragments of adenovirus DNA (134). When crude extracts or purified proteins were used to replicate viral DNA digested with restriction enzymes, only terminal fragments incorporated radioactivity. Further incubation showed the production of radioactive single-stranded DNA fragments from the terminal fragments. Thus, reinitiation takes place at the termini.

4.1.2. Origin of DNA replication. A remarkable feature of the initiation detected by the formation of the pTP-dCMP complex is that template DNA is required for the complex formation (81,87,107). Viral DNA functions as a template, whereas protease-treated viral DNA does not (81,87,107). However, piperidine treatment of the protease-treated viral DNA revived the inactive template, suggesting that residual amino acids interfere with the initiation (81).

Further study to define the origin of DNA replication was facilitated by the use of plasmid DNAs containing terminal sequences of adenovirus DNA. I have shown two such plasmid DNAs, pLA1 DNA and XD7 DNA, in Fig. 3. These plasmid DNAs supported the initiation *in vitro* when linearized with *Eco*RI (81,135-139). However, supercoiled DNA or a linear DNA with the adenovirus terminal sequence located internally did not support the initiation. Thus the adenovirus terminal sequence has to be located very close to the end of the linear DNA. Several nucleotides added on to the adenovirus terminal sequence due to linkers used for the cloning did not influence the initiation. However, addition of about 20 base pairs of dG-dC destroyed the template activity (81). On the linearized pLA1 DNA the initiation took place at the first dC as well as the second dC four nucleotides away from the first dC (81). In addition to the three viral and two host proteins described before, an additional host protein, factor pL, is required for the initiation of the plasmid DNA replication (138,139).

As shown in Fig. 6 a series of deletions was constructed in order to determine the maximum length required for the initiation (137,140). One set of deletions begins from an internal restriction site and extends toward the adenovirus terminus, whereas the other set of deletions starts from the terminus. No decrease in the template activity was observed as long as the plasmid DNA contained adenovirus terminal sequences longer than 50 base pairs. When the deletion extended further, a low level of initiation was observed and no initiation was seen with the plasmid DNAs containing less than 20 base pairs of adenovirus terminal sequence. On the other hand, deletion of several nucleotides of the terminal sequence destroyed the template activity. From these results it was concluded that the terminal 20 base pairs (Domain I) constitute the minimal region absolutely required for the initiation and the adjacent 30 base pairs (Domain II) are required for the efficient utilization of the origin (137). A similar conclusion was obtained by Rawlins et al. (141) and Challberg and Rawlins (136) defined the Domain I as the terminal 18 base pairs.

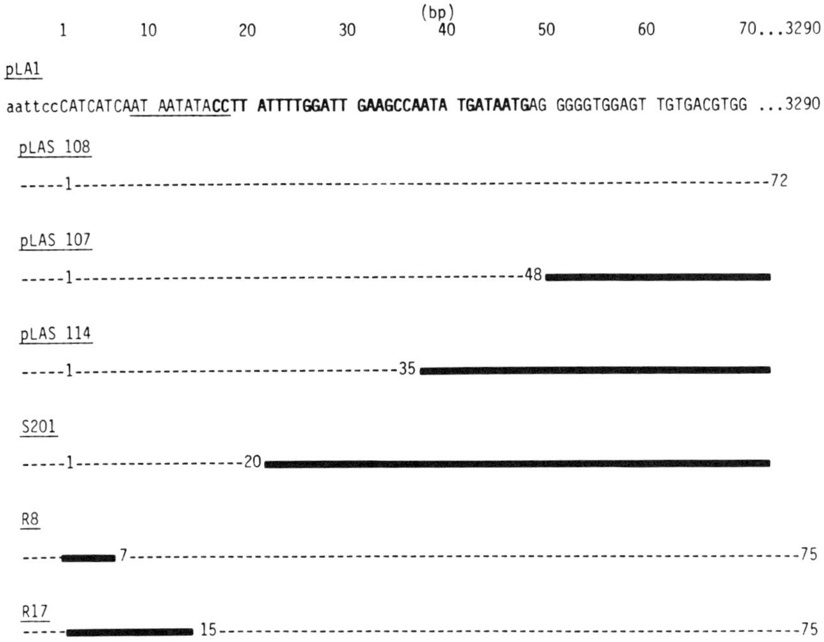

Fig. 6. Deletion mutants used to map the origin of adenovirus DNA replication. DNA sequence at the top indicate the sequences at the 5′ end of the plasmid DNA molecules after digestion with *EcoRI*. Capital letters indicate adenovirus DNA sequences and small letters indicate the *EcoRI* linker DNA sequence. Plasmids, pLAS108, pLAS107, pLAS114 and S201 contain deletions beginning from the internal *PvuII* site and extending to the 5′ terminal *EcoRI* site. Plasmids, R8 and R7, contain deletions beginning at the 5′ *EcoRI* site. ———— deleted sequences, - - - - - sequences present in the various deletion mutants. Numbers represent nucleotides from the adenovirus terminus. 10 bp conserved sequence is underlined and the nuclear factor I binding site is written in bold letters. From Guggenheimer et al. (137).

The terminal 20 base pairs are highly conserved among DNAs from different serotypes of adenovirus (discussed in chapter 2.1.). In particular, a ten base-pair sequence ATAATATACC which is present in the region between the 9th and 18th nucleotide from the terminus is perfectly conserved in all DNAs from human adenoviruses. Point mutations were constructed in order to assess the importance of this ten base pair sequence. Using synthetic oligonucleotides, Tamanoi and Stillman (140) changed the 13th and 14th nucleotides TA to GG. This resulted in a significant decrease of its template activity. Challberg and Rawlins (136) used bisulfite mutagenesis to alter the 17th and 18th nucleotides C to T. This kind of base change also markedly decreased the template activity. However making a base change at the 4th nucleotide which is outside of the conserved sequence did not affect the template activity (135). This suggests that a region between the conserved sequence and the terminal nucleotide merely constitutes a spacer and the alteration within this spacer does not affect the template activity (135,136).

4.1.3. DNA-protein interaction at the origin. The initiation can be carried out by four purified proteins; terminal protein precursor, adenovirus DNA polymerase, adenovirus DNA-binding protein and nuclear factor I. How these proteins interact with the origin described above is of interest in understanding the mechanism of initiation.

The domain II appears to constitute a region where interactions between DNA and nuclear factor I are taking place. As described before, DNase protection experiment has shown that nuclear factor I binds specifically to a region within the origin between the 17th and 48th nucleotide (discussed in chapter 3.2.1.). Guggenheimer et al. (137) further investigated this point using deletion plasmids shown before in Fig. 6. At low concentrations of NaCl all deletion plasmids were retained on nitrocellulose filters after incubation with nuclear factor I. However, specific binding was observed at high concentration of NaCl and only those plasmid DNAs which contained the region between 15 and 48th nucleotide from the adenovirus terminus were bound on nitrocellulose filters (137). This is in excellent agreement with the DNaseI protection experiment (133). Rawlins et al. (141) utilized competition assay to investigate the specific binding of cellular factor (most likely to be nuclear factor I) to the origin. The cellular factor bound to a plasmid DNA, pMDC10, which contained terminal 358 nucleotides of Ad2 genome. A plasmid DNA which contained terminal 67 nucleotides competed as efficiently as the pMDC10 DNA. Plasmid DNAs containing only 36, 31 or 18 nucleotides of the terminal sequence had a greatly reduced ability to compete for the cellular factor. The apparent dissociation constant for binding (Kd) of the cellular factor to the origin was 2×10^{-11}M (141). Further investigation into the interaction between nuclear factor I and the DNA is being carried out by constructing mutant DNAs which have base changes in the binding site.

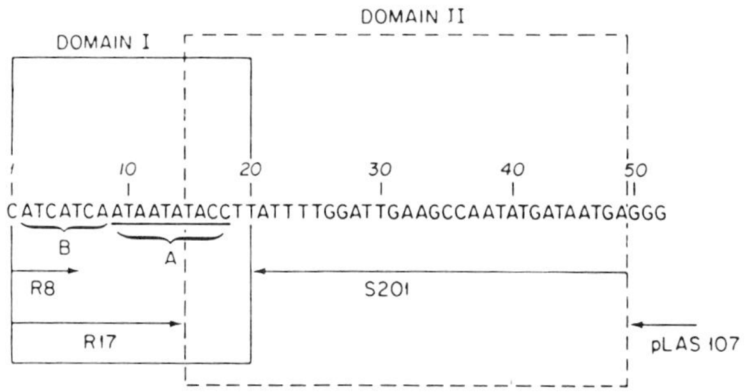

Fig. 7. Structure of the origin of adenovirus DNA replication. The origin can be divided into two domains; domain I and domain II. The domain I can further be divided into two areas; A, 10 bp core sequence and B, a spacer region separating the core and the terminal dC. Arrows represent the boundaries of the deletion plasmids used to define these domains. From Guggenheimer et al. (137).

A similar nitrocellulose filter binding was used to isolate human DNA sequences that bind to nuclear factor I (142). Covalently closed superhelical pLA1 DNA competed for the binding of nuclear factor I to linear plasmid DNA containing adenovirus origin. The competition is specific since neither the RFI forms of øX174, M13mp8, pBR322, SV40, bovine papilloma virus nor several plasmids that contained _AluI_ family and _KpnI_ family DNA sequences showed such competition. Thus this assay provides a means to detect any DNA which contains nuclear factor I binding site. Examination of _HindIII_ digested HeLa cell DNA revealed the presence of such binding sites. Such sequences were isolated and cloned. One nuclear factor I binding site was present about every 100,000 base pairs in the HeLa cell genome and the binding was resistant to high ionic strength. A similar nuclear factor I binding site appears to be present upstream of _myc_ gene (143). Crude extracts from human lymphoblastoid cell nuclei contains an activity to bind a site within the 5′ flanking region of the _myc_ gene. Further precise mapping showed that this region contains a sequence, $TGGAN_6CCAA$, common with nuclear factor I binding site on adenovirus DNA. Another site that has the same common sequence is present downstream of the former site, but this site is bound to a protein factor only under certain conditions which involve the use of 10% glycerol. The protein which binds to the _myc_ DNA is likely to be the nuclear factor I since Ad5 DNA has a higher affinity for the protein than the _myc_ DNA. Extracts of chicken oviduct nuclei also contain similar DNA binding proteins (144). This protein binds to four sites flanking chicken lysozyme gene. Comparison of DNA sequences among the four sites revealed the presence of a symmetrical consensus sequence 5′-TGGCANNNTGCCA-3′ (145).

DNA-protein interaction at the domain I has been investigated by Rijinders et al. (146). They have found that the purified complex of terminal protein precursor and adenovirus DNA polymerase (pTP-Ad pol complex) bound to Ad5 DNA cellulose but not to calf thymus DNA-cellulose. Furthermore by using indirect binding of the pTP-Ad pol complex to DEAE-Sephacel column through the binding to DNA, the binding site was localized to a sequence between the 9th and 22nd nucleotides from the terminus. Thus, pTP-Ad pol complex appears to bind to a conserved sequence in the domain I. However, the DNase I protection experiment to show the direct binding of the pTP-Ad pol to the conserved sequence has failed so far and further experiments are needed to establish this point.

4.2. Elongation of DNA replication

4.2.1. _Properties of elongation reaction._ Three viral proteins (terminal protein precursor, adenovirus DNA polymerase and adenovirus DNA binding protein) and two cellular proteins (nuclear factors I and II) carried out synthesis of full-length DNA on the adenovirus DNA-protein complex (86). The synthesis required 90 -120 min. of incubation at 30°C and started with a marked lag, which depended upon the amount of the pTP-Ad pol added. Without nuclear factor II the synthesis only continues to 25% of the full-length adenovirus DNA. The rate of DNA synthesis obtained by these proteins is 10-20 nucleotides per second, which is close to what is seen *in vivo* for adenovirus DNA replication (147) and for cellular DNA replication (148).

It has been known that aphidicolin inhibits adenovirus DNA replication *in vivo* as well as *in vitro* (110-113). However, the drug concentration required for 50% inhibition of adenovirus DNA replication was 300- 400-fold (*in vivo*) and 130-fold (*in vitro*) higher than for a similar effect on cellular DNA synthesis. These characteristics of aphidicolin sensitivity were reproduced when purified proteins were used to replicate adenovirus DNA (86). Only 50% inhibition was seen at 10μM aphidicolin, but almost 90% inhibition was obtained by 100μM aphidicolin. Products made in the presence of aphidicolin were intermediate-sized DNA chains. Aphidicolin did not inhibit either adenovirus DNA polymerase activity (with nicked salmon sperm DNA), pTP-dCMP complex formation or the 88K dalton complex formation. Furthermore aphidicolin did not affect the topoisomerase I activity of nuclear factor II (86). Since the effects of aphidicolin are demonstrable only in reactions requiring extensive elongation, further study to understand the mechanism of aphidicolin inhibition should be revealing.

Adenovirus DNA-binding protein plays a major role in DNA chain elongation. This has been shown by the use of temperature-sensitive mutant of adenovirus (H5*ts*125) (93-95) or using an antiserum against the DNA-binding protein (149). The DNA binding protein also seems to play a role in early elongation; synthesis of 26 nucleotides of DNA with covalently bound terminal protein precursor (88 kilodalton complex formation) (88,99,101,107). Using H5*ts*125 or H5*ts*107 extracts it was shown that the addition of the DNA-binding protein enhanced the early elongation 2 ~ 4 fold at non-permissive temperature. But little effect was seen with DNA synthesis at permissive temperatures. Dependence of early elongation on the DNA binding protein was incomplete and a residual amount was formed at non-permissive temperature. When terminal protein precursor and DNA polymerase were added to HeLa nuclear extract, similar residual elongation was observed. This may suggest that there is a factor in HeLa nuclear extracts that partially substitute for the adenovirus DNA binding protein. This could be a cellular DNA binding protein.

4.2.2. *Protein-protein interaction during elongation.* How does adenovirus DNA binding protein affect the elongation process? The protein may be needed to protect single-stranded DNA since it is capable of binding to single-stranded DNA. The DNA-binding protein may also play a direct role in the elongation process by interacting with other replication proteins. As described earlier (chapter 3.1.2.), with poly(dT) as template and oligo(dA) or oligo(rA) as primer, DNA synthesis by the adenovirus DNA polymerase (pTP-Ad pol complex was used) was stimulated as much as 100-fold by the addition of adenovirus DNA binding protein. The stimulation is specific for the activity of Ad DNA polymerase since the DNA-binding protein does not stimulate HeLa DNA polymerase α using the same template. Furthermore, E. coli single-stranded DNA-binding protein cannot substitute the adenovirus DNA-binding protein. This suggests that specific interaction between the adenovirus DNA binding protein and adenovirus DNA polymerase is taking place. Interaction between these proteins is also suggested by the coinfection experiment using different serotypes of adenovirus (150).

Whether terminal protein on the template DNA plays any role in the elongation has been investigated by Sussenbach and van der Vliet (151). Addition of antiserum against terminal protein reduced DNA synthesis in infected nuclei to the level of DNA synthesis in uninfected nuclei. When the initiation and partial elongation were carried out and then anti-TP IgG was added, DNA elongation was again reduced. This suggests that the terminal protein on the template DNA or terminal protein precursor at the end of nascent DNA plays a role in DNA elongation. A model which predicts the association of parental terminal protein with adenovirus DNA polymerase has been proposed (151).

5. SUMMARY AND PROSPECTS

Several important findings were obtained from the investigation into adenovirus DNA replication. First of all, a novel DNA polymerase was found. The adenovirus DNA polymerase contains a tightly bound terminal protein precursor which serves as a primer. Association of a priming protein with DNA polymerase may be a general feature since DNA polymerase α also contains a tightly bound protein for primer formation (152-154). However, the priming protein associated with DNA polymerase α forms a short RNA primer of discrete sizes, instead of serving as a protein primer itself as observed with adenovirus DNA polymerase. Undoubtedly further characterization of adenovirus DNA polymerase will be revealing. Defining the region that interacts with the terminal protein precursor might provide information on how active sites are localized. The dCTP binding pocket may be located close to the serine residue on the terminal protein precursor. Furthermore, adenovirus DNA polymerase interacts with adenovirus DNA binding protein and as a result stimulation of DNA synthesis is obtained. This could be obtained by direct association of the DNA binding protein to adenovirus DNA polymerase. Such an association was observed with gene 32 protein and T4 DNA polymerase (155,156). If this were the case, defining the region on adenovirus DNA polymerase for this interaction and investigating the consequence of the interaction will be interesting.

Nuclear factor I is also a novel protein which binds to specific DNA sequences. The concensus sequence appears to be $TGGAN_6CCAA$ which appears within the origin of adenovirus DNA replication and also in cellular DNA. Frequency of the nuclear factor I site in the HeLa cell genome has been calculated to be one site every 100 kb, which is similar to the spacing of origins of replication in various mammalian cells (142). Thus it is possible that nuclear factor I functions in the initiation of cellular DNA replication and the factor I binding sites constitute cellular replication origins. However, nuclear factor I also appears to bind to the regulatory region of *myc* genes, as well as to four sites flanking the chicken lysozyme gene. These results rather suggest that nuclear factor I is involved in a broader activity such as transcriptional activation of genes in general. In this case, adenovirus has, through evolution, gained an ability to utilize a protein originally designed for a different role. Such a case is seen with some bacteriophages. Phage QB makes use of host elongation factors, EF-Tu and EF-Ts (157,158). Phage T7 DNA replication requires thioredoxin from host cells which is tightly associated with the phage DNA polymerase (159,160).

All the studies on enzymology of adenovirus DNA replication have, so far, dealt with type I replication; replication on a double-stranded DNA template. However, very little is known about type II replication which takes place on a displaced single-stranded DNA. It has been suggested that a panhandle structure is formed on the single-stranded DNA thus producing molecular ends identical to those on double-stranded DNA (161-163). The finding that adenovirus genomes lacking part of the left inverted terminal repeat produce infectious progeny which contain two complete inverted terminal repeats (164) supports the idea that a panhandle structure is formed during the replication. How the replication of this panhandle structure is carried out will become clearer by the development of an appropriate in vitro system.

References

1. Van der Eb, A.J. and van Kesteren, L.W. Biochem. Biophys. Acta. 129; 441-444, 1966.
2. Green, M., Pina, M., Kimes, R., Wensink, P.C., Machattie, L.A. and Thomas, C.A. Proc. Natl. Acad. Sci. USA 57; 1302-1309, 1967.
3. Van der Eb, A.J., van Keresteren, L.W. and van Bruggen, E.F.J. Biochem. Biophys. Acta. 182; 530-541, 1969.
4. Robinson, A.J., Younghusband, H.B. and Bellet, A.J.D. Virology 56; 54-69, 1973.
5. Carusi, E.A. Virology 76; 380-394, 1977.
6. Rekosh, D.M.K., Russell, W.C., Bellett, A.J.D. and Robinson, A.J. Cell 11; 283-295, 1977.
7. Winnacker, E.L. Cold Spring Harbor Symp. Quant. Biol. 39; 547, 1975.
8. Schilling, R., Weingartner, B. and Winnacker, E.L. J. Virol. 16; 767, 1975.
9. Tolun, A. and Pettersson, U. J. Virol. 16; 759-766, 1975.
10. Weingartner, B., Winnacker, E.L., Tolun, A. and Pettersson, U. Cell 9; 159, 1976.
11. Horwitz, M.S. J. Virol. 18; 307, 1976.
12. Sussenbach, J.S. and Kuijk, M.G. Virology 77; 149, 1977.
13. Ariga, H. and Shimojo, H. Virology 78; 415, 1977.
14. Stillman, B.W. and Bellett, A.J.D. Cold Spring Harbor Symp. Quant. Biol. 43;729-739.
15. van Wielink, P.S., Naaktgeboren, N. and Sussenbach, J.S. Biochem. Biophys. Acta. 563; 89-99, 1979.
16. Coombs, D.H., Robinson, A.J., Bodnar, J.W., Jones, C.J., Pearson, G.D. Cold Spring Harbor Symp. Quant. Biol. 43; 741-753, 1979.
17. Kelly, T.J. Jr. and Lechner, R.L. Cold Spring Harbor Symp. Quant. Biol. 43; 721-728, 1979.
18. Challberg, M.D., Desiderio, S.V., Kelly, T.J. Jr. Proc. Natl. Acad. Sci. USA 77; 5105-5109, 1980.
19. Lechner, R.L. and Kelly, T.J. Jr. Cell 12; 1007-1020, 1977.
20. Challberg, M.D. and Kelly, T.J. Jr. J. Virol. 38; 272-277, 1981.
21. Stillman, B.W., Lewis, J.B., Chow, L.T., Mathews, M.B. and Smart, J.E. Cell 23; 497-508, 1981.
22. Weissbach, A. Cell 5; 101-108, 1975.
23. Kornberg, A. DNA replication. W.H. Freeman and Co., San Francisco, 1980.
24. Gingeras, T.R., Sciaky, D., Gelinas, R.E., Bing-Dong, J., Yen, C.E., Kelly, M.M., Bullock, P.A., Parsons, B.L., O'Neill, K.E. and Roberts, R.J. J. Biol. Chem. 257; 13475-13491, 1982.
25. Alestrom, P., Akusjarvi, G., Pettersson, M. and Pettersson, U. J. Biol. Chem. 257; 14392-13498, 1982.
26. Smart, J.E. and Stillman, B.W. J. Biol. Chem. 257; 13499-13506, 1982.
27. Kornberg, A. Supplement to DNA replication. W.H. Freeman and Co., San Francisco, 1982.
28. Tomizawa, J-I. and Selzer, G. Ann. Rev. Biochem. 48; 999-1034, 1979.
29. Ogawa, T. and Okazaki, T. Ann. Rev. Biochem. 49; 421-457, 1980.
30. Berns, K.I., Hauswirth, W.W., Fife, K.H. and Lusby, E. Cold Spring Harbor Symp. Quant. Biol. 43, 781-787, 1978.
31. Ortin, J., Vinuela, E., Salas, M. and Vasquez, C. Nature New Biol. 234; 275-277, 1971.
32. Hirokawa, H. Proc. Natl. Acad. Sci. USA 69; 1555-1559, 1972.

33. Salas, M., Mellado, R.P., Vinuela, E. and Sogo, MJ.M. J. Mol. Biol. 119; 269-291, 1980.
34. Arnberg, A.C. and Artmert, F. J. Virol. 18; 783-784, 1976.
35. Ito, J., Kawamura, F. and Yanofsky, S. Virology 70, 37-51, 1976.
36. Kawamura, F. and Ito, J. Virology, 83; 233-245, 1977.
37. Salas, M. In; Current Topics in Microbiology and Immunology (Ed. W. Doerfler), Springer-Verlag, Berlin-Heidelberg, 1983.
38. Hermoso, J.M. and Salas, M. Proc. Natl. Acad. Sci. USA 77; 6425-6428, 1980.
39. Revie, D., Tseng, B.Y., Grafstrom, R.H. and Goulian, M. Proc. Natl. Acad. Sci. USA 76; 5539, 1979.
40. Gerlich, W.H. and Robinson, W.S. Cell 21, 801, 1980.
41. Golini, F., Nomoto, A. and Wimmer, E. Virology 89; 112, 1978.
42. Daubert, S.D., Bruening, G. and Najarian, R.C. Eur.J. Biochem. 92; 45, 1978.
43. Gunge, N., Tamaru, A., Ozawa, F. and Sakaguchi, K. J. Bacteriol. 145; 382-390, 1981.
44. Kikuchi, Y., Hirai, K. and Hishinuma, F. Nuc. Acids Res. 12; 5685-5692, 1984.
45. Watson, J.D. Nature 239; 197-201, 1972.
46. Challberg, M.D. and Kelly, T.J. Jr. Proc. Natl. Acad. Sci. USA, 76; 655-659, 1979.
47. Challberg, M.D. and Kelly, T.J. Jr. J. Mol. Biol. 135; 999-1011.
48. Stillman, B.W. J. Virol. 37; 139-147, 1981.
49. Kaplan, L.M., Ariga, H., Horwitz, M.S. and Hurwitz, J. Proc. Natl. Acad. Sci. USA 76; 5334-5338, 1979.
50. Arrand, J.R. and Roberts, R.J. J. Mol. Biol. 128, 577-594, 1979.
51. Steenbergh, P.H., Maat, J., van Ormondt, H. and Sussenbach, J.S. Nuc. Acids Res. 4; 4371-4389, 1977.
52. Shinagawa, M. and Padmanabhan, R. Proc. Natl. Acad. Sci. 77; 3831-3835, 1980.
53. Sugisaki, H., Sugimoto, K., Takanami, M., Shiroki, K., Saito, I., Shimojo, Y., Sawada, Y., Uemizu, Y., Uesugi, S-I. and Fujinaga, K. Cell 20; 777-786, 1980.
54. Tolun, A., Alestrom, P. and Pettersson, U. Cell 17; 705-713, 1979.
55. Garon, C.F., Parr, R.P., Padmanabhan, R.K. Virology 121; 230-239, 1982.
56. Stillman, B.W., Topp, W.C. and Engler, J.A. J. Virol. 44; 530-537, 1982.
57. Dijkema, R. and Dekker, B.M.M. Gene 8; 7-15, 1979.
58. Tokunaga, O., Shinagawa, M. and Padmanabhan, R. Gene 18; 329-334, 1982.
59. Temple, M., Antoine, G., Delius, H., Stahl, S. and Winnacker, E-L. Virology 109: 1-12, 1981.
60. Shinagawa, M., Ishiyama, T., Padmanabhan, R., Fujinaga, K., Kamada, M. and Sato, G. Virology 125; 491-495, 1983.
61. Alestrom, P., Stenlund, A., Li, P. and Pettersson, U. Gene 18; 193-197, 1982.
62. Shinagawa M. and Padmanabhan, R. Biochem. Biophys. Res. Commun. 87; 671-678, 1979.
63. Tamanoi, F. and Stillman, B.W. Current Topics in Microbiology and Immunology. Vol. 109 75-87.
64. Futterer, J. and Winnacker, E-L. Current Topics in Microbiology and Immunology. Vol. III 41-64.
65. Seif, I., Khoury, G. and Dhar, R. Cell 18; 963-977, 1979.
66. Prage, L. and Pettersson, U. Virology 45; 364-373, 1971.

67. Brown, D.T., Westphal, M., Burlingham, B.T., Winterhof, U. and Doerfler, W. J. Virol. 16; 366-387, 1975.
68. Lischue, M.A. and Sung, M.T. Nature 267; 552-555, 1977.
69. Sung, M.T., Cao, T.M., Coleman, R.T. and Budelier, K.A. Proc. Natl. Acad. Sci. USA 80; 2902-2906, 1983.
70. Brown, M. and Weber, J. Virology 107; 306-310, 1980.
71. Daniell, E., Groff, D.E. and Fedor, M.J. Mol. Cell. Biol. 1; 1094-1105, 1981.
72. Sergeant, A., Tigges, M.A. and Raskas, H.J. J. Virol. 29; 888-898, 1979.
73. Tate, V.E. and Philipson, L. Nuc. Acids Res. 6; 2769-2785, 1979.
74. Reuben, M., Bacchetti, S. and Graham, F. Nature 301; 172-174, 1983.
75. Graham, F.L. EMBO J., in press, 1984.
76. Sambrook, J., Green, R., Stringer, J., Mitchison, T., Hu, S.L. and Botchan, M. Cold Spring Harbor Symp. Quant. Biol. 44; 569-584, 1979.
77. Vardimon, L. and Doerfler, W. J. Mol. Biol. 147; 227-246, 1981.
78. Visser, L., Reemst, A.C.M.B., van Mansfeld, A.D.M. and Roziju, T.H. Nuc. Acids Res. 10; 2189-2198, 1982.
79. Visser, L., Wassenaar, A.T.D.C., van Maarschalkerweerd, M.W. and Rozijn, t.H. J. Virol. 39; 684-693, 1981.
80. van Doren, K., Hanahan, D. and Gluzman, Y. J. Virol. 50; 606-614, 1984.
81. Tamanoi, F. and Stillman, B.W. Proc. Natl. Acad. Sci. USA 79; 2221-2225, 1982.
82. Pearson, G.D., Chow, K-C., Corden, J.L. and Harpst, J.A. ICN-UCLA Symposia on Molecular and Cellular Biology 22; 581-595, 1981.
83. Hanahan, D. and Gluzman, Y. Mol. Cell. Biol. 4; 302-309, 1984.
84. Berkner, K.L. and Sharp, P.A. Nuc. Acids Res. 11; 6003-6020, 1983.
85. Hay, R.T., Stow, N.D. and McDougall, I.M. J. Mol. Biol. 175; 493-510, 1984.
86. Nagata, K., Guggenheimer, R.A. and Hurwitz, J. Proc. Natl. Acad. Sci. USA 80; 4266-4270, 1983.
87. Challberg, M.D., Ostrove, J.M. and Kelly, T.J. Jr. J. Virol. 41; 265-270, 1982.
88. Friefeld, B.R., Krevolin, M.D. and Horwitz, M.S. Virology 124; 380-389, 1983.
89. van der Vliet, P.C., van Dam, D. and Kwant, M.M. FEBS LETTERS 171; 5-8, 1984.
90. Kruijer, W., van Schaik, F.M.A. and Sussenbach, J.S. Nuc. Acids Res. 9; 4439-4457, 1981.
91. Kruijer, W., van Schaik, F.M.A. and Sussenbach, J.S. Nuc. Acids Res. 10; 4493-4500, 1982.
92. Stillman, B.W., Tamanoi, F. and Mathews, M.B. Cell 31; 613-623, 1982.
93. Ensinger, J.M. and Ginsberg, H.S. J. Virol. 10; 328-339, 1972.
94. van der Vliet, P.C., Levine, A.J., Ensinger, M.J. and Ginsberg, H.S. J. Virol. 15; 348-354, 1975.
95. Horwitz, M.S. Proc. Natl. Acad. Sci. USA 75; 4291-4295, 1978.
96. Ginsberg, H.S., Ensinger, M.J., Kaufmann, R.S., Mayer, A.J. and Lundholm, U. Cold Spring Harbor Symp. Quant. Biol. 39; 419-426.
97. Wilkie, N.M., Ustacelibi, S., and Williams, F. Virology 51; 499-503.
98. Galos, R.S., Williams, J., Binger, M.H. and Flint, S.J. Cell 17; 945-956, 1979.
99. Lichy, J.H., Nagata, K., Friefeld, B.R., Enomoto, T., Field, J., Guggenheimer, R.A., Ikeda, J., Horwitz, M.S. and Hurwitz, J. Cold

Spring Harbor Symp. Quant. Biol. 47; 731-740, 1982.
100. Ostrove, J.M., Rosenfeld, P., Williams, J. and Kelly, T.J. Jr. Proc. Natl. Acad. Sci. USA 80; 935-939, 1983.
101. van Bergen, G.M. and van der Vliet, P.A. J. Virol. 46; 642-648, 1983.
102. Desiderio, S.V. and Kelly, T.J. Jr. J. Mol. Biol. 145; 319-337, 1981.
103. Weber, J. J. Virol. 17; 462-471, 1976.
104. Mirza, M.A.A. and Weber, J. Virology 80, 83-97, 1977.
105. Binger, M.H., Flint, S.J. and Rekosh, D.M. J. Virol. 42; 488-501, 1982.
106. Enomoto, T., Lichy, J., Ideda, J-E. and Hurwitz, J. Proc. Natl. Acad. Sci. USA 78; 6779-6783, 1981.
107. Lichy, J.H., Horwitz, M.S. and Hurwitz, J. Proc. Natl. Acad. Sci. USA 76; 2678-2682, 1981.
108. Lichy, J.H., Field, J., Horwitz, M.S. and Hurwitz, J. Proc. Natl. Acad. Sci. USA 79; 5225-5229, 1982.
109. Friefeld, B.R., Lichy, J.H., Hurwitz, J. and Horwitz, M.S. Proc. Natl. Acad. Sci. USA 80; 1589-1593, 19__.
110. Longiaru, M., Ikeda, J., Jarkovsky, Z., Horwitz, S.B. and Horwitz, M.S. Nuc. Acids Res. 6; 3369-3386, 1979.
111. Krokan, H., Schaffer, P. and DePamphilis, M.L. Biochemistry 18; 4431-4443, 1979.
112. Kwant, M.M. and van der Vliet, P.C. Nuc. Acids Res. 8; 3993-4007, 1980.
113. Pincus, S., Robertson, W. and Rekosh, D. Nuc. Acids Res. 9; 4919-4338, 1981.
114. van der Vliet, P.C. and Kwant, M.M. Nature 276; 532-534, 1978.
115. Chen, S., Zubay, G. and Ginsberg, H.S. Eur. J. Biochem. 104; 587-594, 1980.
116. Shaw, C.H., Rekosh, D.M. and Russell, W.C. J. Gen. Virol. 48; 231-326, 1980.
117. Field, J.., Gronostajski, R.M. and Hurwitz, J. J. Biol. Chem. 259; 9487-9495, 1984.
118. van der Vliet, P.C. and Levine, A.J. Nature 246; 170-173, 1973.
119. Fowlkes, D.M., Lord, S.T., Linne, T., Pettersson, U. and Philipson, L. J. Mol. Biol. 132; 163-180, 1979.
120. Schechter, N.M., Davies, W. and Anderson, C.W. Biochem. 19; 2802-2810, 1980.
121. Nass, K. and Frenkel, G.D. J. Virol. 35; 314-319, 1980.
122. Kedinger, C., Brison, O., Perrin, F. and Wilhelm, J. J. Virol. 26; 364-379, 1978.
123. Wohlgemuth, D.J. and Hsu, M-T. J. Mol. Biol. 147; 247-268, 1981.
124. Kruijer, W., Nicolas, J.C., van Schaik, F.M.A. and Sussenbach, J.S. Virology 124; 425-433, 1983.
125. Ariga, H., Klein, H., Levine, A.J. and Horwitz, M.S. Virology 101; 307-310, 1980.
126. Eron, L., Westphal, H. and Khoury, G. J. Virol. 15; 1256-1261, 1975.
127. Klessig, D.F. and Anderson, C.W. J. Virol. 16; 1650-1668, 1975.
128. Klessig, D.F. and Grodzicker, T. Cell 17; 957-966, 1979.
129. Klein, H., Maltman, W. and Levine, A.J. J. Biol. Chem. 254; 11051-11060, 1979.
130. Linne, T. and Philipson, L. Eur. J. Biochem. 103; 259-270, 1980.

131. Kruijer, W., van Schaik, F.M.A., Speijer, J.G. and Sussenbach, J.S. Virology 120: 140-153, 1983.
132. Nagata, K., Guggenheimer, R.A., Enomoto, T., Lichy, J.H. and Hurwitz, J. Proc. Natl. Acad. Sci. USA 79: 6438-6442, 1982.
133. Nagata, K., Guggenheimer, R.A. and Hurwitz, J. Proc. Natl. Acad. Sci. USA 80: 6177-6181, 1983.
134. Horwitz, M.S. and Ariga, H. Proc. Natl. Acad. Sci. USA 78: 1476-1480, 1981.
135. van Bergen, B.G.M., van der Ley, P.A., van Driel, W., van Mansfeld, A.D.M. and van der Vliet, P.C. Nuc. Acids Res. 11: 1975-1989, 1983.
136. Challberg, M.D. and Rawlins, D.R. Proc. Natl. Acad. Sci. USA 81: 100-104, 1984.
137. Guggenheimer, R.A., Stillman, B.W., Nagata, K., Tamanoi, F. and Hurwitz, J. Proc. Natl. Acad. Sci. USA 81: 3069-3073, 1984.
138. Guggenheimer, R.A., Nagata, K., Lindenbaum, J. and Hurwitz, J. J. Biol. Chem. 259: 7807-7814, 1984.
139. Guggenheimer, R.A., Nagata, Kenny, M. and Hurwitz, J. J. Biol. Chem. 259: 7815-7825, 1984.
140. Tamanoi, F. and Stillman, B.W. Proc. Natl. Acad. Sci. USA 80: 6446-6450, 1983.
141. Rawlins, D.R., Rosenfeld, P.J., Wides, R.J., Challberg, M.D. and Kelly, T.J. Jr. Cell 37: 309-319, 1984.
142. Gronostajski, R.M., Nagata, K. and Hurwitz, J. Proc. Natl. Acad. Sci. USA 40134017, 1984.
143. Siebenlist, U., Hennighausen, L., Battey, J. and Leder, P. Cell 37: 381-391, 1984.
144. Nowock, J. and Sippel, A.E. Cell 30: 607-615, 1982.
145. Bergmejer, U., Nowock, J. and Sippel, A.E. Nuc. Acids Res. 12: 4295-4311, 1984.
146. Rijnders, A.W.M., van Bergen, B.G.M., van der Vliet, P.C. and Sussenbach, J.S. Nuc. Acids Res. 11: 8777-8789, 1983.
147. Bodnar, J.W. and Pearson, G.D. Virology 100: 208-211, 1980.
148. Cairns, J. J. Mol. Biol. 15: 372-373, 1966.
149. van der Vliet, P.C., Zandberg, J., Janz, H.S. Virology 80: 98-100, 1977.
150. McDonough, J.S. and Rekosh, D.M. Virology 120: 383-398, 1982.
151. Sussenbach, J.S. and van der Vliet, P.C. In Current Topics in Microbiology and Immunology 109: 53-73. (Ed. Walter Doerfler) Springer-Verlag, Berlin, Heidelberg, 1983.
152. Conaway, R.C. and Lehman, I.R. Proc. Natl. Acad. Sci. USA 79: 4584-4588, 1982.
153. Yagura, T., Kozu, T. and Seno, T. J. Biochem. (Tokyo) 91: 607-618, 1982.
154. Shioda, M., Nelson, E.M., Bayne, M.L. and Benbow, R.M. Proc. Natl. Acad. Sci. USA 79: 7209-7213, 1982.
155. Huberman, J.A., Kornberg, A. and Alberts, B.M. J. Mol. Biol. 62: 39-52, 1971.
156. Alberts, B., Barry, J., Badinger, P., Formosa, T., Jongeneel, E.V. and Kreuzer, K.N. Cold Spring Harbor Symp. Quant. Biol. 47 Part 2, 655-668, 1982.
157. Blumenthal, T., Landers, T.A. and Weber, K. Proc. Natl. Acad. Sci. USA 69: 1313, 1972.
158. Hori, K. Nature 250: 659, 1974.
159. Modrich, P. and Richardson, C.C. J. Biol. Chem. 250: 5515, 1974.

160. Mark, D.F. and Richardson, C.C. Proc. Natl. Acad. Sci. USA 73: 780, 1976.
161. Garon, C.F., Berry, K.W. and Rose, J.A. Proc. Natl. Acad. Sci. USA 69: 2391, 1972.
162. Garon, C.F., Berry, K.W. and Rose, J.A. Proc. Natl. Acad. Sci. USA 72: 3039, 1975.
163. Wolfson, J. and Dressler, D. Proc. Natl. Acad. Sci. USA 69, 3054.
164. Stow, N.D. J. Virol. 37: 171-180, 1981.

4

TRANSCRIPTIONAL AND POST-TRANSCRIPTIONAL REGULATION OF ADENOVIRUS GENE EXPRESSION

MICHAEL J. IMPERIALE and JOSEPH R. NEVINS
The Rockefeller University, New York, N. Y. 10021

1. INTRODUCTION

The phenotype of a cell or organism is determined by the expression of particular genes at specific times and in specific places. Thus, the mechanisms that regulate the expression of protein coding genes are central to our understanding of many diverse aspects of biology. A useful tool to probe these mechanisms has been offered by the viruses that infect animal cells. In the case of the DNA viruses that replicate in the nucleus of infected cells, transcription of the genome and subsequent processing of the viral RNA is usually carried out by host enzymes. Therefore, the study of these "simple systems" allows insight into the more general case of mammalian cell gene control.

Adenovirus has indeed proven to be a fruitful model system. This is most evident by the fact that the initial discoveries of such events as RNA splicing, poly(A) site formation, transcription initiation sites and in vitro transcription systems were made using the adenovirus system. The genome is simple enough to allow its detailed study but sufficiently complex to allow these regulatory interactions. Over the past ten years or so a wealth of information has accumulated concerning the details of adenovirus gene expression. The subject has been reviewed extensively (see refs. 1-3); therefore, this review will focus on recent findings pertaining to transcriptional initiation and RNA processing events.

2. TRANSCRIPTIONAL CONTROL

2.1. Temporal Control of Transcription

It has been known for quite some time that adenovirus gene expression is regulated, just based on the definition of early and late phases of infection. A specific set of genes is expressed prior to DNA replication (early genes) and another set is expressed after replication (late genes). Originally, it was believed that the absence of late gene expression prior to DNA replication was due to transcriptional control; that the major late promoter was not active until after replication of the genome. This turns out not to be the case, at least in its simplest form, but nevertheless transcriptional regulation is a major factor in adenovirus gene control. Perhaps the clearest case of "on/off" control involves the transcription of the genes for protein IX and IVa_2. It appears likely that these genes are not expressed until after DNA replication (4,5).

Much of the current understanding of transcriptional control depends upon two important factors: the availability of mutants in various regulatory functions and the mapping of transcription units and promoters (see 2 for review).

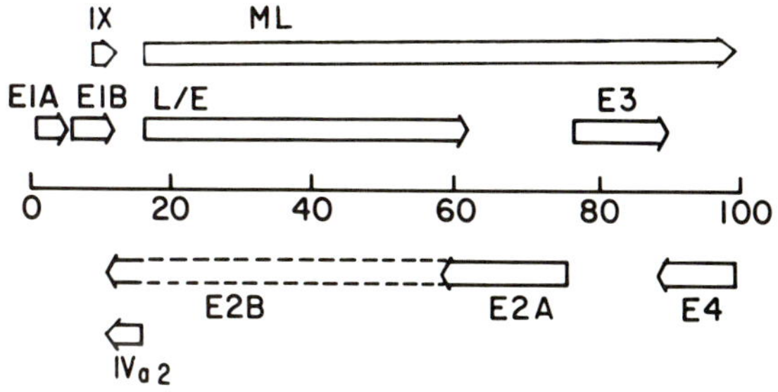

FIGURE 1. Schematic diagram depicting the various transcription units encoded in the adenovirus genome.

Measurement of transcription rates from each of the early regions during the course of an infection revealed differential activation of the early promoters (6,7). The E1A transcription unit is activated first, followed by E1B, E3 and E4, and finally E2. The reason for these temporal differences, for instance the delayed activation of E2, is not clear although it may reflect relative promoter strengths. Possibly the E2 promoter is weak compared to the other promoters and requires a stronger stimulation, perhaps a higher level of a limiting transcriptional factor. Alternatively, there may be more complex interactions that result in differential activation. The elucidation of this activation pathway will clearly require the definition of the exact mechanism of action of E1A and the identification of any other protein factors involved.

As stated above, the initial belief that the major late promoter does not function early turns out not to be correct. Examination of the transcription patterns during early infection revealed transcripts initiating at the late promoter (8-12), at a frequency nearly equal to the other early regions (7,11), and dependent upon the E1A gene product (7). However, there appear to be two distinctions between the transcription of this region early and late. First, there is a quantitative change in transcription such that there is considerably more activity late in infection than early. Whether this is due to an increase in transcription initiation from this promoter, whether it is due to the vastly increased number of templates, or both, is not clear. Second, there is a change in the site of transcription termination in the late transcription unit. During early infection, transcripts terminate somewhere near the middle of the genome, probably between the L3 and L4 late regions (11). Late in infection, transcripts initiating at the same promoter read through this position and terminate near the end of the genome, distal to the L5 region (13). Clearly, transcriptional termination precludes the expression of the L4 and L5 sequences during early infection and thus is a controlling factor in gene expression. The mechanism for this control is not understood but it may represent a _cis_-acting control involving a change in the template after replication of the genome rather than the result of a _trans_-acting factor (14).

2.2. ElA Control of Transcription

2.2.1. The Structure and Function of the ElA Gene. The ElA gene, which maps to the very left end of the viral genome, is the first to be expressed upon infection (7). This gene encodes two early mRNA's, a 13S and a 12S, as well as a 9S mRNA that is produced late in infection (15-17). All of the RNAs have the same 5' and 3' end and differ by the size of an internal splice. All of the proteins are read from the same reading frame and are therefore identical at their N-termini and C-termini, differing only by the sequences in the middle of the protein (3,18).

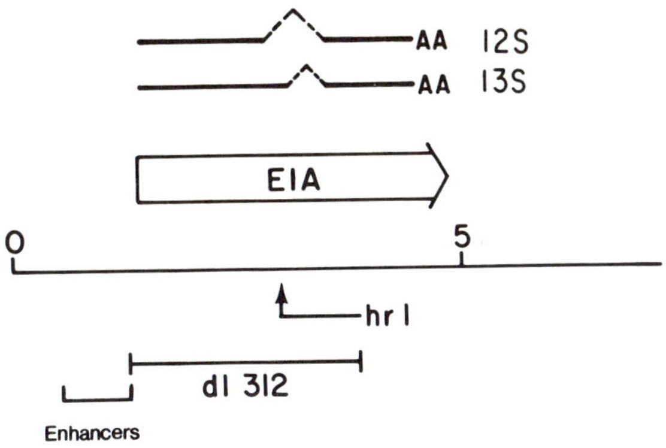

FIGURE 2. Schematic diagram depicting the ElA region. Shown are the mRNA structures, mutant map positons, and locations of the defined enhancer sequences.

The availability of adenovirus transformed cell lines that express the ElA and ElB genes, in particular the 293 human cell line (19,20), allowed the isolation of various viral mutants in these genes (21-23). These cells are especially important since it has not yet been possible to isolate conditional ElA or ElB mutants with the exception of certain cold-sensitive ElA mutants (24-26). The initial definition of the ElA phenotype employed a deletion mutant, dl312 (27), and a single base deletion/frameshift mutant, hrl (28). It was found that in either of these mutants there is no production of the other early viral RNAs. Although there is evidence for post-transcriptional control mediated by ElA (29), the major point of action for ElA control of early viral gene expression appears to be transcriptional initiation, as determined by measurement of transcription rates in intact viral infected cells (7) or, more recently, from nuclei isolated after a co-transfection of the ElA and E2 genes (30). In addition, the ElA-mediated induction of at least one cellular gene, the 70kd heat shock protein gene, is clearly transcriptional (31).

Since the hrl mutation leaves the ElA gene essentially intact, it was possible to assay for ElA transcription from this virus in the absence of a functional ElA protein. ElA transcription in hrl is lower than that seen in wild type infections by about 2-3 fold, indicating that ElA protein also stimulates ElA transcription (7,28). The major difference in the

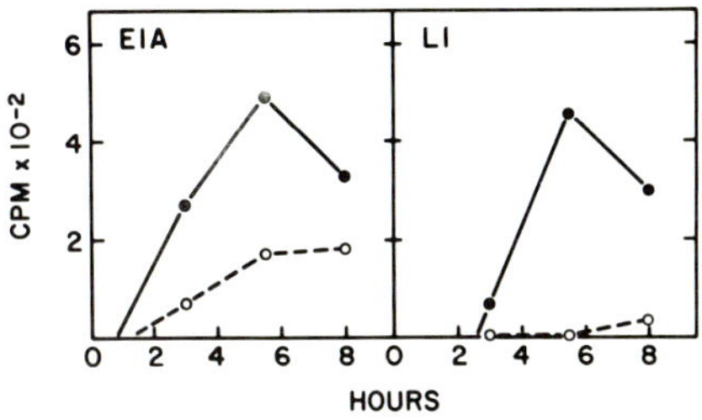

FIGURE 3. Transcription rates of the E1A and L1 sequences in the presence (•) and absence (o) of E1A function. For details, see ref. (7). Reproduced from (7) with permission.

level of expression between E1A and the other early regions appears to be that obtained in the absence of E1A protein. E1A transcription is 10 - 20 fold higher than the other regions in the absence of E1A protein. The end result, in the presence of E1A protein, is nearly the same level of transcription for all of the promoters (7).

Since the hr1 mutation maps to a sequence in the E1A region that is unique to the 13S mRNA and does not affect the 12S product (32), it has been concluded that the product of the 13S mRNA is responsible for transcription activation. Additional mutants have been engineered into this region yielding viruses that produce only the 12S mRNA or only the 13S mRNA (33-35). Consistent with the hr1 results, a virus expressing only the 13S mRNA is capable of transcriptional activation whereas one expressing 12S only is not. A question remains, however, as to whether the 12S product has any capacity to induce transcription, albeit at an inefficient level. This is of particular significance with respect to the role of ElA in transformation (36,37). It appears that a virus that only expresses the 12S mRNA can at least partially transform cells (34,35). The experiments mentioned above have indicated that there was little or no activation of transcription by the 12S mRNA. However, other experiments suggest that although a virus that expresses only the 12S mRNA is inefficient compared to its 13S counterpart, there nevertheless is some activation of transcription (35). Furthermore, in transfection assays, a plasmid expressing only the 12S product was able to efficiently activate transcription of an ElA inducible gene (30). It is somewhat difficult to judge relative efficiencies from the transfection experiments, but these experiments do suggest that the 12S product does possess the ability to induce transcription.

In addition to its role in transformation, another potential role for the 12S product has been suggested from experiments in which the two viruses that express only one of the two early ElA mRNAs have been introduced into quiescent or growing WI38 cells. Although the virus expressing only the 12S mRNA cannot replicate in either case, the virus that expresses the 13S mRNA only gives a lower yield than wild type virus and DNA replication is delayed (34; A.J. Berk, pers. comm.). Thus, the 12S product may be required for some replication function that is missing in cells that are in the G0 state. Alternatively, the 12S gene product may induce the expression of a specific cellular gene that is not expressed when cells arrest in G0, the product of which is essential for exit from the G0 state.

The transcription of the ElA gene itself is apparently controlled by a number of cis-acting elements that appear to be responsible for the high level of transcription in the absence of ElA protein. Indeed, one experiment has shown that ElA upstream sequences inserted into an E2 plasmid can render the expression of the E2 gene independent of ElA induction (38). One might consider two possible explanations for this result. Perhaps the ElA enhancer region utilizes a transcriptional factor that cannot be used by the other early promoters and this factor is present in most cells. Alternatively, the ElA enhancer could utilize the same factor that the other promoters use but with a much higher affinity, allowing ElA to out-compete the other promoters when the factor is limiting. The sequences from -141 to -305, -45 to -188, and -321 to -344 relative to the ElA transcription initiation site have all been shown to act in cis to activate transcription of heterologous genes (38-40; see Figure 2). One of these sequences (40) has homology to the SV40 enhancer core sequence (40a) whereas the other elements do not. Finally, there are apparently cis-acting sequences affecting ElA transcription efficiency that are located within the ElA coding region (41). Thus, the sequences (and possibly factors) that mediate ElA transcription are quite complex.

2.2.2. *E1A Induction: Specificity and Sequence Requirements.* The sequence requirements for the expression of the other early genes have been a subject of study both *in vivo* and *in vitro*. Given the fact that the early genes are positively regulated by a known gene product, studies on the early promoters have focused both on a determination of the sequence requirements for basal and E1A-induced transcription. Among the early promoters, the E2 promoter has been the most intensely studied. Using a series of deletion mutants in the region upstream from the major E2 transcriptional start site, transient expression assays in 293 cells and HeLa cells were used to determine sequence requirements for the expression of the E2 gene (42-44). It has been found that 79 (refs. 42,43) or 94 (ref. 44) nucleotides upstream from the start site were sufficient for wild type levels of expression, although a plasmid containing 70 nucleotides upstream was reduced to 5% of the wild type activity. Furthermore, it has been shown that these same sequence requirements hold when deletion mutants are transcribed *in vitro* in uninfected cell extracts (44; R. Roeder, personal communication). The role of these sequences in E1A induction has been investigated by assaying promoter mutants in HeLa cells in the presence or absence of E1A (43,44). The results indicate that the same sequences required for uninduced transcription are also required for E1A induction. Specifically, a promoter containing 70 nucleotides of upstream sequence is severely impaired both in uninduced expression as well as induced expression. Nevertheless, it clearly is still inducible. In contrast, a promoter containing 59 nucleotides is virtually inactive and is not induced by ElA. Using a different assay system, namely selecting for methotrexate-resistant colonies after transfection of a dihydrofolate reductase gene under the control of either wild type or deleted E2 promoters, it has been suggested that there is no specific sequence requirement for ElA induction (45). It was found that a construction retaining only 18 nucleotides upstream of the E2 transcriptional start site yields an increased number of colonies when co-transfected with an ElA gene. The discrepancy between these two sets of results may be due to the assay system, as no selective pressure is utilized in the transient expression assays described above. Nonetheless, it has been shown that the E2 sequences from -262 to -21 can render a promoter that is normally not ElA inducible responsive to such induction, indicating that there is something in that region that responds to the action of ElA (43).

This finding arose from experiments in which it was attempted to restore function to a -28 E2 mutant by inserting the E2 sequences from -262 to -21 back in at the site of the deletion. It was shown that the -262 to -21 fragment could restore wild type levels of expression in either orientation (42). Furthermore, the same fragment, when inserted approximately 4000 nucleotides from the transcriptional start site, still increased the level of expression to 20% of wild type. More recently, it has been shown that this fragment can provide to a heterologous promoter both a basal promoter function and the ability to be induced by ElA (43). Thus, this upstream element is an enhancer-like element that requires the activity of the E1A protein for full function.

The E1B, E3 and E4 promoter regions have also been analyzed with respect to sequence requirements for expression. An E1B promoter, driving expression of the herpesvirus thymidine kinase gene and retaining 135 nucleotides upstream of the transcriptional initiation site, was functional but one that was deleted in to -83 was inactive (46). Using an _in vitro_ transcription system, it was demonstrated that 325 nucleotides (the entire right end of the genome) are required upstream from the E4 cap site for full expression of the E4 gene in vitro (47). Finally, deletion analysis of the E3 promoter has indicated that sequences up to -105 are essential for expression (N. Jones, pers. comm.). Furthermore, E3 sequences upstream of -80 (thus presumably those sequences between -80 and -105) possess the same properties as the E2 upstream sequences in that they can function independently of position or orientation to activate a heterologous gene.

2.2.3. _Mechanism of E1A Action_. The studies on the sequence requirements for the expression of the early genes have yielded some insight into the mechanism by which E1A activates transcription. As deleting in towards the initiation site has always yielded a decrease in the level of expression, the role of a positive acting factor is indicated. The finding that the sequences necessary for the transcription of the E2 gene cannot be separated from those sequences required for E1A induction does not prove, but does imply, that these upstream regions might be the binding sites for a transcriptional factor(s), and that the role of ElA may be to increase the functional level of such a factor. The enhancer-like nature of the E2 and E3 promoters further implies that the binding of such a factor can occur independently of the orientation of the binding site and, in addition, can sometimes provide a functional interaction regardless of the position of the binding site with respect to the transcriptional start site.

The resolution of the question of proteins that interact with ElA-inducible promoters will require the isolation and identification of these proteins. To date, such proteins have been identified through the use of <u>in vitro</u> transcription systems. Specifically, factors required for the transcription of the early SV40 promoter (48) and the Drosophila hsp70 promoter (49) have been isolated by fractionating cell free extracts. Unfortunately, such may not be an easy task for ElA regulated transcription since extracts prepared from viral infected cells do not show an enhanced activity for the early viral promoters (50,51).

Previous work indicated the involvement of a negative acting component in the ElA mediated activation of viral genes, given the result that the ElA requirement could be at least partially overcome by treating cells with cycloheximide (7,29). Furthermore, the addition of cycloheximide in the presence of ElA results in a superinduction of viral transcription (52). Initially, it was suggested that a cellular repressor might be invovled, namely a cellular protein that recognized the early adenovirus promoters and blocked their activity. The role of the ElA protein would be to block the action of such a repressor. The present data, however, speak strongly for the involvement of a positive factor in the transcription of the ElA inducible promoters. Analysis of the sequence requirements of these promoters, as discussed above, has not yielded any evidence of a repressor interaction. These results would therefore suggest that a negative acting component might operate at a level apart from a direct interaction with specific sequences in the DNA. One might postulate, for instance, that there is a cellular component that is an antagonist of ElA.

Various experiments do, however, indicate that the mechanism of ElA action involves an interaction with the cell and may therefore be analogous to a form of cellular transcription control. First of all, it was found that the immediate early gene of pseudorabies virus, a herpesvirus, can substitute for the ElA gene in both a viral co-infection and in a co-transfection of recombinant plasmids into cells (38,53). The finding that the heterologous activator functioned better than the ElA gene product itself suggested that they function through a common cellular intermediate. The implication that the ElA and immediate early proteins might act through a cellular intermediate in turn suggested that there may be some cellular genes that are analogous to these viral activators as well as other cellular genes that are regulated in the same fashion as the viral early genes. In fact, there is evidence for both types of cellular gene. A number of cellular genes have been shown to be ElA-inducible. Perhaps the clearest studies have been those on the human 70kd heat shock protein gene. It has been demonstrated that the expression of this gene is activated during an infection, and that such induction requires the ElA gene product (54). Furthermore, this induction occurs at the level of transcription and the kinetics of induction of this gene are identical to those seen for the viral early genes (31).

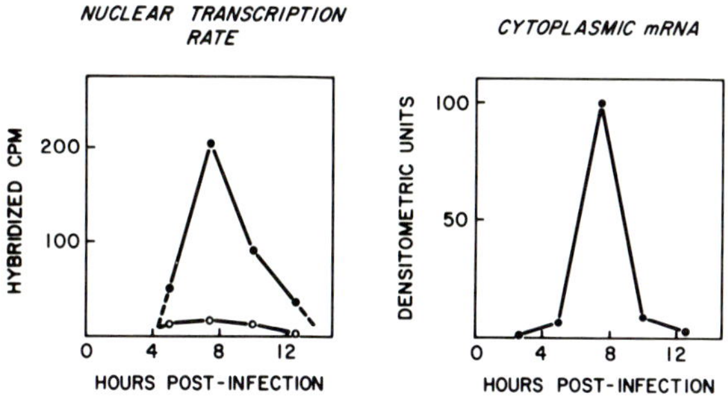

FIGURE 4. Induction of the human HSP70 (heat shock) gene by the adenovirus ElA gene product. Depicted is a graphic representation of transcription rates and mRNA levels during a wild type and dl312 infection. Reproduced from (31) with permission.

A number of other cellular genes have been found to be inducible by ElA, although possibly not in the same manner as the heat shock gene. Initially, it was shown that the human beta-globin gene could be activated by ElA (or the pseudorabies immediate early gene) when transfected into HeLa cells (55,56). In addition, these same studies demonstrated that the SV40 early promoter, without the 72bp repeat enhancer element, could be stimulated by ElA. In the latter case there was a requirement for the 21 bp repeat element, which is known to be an essential component of the SV40 early promoter (57). The beta-globin gene, however, could be stimulated by ElA in the absence of sequences upstream of the TATA element. This is a potentially interesting result in view of recent data suggesting a requirement for sequences internal to the beta-globin gene for activation during mouse erythroleukemia cell differentiation (58). It has now been demonstrated that at least one other gene can be stimulated by ElA when introduced into cells via transfection or as part of the viral genome (59). In all these cases (other than the heat shock gene), however, the endogenous counterparts of these cellular genes are inactive. Presumably this is due to an inaccessibility of the endogenous gene to the ElA function due to chromatin structure.

There is also evidence for an ElA-like activity in certain uninfected cells (60). It was found that a strong correlation exists between the basal level of expression of the 70kd heat shock gene (in the absence of heat or ElA-mediated induction) and the ability of the viral early genes to be expressed independently of ElA. Namely, cells that have a high uninduced level of heat shock gene expression are able to express the early genes in the absence of the E1A gene, whereas cells that have low uninduced levels of heat shock gene expression require E1A function to be provided by the virus in order to obtain early gene expression. The most explicit behavior of this type is exemplified by studies done with the murine teratocarcinoma cell line, F9. The undifferentiated F9 cells express a high level of heat shock protein and are able to support the E1A-independent expression of the viral E2 gene. When these same cells are induced to differentiate with cAMP plus retinoic acid, however, heat shock protein expression becomes undetectable and the E2 gene is only expressed in a wild type infection where the E1A protein is present. In each of these cases, the underlying consistent finding was a correlation of cellular E1A activity with rapid growth rates of cells. In each case, cells that exhibited tight growth control possessed no demonstrable E1A-like activity. This result is, of course, consistent with the role of E1A during transformation in immortalizing cells.

Recent experiments have suggested a possible role for such a cellular E1A-like function. The expression of the heat shock gene in growing cells was investigated to attempt to provide a clue to the manner in which this gene is controlled. It was found that the heat shock gene was subject to cell cycle regulation in HeLa cells and in 293 cells, and that this regulation was strictly transcriptional (61). The timing of the control is such that transcription is highest in the late S-phase or G2 phase of the cell cycle. These results therefore suggest that a cellular

ElA analog may be involved in the regulation of cell cycle transcription.

2.3. Negative Transcription Control

In addition to the negative regulation of the class I MHC genes by the ElA protein (see paper by van der Eb, this volume), there is one other example of negative regulation during adenovirus infection. Carter and Blanton originally described the fact that there was overproduction of early viral RNA under non-permissive conditions for the mutant ts125 (62). This is an E2 mutant that produces a thermolabile 72 kd DNA binding protein (63). The basis for this overproduction of viral RNA was investigated by measuring transcription rates in wild type and mutant infected cells. It was found that transcription of the E4 gene was not shut-off in mutant infected cells as is the case in wild type infection (64).

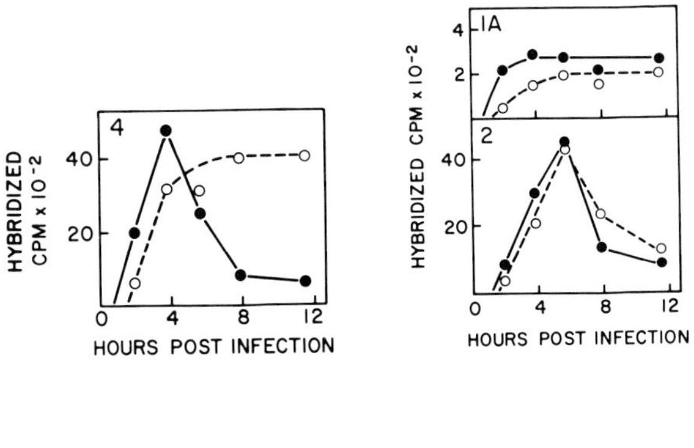

FIGURE 5. Role of the 72kd DNA binding protein in negative tran-

scription control. Transcription rates for E4, ElA and E2 in a wild type infection and a ts125 infection at 41° are depicted. From (64) with permission.

This was the only gene whose expression was affected at the transcriptional level, however, indicating specificity of action on the part of the 72kd DNA binding protein. It thus appears that one function of this protein is to repress transcription of the E4 transcription unit during lytic infection.

These findings have been confirmed in *in vitro* studies using the E4 gene as a template. Transcription from the E4 promoter can be specifically inhibited (as compared to transcription from the major late promoter) in HeLa whole cell extracts upon the addition of purified 72kd protein (65). Left to be determined is the mechanism for the inhibition. Does the protein bind to a specific sequence near the E4 promoter, possibly related to its ability to bind to the termini of double stranded DNA (66), or does it have a more indirect, yet specific, effect?

3. POST-TRANSCRIPTIONAL CONTROL

The pathway of eukaryotic mRNA biogenesis is now well established, as is the fact that certain post-transcriptional processing events are subject to regulation (67,68). Each of the events of RNA processing as well as subsequent metabolism of the mRNA have been shown to be subject to change in adenovirus infected cells. Transcripts from the major late promoter are spliced differently early and late in infection (11,12); poly(A) site selection is altered (11,69); the transport of cellular mRNA is inhibited during late infection (70,71); and, the stability of early viral mRNA is controlled by the 72kd DNA binding protein (72). Thus, a wealth of information has accumulated from the study of adenovirus infection concerning the various events of RNA processing and their regulation. The recent studies on these processes will be discussed here.

3.1. Splicing

The phenomenon of splicing of mRNA precursors was first observed in adenovirus infected cells (73,74). This aspect of gene expression has, of course, received tremendous attention in the time since its discovery, primarily for two reasons. First, the inherent complexity and precision of splicing renders the process intriguing. Second, the fact that pieces of a transcript are reassembled affords an extra degree of regulation of mRNA manufacture.

The splicing of transcripts from the adenovirus major late transcription unit surely must be among the most complex of such processes within the eukaryotic cell. A minimum of 20 separate exons within this single transcription unit can be combined to yield at least 18-20 distinct messenger RNAs (3). Furthermore, the process does appear to be regulated such that different mRNAs are produced under different circumstances. The L1 region of the late transcription unit is spliced differently depending on the time of infection (11,12). During the early phase, a single mRNA of 4.0 kb is the product of processing. However, during the late phase, three mRNAs of 4.3 kb, 3.8 kb, and 2.3 kb are prcduced.

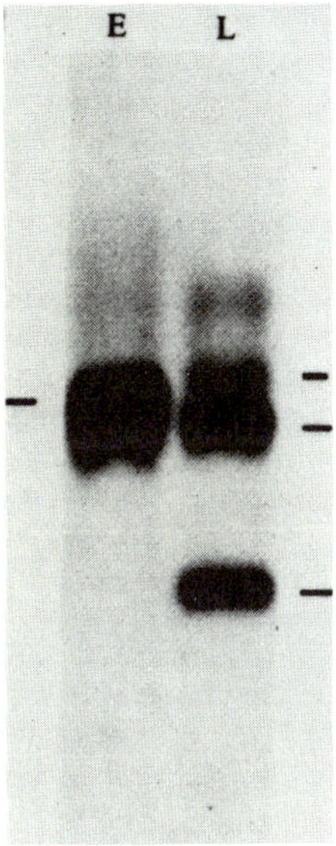

FIGURE 6. Differential splicing of the L1 precursor between early and late infection. Depicted is a Northern analysis of L1-specific RNA prepared from early (E) and late (L) infected cells. From (11) with permission.

An analysis of pulse-labeled early and late mRNA ruled out differential cytoplasmic stability as a contributing factor and argues strongly for splicing control. Presumably a factor is present early and/or late in infection that alters the specificity of splicing.

Recent work using in vitro systems offers the hope that the mechanisms for such splicing selection may soon be elucidated. Systems employing nuclear extracts have exhibited accurate splicing of the L1 and L2 leaders of the late transcription unit (75,76) or the three exons of the E2A transcript (77,78). The intermediates produced in the L1/L2 splicing reaction as well as the requirements of the system have been analyzed in detail. Three aspects are of particular interest. First, the reaction apparently proceeds in two steps. An initial cleavage is made at the splice donor site yielding a free 5' exon sequence, and the free 5' end of the intron forms a covalent linkage to the 2' position of an adenosine residue near the 3' end of the intron, resulting in a lariat type structure (79). Following this, cleavage occurs at the splice acceptor site and the two exons are joined. This may well be a general process for splicing since nearly identical results have been obtained in the analysis of splicing of a human beta-globin precursor in a HeLa extract (80).

The second feature of interest in the splicing reaction is the requirement for the Ul snRNP. This possibility had been postulated earlier based on sequence homologies (81,82) as well as from experiments utilizing antibody introduced into isolated nuclei (83). In the latter case, there was an inhibition of early adenovirus RNA processing in the nuclei receiving the appropriate antisera. The present _in vitro_ experiments seem to clearly show that the Ul snRNP is a requirement for the splicing reaction (84). Furthermore, recent experiments have shown a requirement for the 5' end of the Ul RNA (85). The splicing extract was made dependent on exogenous Ul snRNP by depleting the extract of snRNP with specific antisera. Then, the 5' end of Ul RNA in purified snRNP was removed through a combination of oligonucleotide hybridization and RNase H digestion. Such an snRNP could not restore splicing activity in the depleted extract. Therefore, it would appear that the 5' end of the U1 snRNA is necessary for splicing and very likely interacts with the sequence at the 5' end of the intron. Clearly, the next step in these systems is to extend the analysis to downstream splice sites, including those that are involved in selective splicing under early and late conditions.

Finally, there also appears to be a specific requirement for the 5' cap structures in the splicing reaction, other than just protection against nuclease degradation. Efficient splicing depends upon the presence of the cap and more importantly, splicing of capped precursors can be inhibited by the addition of cap analogs to the reaction (86).

3.2. Poly(A) Addition

Adenovirus early and late mRNAs, like most eukaryotic mRNAs (except the histone mRNAs), are polyadenylated (87,88). Experiments performed several years ago employing the adenovirus late transcription unit as well as with several early adenovirus transcription units demonstrated that the formation of the poly(A) sites [e.g. the RNA sequence to which poly(A) is added] involves an RNA chain cleavage rather than termination of transcription (69,89). Furthermore, the selection of the late poly(A) sites is not equal (69). The L3 poly(A) site is used three times more frequently than the L1 site late in infection, even though transcription across the unit is equimolar. Of considerably more interest is the fact that this frequency of selection can change, since early in infection the L1 site is selected three times more frequently than the L3 site (11). These results suggest that the selection of a poly(A) site can indeed be a regulatory event, resulting in an alteration of gene expression. This type of regulation is not confined to the virus, however, as the same phenomenon apparently contributes to tissue specific gene expression as well as developmentally regulated expression. The calcitonin gene and the calcitonin gene related neuro-peptide (CGRP) are encoded in the same transcription unit. The production of the mRNAs is governed by alternate poly(A) site selection such that calcitonin mRNA is produced in the thyroid and CGRP mRNA is produced in the brain (90-92). Likewise, poly(A) site selection controls the developmental expression of the immunoglobulin heavy chain products (93-97).

What then are the factors that determine poly(A) site selection? Two recent developments in adenovirus systems have potentially brought us closer to an understanding. First, mutational analysis of sequences required for poly(A) site selection have demonstrated a greater complexity than originally believed. It has been known for some time now that all poly(A) containing mRNAs contain the sequence AAUAAA or a close approximation near their 3' terminus (98). Mutational studies have demonstrated that this highly conserved sequence is required for poly(A) site formation (99-101). For instance, recent experiments utilizing the adenovirus ElA transcription unit demonstrated that a single base change in the AATAAA (to AAGAAA) could abolish formation of poly(A)+ ElA RNA (100). The ElA AATAAA sequence was changed to AAGAAA and this altered gene was then reconstructed into virus. RNA produced from this mutant was inefficiently cleaved at the ElA poly(A) site. Interestingly, the RNA that was cleaved was polyadenylated, suggesting that at least this particular mutation did not affect the process of poly(A) addition. It was thus clear from these results that the AATAAA sequence was required for poly(A) site formation.

It has been argued, however, that the AATAAA cannot be the sole determinant of a poly(A) site. First of all, it would be difficult to conceive how widely different efficiencies of utilization could be achieved if this conserved sequence were solely responsible. Second, there are instances where there are transcribed AATAAA sequences (within a transcription unit) that do not serve as poly(A) addition site signals (102-104). Thus, there must be additional sequences specifying poly(A) sites. This has recently been shown to be the case for the adenovirus E2A poly(A) site (105). A plasmid containing the E2A transcription unit was used for the generation of deletions downstream of the poly(A) site. A mutant containing 35 nucleotides of downstream sequence was fully functional whereas one containing 20 downstream nucleotides was not, indicating that the sequence in this region is important for the cleavage/polyadenylation reaction. An analysis of RNA produced from the plasmids revealed that the poly(A) site cleavage was impaired. Once again, this is not a virus-specific phenomenon: similar sequence requirements for proper polyadenylation have been found for the bovine growth hormone and rabbit beta-globin genes (106,107).

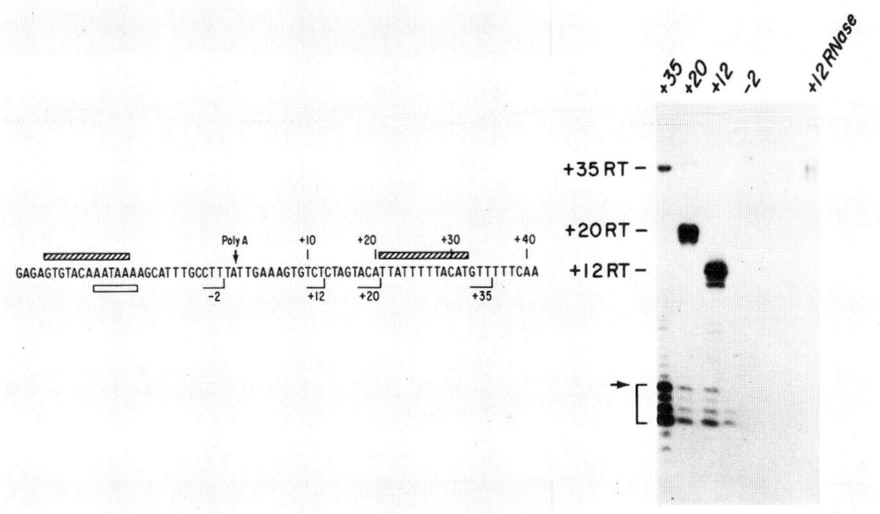

FIGURE 7. The adenovirus E2A poly(A) site and requirement of downstream sequences for function. Depicted is the sequence at the E2A poly(A) site and the end points of several deletion mutants. S1 assays of the RNA produced by these mutants is shown on the right. RT = read-through, unprocessed transcripts. See (105) for details. Reproduced from (105) with permission.

Thus, in addition to the AATAAA requirement there appears to be a requirement for downstream sequences in the processing of RNA at the poly(A) site. It is tempting to speculate that these additional sequences, which do not show conservation among various poly(A) sites, are responsible for the efficiency of usage of a poly(A) site.

The second development concerning poly(A) site formation is the achievement of specific cleavage at a poly(A) site in a cell-free system. A whole cell extract prepared from HeLa cells was able to generate the correct L3 poly(A) site (108). In the initial experiments it was found that the precursor RNA had to be transcribed in the extract for the cleavage to occur, as an exogenous precursor added to the extract was not cleaved. However, recent data indicates that a purified precursor RNA can be processed if added to a nuclear extract rather than the whole cell extract used in the initial experiments (Moore and Sharp, personal communication). Yet to be determined is whether the sequences required for _in vivo_ cleavage are also required for cleavage of a precursor RNA _in vitro_.

Finally, is the mechanism of formation of the poly(A) site really a site-specific endonucleolytic cleavage or is it the result of exonuclease digestion that stops precisely at the poly(A) site? An exonuclease mechanism would afford an explanation for the observation that the downstream cleavage product has never been observed, at least in S1 nuclease assays. However, previous data clearly indicate that an endonuclease must be involved, at least at some point. The timing of formation of the L1 versus the L5 poly(A) sites was examined using very short ^{3}H-uridine labeling (69). Since the polymerase transcribes to the end of the genome (13,69), a purely exonucleolytic mechanism using a terminated transcript would yield a relatively long lag for the formation of the L1 poly(A) site since it is about 30 kb from the end of the transcript. The L5 poly(A) site is only ~3 kb from the transcript terminus and therefore the lag for formation of this poly(A) site should be relatively shorter. However, it was found that each poly(A) site was formed equally fast, within 1-2 minutes after synthesis. Thus, at least for the L1 poly(A) site an endonuclease must cleave the chain shortly after the polymerase has transcribed through the site. It is still possible, however, that this endonucleolytic cleavage is downstream from the poly(A) site and then an exonuclease digests up to the actual poly(A) site.

4. SUMMARY

Adenovirus has proven to be a valuable model system for the study of mRNA biogenesis. The recent development of in vitro systems for studying the steps in RNA processing should open the way to an elucidation of mechanisms regulating these events. It seems certain that within the next few years the examination of these processes, as well as aspects of viral transcription control, will yield a wealth of information that will be applicable to the study of eukaryotic mRNA biogenesis in general.

REFERENCES

1. Ziff, E. B. Nature 287: 491-499. 1980.
2. Nevins, J. R. and Chen-Kiang, S. Adv. Virus Res. 26: 1-35. 1981.
3. Tooze, J. DNA Tumor Viruses. Cold Spring Harbor Laboratory, Cold Spring Harbor, N. Y. 1981.
4. Wilson, M. C. and Darnell, J. E. J. Mol. Biol. 148: 231-251. 1981.
5. Crossland, L. D. and Raskas, H. J. J. Virol. 46: 737-748. 1983.
6. Nevins, J. R., Ginsberg, H. S., Blanchard, J.-M., Wilson, M. C. and Darnell, J. E. J. Virol. 32: 727-733. 1979.
7. Nevins, J. R. Cell 26: 213-220. 1981.
8. Shaw, A. R. and Ziff, E. B. Cell 22: 905-916. 1980.
9. Chow, L.T., Broker, T.R., and Lewis, J.B. J. Mol. Biol. 134:265-304. 1979.
10. Kitchingman, G. R. and Westphal, H. J. Mol. Biol. 137: 23-48. 1980.
11. Nevins, J. R. and Wilson, M. C. Nature 290: 113-118. 1981.
12. Akusjarvi, G. and Persson, H. Nature 292: 420-426. 1981.
13. Fraser, N. W., Nevins, J. R., Ziff, E. B. and Darnell, J. E. J. Mol. Biol. 129: 643-656. 1979.
14. Thomas, G. and Mathews, M. Cell 22: 523-533. 1980.
15. Berk, A. J. and Sharp, P. A. Cell 14: 695-711. 1978.
16. Spector, D. J., McGrogan, M. and Raskas, H. J. J. Mol. Biol. 126: 395-414. 1978.
17. Perricaudet, M., Akusjarvi, G., Virtanen, A. and Pettersson, U. Nature 281: 694-696. 1979.
18. Van Ormondt, H., Maat, J., de Waard, A., and Van der Eb, A. J. Gene 4: 309-328. 1978.
19. Aiello, L., Guiloyle, R., Huebner, K., and Weinmann, R. Virology 94: 460-469. 1979.
20. Graham, F. L., Smiley, J., Russell, W. C. and Nairn, R. J. Gen. Virol. 36: 59-72. 1977.
21. Harrison, T., Graham, F. and Williams, J. Virology 77: 319-329. 1977.
22. Jones, N. and Shenk, T. Cell 17: 683-689. 1979.
23. Graham, F.G., Harrison, T.J., and Williams, J.F. Virology 86:10-21. 1978.
24. Ho, Y.-S., Galos, R., and Williams, J. Virology 122:109-124. 1982.
25. Babiss, L.E., Fisher, P.B., and Ginsberg, H.S. J.Virol. 49:731-740. 1984.
26. Babiss, L.E., Ginsberg, H.S., and Fisher, P.B. Proc. Natl. Acad. Sci. USA 80:1352-1356. 1983.
27. Jones, N. and Shenk, T. Proc. Nat'l. Acad. Sci. USA 76: 3665-3669. 1979.
28. Berk, A., Lee, F., Harrison, T., Williams, J. and Sharp, P. A. Cell 17: 935-944. 1979.
29. Katze, M. G., Persson, H. and Philipson, L. Mol. Cell. Biol. 1: 807-813. 1981.
30. Leff, R., Elkaim, R., Goding, C. R., Jalinot, P., Sassone-

Corsi, P., Perricaudet, M., Kedinger, C. and Chambon, P. Proc. Nat'l. Acad. Sci. USA 81: 4381-4385. 1984.
31. Kao, H.-T. and Nevins, J. R. Mol. Cell. Biol. 3: 2058-2065. 1983.
32. Ricciardi, R. P., Jones, R. L., Cepko, C. L., Sharp, P. A. and Roberts, B. E. Proc. Nat'l. Acad. Sci. USA 78: 6121-6125. 1981.
33. Montell, C., Fisher, E. F., Caruthers, M. H. and Berk, A. J. Nature 295: 380-384. 1982.
34. Montell, C., Courtois, G., Eng, C. and Berk, A. Cell 36: 951-961. 1984.
35. Winberg, G. and Shenk, T. EMBO J. 3: 1904-1912. 1984.
36. Ruley, H. E. Nature 304: 602-606. 1983.
37. Houweling, A., van den Elsen, P. and van der Eb, A. J. Virology 105: 537-550. 1980.
38. Imperiale, M. J., Feldman, L. T. and Nevins, J. R. Cell 35: 127-136. 1983.
39. Hen, R., Borrelli, E., Sassone-Corsi, P. and Chambon, P. Nucl. Acids Res. 11: 8747-8760. 1983.
40. Hearing, P. and Shenk, T. Cell 33: 695-703. 1983.
40a.Weiher, H., Konig, M. and Gruss, P. Science 219: 626-631. 1983.
41. Osborne, T. F., Arvidson, D. N., Tyau, E. S., Dunsworth-Browne, M., and Berk, A. J. Mol. Cell. Biol. 4: 1293-1305. 1984.
42. Imperiale, M. J. and Nevins, J. R. Mol. Cell. Biol. 4: 875-882. 1984.
43. Imperiale, M.J., Hart, R.P., and Nevins, J.R. Proc. Natl. Acad. Sci. USA, in press. 1984.
44. Elkaim, R., Goding, C., and Kedinger, C. Nucl. Acids Res. 11: 7105-7117. 1983.
45. Kingston, R. E., Kaufman, R. J. and Sharp, P. A. Mol. Cell. Biol. 4: 1970-1977. 1984.
46. Bos, J. L. and ten Wolde-Kraamwinkel, H. C. EMBO J. 2: 73-76. 1983.
47. Handa, H. and Sharp, P.A. Mol. Cell. Biol. 4:791-798. 1983.
48. Dynan, W.S. and Tjian, R. Cell 32:669-680. 1983.
49. Parker, C.S. and Topol, J. Cell 36:357-369. 1984.
50. Fire, A., Baker, C. C., Manley, J. L., Ziff, E. B., and Sharp, P. A. J. Virol. 40: 703-719. 1981.
51. Lee, D. C. and Roeder, R. G. Mol. Cell. Biol. 1: 635-651. 1981.
52. Cross, F. R. and Darnell, J. E. J. Virol. 45: 683-692. 1983.
53. Feldman, L. T., Imperiale, M. J. and Nevins, J. R. Proc. Nat'l. Acad. Sci. USA 79: 4952-4956. 1982.
54. Nevins, J. R. Cell 29: 913-919. 1982.
55. Green, M. R., Treisman, R. and Maniatis, T. Cell 35: 137-148. 1983.
56. Treisman, R., Green, M. R., and Maniatis, T. Proc. Nat'l. Acad. Sci. USA 80: 7428-7432. 1983.
57. Fromm, M. and Berg, P.J. J. Mol. Appl. Genet. 1:457-481. 1982.

58. Charnay, P., Treisman, R., Mellon, P., Chao, M., Axel, R. and Maniatis, T. Cell 38: 251-263. 1984.
59. Gaynor, R. B., Hillman, D. and Berk, A. J. Proc. Nat'l. Acad. Sci. USA 81: 1193-1197. 1984.
60. Imperiale, M. J., Feldman, L. T., Kao, H.-T., Nevins, J. R. and Strickland, S. Mol. Cell. Biol. 4: 867-874. 1984.
61. Kao, H-T., Capasso, O., Heintz, N. and Nevins, J. R. Submitted.
62. Carter, T. H. and Blanton, R. A. J. Virol. 25: 664-674. 1978.
63. Van der Vliet, P. C., Levine, A. J., Ensinger, M. J. and Ginsberg, H. S. J. Virol. 15: 348-354. 1975.
64. Nevins, J. R. and Winkler, J. J. Proc. Nat'l. Acad. Sci. USA 77: 1893-1897. 1980.
65. Handa, H., Kingston, R. E. and Sharp, P. A. Nature 302: 545-547. 1983.
66. Fowlkes, D. M., Lord, S. T., Linne, T., Pettersson, U. and Philipson, L. J. Mol. Biol. 132: 163-180. 1979.
67. Nevins, J. R. Ann. Rev. Biochem. 52: 441-466. 1983.
68. Nevins, J. R. In: Processing of RNA (Ed. D. Apirion), CRC Press, Boca Raton, FL., 1984.
69. Nevins, J.R. and Darnell, J.E. Jr. Cell 15:1477-1493. 1978.
70. Beltz, G. A. and Flint, S. J. J. Mol. Biol. 131: 353-373. 1979.
71. Babich, A., Feldman, L. T., Nevins, J. R., Darnell, J. E. and Weinberger, C. Mol. Cell. Biol. 3: 1212-1221. 1983.
72. Babich, A. and Nevins, J. R. Cell 26: 371-379. 1981.
73. Berget, S., Moore, C. and Sharp, P. A. Proc. Nat'l. Acad. Sci. USA 74: 3171-3175. 1977.
74. Chow, L. T., Gelinas, R. E., Broker, T. R. and Roberts, R. J. Cell 12: 1-8. 1977.
75. Padgett, R. A., Hardy, S. F. and Sharp, P. A. Proc. Nat'l. Acad. Sci. USA 80: 5230-5234. 1983.
76. Hernandez, N. and Keller , W. Cell 35: 89-99. 1983.
77. Goldenberg, C. J. and Raskas, H. J. Proc. Nat'l. Acad. Sci. USA 78:5430-5434. 1981.
78. Goldenberg, C. J. and Hauser, S. D. Nucl. Acids Res. 11: 1337-1348. 1983.
79. Padgett, R. A., Konarska, M. M., Grabowski, P. J., Hardy, S. F. and Sharp, P. A. Science 225: 898-903. 1984.
80. Ruskin, B., Krainer, A. R., Maniatis, T., and Green, M. R. Cell 38: 317-331. 1984.
81. Rogers, J. and Wall, R. Proc. Nat'l. Acad. Sci. USA 77: 1877-1879. 1980.
82. Lerner, M. R., Boyle, J. A., Mount, S. M. and Steitz, J. A. Nature 283: 220-224. 1980.
83. Yang, V. W., Lerner, M. R., Steitz, J. A. and Flint, S. J. Proc. Nat'l. Acad. Sci. USA 78: 1371-1375. 1981.
84. Padgett, R. A., Mount, S. M., Steitz, J. A. and Sharp, P. A. Cell 35: 101-107. 1983.
85. Kramer, A., Keller, W., Appel, B. and Luhrmann, R. Cell 38: 299-307. 1984.
86. Konarska, M. M., Padgett, R. A., and Sharp, P. A. Cell 38:

731-736. 1984.
87. Philipson, L., Wall, R., Glickman, G., and Darnell, J.E. Proc. Natl. Acad. Sci. USA 68:2806-2809. 1971.
88. Lindberg, U., Persson, T., and Philipson, L. J.Virol. 10:909-919. 1972.
89. Nevins, J. R., Blanchard, J.-M., and Darnell, J. E. J. Mol. Biol. 144: 377-386. 1980.
90. Amara, S. G., Jonas, V., Rosenfeld, M. G., Ong, E. S., and Evans, R. M. Nature 298: 240-244. 1982.
91. Rosenfeld, M. G., Mermod, J. J., Amara, S. G., Swanson, L. W., Sawchenko, P. E., Rivier, J., Vale, W. W., and Evans, R. M. Nature 304: 129-135. 1983.
92. Amara, S. G., Evans, R. M. and Rosenfeld, M. G. Mol. Cell. Biol. 4: 2151-2160. 1984.
93. Early, P., Rogers, J., Davis, M., Calame, K., Bond, M., Wall, R. and Hood, L. Cell 20: 313-319. 1980.
94. Rogers, J., Early, P., Carter, C., Calame, K., Bond, M., Hood, K., and Wall, R. Cell 20, 303-312. 1980.
95. Alt, F. W., Bothwell, A. L. M., Knapp, M., Siden, E., Mather, E., Koshland, M. and Baltimore, D. Cell 20: 293-301. 1980.
96. Maki, R., Roeder, W., Traunecker, A., Sidman, C., Wabl, M., Raschke, W. and Tonegawa, S. Cell 24: 353-365. 1981.
97. Mather, E. L., Nelson, K. J., Haimovich, J. and Perry, R. P. Cell 36: 329-338. 1984.
98. Proudfoot, N. J. and Brownlee, G. G. Nature 263: 211-214. 1976.
99. Fitzgerald, M. and Shenk, T. Cell 24: 251-260. 1981.
100. Montell, C., Fisher, E. F., Caruthers, M. H. and Berk, A. J. Nature 305: 600-605. 1983.
101. Higgs, D.R., Goodbourn, S.E.Y., Lamb, J., Clegg, J.B., Weatherall, D.J., and Proudfoot, N.J. Nature 306:398-400. 1983.
102. Reddy, V.E., Ghosh, P.K., Lebowitz, P., Piatek, M., and Weissman, S. J.Virol. 30:279-296. 1979.
103. Perricaudet, M., LeMoullec, J.M., Tiollais, P., and Petterson, U. Nature 288:174-176. 1980.
104. Aho, S., Tate, U., and Boedtker, H. Nucl. Acids. Res. 11:5443-5450. 1983.
105. McDevitt, M. A., Imperiale, M. J., Ali, H. and Nevins, J. R. Cell 37: 993-999. 1984.
106. Woychik, R.P., Lyons, R.H., Post, L., and Rottman, F.M. Proc. Natl. Acad. Sci. USA 81:3944-3948. 1984.
107. Gil, A. and Proudfoot, N.J. Nature, in press.
108. Moore, C. L. and Sharp, P. A. Cell 36: 581-591. 1984.

5

GENETIC ELEMENTS GOVERNING ADENOVIRAL GENE EXPRESSION

N. C. JONES

Department of Biological Sciences, Purdue University,
West Lafayette, Indiana 47907

1. INTRODUCTION

The adenoviruses have long been recognized as providing a convenient model system for studying eukaryotic gene expression and its regulation. Advances over the last few years have reinforced this position; a number of features of eukaryotic gene expression were first discovered in the adenovirus system, RNA splicing being the classic example. Some of the principal features that make this system so attractive are: (1) The ease by which the adenoviruses can be grown and purified in large amounts. (2) The complex, temporal regulation of adenovirus gene expression; different viral genes are expressed at different times of the productive infection (1, 2). (3) The complete sequence of Adenovirus type 2 (Ad2) is known. In addition there is significant sequence information on a number of other adenovirus serotypes (3, 4). Cloned viral genes can be manipulated in vitro and subsequently incorporated into an intact, infectious viral genome.

A detailed description of the structure and expression of the adenovirus genome is given in other chapters of this volume (1, 2). Briefly, the various adenovirus genes can be categorized as early, intermediate or late depending on when during the productive cycle their expression begins. The early genes are expressed before the onset of viral DNA replication which begins 6-8 hours post-infection. They can be subdivided into immediate early and early, the immediate early gene being the ElA gene. A product encoded by this gene trans-activates expression of all the early viral genes (4, 5). The mechanism of this trans-activation and the recognition of early gene sequences that render these genes sensitive to the action of the ElA protein has been the focus of a great deal of work over the last few years. Our understanding of this process however is still minimal. Expression of the early genes is clearly regulated at the transcriptional level (7, 8).

However, regulation of gene expression at a number of post-transcriptional levels as well as the translational level has also been found during the adenovirus productive infection (8). This includes regulation at the level of splicing, polyadenylation, message stability and selective transport of processed message to the cytoplasm.

In order to understand these regulatory mechanisms in great detail it is necessary first to identify and characterize the genetic elements responsible for mRNA biosynthesis. Over the last two to three years there has been an explosion in studies aimed at identifying such elements in particular the elements responsible for efficient transcription initiation. The aim of this review is to briefly summarize these studies and describe how they have contributed to a better understanding of how eukaryotic genes are expressed and regulated.

2. EARLY REGION 1A

One of the most extensively studied adenoviral genes is the E1A gene. The products encoded by this gene are of great interest and importance because they are involved in regulating the expression of the adenovirus genome (4-6) as well as the process of oncogenic transformation by this virus (9). In addition viral mutants containing alterations in this region can be readily constructed and propagated. Such mutants are propagated in 293 cells (10) which are transformed human embryonic kidney cells that contain and express the Ad5 E1 region. The alterations in the majority of E1A mutants were first introduced into cloned E1A sequences in vitro. Expression of the altered gene can be assayed following transient expression in appropriate human cell lines. Alternatively, complete viral genomes containing the alterations can be constructed by one of two methods: 1) Ligation of a modified 0-3.8 map unit fragment to the 3.8-100 map unit fragment from the viable mutant dl309 (11) obtained by cleaving the viral DNA with XbaI (12). 2) Transfection of cells with DNA fragments that overlap and consequently relying on recombination events within the cell to regenerate an intact genome (13).

The E1A gene is expressed immediately after infection of the host cell by the virus, even in the presence of protein synthesis inhibitors (4, 6). Therefore unlike the other early genes it does not require the synthesis of a positive transcriptional factor for its expression. What are the fundamental features which render the E1A gene self-sufficient with respect to it's transcriptional activity? The characterization of mutants within the sequences upstream of the E1A major cap site has indicated the presence of elements that behave as cis-acting transcriptional enhancers. These elements are analogous to similar elements present in other viral genomes (for Refs., see 14) and in certain cellular genes (15-17). Enhancers function to potentiate gene transcription in a manner relitavely independent of position and orientation with respect to the nearby gene.

It is essential that the E1A gene is efficiently expressed early in the productive cycle. A protein product of the gene is necessary for efficient expression of the other early genes some of which encode products absolutely necessary for viral DNA replication. The overall scheme for adenovirus early gene expression appears logical and efficient; by containing a gene that is efficiently expressed such as the E1A gene which encodes a trans-acting transcriptional activator it ensures that all the early genes are expressed well without the necessity for cis-acting enhancers within the upstream region of each gene. In addition, regulation of the effective concentration of E1A gene product allows coordinate regulation of expression of all the early genes. This scheme is not restricted only to the adenoviruses since a similar scheme operates for regulation of expression of the early genes of pseudorabies virus (18), herpes simplex virus (19-20) and possibly cytomegalovirus (22). For example, an immediate early gene of HSV encodes a trans-acting factor that stimulates early gene expression; a cis-acting enhancer element ensures efficient expression of the immediate early gene (23).

In an effort to define which sequences are necessary for efficient expression of E1A, a number of groups have introduced specific mutations into the upstream region. However despite the effort some disagreement exists particularly with respect to which sequences constitute the cis-acting enhancer element. The most extensive mutational analysis has been carried out by Hearing and Shenk (24, 25). They constructed a set of small deletion mutations in the region surrounding the cap site; the mutations were introduced into an E1A containing plasmid using a modified D-loop mutagenesis procedure (24). When the effect of these deletions on transcription of the E1A gene _in vitro_ using cell free extracts was tested the following observations were made: 1) The sequence 5'TATTTATA3' which is located approximately 30bp upstream of the major cap site was essential for transcription; a 13bp deletion that completely removed this sequence abolished in vitro transcription entirely whereas a deletion that removed only a portion of this sequence reduced the level of transcription about eightfold. The sequence is similar to the 'TATA' consensus sequence found at about this position in most eukaryotic genes (26). In the majority of the cases where it has been examined, the TATA consensus has been found to be essential for in vitro transcription (26, 27). In addition the 'TATA' consensus has been found to serve as a reference point from which the transcriptional start site is determined. The same is true for the 'TATA'-like sequence of E1A as deletions between this sequence and the normal cap site resulted in new start sites displaced downstream by about the size of the deletion (24). No sequences upstream of position -35 were found to be necessary.

Some of the deletion containing E1A sequences were used to reconstruct complete viral genomes and the effect of the mutations on in vivo expression of E1A was tested. Removal of the 'TATA' like sequence also lowered in vivo expression but only by a factor of two. However, the two most surprising results from the in vivo studies were: 1) Sequences between positions -35 and -168 were dispensible. 2) Sequences upstream of position -168 were essential for efficient expression.

Within the -35 to -168 region is a sequence 5'CAAAGT3' located at about position -70 which is homologous to the 'CAAT' consensus found at approximately this position in many but not all eukaryotic genes (26, 27). These sequences appear to be important for efficient in vivo expression of some of these genes (28, 29) but clearly this is not the case with E1A. The essential sequences at least for in vivo expression, are far upstream (beyond) position -168). It is somewhat unusual to have regions so far upstream as being critical for expression but in many of the cases where this has been found the far upstream sequences constitute cis-acting enhancer elements. The same is true for E1A; the far upstream sequences could enhance expression of the E1A gene irrespective of whether they were located at the 5' or the 3' end of the E1A coding sequence and in either possible orientation (25). In addition, the sequences were shown to enhance transcription from a heterologous promoter such as the HSV thymidine kinase (tk) promoter (25). By examining mutants with deletions in this far upstream region, Hearing and Shenk concluded that the enhancer element was located between positions -305 and -168 and was subdivided into two domains both of which were critical for enhancer function. Deletion of the entire -305 to -168 region resulted in an approximate 15-fold reduction in ElA expression whereas two smaller deletions between -305 and -226 and between -209 and -144 only resulted in reductions of 2-3 fold. The two regions contain the 11bp repeat sequence 5'AGGAAGTGACA3' repeated with only one bp change centered at approximate positions -300 and -200. Any combination of mutations that altered both repeat sequences gave very low expression levels although some residual expression was evident. Surprisingly a small amount of residual expression was also seen with a mutant (dl340-11) which contained a deletion extending from position -393 to +10. Thus the deletion removed both the promoter sequences and the upstream cis-acting enhancers described above. The reason for residual expression in this mutant is not clear but may be a consequence of other regulatory regions that have been identified in the ElA transcriptional unit that are located outside of the deletion end points (see below). The adenovirus repeat sequence is found within enhancer regions of other viral genomes such as the polyoma virus early gene (30), the BKV early gene (31) and the long terminal repeats of ASV (32) and MMTV (35).

Additional studies by other groups however, suggest that the organization of the ElA upstream region is more complex and probably contains multiple elements that influence expression of ElA and which can behave as enhancers. Sassoni-Corsi et al. (34) introduced deletions into the far upstream region of ElA and measured their effect on transcription of ElA following transient expression in transfected HeLa cells. In agreement with the results described above, they found sequences upstream of position -168 to be essential. Examination of two series of converging deletions revealed that although sequences centered around -300 had some importance with respect to enhancement activity (about four-fold), the critical sequences were localized between positions -320 and -344 (35), clearly outside of the region defined by Hearing and Shenk. This critical region contains the sequence 5'TGTGGTAAAAG3' which resembles the SV40 core sequence first derived by Weiher et al. (36). This core sequence, $\mathrm{5'GXTGTGG}^{\mathrm{TTT}}_{\mathrm{AAA}}\mathrm{G3'}$, is found within a number of different enhancer elements and at least in the case of the SV40 enhancer, is crucial for function. The element containing the SV40 core sequence may not be strictly analogous to other enhancer elements in that it does not demonstrate bidirectionality; it works much better in one direction (its direction in the intact adenovirus genome) than the other.

It is somewhat preplexing that two different groups have defined an enhancer within the ElA upstream region but that the location and critical sequences are not the same. It should be emphasized however that a very different assay system was used for each study. Therefore an explanation for such apparently conflicting results is that multiple, independent enhancers are present within the E1A upstream region and that the relative importance of each with respect to stimulation of nearby promoters is dependant upon the experimental system used. Indeed, the existence of multiple enhancer domains that contain either the adenovirus repeat sequence (Ad-core) or the SV40 core sequence (SV-core) appears to be a fairly common occurence. A very nice parallel to E1A is found in polyoma virus (30). At least two distinct non-overlapping enhancer elements have been defined within the upstream region of the polyoma early gene. Interestingly, one of these elements contains a sequence resembling the Ad-core and the other contains a sequence resembling the SV-core. The relative strength of each element varies depending on the cell type used for the assay. Other possible parallels are found with the mouse immunoglobin enhancer region (15), the enhancer element of the human BK virus (31) and the enhancer element of rous sarcoma virus (37). Both the immunoglobin and BK enhancers contain sequences homologous to the Ad- and SV-core sequences. In the case of RSV (Prague strain) the enhancer element consists of three separate domains and a combination of any two of the three domains is sufficient for full enhancer activity (37). Two of the domains contain the Ad-core sequence and the third contains the SV-core sequence.

Yet a third possible upstream enhancer element may be positioned between nucleotides -44 and -188 (38). Inserting this sequence upstream or downstream of the adenovirus E2A gene resulted in higher levels of E2A expression in transfected cells. This region contains neither the SV-core nor the Ad-core sequence. In addition, Hearing and Shenk (24) showed that these sequences were completely dispensible for *in vivo* expression of the E1A gene.

Questions concerning the E1A upstream enhancer element remain. Clearly the question of how many upstream enhancer elements exist has not yet been resolved. More importantly perhaps is the question of why there is a need for multiple enhancers.

To complicate matters even further, it has been reported by Osborne et al. (39) that deletions within the E1A coding region depressed E1A transcription to 2% of the wild-type rate. Part of this decrease (about 5-fold) could be complemented *in trans* and presumably resulted from inactivation of the E1A protein which is known to stimulate expression of its own gene by 4-5 fold. However the remaining decrease (about 10-fold) was not complemented in trans and thus suggested the existence of important cis-acting sequences within the E1A protein coding region. A single base pair deletion at position +399 was sufficient to produce this phenotype. It is intruiging that the deletion lies within a sequence that shows extensive homology to the SV-core sequence. Although the nature of these internal sequences is not completely understood, the possibility remains that they constitute an enhancer-like element. Indeed, a fragment containing this internal region has been found to stimulate transcription from three different heterologous promoters independent of orientation and position. For example, insertion of sequences extending from position +126 to +421 downstream of the rabbit β-globin gene, resulted in significant expression in transfected HeLa cells. Without the insert β-globin expression was too low to be detected. It remains a puzzle however why such a reduction in E1A expression (10-fold) is seen when the internal sequences are removed given the fact that at least one and probably two or more enhancer elements are located in the upstream region. With this in mind it might be informative to construct a viral mutant that lacks both the upstream and internal control sequences and compare such a mutant to ones that lack either the upstream or the internal regions. The various regions of the E1A gene reported to contain enhancer elements are indicated in Figure 1.

The effect of upstream or internal deletions is not apparent when E1A transcription is measured late in infection (25, 39); E1A mRNA levels are the same in mutant and wild-type infected cells. This indicates that sequences required for E1A transcription at late infection times may be different from those required during the early phase. The nature of these differences is not known. Of probable relevance is the lowered rate of E1A transcription during the last phase; the increase in E1A mRNA at late times is very modest (about 5-10 fold) compared to the enormous (greater than 10^4) increase in gene copy number due to viral DNA replication (41, 42). A very provocative finding recently reported by Borrelli et al. (43) shows that under the right conditions E1A protein can down regulate or repress enhancers. Most of the work was carried out using the polyoma early gene enhancer; stimulation of transcription of a linked rabbit β-globin gene was repressed in cells expressing E1A. Other enhancers were also shown to be probable targets of repression, including the upstream E1A enhancers. We are therefore faced

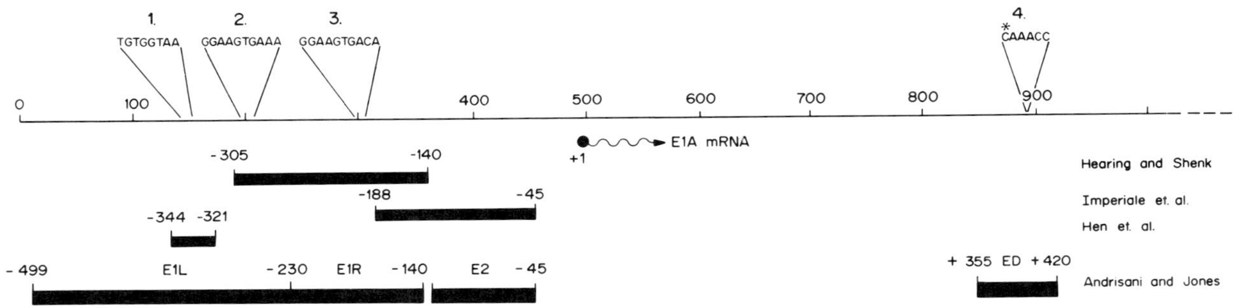

Figure 1: Location of enhancer elements within the left-hand end of Ad2 and Ad5. The left-hand end 1000 bp is depicted. The initiation position and direction of transcription of the E1A gene is indicated by +1 and the arrow. The fragments containing enhancer activity are shown by the solid bars; their location relative to the E1A cap site is indicated. At the top of the figure, sequences implicated as being important components of enhancer elements, are shown. Sequence 1, pointed out by Hen et al. (35) is homologous to the SV40 core sequence. Sequences 2 and 3 are the repeats shown by Hearing and Shenk (25) to be vital for enhancer acticity localized in the -305 to -140 region. Sequence 4 is within the internal enhancer region (39-40) and is homologous to the SV40 core sequence. The asterisk indicates the position of the single base pair deletion in the Ad5 mutant 15606, which leads to decreased E1A transcription (39).

with the situation that the E1A proteins can both stimulate (by an unknown mechanism) and repress (by inactivation of the enhancers) expression of its own gene. There probably exists a very complex form of autoregulation of E1A transcription where parameters such as concentration of cellular factors or ratio of gene copy number to protein, may determine whether repression or activation would be the predominant activity.

3. EARLY REGION 2A

The early region 2 transcription unit is one of the more interesting early genes because of its complex organization. At early times of infection multiple mRNAs are produced by differential splicing of a precursor RNA that initiates at approximately 76.0 map units. There are in fact two promoters that control transcription of the E2 region (44, 45). The one initiates at 76.0 map units and is referred to as the E2A early promoter. The second promoter approximately 1100 bp downstream is predominantly used to express the E2A gene sequences at intermediate and late times of infection. Both the early and late promoters of E2A are unusual in that they lack a typical 'TATAA'-like sequence normally located 20-30 bp upstream of the initiation site (45). These promoters can be transcribed in vitro (46, 47), but at very low efficiencies, a common finding with promoters that lack the typical 'TATAA' motif. In contrast the promoters are very active in infected cells and are clearly as active as the other early gene promoters that do contain the 'TATAA'-like sequences. The E2A early promoter is positively regulated by an E1A gene product (4, 5). This does not seem to be the case with the late promoter and in fact there is some evidence that expression from this promoter can be repressed by E1A encoded products (48).

The E2A early promoter has been the subject of intense scrutiny from the point of view of determining which sequences are vital for promoter activity both in the absence and in the presence of E1A products. Both deletions and point mutations have been introduced into the upstream sequences and promoter activity subsequently assayed either in transient expression systems (49, 50, 52) or stable transformation assays (51). In the majority of cases expression of chimeric genes has been assayed, where the E2A early promoter was fused to a variety of coding sequences such as those for the enzymes CAT (52) or dihydrofolate reductase (51).

Although the E2A promoter requires the presence of E1A for efficient activity, a basal level of expression (or uninduced level) can be detected even in its absence. The E1A products do not change a non-functional promoter to one that is functional but instead increases the efficiency of use of a promoter that is already partially active. The ability to detect uninduced transcription from the E2A promoter has allowed examination of which upstream sequences define the actual promoter. Examination of a series of deletion mutants (49-52) showed that sequences upstream of position -79 were not important for promoter activity. However deletion of additional sequences to position -59 reduced activity 5-10 fold. A critical element of the promoter is therefore located within the -79 to -59 region. This conclusion has been confirmed by examining a number of linker-scanning mutants in this area (52). Linker-scanning mutagenesis, first used by McKnight and Kingsbury (53) to examine in detail the HSV tk promoter, introduces clusters of point mutations into the regions of interest. Such mutants provide more information than simple deletion mutants and do not disturb the spacing of important control signals. Point mutations within the region -66 to -82 reduced promoter activity 9-12 fold (52). The sequence located between -66 to -82 is 5'TGGAGATGACGTAGTTT3' and is clearly different from the most distal control signals of other promoters such as the tk and β-globin promoters. Part of the above sequence, namely 5'AGATGACGTA3' is very similar to sequences found to be important for activity of other early promoters (see below). Point mutations in the region between -35 and -66 had no affect on promoter activity. However point mutations in the region between -21 and -35 lowered promoter activity 6-7 fold and hence define a second control region. In addition variability in position of the cap sites was seen with mutants in this area. Hence there seems to be an element which in terms of position and function is analogous to the 'TATAA'-element found in the upstream region of most RNA polymerase II transcribed genes. This region does contain the sequence 5'TTAAGA3' which has some resemblence to the 'TATAA' motif. It would be interesting to know whether in fact this sequence of the E2A promoter is functioning in an identical fashion to the 'TATAA' element of other genes, especially in terms of interaction with transcriptional factors. Point mutations located between the cap site and position -21 had no effect on activity. Therefore at least two distinct control elements are present in the E2A upstream region and alteration of either significantly reduces uninduced expression (Figure 2).

It is obviously interesting and important to not only define the upstream sequences necessary for uninduced expression but also the sequences necessary for expression induced by E1A. In particular it is important to know whether there are specific sequences that are required for induced expression. Such information would provide valuable insight as to how the E1A proteins activate transcription. Presently, the majority of the data support the view that there are no sequences that are specifically required for E1A induction. This view is based on the observation that most of the mutants containing alterations within the E2A upstream sequences show enhanced expression in the presence of functional E1A (51, 52). This is true even for the constructs where basal level or uninduced expression is lowered; the induced level with these constructs is also lower but the induction ratio approaches that seen with the wild-type promoter. Accurate measurement of the extent of induction is difficult with these mutants because of the very low level of uninduced expression which is hardly detectable and therefore difficult to gauge. Nevertheless, even though the ratios in some cases may not be as high as wild-type, the important conclusion is that induction can be observed and hence there is no specific sequence that is absolutely required for this induction to take place. The strongest support for this conclusion comes from the study employing the linker-scanning mutants where almost every nucleotide of the E2A promoter is mutated (52). All the mutant promoters were activated to some extent by E1A. The sequences required for uninduced and induced expression are linked and in fact may not be separable.

Imperiale et al. have reported findings that may not fully support the above conclusions (50). Constructs with deletion end points at -59, -45, and -17 all show greatly reduced levels of uninduced expression. However, unlike the studies described above induced expression is extremely low and not detectable. The authors claim therefore that in the absence of sequences between -59 and -79 induction by E1A can not take place. It is not easy to reconcile these findings with those of others described above, although as pointed out earlier, because of the very low expression levels obtained with mutants that lack the -59 to -79 region, some induction by E1A may have gone unnoticed. Imperiale et al. (50) also demonstrated a very interesting and intriguing feature of the E2A upstream region. A mutant E2 promoter lacking all sequences upstream of position -28 was inactive both in the presence and absence of E1A. Not surprisingly, they showed that promoter activity could be restored by inserting upstream of the deletion end point a restriction fragment containing sequences from -21 to -262 of the wild-type gene. However what was surprising was the finding that this fragment could restore activity regardless of its orientation and even at a distance of about 4kb. The region therefore had properties that we normally associate with enhancer elements. They also showed that the fragment behaved in a similar fashion when inserted adjacent to heterologous promoters that also lack their upstream, distal control signals. So far, this property shown by the E2A upstream control element has not been shown with upstream sequences of the other adenovirus early genes. Such flexibility with regards to position and orientation has been seen with promoter elements of other non-adenovirus genes such as the early SV40 gene (54), the major heat shock gene of Drosophila (55) and the HSV tk gene (56, 57). However in these cases the flexibility if not as extensive as that seen with the E2A promoter element.

4. EARLY REGION 3

A similar but less detailed analysis of the E3 promoter has been performed by more than one group to investigate the sequence requirements for E1A induced and uninduced transcription (58, 59). Qualitatively the same results were obtained as those described for the E2A promoter. At least two and probably three distinct elements are required for function; one of these elements is centered around position -25 and contains the typical 'TATAA' motif. No specific sequence has been recognized that is absolutely required for E1A induction. Some induction could be seen with all the mutants tested.

Leff et al. (59) constructed a series of mutants that contained small, internal deletions within the E3 upstream region. The activity of the mutant promoters was assayed following transient expression in transfected cells both in the presence and absence of E1A. Deletions that removed the 'TATAA' sequence entirely, lowered promoter efficiency about 5-fold. Surprisingly however, and in contrast to observations made with other eukaryotic genes, removal of this sequence did not result in 5' end heterogeneity; the wild-type transcriptional start-sites were utilized. With this gene, therefore, structural features other than the 'TATAA' element must determine the start-site. A second important element was found to be centered around positions -57/-55. An internal deletion between positions -39 and -55 had little effect on promoter activity. In contrast however extending the deletion by two nucleotides to position -57 resulted in a 5-10 fold reduction in promoter efficiency. Leff et al. also suggest that

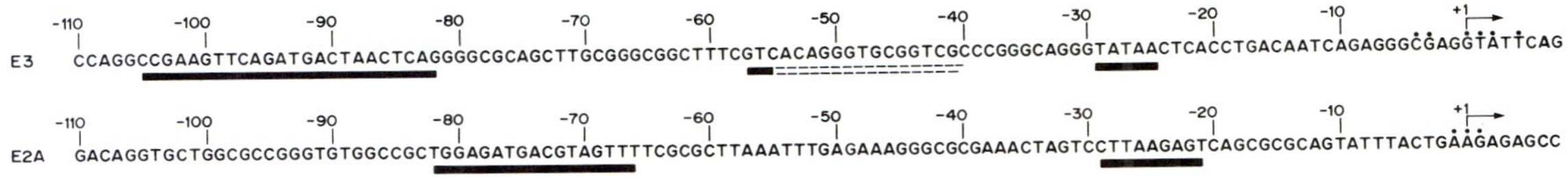

Figure 2: Sequences upstream of the transcriptional start site of the E2A early and E3 genes are shown. Sequences found to be essential for uninduced and E1A-induced expression (49-52; 58, 59) are underlined. The exact limits of the first distal signal of the E3 promoter are not known but at least some of the sequences indicated by the dotted line are probably important (59).

5'-AGATGAC-TA E3 -96 to -88

5'-AGATGACGTA E2 -79 to -70

5'-AG-TGACGTA E4 -170 to -162

5'-AGA-GACG E3 -118 to -112

Figure 3: Comparison of sequences within the distal control elements of the E3, E2A and E4 promoters. The position of the sequences with respect to the transcription initiation sites is indicated.

there is a second, more distal element that is critical for function which is located between positions -111 and -233. This conclusion is based on results obtained with three mutants; one contains a deletion between -39 and -57, the other has a deletion between -57 and -111 and the third deletes the entire upstream region between -37 and -233. The -39/-57 deletion shows about a 10-fold reduction in activity, the -57/-111 deletion only a slight additional decrease but the -37/-233 deletion, complete loss of activity. Their conclusion therefore is that sequences between -57 and -111 do not contain a second distal signal but that sequences between -111 and -233 do. When each of the mutants was examined in the presence of E1A, the authors found that the effect of the deletion was approximately the same implying that in each case activation by E1A could take place.

This positioning of the second, most distal control sequence is in conflict with the results of Weeks and Jones (58). Deletions were introduced into the E3 sequences of the chimeric gene E3CAT (7) which contains sequences from the 5' end of the E3 gene fused to the CAT coding region. Removal of sequences upstream of position -105 has little effect both on the uninduced level of expression and on induction by E1A. In contrast, at least a 10-fold reduction was observed when the deletion was extended to position -82. These results would therefore place an important control sequence within the -82 to -105 region (Figure 2). Very little activity was seen with the -82 deletion even in the presence of E1A. However, because of the difficulty in consistantly and accurately measuring the uninduced level of expression with this mutant, one could not confidently conclude that E1A induction could not take place at all. Within the region between -82 and -105 is the sequence 5'AGATGACTA3' which is highly homologous to a sequence found within the region of the E2A promoter that harbours a critical distal signal (see earlier). Such a sequence is also found within an important region of the E4 promoter (see next section; Figure 3). Examination of the sequences of the E3 region just upstream of position -111 reveals a second sequence homologous to the conserved region described above; this sequence 5'AGAGACG3' is located between positions -118 and -112 (Figure 3). Therefore in the E3 promoter two copies of the conserved region are present and it is conceivable that both serve as regulatory elements. If this were the case it would help explain the apparent contradiction between the results of Leff et al. and Weeks and Jones. Deletion of one or the other (as in the -105 mutant of Weeks and Jones and the -57 to -111 mutant of Leff et al.) might have only a small effect on transcription from the E3 promoter whereas deletion of both (as in the -82 mutant of Weeks and Jones and the -37 to -233 mutant of Leff et al.) could result in drastic reduction of promoter efficiency.

Although there is no evidence that sequences other than those that constitute the basic promoter are required for the action of E1A, not all promoters are equally sensitive to activation by E1A products. It follows therefore that there must be some feature of the early gene promoters that render them particularly sensitive to E1A action. Some results reported by Weeks and Jones (58) suggest that the difference between an insensitive and sensitive promoter may lie in the distal control region. Chimeric promoters were constructed that consisted of both E3 and HSV tk sequences. Expression of a gene where the tk promoter was fused to CAT coding sequences, was not induced to any appreciable extent by E1A. However, by substituting the tk promoter sequences upstream of position -79 with similarily positioned sequences of the E3 promoter, a hybrid promoter was formed that was sensitive to E1A action. The reverse hybrid construct containing E3 promoter sequences from the start site to -82 and fused to upstream tk promoter sequences remained insensitive to E1A.

4. EARLY REGION 4

The E4 promoter provides a very convenient and interesting system for studying the action of both positive and negative trans-acting factors. Like all the other early genes, transcription from the E4 promoter is stimulated by E1A gene products. However, transcription is repressed by the 72,000 molecular-weight DNA binding protein encoded by early region 2A (60, 61) and indeed addition of this protein to an _in vitro_ reaction specifically suppresses E4 transcription (62). E4 promoter sequences can thus be investigated with respect to both forms of regulation.

The only extensive analysis of E4 promoter sequences has been done using _in vitro_ transcription systems (63). A series of deletion mutants spanning the E4 promoter region were tested for _in vitro_ activity. Three elements were found to be important for maximal transcription; the 'TATAA'-like sequence centered around position -25, an element located between positions -44 and -58 and an element located upstream of position -140. This distal upstream element was particularily important and it's removal resulted in at least a 10-fold drop in transcription.

The importance of a far upstream element has been confirmed for in vivo expression from the E4 promoter (64). A series of deletion mutants of the chimeric gene E4-CAT were tested for in vivo activity following transient expression in HeLa cells with or without the E1A gene. A critical element was located between positions -179 and -158; removal of these sequences resulted in an 8-10 fold drop in induced expression. The removal of these sequences, however, appeared to have much less effect on uninduced expression although again, because of the very low levels of uninduced expression observed, these results should be viewed with some caution. In addition they would be in conflict with the in vitro results discussed above which clearly show a requirement for these distal sequences; the in vitro system is presumably measuring the equivalent of uninduced expression because the assay is carried out in the absence of E1A and because the in vitro systems fail to show an E1A response. If the in vivo results are confirmed, it would clearly distinguish the E4 promoter from other early promoters in that sequences vital for induction by E1A are separate from sequences that constitute the basic promoter.

The -158 to -179 region contains a sequence that has striking homology to sequences in the most distal control elements of the E2A and E3 promoters (Figure 3).

Despite a great deal of effort over the last few years, very little is known concerning the mechanism of E1A trans-activation. However, it is unlikely that E1A protein acts by directly binding to specific nucleotide sequences within the upstream regions of responsive genes. The studies described above, with the possible exception of the E4 gene, suggest that specific binding sequences do not exist. Rather, sequences required for uninduced and E1A-induced transcription are not separable. In addition, evidence is accumulating that E1A protein is not capable of binding to DNA (65). How does E1A trans-activate? At present very few clues exist although a number of models can be proposed. It is possible that E1A protein is a transcriptional factor capable of associating with the transcriptional machinery and modifying it such that certain promoters are recognized and utilized more efficiently. It would thus be analogous to the sigma factors that modulate transcription in prokarytic cells. Alternatively, E1A protein could in some way increase the effective concentration of a limiting cellular transcriptional factor that is required for the efficient transcription of the E1A responsive genes. Given the large number of laboratories actively investigating the mechanism, it is anticipated that significant insights will be obtained within the next year or two.

The biogenesis of mature eukaryotic mRNA is a complex event that not only requires transcription of the gene but also correct post-transcriptional processing of the primary transcript (66). There are three basic processing events involved during the synthesis of most mRNAs namely the addition of the 'cap' at the 5' end, the removal of internal sequences via RNA splicing and the formation of a poly(A) tail at the 3' end. The adenovirus system has proved to be a valuable model system for investigating each of these processing events, both the mechanism of processing and their possible role in regulating gene expression.

5. TRANSCRIPTION TERMINATION AND POLY(A) ADDITION

How mature, polyadenylated 3' ends are formed is not yet known. However, an important observation first described for the adenovirus late transcriptional unit, clearly demonstrates that the mechanism does not rely on specific termination of transcription (67, 68). In all cases where it has been examined (67-72), transcription does not terminate at the site of poly(A) addition but rather at some point downstream. Formation of the mature 3' end therefore has to involve a very specific endonucleolytic cleavage of a larger precursor; the poly(A) tail is subsequently added at the cleavage site. There have to be signals therefore within the precursor RNA that specify where cleavage is to take place. An examination of nucleotide sequences close to the 3' end of all poly(A)-containing mRNAs have revealed the highly conserved sequence AAUAAA situated 10-30 nucleotides upstream of the poly(A) addition site (73). All adenovirus messages both early and late contain this conserved hexanucleotide sequence with one exception. One of the mRNAs of the adenovirus E3 transcriptional unit does not contain this sequence but instead has the sequence AUUAAA at the approximate same position (74); it is presumed that this varient is functionally analogous. Indeed, the same varient is found at the 3' end of the chicken lysozyme mRNA (75) and yet another varient AAUAUA is found in one of the mRNAs of the pancreatic alpha amylase gene (76). Because the sequence is so highly conserved it was speculated that it played a critical role in the process of poly(A) addition either for the specific endonucleolytic cleavage event or the subsequent poly(A) addition by the poly(A) polymerase. This speculation has been tested by specific mutagenesis studies. Deletion of this sequence at the 3' end of the SV40 late gene eliminates the synthesis of messages that are polyadenylated at the normal site (77). In addition, a point mutation in the Ad5 E1A AAUAAA sequence to create AAGAAA resulted in defective E1A mRNA production (78). Only 10% of the nuclear transcripts were cleaved at the normal site but all of them were polyadenylated. These studies confirm the importance of the hexanucleotide sequence to 3' end formation and further suggest that the sequence is important for the cleavage reaction rather than poly(A) addition itself.

In addition to the obvious importance of the hexanucleotide sequence it is equally obvious that this sequence cannot be the only signal required for formation of the correct 3' terminus. The hexanucleotide is often found in transcribed regions far removed from any known poly(A) addition sites. For example, within the Ad12 E1A gene there are two sets of AAUAAA sequences at positions 1397 and 1241 that appear to be non-functional in that there is no evidence for poly(A) addition just downstream of these sites (79). Another argument in favor of additional signals being important, stems from the observation that many genes possess multiple poly(A) addition sites and that different mRNAs can be produced from the precursor RNA by choosing which poly (A) site to utilize. A classic example of this is the adenovirus major late transcription unit which extends from map unit 16·4 to close to the right-hand end terminus of the genome. The primary RNA can be processed by cleavage and poly(A) addition at any one of five different sites (to give the L1 through L5 series of message families)(67). The choice of poly(A) addition site can and probably does represent a mode of regulation of gene expression. The different sites do not appear to be utilized with equal efficiency, an observation that would be difficult to reconcile with the hypothesis that the hexanucleotide sequence is the sole signal for generation of mature 3' ends.

The above arguments suggest that the context of the AAUAAA sequence is important i.e., the nature of the flanking sequences. Possibly these flanking regions contain additional critical signals either in the form of primary sequence recognized by the cleavage machinery or by contributing to a secondary or tertiary structure that may constitute an additional feature of the cleavage signal. It can be argued that the latter possibility is probably the case because by examining sequences at the 3' end of eukaryotic genes (73) no other obvious universal sequence (other than AAUAAA) is present.

Clues as to where these other signals might be located has come from specific mutagenesis studies of the cloned Ad2 E2A gene (80). Sequences just distal to the E2A poly(A) addition site are essential for normal 3' end formation. Specific deletions were introduced into the region surrounding the E2A poly(A) addition site and the consequence of the deletions on expression of E2A measured following transient expression in 293 cells. Deletion of sequences downstream of position +32 (where +1 is the poly(A) addition site and sequences 3' to this site are given the plus notation) had no affect on synthesis of authentic E2A message whereas deletion of additional sequences to position +20 resulted in a 10-20 fold reduction in message production (Figure 4). S1 mapping confirmed that the defect was at the level of 3' end formation. So sequences between +20 and +35 contain a critical element for 3' end formation. There may even be additional important sequences between the poly(A) addition site and position +20 that would have gone unnoticed in this study. The role of these sequences is completely unknown but they do demonstrate one interesting feature; the +20 to +35 region contains a sequence that is complementary to the AAUAAA sequence and flanking region suggesting the possibility of the formation of a stem and loop structure. Unfortunately the presence of such complementary sequences downstream of the poly(A) addition site is not a common feature of all the adenovirus messages or the majority of all eukaryotic messages whose sequence is known. In fact, a similar study to that performed on the E2A gene, has been carried out with the rabbit β-globin gene (81). The results were similar in that sequences between the poly(A) addition site and position +31 were found to be critical for normal 3' end formation. However, these downstream sequences and those of the E2A gene are not alike. Nothing is known about the downstream sequences of any of the other adenovirus messages. Hopefully such studies will be forthcoming and that by comparing the essential regions of a number of these genes a better understanding of their possible role in 3' end formation might be revealed.

The development of an efficient <u>in vitro</u> system for synthesizing RNA with correct 3' termini will also be essential in order for the mechanism of polyadenylation to be understood. Considerable progress has been made in this direction (82). A cell free system has been described that generates RNA accurately polyadenylated at the L3 site of the major late transcript of Adenovirus. The system is a coupled transcription/polyadenylation system and results in about 10% of the transcripts being correctly processed. The coupling of transcription and polyadenylation was found to be essential; addition of presynthesized precursor RNA did not result in proper 3' end formation due to lack of the specific endonucleolytic cleavage step. In contrast to the *in vivo* studies on the E2A gene, no sequence downstream of +10 were found to be necessary for *in vitro* poly(A) addition at the L3 site. Therefore either the position of putative downstream signals is different in this transcription unit or the requirements for 3' end formation are different in the *in vitro* and *in vivo* situations. The results on the *in vitro* polyadenylation of other adenovirus transcription units, in particular the E2A unit, will be eagerly awaited.

In terms of understanding the biochemistry of 3' end processing, the cell free system will be invaluable. Initial studies have indicated that some snRNP's may be involved as a number of anti snRNP antisera inhibited *in vitro* 3' end formation, including anti(U1)RNP that has also been found to inhibit *in vitro* splicing.

Clearly, 3' end formation involves processing of a longer precursor RNA. What is not clear is whether transcription actually terminates at a specific downstream location i.e., is there such a thing as a termination site? No good example of a termination site has been established for any polII transcribed gene. Often termination can be mapped to a certain region of the genome which is usually a few hundred nucleotides in length (68, 83, 84) but whether there are specific termination points within that region has not been established. Very little work has been done on termination of adenovirus gene transcription. It has been shown that there is no transcriptional termination site within at least a 4kb region downstream of the E1A poly(A) addition site (78). In the E1A mutant where the AAUAAA sequence was altered to AAGAAA, only one in ten transcripts were cleaved and polyadenylated at the normal E1A poly(A) addition site. The other transcripts continued through the E1B gene and at least 1·5kb downstream of the E1B poly(A) addition site; some of these long transcripts were subsequently cleaved and polyadenylated at the E1B site.

The only other studies using the adenovirus system to study transcription termination have involved the major late transcription unit. Several different assays have shown that transcription can extend to the region between map units 98-100, well beyond the final poly(A) addition site at map unit 91·4 (68, 83). None of these different methods have convincingly shown that termination stops at specific sites within this extreme right-hand end region. However, progress in this regard may soon be forthcoming. Insertion of Ad2 sequences extending from map units 97·1-100, between the mRNA cap site and translational start site of a chimeric RSVLTR-CAT gene, inhibits expression (85). Intruigingly inhibition by the insert is specific for the orientation of the major late transcription unit. The most likely explanation for these results is that transcription termination within the insert has taken place. If this proves to be correct, deletion analysis of the insert should provide valuable information regarding the nature of the termination signals.

6. ROLE OF NUCLEOTIDE SEQUENCES IN SPLICING OF ADENOVIRUS MESSAGES

More than 40 individual mRNAs are synthesized from the adenovirus genome throughout the productive infection. Out of these only one, the message that codes for protein IX (86), does not require the removal of internal sequences by RNA splicing for its biogenesis. Many of the messages share overlapping sequences and derive from a common precursor by differential splicing (1, 2). This of course allows the synthesis of many different viral products from a limited amount of genetic information; because all viruses are faced with the problem of limited genetic information, differential splicing of viral precursor RNAs is found to be common. Thus the mechanism and regulation of RNA splicing is very complex particularily in the cases where choices between different processing pathways have to be made. What governs these choices is completely unknown but obviously represents a stage of mRNA biogenesis where regulation can be exercised. There is evidence that the synthesis of some adenovirus mRNAs is regulated at the level of differential splicing (87-90).

Despite the attractiveness of the adenovirus system for studying splicing and its regulation very few detailed studies have been carried out particularily with respect to investigating the nucleotide signals vital for correct splicing to take place. The majority of studies of this type have been done with the globin genes where the necessity for the sequences around the splice junctions and within the introns have been characterized (91, 92).

Nucleotide sequences at the exon/intron junctions of many of the Ad2 genes is shown in Fig. 5. The consensus sequences of ${}^{C}_{A}AG/GT^{A}_{G}AGT$ for the donor splice site and ${}^{C}_{T}AG/G$ at the acceptor splice site (93) are conserved for the adenovirus genes with strict conservation (no exceptions) of the GT dinucleotide and AG dinucleotide at the 5' and 3' ends of the intron respectively. The importance of these conserved sequences to the function of splice sites has been demonstrated in the E1 region of Ad2 by the construction and analysis of the following mutants: (1) A T to G transversion at nucleotide

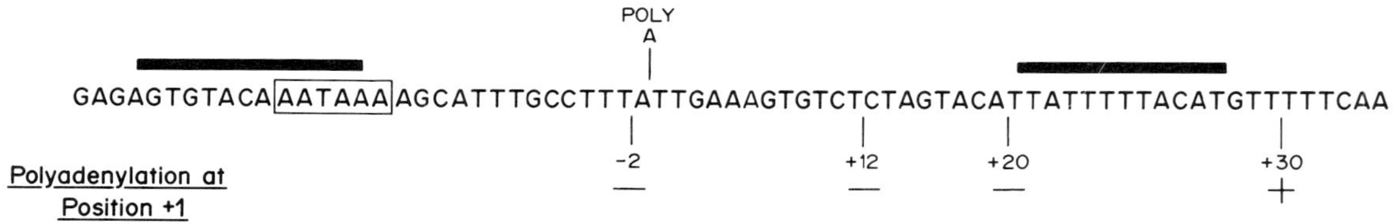

Figure 4: Sequence of the E2A poly(A) addition site (adapted from McDevit et al. (80)). The poly(A) addition site is indicated and is designated +1. Below the sequence are indicated the positions of the end-points of deletion mutants examined (+35, +20, +12, -2) and whether polyadenylation at the +1 position does (+) or does not (-) occur. The box indicates the conserved AATAAA hexanucleotide and the bars the 12bp inverted repeat.

	DONOR		ACCEPTOR
E1A	AG/GTACTG		
	GG/GTGAGG		TAAAAG/GT
	CA/GTAAGT		
E1B	CT/GTATCC		TTGCAG/CA
	AG/GTACTG		
E2A	GG/GTGAGG		TTGCAG/AG
	AG/GTATGA		CTATAG/GA
E3			CAACAG/TT
	CG/GTGAGT		ACACAG/GT
			AAAGAG/GT
E4			ATGCAG/GA
	TG/GTAAGG		ATACAG/GT
			CAAGAG/TT
			CTTCAG/GA
$\overline{\text{IV}}a_2$	AG/GTAAGA		GCGCAG/GG
MAJOR	GG/GTGAGT		CCACAG/CT
LATE	CG/GTAAGA		GTGTAG/GT
			TCCCAG/AT
			TTTGAG/CA
			TTGTAG/AA
			CGCCAG/AG
	AG/GTAGGC		TTGAAG/TC
			CTATAG/GC
			CTACAG/CC
			TTGCAG/AT
CONSENSUS	AG/GTAAGT		$^{CC}_{TT}$NCAG/GT

Figure 5: Ad2 sequences at various donor and acceptor splice junctions. The dash indicates the splice point.

position 975 of the Ad2 genome which changes the GT dinucleotide at the 5' end of the E1A 12s intron to GG (94). (2) A T to A change at nucleotide position 1114 of the Ad2 genome which is the third nucleotide from the 5' end of the E1A 13s intron (95). (3) Single nucleotide substitutions at the fifth and sixth positions of the E1A 12s intron (nucleotide position 978 and 979) (96). (4) A G to C transversion at the first position of the E1B 13s intron (nucleotide 2250) which alters the GT dinucleotide to CT (90). In all four cases the changes resulted in inactivation of the splice junction. The results therefore confirm the importance of the GT dinucleotide at the 5' end of the intron and also indicate that other positions of the consensus sequence are also important. In most cases the mutants were designed and constructed to allow functional characterization of overlapping viral genes. When the E1A 12s donor junction is inactivated all the precursor RNA is processed to give 13s message and vice versa (94, 95). Interesting results were obtained with the E1A mutant that has the single base pair change at position 2250 (90). As a result of the change the 13s splice junction was inactivated and consequently at early times of infection all the E1B precursor RNA was processed to give 22s mRNA. However, at late infection times, in addition to the 22s mRNA, two novel messages were formed which utilized cryptic donor splice sites close to the inactivated 13s site. The surprising result led to the speculation that at late times there is some inactivation of the 22s donor junction and an increase in the propensity to utilize donor junctions close to the position of the 13s junction. In other words splicing of E1B precursor is regulated during infection and probably contributes to the higher ratio of E1B 13s/22s mRNA at late times versus early times.

Apart from these examples no other mutants have been described that investigate the sequence requirements for splicing of adenovirus messages. As the mutants described above convincingly show, the junction consensus sequences are important for defining splicing specificity. However, the consensus sequences can not be the sole determinants for specificity. There are numerous examples of consensus sequences throughout genes that are not used for splicing; in some cases, as was found for the E1B gene, these cryptic sites can be used when the normal site is inactivated (90). There must be other features of the precursor RNA that define splicing specificity and distinguish between active and cryptic sites. What these other features are have not been investigated in adenovirus genes.

Until recently very little was known about the mechanism and biochemistry of the splicing reaction. However, due to the development of efficient *in vitro* splicing extracts, very exciting and significant advances have been made within the last year (97-100). Pre-mRNAs can be added to these extracts and in the presence of ATP, $MgCl_2$ and monovalent cations, correct splicing of a significant portion of the added molecules occurs. Two different adenovirus pre-mRNAs have been investigated using the *in vitro* system. In one case pre-mRNA containing the E2A sequences was synthesized *in vitro* from a plasmid in which the E2A coding region is fused to the Sp6 bacteriophage promoter (100). In the other case, a substrate was prepared that contained the first and second exons of the major late transcription unit together with a shortened version of the intron that separates these exons (98). In both cases efficient splicing of the pre-mRNAs took place. One of the next steps will presumably be an investigation by mutational analysis, of the contribution that different sequences within the pre-mRNAs make towards splicing specificity.

The *in vitro* systems have provided important information regarding the mechanism of splicing. The excised introns as well as RNA species that are probably intermediates in the splicing reaction have the unusual configuration of a circle with a tail where the branch point consists of a 2'-5' phosphodiester bond (101-103); this structure is often referred to as a 'lariat' structure. The bond is between the G residue at the 5' end of the intron and an A residue located just upstream of the 3' end of the intron. In the case of the first intron of the major late transcription unit this A residue is situated 24 nucleotides upstream from the intron 3' end point (104). It has been hypothesized and suggested that the A residue lies within a sequence PyXPyTPuAPy that is conserved in introns of higher eukaryotes (103). It would of course be interesting to mutate this residue and flanking sequences within the L1-L2 intron and investigate the consequences of such changes on the L1-L2 splice.

7. SEQUENCES REQUIRED FOR EXPRESSION OF THE VA1 AND VA11 GENES

During a productive infection of cells by Adenovirus, two distint low molecular weight RNAs are synthesized in very large amounts (105, 106). These RNAs which are transcribed by RNA polymerase III (106, 107) are designated VA1 and VA11. The respective genes are located between 29.0 and 31.0 map units on the Ad2 or Ad5 genome. Since the development of efficient *in vitro* transcription systems a number of RNA polymerase III-transcribed genes including the VA1 gene have been genetically analyzed in detail to define the sequences necessary for accurate and efficient expression (Figure 5).

Deletion mutants of the cloned VA1 gene have been constructed and the ability of the altered templates to direct *in vitro* transcription assayed (108, 109). Sequential removal of sequences at the 5' and 3' ends of the gene led to the identification of an internal control region (ICR) situated between nucleotides +10 and +69. Deletion of this ICR reduces transcription at least 100-fold. The presence of control regions within the coding region is typical of RNA polymerase III transcribed genes; ICR's have been described for the 5S (110, 111) and numerous tRNA genes (112-114). Sequences upstream of the VA1 transcriptional start site were found not to affect the level of transcription in a significant way but were found to influence the position of the start site (108, 115). Normally, two VA1 RNA species are synthesized and arise by initiation of transcription occuring at a G residue or an A residue that are 3 nucleotides apart. A 2 base pair deletion between nucleotide positions -22 and -25 results in the synthesis of the VA1 (G) species only (115). Comparison of the ICR of VA1 and numerous tRNA genes, reveals two blocks of conserved sequences (block A and B) at either end of the control region (108). The conserved regions have the consensus sequence PuPuPyNNAPuPyGG for element A at the 5' end of the ICR and G^{T}_{A}TCPuANNC for element B at the 3' end of the ICR. These two conserved elements have been shown to be critical for efficient expression of the *X. laevis* tRNA met gene (112). The same appears to be true for the VA1 gene (116). Specific small alterations such as substitutions and insertions within the +18 to +53 region of the VA1 ICR had little affect on transcription. These results suggested therefore that the critical regions for transcription were between nucleotides +10 and +18 and between +54 and +69. (The +10 to +18 block could possibly extend to +25 since substitutions in the region between +18 and +25 resulted in an approximate 5-fold reduction in transcription levels.) These two critical regions coincide with the conserved elements A and B. The nucleotide sequence of the inter-element region (C) does not appear to be important. However the size of C is important (116). The region could be enlarged without affecting expression; expansion of C from 35 to 50 nucleotides had no affect, expansion to 75 nucleotides resulted in a 5-fold drop in expression and expansion to 115 nucleotides abolished expression. There was found to be less leeway in terms of shortening C since

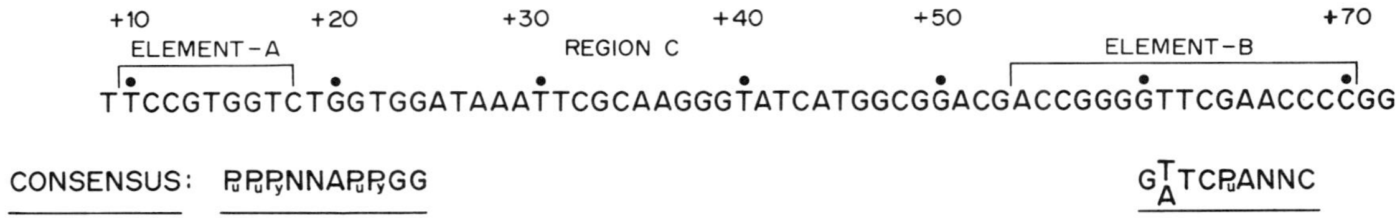

Figure 6: Location of the transcription control elements on the VA1 RNA gene (taken from Bhat et al (116)). The single strand shown reads the sense strand for the VA1 RNA. The conserved A and B elements are indicated and compared to the concensus sequence described by Ciliberto et al. (114).

deletions of 4 and 6 nucleotides resulted in 2% to 1% of wild-type activity respectively. Thus, the C region appears to function as a spacer between the conserved A and B elements with the optimum spacing being in the range of 31-75 base pairs. Similar results have been obtained with certain tRNA genes (117). The VAII gene of Ad2 and Ad5 also contains an ICR which encompasses the consensus A and B sequences. Detailed genetic analysis of the VAII ICR region has not been carried out but there is every reason to believe that it is organized in an indential fashion to that of VAI.

The results clearly suggest a functional similarity in the A and B elements of the VAI and tRNA genes. This was elegantly tested by constructing hybrids between VAI of Ad2 and the *X. laevis* tRNA met gene (116); the A and B elements of VAI were substituted with the corresponding elements of the tRNA gene. In both cases close to wild-type levels of transcription were obtained.

The exact role of the A and B elements are not yet known but are hypothesized to interact with one or more essential transcription factors or even RNA polymerase III itself or both (118). In addition to the polymerase two critical factors IIIB and IIIC have been described. Preliminary evidence suggests that a stable complex is formed between factors IIIC and the B element of VAI (118). Incubation of the VAI gene with factor IIIC protects the DNA from digestion with endonuclease BamHI the recognition site of which is located 1 nucleotide downstream of the B consensus sequence.

REFERENCES:

1. Akusjarvi G, Pettersson U: Chapter 2, this volume, 1985.
2. Nevins, JR, Imperiale MJ: Chapter 4, this volume, 1985.
3. VanOrmondt H, Galibert F: Current Topics in Microbiology and Immunology (110):73-142, 1984.
4. Berk AJ, Lee F, Harrison T, Williams J, Sharp PA: Cell (17):935-944, 1979.
5. Jones NC, Shenk TE: Proc. Natl. Acad. Sci. USA (76):3665-3669, 1979.
6. Nevins JR: Cell (26):213-220, 1981.
7. Weeks DL, Jones NC: Mol. Cell Biol. (3):1222-1234, 1983.
8. Nevins JR: Ann. Rev. Biochem. (52):441-466, 1983.
9. Van der Eb AJ, Bernards R: Current Topics in Microbiology and Immunology (110):23-51, 1984.
10. Graham FL, Smiley J, Russell WC, Nairn R: J. Gen. Virol. (36):59-94, 1977.
11. Jones NC, Shenk T: Cell (17):683-689, 1979.
12. Stow ND: J. Virol. (37):171-180, 1981.
13. Chinnadurai G, Chinnadurai S, Brusca J: J. Virol. (32):623-628, 1979.
14. Khoury G, Gruss P: Cell (33):313-314, 1983.
15. Banerji J, Rusconi S, Schaffner W: Cell (27):299-308, 1981.
16. Gillies SD, Morrison SL, Oi VT, Tonegawa S: Cell (33):717-728, 1983.
17. Queen C, Baltimore D: Cell (33):741-748, 1983.
18. Feldman LT, Imperiale MJ, Nevins JR: Proc. Natl. Acad. Sci. USA (79): 4952-4956, 1982.
19. Preston CM: J. Virol. (29):275-284, 1979.
20. Watson RJ, Clements JB: Virology (91):364-379, 1978.
21. Watson RJ, Clements JB: Nature (285):329-330, 1980.
22. Spector D, personal communication.
23. Lang JC, Spandidos DA, Wilkie NM: EMBO J. (3):389-395, 1984.
24. Hearing P, Shenk T: J. Mol. Biol. (167):809-822, 1983.
25. Hearing P, Shenk T: Cell (33):695-703, 1983.
26. Breathnach R, Chambon P: Ann. Rev. Biochem. (50):349-383, 1981.
27. Shenk T: Current Topics in Microbiology and Immunology (93):25-46, 1981.
28. Grosveld GC, deBoer E, Shewmaker CK, Flavell RA: Nature (295):120-126, 1982.
29. Dierks P, Van Ooyen A, Cochran MD, Dobkin C, Reiser J, Weissmann C: Cell (32):695-706, 1983.
30. Herbomel P, Bourachot B, Yaniv M: Cell (39):653-662, 1984.
31. Rosenthal N, Kress M, Gruss P, Khoury G: Science (222):749-755, 1983.
32. Czernilofsky AP, DeLorbe W, Swanstrom R, Varmus HE, Bishop JM, Tischer E, Goodman HM: Nucl. Acids Res. (8):2967-2984, 1980.
33. Donehower LA, Huang AL, Hager GL: J. Virol. (37):226-238, 1981.
34. Sassone-Corsi P, Hen R, Borrelli E, Leff T, Chambon P: Nucl. Acids Res. (11):8735-8745, 1983.
35. Hen R, Borrelli E, Sassone-Corsi P, Chambon P: Nucl. Acids Res. (11): 8748-8760, 1983.
36. Weiher H, Konig M, Gruss P: Science (219):626-631, 1983.
37. Laimins LA, Tsichlis P, Khoury G: Nucl. Acids Res. (12):6427-6442, 1984.
38. Imperiale MJ, Feldman LT, Nevins JR: Cell (35):127-136, 1983.
39. Osborne TF, Arvidson DN, Tyau ES, Dunsworth-Browne M, Berk AJ: Mol. Cell. Biol. (4):1293-1305, 1984.

40. Andrisani O, Jones NC: submitted for publication
41. Shaw AR, Ziff EB: Cell (22):905-916, 1980.
42. Spector D, McGrogan M, Raskas H: J. Mol. Biol. (126):395-414, 1978.
43. Borrelli E, Hen R, Chambon P: Nature (312):608-612, 1984.
44. Stillman BW, Lewis JB, Chow LT, Mathews MB, Smart JE: Cell (23): 497-508, 1981.
45. Baker CC, Ziff E: J. Mol. Biol. (149):189-221, 1981.
46. Mathis DJ, Elkaim R, Kedinger C, Sassone-Corsi P, Chambon P: Proc. Natl. Acad. Sci. USA (77):7102-7106, 1981.
47. Matsui T: Nucl. Acids Res. (10):7089-7101, 1982.
48. Rossini M: Virology (131):49-58, 1983.
49. Imperiale MJ, Nevins JR: Mol. Cell Biol. (4):875-882, 1984.
50. Imperiale MJ, Hart RP, Nevins JR: Proc. Natl. Acad. Sci. USA (82), 381-385, 1985.
51. Kingston RE, Kaufman RJ, Sharp PA: Mol. Cell Biol. (4):1970-1977, 1984.
52. Murthy SCS, Bhat GP, Thimmappaya B: Proc. Natl. Acad. Sci., in press.
53. McKnight SL, Kingsbury R: Science (217):316-324, 1982.
54. Everett RD, Baty D, Chambon P: Nucl. Acids Res. (11):2447-2464, 1983.
55. Pelham HRB: Cell (30):517-528, 1982.
56. McKnight SL: Cell (31):355-365, 1982.
57. McKnight SL, Kingsbury RC, Spence A, Smith M: Cell (37):253-262, 1984.
58. Weeks DL, Jones NC: submitted for publication.
59. Leff T, Corden J, Elkaim R, Sassone-Corsi P: submitted for publication.
60. Blanton RA, Carter T: J. Virol. (29):458-465, 1979.
61. Nevins JR, Winkler JJ: Proc. Natl. Acad. Sci. USA (77):1893-1897, 1980.
62. Handa H, Kingston RE, Sharp PA: Nature (302):545-547, 1983.
63. Handa H, Sharp PA: Mol. Cell Biol. (4):791-798, 1984.
64. Gilardi P, Perricaudet M: Nucl. Acids Res. (12):7877-7888, 1984.
65. Ferguson B, Krippl B, Westphal H, Jones N, Richter J, Rosenberg M: *In* Cancer Cells 3: Growth Factors and Transformation. Cold Spring Harbor Laboratory, 1985.
66. Nevins JR: Ann. Rev. Biochem. (52):441-466, 1983.
67. Nevins JR, Darnell JE: Cell (15):1477-1493, 1978.
68. Fraser NW, Nevins JR, Ziff EB, Darnell JE: J. Mol. Biol. (129):643-656, 1979.
69. Ford JR, Hsu MT: J. Virol. (28):795-801, 1978.
70. Lai CJ, Dhar R, Khoury G: Cell (14):971-982, 1978.
71. Hofer E, Darnell JE: Cell (23):585-593, 1981.
72. Weintraub H, Larsen A, Groudine M: Cell (24):333-344, 1981.
73. Proudfoot NJ, Brownlee GG: Nature (263):211-214, 1976.
74. Ahmed CMI, Chandra RS, Stow ND, Zain BS: Gene (20):337-346, 1983.
75. Jung A, Sippel AE, Grez M, Schutz G: Proc. Natl. Acad. Sci. USA (77):5759-5763, 1980.
76. Tosi M, Young RA, Hagenbuchle O, Schibler U: Nucl. Acids Res. (9): 2313-2323, 1981.
77. Fitzgerald M, Shenk T: Cell (24):251-260, 1981.
78. Montell C, Fisher EF, Caruthers MM, Berk AJ: Nature (305):600-605, 1983.
79. Perricaudet M, LeMoullec JM, Tiollais P, Pettersson U: Nature (288): 174-176, 1980.
80. McDevitt MA, Imperiale MJ, Ali H, Nevins JR: Cell (37):993-999, 1984.
81. Gil A, Proudfoot NJ: Nature (312):473-474, 1984.
82. Moore CL, Sharp PA: Cell (36):581-591, 1984.
83. Fraser NW, Tsu MT: Virology (103):514-516, 1980.

84. Hofer E, Hofer-Warbinek R, Darnell JE: Cell (29):887-893, 1982.
85. Dressler GR, Fraser NW: personnal communication.
86. Alestrom P, Akusjarvi G, Perricaudet M, Mathews MB, Klessig DF, Pettersson U: Nature (19):671-681, 1980.
87. Nevins JR, Wilson MD: Nature (290):113-118, 1981.
88. Akusjarvi G, Persson H: Nature (292):420-426, 1981.
89. Chow LT, Broker TR, Lewis JB: J. Mol. Biol. (134):265-303, 1979.
90. Montell C, Fisher EF, Caruthers MH, Berk AJ: Mol. Cell. Biol. (4): 966-972, 1984.
91. Wieringa B, Meyer F, Reiser J, Weissman C: Nature (301):38-43, 1983.
92. Wieringa B, Hofer E, Weissman C: Cell (37):915-925, 1984.
93. Mount SM: Nucl. Acids Res. (10):459-472, 1982.
94. Montell C, Fisher EF, Caruthers MH, Berk AJ: Nature (295):380-384, 1982.
95. Montell C, Berk AJ: Nucl. Acids Res. (12):3821-3927, 1984.
96. Solnick D: Nature (291):508-510, 1981.
97. Hernandez N, Keller W: Cell (35):89-99, 1983.
98. Hardy SF, Grabowski PJ, Padgett RA, Sharp PA: Nature (308):375-377, 1984.
99. Krainer AR, Maniatis T, Ruskin B, Green MR: Cell (36):993-1005, 1984.
100. Goldenberg C: Proc. Natl. Acad. Sci. USA (81):4707-4711, 1984.
101. Grabowski PJ, Padgett RA, Sharp PA: Cell (37):415-427, 1984.
102. Padgett RA, Konorska M, Grabowski PJ, Hardy SF, Sharp PA: Science (225): 898-903, 1984.
103. Ruskin B, Krainer AR, Maniatis T, Green MR: Cell (38):317-331, 1984.
104. Konarska MM, Grabowski PJ, Padgett RA, Sharp PA: Nature (313):552-557, 1985.
105. Reich PR, Forget B, Rose J, Weissmann SM: J. Mol. Biol. (17):428-438, 1966.
106. Soderlund H, Pettersson U, Vernstrom B, Philipson L, Mathews MB: Cell (7):585-593, 1976.
107. Weinmann R, Raskas HJ, Roeder RG: Proc. Natl. Acad. Sci USA (71):3426-3430, 1974.
108. Fowlkes DM, Shenk T: Cell (22):405-413, 1980.
109. Guilfoyle R, Weinmann R: Proc. Natl. Acad. Sci. USA (78):3378-3382, 1981.
110. Sakonju S, Bogenhagen DF, Brown DD: Cell (19):13-25, 1980.
111. Bogenhagen DF, Sakonju S, Brown DD: Cell (19):27-35, 1980.
112. Hofstetter H, Kressman A, Birnsteil M: Cell (24):573-585, 1981.
113. Ciliberto G, Castagnoli L, Melton DA, Cortese R: Proc. Natl. Acad. Sci. USA (79):1195-1199, 1982.
114. Ciliberto G, Rongei G, Costanzo F, Dente L, Cortese R: Cell (32): 725-733, 1983.
115. Thimmappaya B, Jones NC, Shenk T: Cell (18):947-954, 1979.
116. Bhat RA, Metz B, Thimmappaya B: Mol. Cell. Biol. (3):1996-2005, 1983.
117. Ciliberto G, Traboni C, Cortese R: Proc. Natl. Acad. Sci. USA (79): 1971-1975.
118. Lassar AB, Martin PL, Roeder RG: Science (222):740-748, 1983.

6

INACTIVATION OF VIRAL PROMOTERS BY IN VITRO METHYLATION: STUDIES ON THE ADENOVIRUS AND BACULOVIRUS SYSTEM

W. DOERFLER, K.-D.LANGNER, D. KNEBEL, H. LÜBBERT, B. KNUST-KRON
Institute of Genetics, University of Cologne, Cologne, Germany

INTRODUCTION

Viruses as tools in molecular biology

Frequently, viruses have served a very important role in investigations on the mechanisms involved in gene expression in eukaryotic cells. For example, the principle of RNA splicing was first recognized during the course of studies on gene expression in adenovirus- and simian virus 40-infected cells (e.g. 1, 2). Using the same viral systems, decisive progress was made in developing methods to map viral transcripts and to employ cell-free transcription (3) and splicing assays (4) in order to elucidate details of these mechanisms. For several years, we have studied the functional significance of site-specific DNA methylations, particularly in the promoter part of adenovirus genes by a number of different experimental approaches. Again, the adenovirus system offered several advantages as an experimental tool, not only due to the manageable size of its genome and the wealth of information available on the molecular biology of these viruses (5). In addition, the interaction of the adenoviral genome with host cell factors could be analyzed in quite different biological environments, i.e. in productively or abortively infected cells, as well as in adenovirus-transformed cells in whose genome the adenovirus genome was stably integrated (6 - 9).

From a functional point of view, and in particular with respect to the state of adenovirus DNA methylation, different compartments can be distinguished in the nuclei of adenovirus-infected or -transformed cells (Fig. 1). The DNA inside adenovirions was not found to be methylated (10 - 12). Similarly,

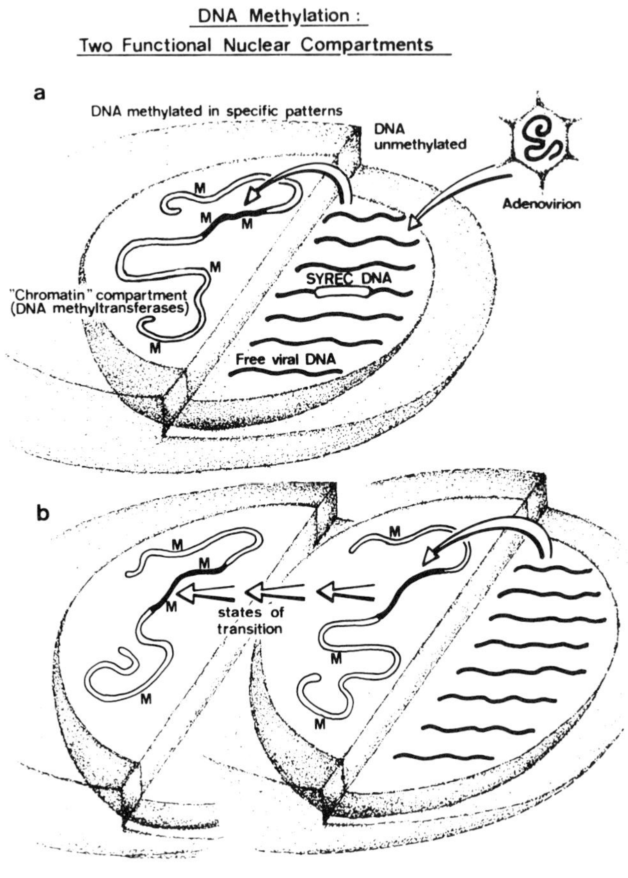

Fig. 1. DNA methylation: two functional nuclear compartments. From a functional point of view, the nucleus of an adenovirus-infected or -transformed cell is subdivided into two compartments. One compartment contains free viral and SYREC DNA replicating and remaining unmethylated. The second compartment - termed "chromatin" - contains nuclear DNA and integrated viral genes which are methylated in specific patterns. These latter patterns are attained only after viral DNA has become fixed in a special way.

free viral DNA in the nuclei of productively or abortively infected cells was not methylated either (13, 14). Even the cellular DNA in symmetric recombinants (SYREC) containing the left terminal 1000 - 2081 nucleotide pairs of adenovirus type 12 DNA and about 26 - 28 kilo base pairs (kbp) of human cell DNA was not methylated (15, 16). The DNA of SYREC could be encapsidated into virions and appeared to replicate in the same nuclear compartment as authentic adenovirus DNA (Fig. 1). In contrast, adenovirus type 2 or adenovirus type 12 DNA stably integrated into the host genome, as in many adenovirus type 2- or adenovirus type 12-transformed cell and tumor lines (8, 9), was found to be methylated in highly specific patterns (13, 17 - 19). Thus, viral DNA incorporated into the "chromatin compartment" (Fig. 1), as it were, was subject to different rules of DNA methylation, probably identical to those that applied to cellular DNA. These findings also demonstrated unequivocally that adenovirus DNA had to be de novo methylated at some time between infection of the cell and its stable fixation in the host genome. However, the situation proved to be more complicated in that at relatively early times after integration of adenovirus type 12 DNA, e.g. in DNA directly extracted from adenovirus type 12-induced tumors, viral DNA was not methylated or hypomethylated (20). Upon explantation of cells from these tumors and continued cell passage in culture, a gradual increase of viral DNA methylation was observed (20, 21), and viral DNA sequences were then methylated in a non-random fashion (22). At present, we do not understand at all how levels of DNA methylation are regulated, nor how such shifts in specific viral DNA methylation can occur or what their consequences may be. The adenovirus system and the study of levels of methylation of well characterized viral DNA sequences in different mammalian cell environments may help to elucidate some of the still enigmatic features of DNA methylation. We have also begun to investigate in detail DNA methyltransferases in uninfected and adenovirus type 2-infected mammalian cells or tissues (23 - 25). In short, the viral system chosen for these investigations offers a wide gamut of problems and, hopefully, the approaches to solve some of these problems.

Site-specific DNA methylations and gene inactivation: Brief synopsis of data

The data on DNA methylation in eukaryotic systems and their significance in long-term gene inactivation have been extensively reviewed in articles (26 - 34) and books (35 - 37). Hence, it will be superfluous to repeat much of the detail covered elsewhere. In this review we shall summarize data recently obtained in the adenovirus system, and in this section restate some of the major conclusions gleaned from a number of different biological systems.

i) The major modified nucleotide in eukaryotic DNA is 5-methyl deoxycytidine (5-mC), it is not certain whether N^6-methyl deoxyadenosine (N^6-mA) occurs to any appreciable extent in mammalian DNA. The nucleotide 5-mC is found frequently, but not exclusively, in the dinucleotide combination 5'-CG-3'.

ii) DNA methylation appears to be a postreplicative event. Upon one round of DNA replication, DNA methylated in both strands of the duplex molecule is converted to hemimethylated DNA which is the preferred substrate for maintenance DNA methyltransferases. De novo DNA methyltransferases are thought to use unmethylated DNA as substrates. It is not clear whether the two enzymatic functions reside in separate enzymes or whether one enzyme can accomplish both activities. Active enzymatic demethylation of DNA has not been unequivocally demonstrated. The cytidine analog 5-azadeoxycytidine (5-azaC) inhibits DNA methyltransferases of the cell after the compound has been incorporated into DNA. Therefore, patterns of DNA methylation can be decisively altered by growing cells in the presence of moderate concentrations (μM) of 5-azaC.

iii) It has been established by a number of different experimental approaches in several systems that methylations of specific DNA sequences, notably in the promoter and 5' parts of eukaryotic genes, can be causally related to inactivation of these genes. The biochemical mechanism of this inactivation is not yet understood, it probably has to be sought in alterations of specific protein-DNA interactions due to the presence of 5-mC in highly specific sequences which are involved in these interactions, in alterations in the structure of DNA or in a combination of both events. The main lines of evidence supporting the notion of gene inactivation by site-specific DNA methylation are the following: 1. Inverse correlations between the levels of DNA methylation in the promoter and 5' parts of eukaryotic genes. 2. Reactivation of previously dormant cellular genes by exposure of cells to 5-azaC. Due to the inhibition of cellular DNA methyltransferases by the drug, DNA is "demethylated" at certain sites, and gene activation can accrue as a consequence of altered promoter methylations. 3. A number of viral and non-viral eukaryotic promoters (cf. Table 3) have been in vitro methylated at highly specific sites. Upon testing the activity of these promoters in eukaryotic cells, it was found that the methylated promoters were inactive, whereas unmethylated promoters remained active.

iv) Since alterations in levels of DNA methylation require one or several rounds of DNA replication, it is reasonable to postulate that the inactivation of eukaryotic genes by site-specific promoter methylations is a long-term event. Once a gene has to be permanently inactivated, DNA methylation may be the signal of choice. However, even if a gene has to be turned on very rarely or for extremely short periods of time, e.g. during certain periods in the regular cell cycle or in certain cells of an organ or organism, it would be futile to use DNA methylation as the inactivating mechanism, since it would be difficult to accomplish short-term and transient reactivation. We hold this to be a very important consideration in interpreting data on DNA methylation.

v) Very scanty information is available on the nature and specificity of eukaryotic DNA methyltransferases (38) and their regulation. If DNA methylation was in fact an important mechanism for long-term gene inactivation, the study of DNA methyltransferases should be given very high priority in the future.

vi) With DNA methylation being a widely observed and generally applicable mechanism of regulating genetic activity, particularly on a long-term scale, it will be very interesting to assess the importance of this mechanism in basic biological phenomena such as differentiation, mutagenesis, oncogenesis and the maturation of the immune system. There is already ample evidence for the notion that the regulation of these processes can be related to site-specific DNA methylations.

Rigid control of viral genes in virus-transformed cells

When inverse correlations between levels of DNA methylation and gene expression were first observed, viral genes in a number of virus-transformed cells represented particularly clear-cut examples (27). In some of the adenovirus type 12-transformed hamster cell lines, which had been investigated in our laboratory (39, 40), the total adenovirus type 12 genome persisted in an integrated form. The late viral genes were not expressed, most of the early viral genes were expressed (41). The promoters of the inactive adenovirus type 12 genes were methylated at many 5'-CCGG-3' (HpaII) and 5'-GCGC-3' (HhaI) sequences, those of the expressed adenovirus type 12 genes were hypomethylated (13, 17 - 19). Similarly well documented examples could be quoted from several other viral systems (for review 27). Virus-transformed cell populations do in fact constitute highly selected systems which have been optimally conditioned to the culture conditions employed in their propagation. This accomodation to certain culture conditions entailed an expression pattern of integrated viral genes which seems to be unique to each transformed cell line and which has to be optimally attuned to the growth requirements of these cells. This expression pattern is rigidly controlled, since survival of the cell line under the actual culture conditions depends on the maintenance of these patterns. This rigid control is apparently perpetuated by a highly specific pattern of DNA methylation. Thus, the stringency of selection and of the control of gene expression might be interdependent, and these interrelationships could perhaps explain the very striking inverse correlations between DNA methylation and extent of gene expression which has been described in many viral systems.

Survey of results obtained with the adenovirus and baculovirus systems

In addition to the adenovirus system, insect baculoviruses were used for some of our studies. The rationale for including this latter system in the analyses was to investigate a DNA insect virus and its promoters for their sensitivity to DNA methylation. Drosophila DNA and the DNA of Spodoptera frugiperda insect cells, on which the baculovirus Autographa californica is routinely grown (42 - 45), do not contain 5-mC in easily detectable amounts (12, 43, 46 - 48). There is one report that claims one 5-mC residue to be present in 12.500 nucleotides in Drosophila DNA (49). It was therefore interesting to test whether baculovirus promoters (45, 50, 51) could be inactivated by site-specific DNA methylations. It will not be possible to describe the baculovirus system in detail. A summary of current research in this field will be shortly published in a book (52).

Virion DNA and free intracellular adenovirus DNA are unmethylated

Adenovirions, which had replicated in the nuclei of human cells growing in tissue culture, can be highly purified. The virions can be freed of cellular DNA contaminants by treatment with DNase. Viral DNA extracted from virions purified in this way has been shown to be devoid of detectable amounts of the methylated bases 5-mC and N^6-mA (10, 12). Other animal viruses like baculoviruses or simian virus 40 which replicate and are assembled in the nuclei of their host cells, are likewise deficient in the methylation of their genomes (12, 43, 53). In contrast, the iridovirus, frog virus 3, with a complicated pathway of replication involving a nuclear phase but assembly in the cytoplasm, carries > 20 % of 5-mC in its DNA (54).

Human cellular DNA in whose immediate environment adenovirus DNA replicates at an extremely high rate contains 3.5 % - 4.4 % of 5-mC relative to the total C content of its DNA. It is still unknown how adenovirus DNA is capable of escaping the modifying activity of the DNA methyltransferase system of the host. We have been able to show that both adenovirus-infected and uninfected human cells exhibit the same DNA methyltransferase activities

(23). Hence, absence of 5-mC in adenovirus DNA is probably not due to a lack or inactivity of the cellular DNA methyltransferase system. Since active enzymatic demethylation of DNA does not appear to be a frequent mechanism, the main alternative remaining to enact changes in the patterns of DNA methylation seem to be DNA replication followed by specific modulations of the maintenance DNA methyltransferase. For this reason, DNA methylation is a very stable signal and, if used in the regulation of gene expression, will impose a long-term inactivating effect on gene expression. In this sense, it would be highly inopportune for adenovirus DNA to succumb to this functional control of the host cell. The question remains of how adenovirus DNA manages to avoid this modification. Compartmentalization of the DNA methyltransferase system in the host's chromatin (Fig. 1) and the high rate of viral DNA replication are probably among the salvaging factors.

It is interesting to mention in this context that human cell DNA that has become integrated into the DNA of a symmetric adenovirus-host cell DNA recombinant, SYREC for symmetric recombinant, which is encapsidated into virions (15, 16), contains human cell DNA sequences that are unmethylated at their 5'-CCGG-3' sites. The same cellular DNA sequences are strongly methylated at the same sites when investigated within the compartment of the human cellular chromatin (15). Hence, methylation of DNA in mammalian cells does not seem to depend on the DNA sequence but rather on the location of DNA in apparently functionally different nuclear compartments (Fig. 1).

We have also investigated the state of DNA methylation in free intracellular adenovirus DNA in cells infected productively (e.g. human cells infected with adenovirus type 2) or abortively (e.g. hamster cells infected with adenovirus type 12) (6, 7, 56) with adenoviruses. As will be expected from the aforementioned reasoning, we have not found any evidence for the presence of 5-mC residues in free intracellular adenovirus DNA (13, 14). This statement is based on the results of analyses of intracellular viral DNA using several restriction endonucleases which are sensitive to site-specific DNA methylations. In this study (14), evidence for the methylation of cytidine residues in 5'-CCGG-3' (HpaII) and 5'-GCGC-3' (HhaI) sequences or the methylation of adenosine residues in 5'-GATC-3' (DpnI) and 5'-TCGA-3' (TaqI) sequences in free intranuclear adenovirus type 2 DNA isolated and analyzed early (5 h) or late (24 h) after infection could not be obtained. In Ad2 DNA 22.5 % of all 5'-CG-3' dinucleotides reside in 5'-CCGG-3' and 5'-GCGC-3' sequences. Intranuclear viral DNA was examined by restriction endonuclease cleavage using HpaII, MboI, HhaI, DpnI or TaqI and Southern blot hybridizations. The HindIII fragments of Ad2 DNA served as hybridization probes. These data (14) rendered it very unlikely that DNA methylation could play a role in the regulation of adenovirus DNA expression in productively infected cells.

Integrated adenovirus DNA is part of the host genome and exhibits specific, functionally significant patterns of DNA methylation

We will have to distinguish (Fig. 1, Table 1) free intracellular adenovirus DNA from adenovirus DNA which has become integrated into host DNA (6 - 9), e.g. in adenovirus-transformed cells or adenovirus-induced tumor cells. These viral DNA sequences - from a functional point of view - have been rendered part of the chromatin of the host cell, and within this compartment (Fig. 1) are subject to the methylation strategies of the host. We have shown that integrated adenovirus DNA exhibits unique and functionally highly significant patterns of DNA methylation, and that inverse correlations exist between the extent of DNA methylation and the degree to which the same viral genes are expressed (13, 17 - 19, 27, 30, 39, 55).

A particularly clear cut but, by no means, the only example of such inverse correlations between DNA methylation at specific sites and gene expression (27, 30) is offered by the E2A region of adenovirus type 2 (13). The data are schematically summarized in Table 2. The E2A region, one of the genes that is expressed early in adenovirus infection, codes for the 72,000 molecular weight (72K) DNA binding protein (62) which has now been recognized as a highly interesting protein exerting several dif-

Table 1. DNA methylation of the genome of human adenoviruses[a]

Type of adenovirus DNA	State of methylation	References
DNA packaged into virus particles	Unmethylated[b]	10, 12
Free viral DNA in infected mammalian cells (productive and abortive systems) both parental and newly synthesized	Unmethylated	13, 14
Integrated adenovirus DNA in established cell lines	Methylated in highly specific patterns: Active genes are hypomethylated; inactive genes are hypermethylated.	13, 17 - 19, 39
Integrated adenovirus DNA in recently produced tumor cells	Hypomethylated. Gradual increase in DNA methylation with passage of tumor cells. Shift in methylation is non-random.	21, 22

[a]The only methylated base detectable is 5-mC (10, 12)

[b]Limit of detectability $\geqq$ 0.04 % 5-mC (12)

Table 2. Inverse correlations between DNA methylation and E2A expression in cell lines HE1, HE2, and HE3

Cell line (58)	E2A gene and late promoter persisting in integrated form (59)	mRNA (60)	E2A protein (61)	Methylation of 14 CCGG sites in E2A gene (13)	Methylation of GGCC sites in E2A gene (63)
HE1	+	+	+	-	-
HE2	+	-	-	+	-
HE3	+	-	-	+	-

ferent functions. Perhaps for this reason, the cell wants to control the expression of this protein very tightly once the corresponding gene has been inserted into its own genome. In this functional analysis we have availed ourselves of a number of adenovirus type 2-transformed hamster cell lines (57, 58). Cell line HE1 expresses the 72K protein (60, 61), and all fourteen 5'-CCGG-3' sites in the E2A region of the integrated adenovirus type 2 DNA are unmethylated (13). In contrast, cell lines HE2 and HE3 fail to express the 72K protein (60, 61), and the fourteen 5'-CCGG-3' sites in the E2A region are all completely methylated (13). We have also investigated the 5'-GGCC-3' sites in the E2A region of the integrated adenovirus type 2 genomes in the same cell lines and found them unmethylated irrespective of the state of genetic activity of the E2A region (63). These data suggested that highly specific sequences might be involved in gene regulation via DNA methylation. Many other examples of such inverse correlations have been described in viral systems (27).

The type of analysis described here in an exemplary fashion predicted a role for DNA methylation in the regulation of gene activity. However, the results did not permit one to distinguish between DNA methylation being the cause or the consequence of specific gene inactivations.

<u>Site-specific DNA methylations in the promoter and 5' part of a gene correlate with its inactive state</u>

Before we attempt to accomplish this distinction, we have to address the problem of the significant sites for DNA methylations in inactivated genes. Results from a variety of eukaryotic systems (for reviews 30, 33) support the concept that the promoter and 5' part of a gene are the functionally decisive parts of a gene with respect to DNA methylation. This notion is documented by the results of analyses in which the 5'-CCGG-3' and 5'-GCGC-3' sites in integrated adenovirus type 12 DNA in three adenovirus type 12-transformed hamster cell lines have been investigated for the presence of 5-mC. Methylations of these sites at the promoter and 5' parts of the integrated adenovirus genes correlated with the inactive states of the genes, whereas the extent of DNA methylation in the 3' parts, the main bodies of these genes, seemed to be functionally less important (19). Frequently, but not always, the early viral genes in the integrated adenovirus type 12 genomes were active, the late genes were inactive (41). Correspondingly, the promoters of each group of these genes in integrated adenovirus type 12 DNA were unmethylated or methylated, respectively. In one instance, cell line HA12/7, one of the early regions, E3, was not expressed, although the gene persisted in an integrated form in this cell line. In the promoter and 5' region of the E3 segment of integrated adenovirus type 12 DNA all 5'-CCGG-3' sites were methylated (19). It is still unknown what the functional meaning of DNA methylations in the 3' main parts of genes could be, if promoter and 5' methylations sufficed to effect gene inactivation. Do methylations in the 3' part of a gene constitute a safeguard or redundancy, do they have a function unrelated to the control of gene expression or are these methylated sites a consequence of the peculiarities of the enzymatic mechanism of methylation? These and other questions cannot yet be answered.

The results quoted in this and the preceding sections were mainly based on restriction enzyme analyses of patterns of DNA methylation. It is important to recall that, using this approach, one would analyze only a subfraction of the total number of CpG dinucleotide pairs in DNA. This dinucleotide is the main, albeit not the only, target of the DNA methyltransferase system. Hence, at present a complete description of the true patterns of DNA methylation including all 5-mC residues in the sequence of an inactivated promoter is not yet available. It is likely that the recently developed genomic sequencing technique (64) will help to alleviate some of these shortcomings.

It is also interesting to recall that the CpG dinucleotide is underrepresented by a factor of four in comparison to statistical expectation in the DNA of eukaryotes (65, 66). In adenovirus DNA, however, the CpG dinucleotide is only slightly underrepresented (67). Superficially, this underrepresentation can be explained by the propensity of 5-mC for deamination and the subsequent conversion of a 5-mCpG dinucleotide to a TpG sequence.

More significantly, the distribution of certain sequences in DNA is the result of long standing functional selection, and in this context it will be meaningful to recognize that the CpG sequence that is relatively rare in eukaryotes, is also very important in long-term gene inactivation. In spite of the relative underrepresentation of CpG dinucleotides, a clustering of this sequence has been recognized in the promoter parts of several eukaryotic genes (23, 27, 68, 69).

Genes *in vitro* methylated at specific sites are inactive upon microinjection or transfection into eukaryotic cells

In order to resolve the problem of whether DNA methylation is the cause or the consequence of gene inactivation, several laboratories have devised experiments in which cloned viral or non-viral eukaryotic genes were *in vitro* methylated using DNA methyltransferases of known sequence specificities. These enzymes were usually of prokaryotic origin. The methylated or the unmethylated gene was subsequently tested for activity after microinjection or transfection into eukaryotic cells. The outcome of many of these experiments demonstrated methylated genes to be inactive and hence supported the notion that sequence-specific DNA methylations led to gene inactivation (cf. 30 - 34).

In designing *in vitro* methylation experiments in the adenovirus system, we soon recognized the absolute necessity to base the designs of these experiments on findings gleaned from investigations on patterns of methylation of adenovirus genes which were permanently fixed in the genomes of virus-transformed cells. In planning one of these experiments the data summarized in Table 2 were considered. The results of analyses on patterns of DNA methylation of the E2A gene in adenovirus type 2-transformed cells suggested that the fourteen 5'-CCGG-3' sequences in this gene might be functionally important. We therefore chose to clone the E2A gene including its late promoter and 5' region into a prokaryotic vector. Subsequently, the E2A region was left unmethylated or was *in vitro* methylated with the prokaryotic HpaII DNA methyltransferase which methylates the internal C of the sequence 5'-CCGG-3'. Complete methylation was ascertained by diagnostic cleavage with the isoschizomeric restriction endonuclease pair HpaII and MspI (70), by electrophoresis, Southern blotting (71) and hybridization to a [^{32}P]-labeled adenovirus type 2 DNA probe followed by autoradiography. It was essential to assure complete methylation prior to the performance of microinjection experiments. The methylated or the unmethylated DNA was then microinjected into the nuclei of Xenopus laevis oocytes. The completely methylated E2A gene was transcriptionally inactive, the unmethylated gene was expressed into E2A-specific messenger RNA which was - at least in part - initiated at the same (late) promoter as in adenovirus type 2-infected human cells (72, 73). Unmethylated histone genes coinjected with the methylated E2A gene were expressed (73). E2A gene methylated at the 5'-GGCC-3' sites which were not methylated in transformed cells, continued to be transcribed after microinjection into the nuclei of Xenopus laevis oocytes (63). These data and those from other systems (for review 30, 33) clearly argued in favor of the concept that site-specific DNA methylations were the cause and not the consequence of gene inactivation.

Methylations of one or a few sequences in the promoter of viral or cellular genes suffice to inactivate these genes

We have recently refined the studies designed to methylate *in vitro* viral genes by restricting methylations to very few sequences, notably in the promoters of viral genes.

We have recently described experiments in which partly methylated clones of the E2A gene were constructed (74). In the promoter (5')-methylated construct, three 5'-CCGG-3' sequences at the 5' end of the subclone were methylated. One of these sites is located 215 base pairs (bp) upstream (bp 26,169 of adenovirus type 2 DNA), and two sites are located 5 and 23 bp downstream from the cap site (bp 25,931 and 25,949 of adenovirus type 2 DNA) of the E2A gene (Fig. 2). The bp numbers refer to the nucleotide sequence of adenovirus type 2 DNA published in this volume (75). The construct was transcriptionally inactive upon microinjection into nuclei of Xenopus laevis oocytes. In the gene (3')-methylated construct, eleven 5'-CCGG-3' sequences in the main part of the transcribed gene region were methylated _in vitro_. This construct was actively transcribed in Xenopus laevis oocytes, and at least a part of the adenovirus type 2-specific RNA synthesized was initiated at the same sites as in adenovirus type 2-infected human cells in culture (74). Mock methylation experiments were carried out similarly to methylation experiments, except that the methyl donor (S-adenosylmethionine) was omitted. Both mock methylated constructs were transcribed into adenovirus type 2-specific RNA in Xenopus laevis oocytes. These results demonstrate that DNA methylations at or close to the

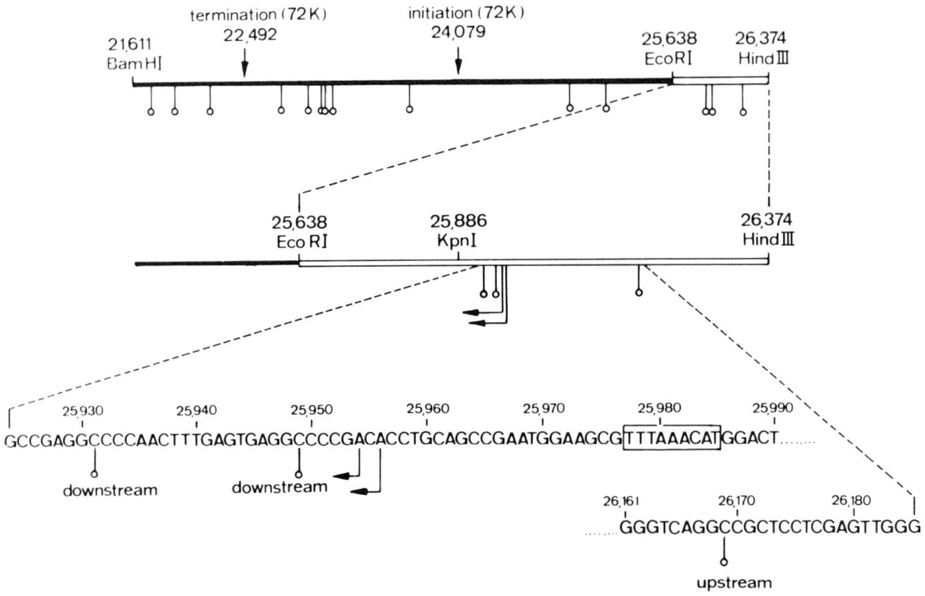

Fig. 2. The structure of the E2A region of adenovirus type 2 DNA. Details of the 5' and promoter region are also shown. Arrows indicate sites of initiation of transcription, ♁ symbols refer to 5'-CCGG-3' sequences, figures designate nucleotide numbers in the viral DNA sequence (75).

promoter and 5' end of the E2A gene cause transcriptional inactivation. Perhaps only one methyl group would be adequate for inactivation. The construct retained its activity when it was grown on an E. coli strain deficient in adenosine methylation (dam$^-$) (76) and subsequently microinjected into Xenopus laevis oocytes (74). These data indicated that adenosine methylation in 5'-GATC-3' sequences did not play a role in the regulation of E2A expression.

The E2A promoter of adenovirus type 2 DNA was also inserted into the pSVO-CAT construct (77) (see below), and the activity of the unmethylated or the methylated DNA was tested after transfection into human HeLa cells. The plasmid construct pSVO-CAT (77), which contained the chloramphenicol acetyl transferase (CAT) gene and a HindIII site immediately in front of it for experimental insertion of eukaryotic promoters, was used to test adenovirus promoter activities in the unmethylated and methylated states. We now observed that the E2A late promoter of adenovirus type 2 DNA also activated the CAT gene upon transfection into mammalian cells, and this E2A promoter was inactivated by specific methylations of three 5'-CCGG-3' sites (78). Similar results had been reported previously after microinjecting promoter-methylated constructs into oocytes of Xenopus laevis (74). Surprisingly, it was observed that the pSVO-CAT construct, which lacks eukaryotic promoter sequences, was able to express the CAT gene upon transfection into human or hamster cells which harbored the El region of adenovirus type 2 or of type 5 (78). In HeLa or BHK21 cells devoid of adenovirus DNA sequences, the pSVO-CAT construct was usually not expressed. The adenovirus type 2-transformed hamster cell lines HE5 and HE7, the adenovirus type 5-transformed hamster cell line BHK297-131 or the adenovirus type 5-transformed human cell line 293 were all capable of expressing the pSVO-CAT construct. We adduced evidence that these latter activities were mediated by prokaryotic promoter-like sequences in the pBR322 section of the construct, and presumably functioned via transactivation by the El region, perhaps by the ElA functions of adenoviruses (78). Low levels of expression of the pSVO-CAT construct in HeLa or BHK21 cells devoid of adenovirus sequences were occasionally found and might be due to cellular functions akin to the E1 functions of adenoviruses. The trans-activating effect was not sensitive to DNA methylation. This finding was commensurate with the notion that trans-activation involved prokaryotic promoter-like sequences which in general were not affected by the methylation of DNA sequences. In cells carrying the left terminal sequences of adenovirus type 2 DNA or of adenovirus type 5 DNA, the expression of the pAd2E2A-CAT construct was enhanced (78).

The effect of DNA methylations at specific promoter sites of adenovirus type 12 DNA on gene expression was also tested by using the plasmid pSVO-CAT. Upon insertion into pSVO-CAT, the ElA or protein IX gene promoter from adenovirus type 12 DNA was capable of mediating CAT expression upon transfection into mouse cells (55). The CAT-promoting activity of the early simian virus 40 promoter in plasmid pSV2-CAT is refractory to methylation by the HpaII or HhaI DNA methyltransferase at 5'-CCGG-3' or 5'-GCGC-3' sequences, respectively, because this promoter lacks such sites. The CAT coding sequence of this plasmid carries three HpaII and no HhaI sites. Methylation of the HpaII sites in the coding region does not affect expression. The ElA promoter of adenovirus type 12 DNA comprising the leftmost 525 base pairs of the viral genome carries two 5'-CCGG-3' and three 5'-GCGC-3' sites immediately upstream from the leftmost TATA signal. Methylation of the HpaII or HhaI sites incapacitates this promoter in mouse cells (55). Similar results were obtained with the ElA promoter in human HeLa and hamster BHK21 cells. Again, methylation inactivated this promoter (D. Knebel, U. Weyer, and W. Doerfler, unpublished results). The promoter of protein IX gene of adenovirus type 12 DNA contains one 5'-CCGG-3' and one 5'-GCGC-3' site downstream and two 5'-GCGC-3' sites > 300 base pairs upstream from the TATA motif, and probably outside the promoter. The protein IX promoter is not inactivated by methylation of these sites (55). These data demonstrate that critical 5' methylations in the promoter region decrease or eliminate transcription; methylations of sites too far upstream and probably sites downstream from this TATA site do not seem to affect expression.

We have also *in vitro* adenine-methylated a single 5-GAATTC-3' (EcoRI) site upstream from the E1A promoter of the pAd12E1A-CAT construct using the EcoRI DNA methyltransferase from Escherichia coli. The *in vitro* methylation of this single site led to a very striking reduction in CAT activity, and this reduction was nearly as significant as upon HpaII methylation of the same construct. These data suggest that modifications of bases other than cytosine in the promoter or 5' region of a gene can also lead to gene inactivation (79).

The complete methylation of CpG sequences in the 5' region of the γ-globin gene, between nucleotides -760 to +100, eliminates transcription of this gene (80). There are 11 CpG dinucleotides in this stretch of 860 nucleotides. It has not yet been elucidated which of these nucleotides are the decisive ones in shutting off gene expression. The authors have used the *in vitro* methylation procedure based on replicating a gene, in this case the γ-globin gene, in the single-stranded vector M13 in the presence of 5-methyl dCTP (81). By choosing the appropriate primer fragment for this *in vitro* replication reaction, specific segments of the gene and its flanking sequences can be methylated. In this way, the cytidine residues in a certain segment will be methylated; this method will therefore not permit identification of the functionally significant nucleotides. The cloned γ-globin gene thus hemimethylated in various regions was then cotransfected into Ltk^- mouse cells, an established mouse cell line, with the thymidine kinase gene of herpes simplex virus. Cell clones containing both the γ-globin and thymidine kinase genes were selected. It was found that the CpG methylation pattern introduced *in vitro* was maintained in both strands of the gene and was faithfully inherited in this system as previously reported for other genes (80, 81). Methylation in the region from -760 in the 5' flanking region to +100 in the γ-globin gene suppressed globin expression, methylation of C residues from +100 in the γ-globin gene to +1950 in the 3' flanking region did not affect globin gene transcription. These data are in excellent agreement with those described in the adenovirus system.

The results discussed attest to the methylation sensitivity of two adenovirus promoters and of the γ-globin promoter. The data on the viral systems demonstrated that methylation had to occur at specific sites to elicit gene inactivation. One of the viral promoters (E2A late of adenovirus type 2) was tested in amphibian oocytes and in mammalian cells; sensitivity toward methylation was the same in both vertebrate systems. Gene activity and its shut-down by site-specific promoter methylations were investigated with the viral genes in the extrachromosomal state or with the globin gene integrated into the recipient cell DNA. Work is presently in progress in this laboratory to fix methylated or unmethylated adenovirus promoter constructs in the genome of the recipient cell and to follow stability of the state of methylation and the expression of genes governed by these promoters. In conjunction with the bulk of evidence derived from correlative experiments, the data presented here make a strong case for the regulatory function of site-specific promoter methylations in higher eukaryotes.

<u>An insect virus promoter is also sensitive to site-specific DNA methylation in insect cells</u>

It has been documented by many investigators that Drosophila DNA is devoid of 5-mC (12, 46, 47) or contains only minute amounts of the modified nucleotide (49). It has also been shown recently that the DNA of the insect cell line (Spodoptera frugiperda), that serves as host for the propagation of Autographa californica Nuclear Polyhedrosis Virus (AcNPV), does not contain detectable amounts of methylated nucleotides (48, 82). Similarly, methylated nucleotides cannot be detected in AcNPV DNA (12, 43). One could, therefore, reason that 5-mC as a regulatory signal may be a late addition in phylogeny to the arsenal of regulatory functions. Although DNA methylation might not play a major role in insect cells, it was nevertheless conceivable that the expression machinery in insect cells was sensitive to site-specific methylations. We investigated this possibility by using a conspicuously strong insect virus promoter in such studies, viz. the p10 gene promoter of AcNPV DNA (45).

In lepidopteran insect cells infected with the baculovirus AcNPV, two major late viral gene products are expressed: The polyhedrin which makes up the mass of the nuclear inclusion bodies, and a 10,000 molecular weight protein (p10) whose function is unknown. The structures of the promoters of these genes have been determined (45, 50, 83). These structures are similar to those of other eukaroytic promoters, except that the nucleotide sequences of these promoters are rich in adenine-thymine base pairs.

We have used the pSVO-CAT construct (77) containing the prokaryotic gene chloramphenicol acetyl transferase (CAT) to study the function of the p10 gene promoter in insect and mammalian cells. Upon transfection of the pAcp10-CAT construct, which contains the p10 gene promoter of AcNPV DNA in the HindIII site of pSVO-CAT, CAT activity was determined. The p10 gene promoter is inactive in human HeLa cells. The same promoter has proved active, however, in AcNPV-infected Spodoptera frugiperda cells and exhibits optimal activity when cells are transfected 18 h after infection (82). This finding demonstrates directly that the p10 gene promoter requires additional viral gene products for its activity in Spodoptera frugiperda cells. The nature of these products is unknown at present.

The p10 gene promoter sequence contains one 5'-CCGG-3' (HpaII) site 40 bp upstream from the cap site of the gene and two such sites far downstream in the coding sequence of the gene. In view of the fact that Drosophila DNA has been reported to contain no 5-mC or extremely small amounts of it, we were interested in determining the effect of site-specific methylation on the insect virus promoter of the p10 gene. Methylation at the 5'-CCGG-3' site led to inactivation of the promoter or to a drastic reduction in its activity (82). These data show that an AcNPV insect virus promoter can be inactivated by site-specific methylation even in insect cells.

The data presented are consistent with the idea that insect cells have the capacity to recognize methylated nucleotides in specific promoter sequences and respond to this signal in a similar way as vertebrate systems do. This finding has been surprising in view of the fact that 5-mC is not a nucleotide modification that abounds in insect cell DNA. We conclude that 5-mC as a regulatory signal, although apparently not extensively used in insect cells, can nevertheless be recognized by the transcriptional machinery in insect cells. This signal might in fact be used in crucial positions in the insect cell genome.

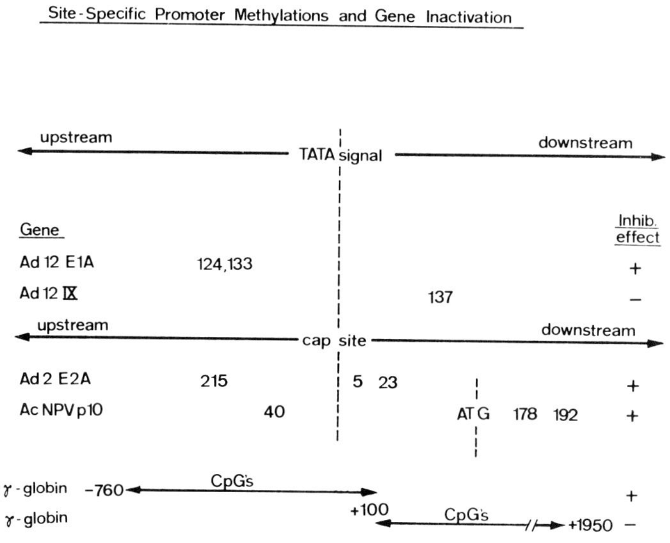

Table 3. The data on promoter inactivation by site-specific DNA methylations are summarized. The data are taken from references 55 (Ad12-E1A, Ad12-IX), 74 (Ad2-E2A), 82 (AcNPVp10), and from 80 (γ-globin).

Summary of in vitro methylation experiments modifying specific sites in viral promoters

The data summarized in Table 3 indicate that promoter sites upstream from the TATA or cap signal are the most sensitive ones with respect to DNA methylation. As stated previously, we do not understand yet the biochemical mechanism by which site-specific promoter methylations lead to gene inactivation.

DNA methylation as a long-term inactivating signal is suitable only for certain genes

Among the extensive literature on DNA methylation in eukaryotes and its biological functions, there are reports claiming that some eukaryotic genes are not methylation sensitive. Obviously, one mechanism for the shut-off of certain genes does not necessarily have to be suitable for the regulation of expression in all genes. Since DNA methylation can serve as a long-term signal in the inactivation of genes, it is not likely that this signal should be selected in the transient inactivation of gene functions. It will be useful in this context to operationally subdivide genes into at least three classes:

1. Genes that become permanently inactivated. In this class of genes, DNA methylation can play an important role.
2. Genes that are intermittently active, but inactive for long periods, or genes that can be reactivated, e.g. by hormone treatment. These genes will probably not be inactivated by site-specific promoter methylations, since it would then be difficult to instantaneously reactivate them.
3. Active genes. The promoters of these genes are unmethylated in most cases.

Of course, this schematic subdivision of gene classes does not take into account the possibility that the sites of promoter methylation leading to gene inactivation may have to be highly specific. Thus, the occurrence of 5-mC in a promoter sequence may not *per se* be associated with the inactive state of a gene. Moreover, this simple classification presupposes that rapid alterations of levels of DNA methylation may be difficult to achieve.

Shifts in the patterns of DNA methylation

It is still unknown how highly specific patterns of DNA methylation are generated in eukaryotic cells. Such elaborate patterns may have functional significance in more than one respect. From the adenovirus system, we should like to describe briefly a set of data which document the generation of complicated patterns. In adenovirus type 12-induced hamster tumors varying amounts of adenovirus DNA become fixed in the cellular genome by integration (21). These patterns of viral DNA integration have been analyzed in great detail. In these tumor cells only few viral genes are being expressed as messenger RNA. It has, therefore, been surprising to find that the integrated adenovirus DNA is strikingly undermethylated at 5'-CCGG-3' and 5'-GCGC-3' sites (20 - 22). These findings therefore demonstrate that the absence of DNA methylation is a necessary but not a sufficient precondition for gene expression.

Upon subcultivation of cells from some of the tumors, the viral genomes were eliminated, apparently in a stepwise process, with segments of the left termini of adenovirus type 12 DNAs persisting the longest. Morphological variants of these tumor cells lost all viral DNA and yet retained the oncogenic phenotype (21). All 13 independently isolated clones from one revertant were devoid of adenovirus type 12 DNA. The extent of viral DNA methylation was minimal in adenovirus type 12-induced tumors, although the viral genome was not extensively expressed, if at all. Upon passage in culture, however, the levels of viral DNA methylation increased. It was interesting that establishment of the final methylation patterns of integrated adenovirus type 12 DNA required many cell generations after the fixation of foreign DNA in the host genome. The shift in methylation was nonrandom. The late parts of the inserted viral genomes became methylated more extensively than did early segments (22).

Studies on DNA methyltransferases in uninfected and adenovirus type 2-infected human cells

We have initiated investigations on DNA methyltransferase activities in mammalian cells. These enzymes and the regulation of their activities must play a crucial role in determining specific DNA methylation patterns and thus regulate genetic activities. In the course of studies on the nature of DNA methyltransferases in uninfected or adenovirus type 2-infected human cells (partly reported in ref. 23), enzymatic activities were detected that methylated adenovirus type 2 DNA *de novo* (24). The activities in uninfected and adenovirus type 2-infected human cells were approximately equal. We have purified DNA methyltransferase activities also from calf thymus. Much more extensive work on these interesting enzymes will be required before one can hope to grasp the significance of interrelationships between DNA methylation and the regulation of a number of important genetic activities.

Current work involving the adenovirus system

As will be apparent from the previously reported findings and our attempts to formulate general concepts about the biological significance of DNA methylation, a great deal of decisive experimentation remains to be done. Our current work is directed toward the following problems.

i) The biochemical mechanism by which specific promoter methylations modulate gene expression is not known. We have started to use an *in vitro* transcription system prepared from HeLa cell extracts similar to the one developed by Manley and colleagues (3) to test the activities of the E1A and E2A promoters. The E1A or the E2A promoter was used in the unmethylated or HpaII methylated form. Preliminary results (P. Dobrzanski and W. Doerfler, in preparation) indicate that under certain conditions both promoters methylated at 5'-CCGG-3' sequences were inactivated. This system might open the way to a careful biochemical analysis of the effect of promoter methylations.

ii) Furthermore, we are trying to narrow down the decisive sequence in a promoter which is responsible for shut-off upon methylation. In different experiments, one or two of the three 5'-CCGG-3' sequences within the E2A promoter region (Fig. 3) will be modified by site directed mutagenesis to convert these sites to render them insensitive to the HpaII DNA methyltransferase. In the modified promoter construct only one sequence could then be methylated by this DNA methyltransferase. Upon coupling the thus modified promoter to the CAT gene, the influence of methylation of a single 5'-CCGG-3' site could then be assessed.

iii) Similarly, we are attempting to generate an E2A promoter sequence that is hemimethylated in one or the other DNA strand. It is conceivable but unknown at present whether methylation of only one strand will be sufficient to accomplish complete shut-off of promoter activity. The CAT gene will again be used as the indicator. Since hemimethylated DNA is thought to arise at least transiently during DNA replication, it will be of general importance to clarify the issue of genetic activities in hemimethylated promoter sequences.

iv) When restriction endonucleases and Southern blot hybridization techniques are used to determine patterns of DNA methylation, e.g. in specific promoter sequences of integrated adenovirus DNA, only a limited subset of all the possible 5'-CG-3' dinucleotide combinations can be scanned for the presence of 5-mC (14). In order to determine complete patterns of methylation of active or inactive promoters of integrated adenovirus promoters the technique of genomic sequencing (64) will have to be adapted.

v) Lastly, experiments have been initiated to fix permanently in the host genome constructs which consist of unmethylated or methylated adenovirus promoters and the CAT gene. Subsequently, the levels of expression of these fixed constructs will have to be assayed. Simultaneously, it will be assessed whether the extents of experimentally imposed promoter methylations will be maintained after integration of DNA constructs introduced into mammalian cells by DNA transfection.

The questions raised are only some among a number of problems that need to be solved before we can understand in detail the mechanisms by which sequence specific promoter methylations can influence gene expression. It is obvious today that DNA methylation is an important signal in eukaryotic cells. Gene expression may be only one of several processes being modulated by this signal. Moreover, via gene expression, specific DNA methylations are thought to affect complicated biological phenomena such as differentiation, oncogenesis, mutagenesis and perhaps the immune response.

ACKNOWLEDGMENTS

We are indebted to Petra Böhm and Gertrud Deutschländer for expert editorial work. Research in the authors' laboratory was supported by the Deutsche Forschungsgemeinschaft through SFB74-C1, by the Ministry of Science and Research of the State North-rhine-Westfalia (IV B5 FA 9227) and by a donation of the Hoechst Comp.).

REFERENCES

1. Berget, S.M., Moore, C. and Sharp, P.A. Proc. Natl. Acad. Sci. USA 74: 3171-3175, 1977.
2. Aloni, Y., Dhar, R., Laub, O., Horowitz, M. and Khoury, G. Proc. Natl. Acad. Sci. USA 74: 3686-3690, 1977.
3. Manley, J.L., Fire, A., Cano, A., Sharp, P.A. and Gefter, M.L. Proc. Natl. Acad. Sci. USA 77: 3855-3859, 1980.
4. Hernandez, N. and Keller, W. Cell 35: 89-99, 1983.
5. The Molecular Biology of Adenoviruses 1 - 3, ed. W. Doerfler. In: Current Topics Microbiol. Immunol. 109 (1983), 110 (1984), 111 (1984).
6. Doerfler, W. Proc. Natl. Acad. Sci. USA 60: 636-643, 1968.
7. Doerfler, W. J. Virol. 6: 652-666, 1970.
8. Doerfler, W. Current Topics Microbiol. Immunol. 101: 127-194, 1982.
9. Doerfler, W., Gahlmann, R., Stabel, S., Deuring, R., Lichtenberg, U., Schulz, M., Eick, D. and Leisten, R. Current Topics Microbiol. Immunol. 109: 193-228, 1983.
10. Günthert, U., Schweiger, M., Stupp, M. and Doerfler, W. Proc. Natl. Acad. Sci. USA 73: 3923-3927, 1976.
11. Von Acken, U., Simon, D., Grunert, F., Döring, H.A. and Kröger, H. Virology 99: 152-157, 1979.
12. Eick, D., Fritz, H.-J. and Doerfler, W. Anal. Biochem. 135: 165-171, 1983.
13. Vardimon, L., Neumann, R., Kuhlmann, I., Sutter, D. and Doerfler, W. Nucleic Acids Res. 8: 2461-2473, 1980.
14. Wienhues, U. and Doerfler, W. J. Virol. 57: 000-000, 1985.
15. Deuring, R., Klotz, G. and Doerfler, W. Proc. Natl. Acad. Sci. USA 78: 3142-3146, 1981.
16. Deuring, R. and Doerfler, W. Gene 26: 283-289, 1983.
17. Sutter, D. and Doerfler, W. Cold Spring Harbor Symp. Quant. Biol. 44: 565-568, 1979.
18. Sutter, D. and Doerfler, W. Proc. Natl. Acad. Sci. USA 77: 253-256, 1980.

19. Kruczek, I. and Doerfler, W. EMBO J. 1: 409-414, 1982.
20. Kuhlmann, I. and Doerfler, W. Virology 118: 169-180, 1982.
21. Kuhlmann, I., Achten, S., Rudolph, R. and Doerfler, W. EMBO J. 1: 79-86, 1982.
22. Kuhlmann, I. and Doerfler, W. J. Virol. 47: 631-636, 1983.
23. Doerfler, W., Kruczek, I., Eick, D., Vardimon, L. and Kron, B. Cold Spring Harbor Symp. Quant. Biol. 47: 593-603, 1982.
24. Knust-Kron, B., Hübscher, U., Eick, D., Spadari, S. and Doerfler, W., submitted, 1985.
25. Hübscher, U., Pedrali-Nay, G., Knust-Kron, B., Doerfler, W. and Spadari, S. Anal. Biochem., submitted, 1985.
26. Razin, A. and Riggs, A.D. Science 210: 604-610, 1980.
27. Doerfler, W. J. gen. Virol. 57: 1-20, 1981.
28. Ehrlich, M. and Wang, R.Y.-H. Science 212: 1350-1357, 1981.
29. Razin, A. and Friedmann, J. Progr. Nucleic Acids Res. Mol. Biol. 25: 33-52, 1981.
30. Doerfler, W. Ann. Rev. Biochem. 52: 93-124, 1983.
31. Cooper, D.N. Human Genetics 64: 315-333, 1983.
32. Riggs, A.D. and Jones, P.A. Adv. Cancer Res. 40: 1-30, 1983.
33. Doerfler, W. Adv. Viral Oncol. 4: 217-247, 1984.
34. Doerfler, W. Current Topics Microbiol. Immunol. 108: 79-98, 1984.
35. Trautner, T.A. (ed.) Methylation of DNA. Springer, Berlin/Heidelberg/New York/Tokyo, 1984.
36. Taylor, J.H. DNA Methylation and Cellular Differentiation. Springer, Wien/New York, 1984.
37. Razin, A., Cedar, H., Riggs, A.D. (eds.) DNA Methylation. Springer, New York, 1984.
38. Hattman, S. DNA Methylation. In: The Enzymes (Ed. P.D. Boyer), Academic Press, New York, 1984.
39. Sutter, S., Westphal, M. and Doerfler, W. Cell 14: 569-585, 1978.
40. Stabel, S., Doerfler, W. and Friis, R.R. J. Virol. 36: 22-40, 1980.
41. Ortin, J., Scheidtmann, K.-H., Greenberg, R., Westphal, M. and Doerfler, W. J. Virol. 20: 355-372, 1976.
42. Carstens, E.B., Tjia, S.T. and Doerfler, W. Virology 99: 386-398, 1979.
43. Tjia, S.T., Carstens, E.B. and Doerfler, W. Virology 99: 399-409, 1979.
44. Lübbert, H. and Doerfler, W. J. Virol. 50: 497-506, 1984.
45. Lübbert, H. and Doerfler, W. J. Virol. 52: 255-265, 1984.
46. Smith, S.S. and Thomas, C.A., Jr. Gene 13: 395-408, 1981.
47. Urieli-Shoval, S., Gruenbaum, Y., Sedat, J. and Razin, A. FEBS Letters 146: 148-152, 1982.
48. Müller, T. and Doerfler, W., unpublished results, 1985.
49. Achwal, C.W., Ganguly, P. and Chandra, H.S. EMBO J. 3: 263-266, 1984.
50. Hooft van Iddekinge, B.J.L., Smith, G.E. and Summers, M.D. Virology 131: 561-565, 1983.
51. Rohel, D.Z., Cochran, M.A. and Faulkner, P. Virology 124: 357-365, 1983.
52. The Molecular Biology of Baculoviruses. (ed.) W. Doerfler. In: Current Topics Microbiol. Immunol., in preparation.

53. Kaye, A.M. and Winocour, E. J. Mol. Biol. 24: 475-478, 1967.
54. Willis, D.B. and Granoff, A. Virology 107: 250-257, 1980.
55. Kruczek, I. and Doerfler, W. Proc. Natl. Acad. Sci. USA 80: 7586-7590, 1983.
56. Doerfler, W. Virology 38: 587-606, 1969.
57. Vardimon, L. and Doerfler, W. J. Mol. Biol. 147: 227-246, 1981.
58. Cook, J.L. and Lewis, A.M., Jr. Cancer Res. 39: 1455-1461, 1979.
59. Vardimon, L., Renz, D. and Doerfler, W. Recent Results Cancer Res. 84: 90-102, 1983.
60. Esche, H. J. Virol. 41: 1076-1082, 1982.
61. Johansson, K., Persson, H., Lewis, A.M., Pettersson, U., Tibbetts, C. and Philipson, L. J. Virol. 27: 628-639, 1978.
62. Van der Vliet, P.C. and Levine, A.J. Nature New Biol. 246: 170-174, 1973.
63. Vardimon, L., Günthert, U. and Doerfler, W. Mol. Cell. Biol. 2: 1574-1580, 1982.
64. Church, G.M. and Gilbert, W. Proc. Natl. Acad. Sci. USA 81: 1991-1995, 1984.
65. Josse, J., Kaiser, A.D. and Kornberg, A. J. Biol. Chem. 236: 864-875, 1961.
66. Subak-Sharpe, H., Burk, R.R., Crawford, L.V., Morrison, J.M., Hay, J. and Keir, H.M. Cold Spring Harbor Symp. Quant. Biol. 31: 737-748, 1966.
67. Morrison, J.M., Keir, H.M., Subak-Sharpe, H. and Crawford, L.V. J. gen. Virol. 1: 101-108, 1967.
68. Felsenfeld, G., Nickol, J., Behe, M., McGhee, J. and Jackson, D. Cold Spring Harbor Symp. Quant. Biol. 47: 577-584, 1982.
69. Tykocinski, M.L. and Max, E.E. Nucleic Acids Res. 12: 4385-4396, 1984.
70. Waalwijk, C. and Flavell, R.A. Nucleic Acids Res. 5: 3231-3236, 1978.
71. Southern, E.M. J. Mol. Biol. 98: 503-517, 1975.
72. Vardimon, L., Kuhlmann, I., Cedar, H., and Doerfler, W. Eur. J. Cell Biol. 25: 13-15, 1981.
73. Vardimon, L., Kressmann, A., Cedar, H., Maechler, H. and Doerfler, W. Proc. Natl. Acad. Sci. USA 79: 1073-1077, 1982.
74. Langner, K.-D., Vardimon, L., Renz, D. and Doerfler, W. Proc. Natl. Acad. Sci. USA 81: 2950-2954, 1984.
75. Roberts, R.J., Akusjärvi, G., Aleström, P., Gelinas, R.E., Gingeras, T.R., Sciaky, D. and Pettersson, U. Adenovirus DNA: The Viral Genome and its Expression, ed. W. Doerfler. In: Developments in Molecular Virology, Martinus Nijhoff Publishers, Boston, 1985.
76. Marinus, M.G., Morris, N.R. J. Bacteriol. 114: 1143-1150, 1973.
77. Gorman, C.M., Moffat, L.F. and Howard, B.H. Mol. Cell. Biol. 2: 1044-1051, 1982.
78. Langner, K.-D., Weyer, U. and Doerfler, W. Proc. Natl. Acad. Sci. USA 82: 0000-0000, 1985.
79. Knebel, D. and Doerfler, W. J. Virol., submitted, 1985.

80. Busslinger, M., Hurst, J. and Flavell, R.A. Cell 34: 197-206, 1983.
81. Stein, R., Gruenbaum, Y., Pollak, Y., Razin, A. and Cedar, H. Proc. Natl. Acad. Sci. USA 79:61-65, 1982.
82. Knebel, D., Lübbert, H. and Doerfler, W. EMBO J. 4: 000-000, 1985.
83. Kuzio, J., Rohel, D.Z., Curry, C.J., Krebs, A., Carstens, E.B. and Faulkner, P. Virology, 139: 414-418, 1984.

7

EARLY PROTEINS CODED BY THE ADENOVIRUS GENOME: FUNCTIONAL ASPECTS

H. ESCHE

Institute of Molecular Biology (Cancer Research), University of Essen, D-4300 Essen 1, Germany

1. INTRODUCTION

The adenovirus early proteins are defined as those virus-coded proteins that are synthesized prior to and in the absence of viral DNA replication. The functions of these early viral proteins provide new insights into regulatory mechanisms of gene expression and DNA replication in higher eukaryotic cells. At present, there are at least 25 early adenovirus proteins identified so far. Some of these early adenovirus gene products regulate transcription in both a positive (1,2,3) and a negative (4,5,6) mode. Other early proteins may be involved in RNA processing (7,8), while at least three early proteins are required for DNA replication (9,10,11,12). A subset of the early viral proteins are both necessary and sufficient for virus-induced cellular transformation (13,14,15,16). Other early gene products can affect the frequency of transformation (17,18,19). Finally, there are indications that some of the early viral functions may influence the host range of the human adenoviruses (7).

Today, the adenovirus early gene products are classified into at least three groups on the basis of their temporal appearance and regulation during virus infection: (A) The 'immediate early' genes encode the E1A proteins, which are synthesized shortly after infection and are required for expression of the remaining early mRNAs. (B) The 'delayed early' gene blocks (regions E1B, E2 (A and B), E3, E4) encode most of the conventionally recognized early proteins and are dependent on functions of the E1A gene products (1,2). (C) Several viral genes, which are predominantly transcribed at late times, after the onset of DNA replication, are also expressed at early times (IVa2, IX, L1, late leader). Investigations of the early viral gene expression after inhibition of macromolecular synthesis during lytic infection by different drugs have indicated a more complex regulation of the 'delayed early' and 'late' genes expressed at early times (20).

Several structural and functional data concerning adenovirus early gene products have been collected in the past years, however, still little is known about some of these early proteins. A better characterization of their functions may lead to a clearer understanding of the biology of these viruses in their natural host.

This review of the adenovirus early proteins will focus upon the structure and function of these proteins, in particular their roles in gene regulation, DNA replication and modulation of virus host cell interaction.

2. METHODS FOR DETECTION OF ADENOVIRUS PROTEINS

The identification, purification and functional analysis of most of the early adenovirus proteins is difficult, mainly because they are synthesized at relatively low levels at early times after infection, when host cell proteins are synthesized at high levels. Several approaches have been used to identify adenovirus proteins. The E2A DNA-binding protein of Ad2 (21,22) and of Ad12 (23) has been purified from lysates of infected cells. A comparison of proteins prepared from infected and uninfected cells by two-dimensional gel electrophoresis has also led to the identification of a number of virus proteins, most of them expressed late in viral infection (for review, see 24). A more specific assignment of the map positions of the adenovirus gene products was obtained by selecting mRNA with genomic DNA restriction fragments, followed by in vitro translation of these mRNAs (25,26,27). This approach has the advantage of establishing the viral origin of the proteins (by mRNA hybrid selection) and assigning map positions to individual genes on the viral DNA. Finally, serum from animals bearing adenovirus-induced tumors has provided a source of antibodies directed against several adenovirus proteins (28,29,30). The latter approach, however, is dependent on the quality of the sera obtained. More recently, monoclonal antibodies and antisera directed against synthetic peptides corresponding to specific parts of the viral proteins have been prepared against several adenovirus proteins (31,32,33, 34). The experimental procedures outlined above have been used collectively to identify and map adenovirus proteins. The functions of some of these proteins have been deduced either by using mutants that alter the function of a gene product (2,17,18) or by developing functional assays, such as in vitro DNA replication (35,36) or in vitro transcription (37,38,39).

3. IDENTIFICATION, STRUCTURE AND FUNCTIONS

Using the methods described in the previous section, at least 25 adenovirus early proteins have been detected so far in cells productively infected with adenovirus types 2, 5 or 12. Six regions of the viral genome are transcribed during the early

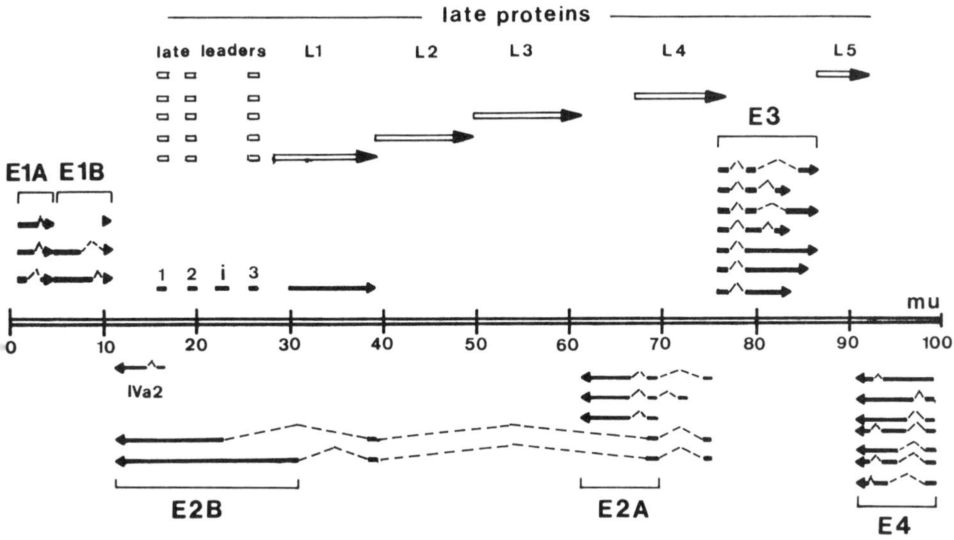

Figure 1. Tanscription map of the Ad2 genome. Early viral mRNAs are shown as dark arrows. Arrowheads indicate 3' ends and dotted lines show spliced-out introns. Each of the late gene blocks (L1-L5) consits of several different spliced mRNAs.

phase: early regions 1A (E1A), E1B, E2 (A and B), E3, E4 and late region 1 (L1) (see Figure 1 and Table 1). Regions E1, E3 and L1 are transcribed from the r-strand, whereas the coding sequences for regions E2 and E4 are located on the l-strand. Each transcription unit generates a complex spectrum of spliced mRNAs containing overlapping sequences from a single promoter.

Table 1. The adenovirus type 2 early proteins

Region	Coordinates (map units)	Transcripts (5'3')(size)	Proteins (MW)
E1A	1.3-4.6	r, 13S r, 12S r, 9S	56K, 46K 47K, 40K 28K
E1B	4.6-11.2	r, 22S r, 13S	58K, 19K 19K
E2A	61.7-66.6	l	72K
E2B	11-31	l	87K, 140K
E3	76.6-86	r	gp19K, 13K, 14K
E4	91.3-99	l	11K, 14K 25K, (34K)
Late genes expressed to some extent at early times			
IVa2	11.2-16	l	54K
L1	30.6-39	r	52K, 55K
late leader	17-22	r	13.5K, 13.6K
IX	9.8-11.2	r, 9S	12.5K

As a consequence, proteins translated from these mRNAs may also show partial homology. Many early regions of the adenovirus DNA indeed specify proteins that are related to one another (40, 41,42).

The structure and functions of the early adenovirus gene products will now be reviewed.

3.1 Region E1 Proteins: Lytic and Transforming Functions

The E1 region of adenoviruses is composed of two distinct transcription units (43,44,45) and two genetic complementation groups (2,46,47,48). Mutants obtained in both genes (2,46) are defective for virus growth and are propagated in line 293 human cells (49), which are transformed by adenovirus and express the E1A and E1B functions. These E1A and E1B mutants also fail to transform cells in culture (49,50,51), indicating that the functions of both groups are necessary for transformation. For clarity these proteins and their possible functions will be considered seperately.

3.1.1 Region E1A Proteins. The E1A transcription unit (1.3-4.6 map units) encodes at least three mRNA species, which have common 5' and 3' ends but differ in their splicing patterns (45,51). The three mRNAs (designated 13S, 12S and 9S) contain coding regions which should give rise to 31.8Kd, 26.5Kd and 13.3Kd polypeptides, respectively (52,53). However, it was found that the 13S and 12S mRNAs each produce two to three polypeptides in vivo with apparent molecular weights of 4oK-58K (41,54,55). The 9S mRNA encodes a 28K protein (41,56). All proteins derived from the three E1A mRNAs are translated in the same reading frame (57). Thus, the mRNAs of region E1A produce sets of related proteins which share common peptides (41,42). The reasons for the heterogeneity of the E1A proteins are not yet clear. Possibly a single primary translation product is produced from each E1A mRNA species and then some of these proteins are modified specifically by a proteolytic event and phosphorylation. That the molecular weights of the E1A proteins calculated from the known nucleotide sequence are all lower than those of the corresponding gene products detected by electrophoresis in SDS-polyacrylamide gels might be due to the high proline and glutamic acid content of these polypeptides. It can cause alterations in the migration of proteins in this gel system. The E1A proteins have relatively short half-lifes (about 2 hr; 3,58) and are localized in both nucleus and cytoplasm of the infected cell.

The functions of the E1A proteins have been elucidated by

analysis of point and deletion mutants in region 1A (2,8,46,59). Viruses carrying mutations within E1A do not transcribe the other early regions (E1B, E2A, E2B, E3, E4, L1) of the genome (50,58,60). However, there are indications that the E1A products are not absolutely essential for the expression of additional viral transcription units. Nearly normal virus yields are produced when KB or Hela cells are infected at very high multiplicities with the deletion mutant dl312 of Ad5 (e.g. 800 plaque forming units per cell; 2) and additional transcription units are turned on slowly after long periods of time, even when cells are infected at relatively low multiplicities (58,61). Experiments with a host range mutant designated hr440 (8,62), which carries several base pair changes in E1A, as well as with inhibitors of protein synthesis during infection (58,63,64) have further indicated that not all of the early transcription units are equally dependent upon the E1A proteins. Regions E2 and E3 seem to be more tightly regulated by E1A proteins than E1B and E4 transcripts.

The regulatory gene product is encoded by the 13S mRNA species, which can be concluded from at least two observations: First, mutants hr1 and in500 (65), which exhibit the regulatory defect, carry lesions within the region unique to the 13S mRNA. Second, a mutant (pm975) with a defective splice donor site in the 12S mRNA expresses only the 13S mRNA species and grows normally in Hela cells (66). The 9S mRNA products seem to be also nonessential, since the mutant dl312 grows well on human 293 cells, which do not synthesize detectable levels of this mRNA.

The mechanism mediating the E1A positive regulation of transcription is still unclear. Interaction with RNA polymerase II or DNA nucleotide sequences, counteraction of a cellular repressor bound to the viral genome (67) or post-transcriptional effects (stabilisation of viral mRNA; 68) are possibilities. However, it is likely that a variety of cellular genes are also regulated by the E1A functions. It has been recently shown that infection of Hela cells with wild-type Ad5, but not with the mutant dl312, induces synthesis of a cellular 70K heat shock protein (58).

3.1.2 Region E1B Proteins. The E1B transcription unit (4.6-11.3 map units) encodes two mRNAs (22S and 13S) early after infection. The 22S mRNA directs the synthesis of two major polypeptides, a 58Kd and 19Kd protein (29,69,70), which are translated from different, but overlapping reading frames (71). The 13S mRNA specifies only the 19K protein (71). In addition, region E1B produces a number of smaller polypeptides that seem to be related to the 58K protein (56,72). The 58K protein is located in the nucleus and the cytoplasm (34,55), while the 19K polypeptide is found in the plasma membrane of infected cells (73).

As in the case of E1A, point and deletion mutants have been imployed to study the E1B functions (46,49,60,74,75). A large deletion mutant, dl313, produces no 58K and 19K proteins (2,41, 56). This deletion mutant lacks the C-terminus of the E1A proteins, the entire E1B 19K and 58K coding regions and the N-terminus of the protein IX (75). The mutant dl313 complements dl312 (effecting only region E1A), which suggests that it expresses functional E1A products. Thus the phenotype of dl313 should result from its inability to produce E1B products. This mutant produces normal levels of the E3 and E4 proteins but reduced levels of the E2A 72K protein at low multiplicities of infection (48). The mutant dl313 cannot synthesize viral DNA in Hela cells (75). The latter observation, however, cannot be unambiguously assigned to the function of E1B, since E1A polypeptides were also altered. The point mutants hr6 and hr7 (46,49), whose defects are localized within the 58K protein coding region, do not synthesize the 58K polypeptide and a reduced level of the 19K protein (48). In addition, hr7 exhibits a DNA replication defect (36). Viral DNA replication does occur in hr7-infected Hela cells but its onset is delayed by about 10 hr (36). Most other 58K-specific mutants synthesize DNA normally.

At least part of the E1B 58K protein has been found in association with a cellular phosphoprotein termed p53K in Ad5-transformed mouse, rat, hamster and human 293 cells (76).

The cellular p53K protein has previously been shown to be associated with the SV40 large T antigen (77,78). In lytically infected cells, however, the Ad5 E1B 58K protein is not associated with this cellular p53K protein. Instead, the 58K protein forms a complex with a virus-coded E4 25K polypeptide (79). That the same cellular protein is associated with the SV40 large T antigen and the large E1B tumor antigen of Ad5 in their respective transformed cells indicates that the two viral proteins may share some common functions. The p53K cellular protein occurs in increased amounts in a variety of transformed cells compared to their nontransformed counterparts (77,80,81,82). So, the elevated concentration of the p53K protein may be significant for the transformed state. The p53K protein has a very short half-life (about 20 min) in nontransformed cells, whilc it has a long half-life (about 24 hr) when associated with large T antigen in SV40-transformed cells (83). This observation has let to speculations about the physical association of the SV40 large T antigen or E1B 58K protein with the p53K protein which seems to stabilize this cellular protein by reducing its turnover or rate of degradation (83). There are some indications that the cellular p53K protein and the SV40 large T antigen are involved in regulating events in the cell cycle (84,85,86) which results in the cellular DNA replication associated with transformation.

Relatively little is known about the function of the E1B 19K protein during productive infection. Mutants exclusively effecting the E1B 19K gene have recently been isolated (87). Their phenotype is presently under investigation.

3.1.3 Region E1 and Transformation. Two lines of evidence suggest that the functions necessary and sufficient for transformation are encoded by regions E1A and E1B: First, transfection of rodent cells in vitro with viral DNA fragments has shown that a fragment carrying region E1 alone is sufficient to induce neoplastic transformation of primary cells (13). A left-end fragment, about 8% of the total E1 region, induces apparently complete transformation of primary rat kidney cells, while a smaller 4.5% fragment from the extreme left leads to an only partially transformed phenotype (15,16,88,89,90). The latter cells are immortalized but neither they grow under soft agar nor produce tumors in animals. Second, host range mutants with mutations mapping to either E1A or E1B are transformation defective (46,49,62,91,92). As both mutant groups are defective in neoplastic transformation, both genes appear to be involved (49). Since at least one of the products of the region E1A is necessary for the activation of the other early transcription units including the E1B region, the main role of region E1A in transformation could be to enhance the expression of region E1B. However, this possibility appears to be ruled out by the discovery that an Ad5 host range mutant (hr440), defective in the expression of E1A but normal in the transcription of E1B, does not transform baby rat kidney cells (62). This indicates that the expression of E1B alone is not sufficient for neoplastic transformation and supports the view that the E1A expression may have its specific function in the transforming process. The analysis of mutants carrying mutations within the E1B coding region has shown that the function of the E1B 19K gene product is required for the production of transformed cell foci, when it is accompanied by the E1A products (93,94). Therefore at least two distinct functions, the E1A and E1B gene products, are required for a complete transformation event in cultured cells (95). The E1B 58K gene function seems to be required for tumorigenesis in animals (90,93).

More recently, cold-sensitive host range mutants of Ad5 have been isolated (91) and assigned to both complementation groups, E1A and E1B. The hr^{cs} mutants mapping in the E1A region are cold-sensitive both for the induction of transformation and for the expression of the transformed phenotype in primary rat embryo cells. The hr^{cs} mutants mapping in region E1B are defective in morphological transformation at both low and high temperature. These findings indicate that regions E1A and E1B are simultaniously required both to initiate the morphological transformation and to express the transformed phenotype.

3.2 Region E2: DNA Replication Proteins

Messenger RNA from the E2 transcription unit initiates at

about 75.2 map units at early times and at 72.0 map units at later times in infection (51). Splicing of this leftward primary transcript produces mRNAs derived from the E2A coding region (61-66 map units) that synthesize the single-stranded DNA-binding protein (DBP; 9) and mRNAs complementary to the E1B coding region (11-31 map units) that produce an 87K and a 140K polypeptide. The 87K polypeptide is the precursor of the 55K terminal protein (pTP) covalently bound to the 5' ends of the adenovirus DNA (11,96,97). It is required for the initiation of viral DNA replication, where it presumably acts as a primer for new DNA strands (11.98). The 140K polypeptide has been identified as a DNA polymerase (98,99,100). The latter two proteins can be found and isolated as a pTP-Pol complex (98), which binds specifically to the termini of adenovirus DNA between nucleotide positions 9 and 22 (101). The structual protein IVa2 (11-15 map units) predominantly synthesized at late times after infection, is also produced in small amounts at early times as part of the E2 transcription unit (Table 1; 41). The three E2 early proteins E2A 72K, E2B 87K and E2B 140K are all required for viral DNA replication.

The DBP is functionally the best characterized adenovirus early protein. Much of what we know about it has come from studies of two temperature-sensitive mutants, H5ts125 and H5ts107 (17, 102) which map in the structural gene of DBP. The DBP binds efficiently, but not covalently, to single-stranded DNA (21,103) and poorly, if at all, to double-stranded DNA. However, it does bind to the ends of the linear double-stranded genome (104). The DBP is a phosphoprotein (105) and composed of 529 amino acids (106) as deduced from the nucleotide sequence of this gene. It can be cleaved by mild chymotrypsin treatment into two structural domains, a 26K N-terminal fragment and a 45K C-terminal peptide (107). The 45K domain alone binds to single-stranded DNA (108), while the 26K fragment, which carries nearly all phosphorylation sites, has no DNA-binding capacity. The degree of phosphorylation of the DBP in the N-terminal domain, however, has an effect upon the binding intensity of the C-terminal fragment to single-stranded DNA (107).

The two temperature-sensitive mutants mentioned above, H5ts125 and H5ts107, both contain a proline to serine change at amino acid residue 413, located in the C-terminal domain (106). This alteration diminishes the DNA-binding activity of the DBP in vitro (102,109) and inhibit DNA replication in vivo (9,110). Experiments with extracts from cells infected with H5ts125 have shown that the DBP is not essential for the initiation reaction, e.g. for the formation of the 80K pTP-dCMP/Pol complex (110). Rather it is necessary in elongation of Ad DNA synthesis (112, 113). Although the synthesis of an early elongation product, a short oligonucleotide 26 nucleotides long, can proceed in the absence of DBP, it is stimulated significantly by the presence of a functional DBP (114,115,116). In this case, a host DBP may substitute for the adenovirus DBP. More extended elongation reactions, however, are strictly dependent on adenovirus DBP.

The H5ts125 and H5ts107 mutants of the DBP express several additional phenotypes. Cells infected with H5ts125 at non-permissive temperature contain elevated levels of early DBP mRNA and other early mRNAs (117,118), suggesting that the DBP autoregulates its own production. Further studies have shown that the accumulation of the early mRNAs probably results from increased stability of these mRNAs (4,119). An inhibitory effect of the DBP on adenovirus transcription has been specifically related to region E4 (4,120). The mechanism of this repression has not yet been sufficiently understood, but it is believed to act on the level of the initiation of transcription (4,120).

That the DBP may involved in some way in RNA processing, has been suggested by the study of adenovirus host range mutants that replicate in monkey cells (7). Monkey cells infected with human wild-type adenovirus produce normal levels of early proteins, replicate their viral DNA but synthesize reduced levels of some late gene products, particulary the fiber protein (121,122). About half of the fiber mRNA that is made in these cells is defectively spliced (123). This defect can be overcome by certain mutants (e.g. H5hr404), whose mutations map in the N-terminal domain of the DBP gene (7). It is interesting to note in this context that SV40 large T antigen or even a small subfragment from its C-terminus, can complement the deficiency of adenovirus growth in monkey cells (124,125,126,127). This observation suggest a functional relationship between the adenovirus DBP and SV40 large T antigen in respect to the efficient expression of late adenovirus genes in these cells. Further evidence for possible functional relatedness of SV40 large T antigen and E2A DBP arises from the observation that SV40 complements the H5ts125 defect in monkey cells co-infected at restricted temperature and allows adenoviral DNA synthesis to proceed (18,128,129).

Finally, the DBP seems also to be involved in some way in virus-induced cellular transformation. At nonpermissive temperature the ts mutants transform rat cells in culture with higher efficiency than wild-type virus (17,18,19). At present, two hypotheses can be proposed to explain the role of the DBP in the process leading to morphological transformation: First, the DBP may influence the frequency of transformation by altering the stability of E1A and E1B mRNAs directly (119). This would lead to an increased level of these proteins involved in transformation. Second, as the DBP autoregulates its own cellular concentation, it may also regulate the level of the E2B gene products. It is known that mutant H5ts36, which maps in the E2B region, effects the frequency of transformation at the nonpermissive temperature.

3.3 Region E3 Proteins

The E3 transcription unit, located at 76 to 86 map units expresses at least eight mRNAs using a common promoter at 76 map units. The mRNAs differ both in their splicing patterns and in poly A site utilization (45,51). From the DNA sequence of region E3, 11 open coding regions in all three reading frames have been identified that could encode polypeptides of 6.8K to 18.5K (131,132). However, only three E3-coded proteins are known, the E3 13K, the E3 14K (133,134) and the E3 19K glycoprotein which is associated with the cell membrane (22,48,133,134,135).

Deletion mutants in which the entire E3 region is eliminated and substitution mutants replicate normally in cell culture, indicating that E3 gene products are not required for growth of adenovirus in cultured cells. It has been shown that cytotoxic T lymphocytes can recognize the E3 19K glycoprotein on the surface of infected and transformed cells and reject these cells via T cell-mediated killing (137). This killing requires H2 (major histocompatibility antigen) identity between killer T cells and their targets. These findings support speculations that the E3 19K glycoprotein may have useful function for the virus in vivo.

Little is known about the functions of the other E3 proteins.

3.4 Region E4 Proteins

The E4 transcription unit is located between 91 and 100 map units and is transcribed in the leftward direction. As most other adenovirus transcription units, it encodes a variety of different mRNAs (9 to 14 transcripts; 45,51,130,138), which are formed by differential splicing from a common nuclear precursor. The complete nucleotide sequence of this region has been determined (132) and contains open coding regions in all three reading frames (53,132). At least six different E4 proteins have been observed by in vitro translation (11K, 13K, 17K, 19K, 21K, 24K) (26). Three polypeptides have been identified in vivo and unambiguously assigned to E4 coding regions. These polypeptides have apparent molecular weights of 11K, 14K and 25K (139,140).

The E4 25K polypeptide exists at least partially in a physical complex with the E1B 58K polypeptide in infected cells (79). The E4 11K protein has been found to be associated with the nuclear matrix (139) and appears immunologically to be highly conserved in comparison to different human adenovirus serotypes (139).

The expression of the E4 region is positively and negatively regulated by products of the regions E1A (1,2) and E2 (4) respectively.

The functions of the E4 proteins are unknown. Several deletion mutants in this region seem to have no effect on DNA replication, while some, but not all, late proteins were produced in reduced amounts (141). These observation may indicate possible functions of at least some of the E4 proteins in the regulation of the levels of early (?) and late viral proteins.

4. CONCLUSIONS

To date, about 25 early adenovirus proteins have been detected in productively infected cells (Table 1). Their coding regions have been mapped on the viral genome and mutants have been isolated within these genes. The early viral proteins fall into three groups with regard to their appearance after infection: (a) immediate early proteins (E1A), which are required to turn on the transcription of (b) the delayed early proteins (E1B, E2, E3, E4) and (c) proteins synthesized at early times but expressed in considerably higher amounts at late times in infection (IVa2, L1, protein IX). The E1 gene products are involved in a positive and possibly negative regulation of viral as well as cellular gene expression after infection. In addition, functions of both E1 cistrons, E1A and E1B, are necessary and sufficient for neoplastic transformation. Proteins of region E2, the single-stranded DNA-binding protein (E2A), the DNA terminal protein (E2B) and the DNA polymerase (E2B) , all are essential for viral DNA replication. The E2A DBP with its separate functional domains is an excellent example of versatile, multifunctional proteins reminiscent to the situation of SV40 large T antigen or even the bacterophage T4 gene 32 product (142). Less is known about the functions of the E3 and E4 gene products. All E3 proteins and some of the E4 proteins are nonessential for viral growth in cultured cells. Yet this fact does not mean that these genes fail to perform an useful function for the virus in nature.

Although extensive data have been collected in recent years especially on the proteins encoded by regions E1A, E1B, E2A and E2B, little is known about the mechanism of their functions. How do these proteins act in the regulation of transcription and RNA processing? How do they interact with cellular proteins and functions? Another important question, namely the question of how viral gene products of region E1 are involved in oncogenic cellular transformation has obviously not been solved so far. Therefore the adenovirus/cell system will surely continue to be one of the attractive fields in molecular biology research.

REFERENCES

1. Berk AJ, Lee F, Harrison T, Williams J, Sharp PA: A pre-early Ad5 gene product regulates synthesis of early viral mRNAs. Cell (17):935-944,1979.
2. Jones N, Shenk T: An adenovirus type 5 early gene function regulates expression of other early viral genes. Proc Natl Acad Sci USA (76):3665-3669,1979.
3. Nevins J: Induction of the synthesis of a 70,000 dalton mammalian heat shock protein by the adenovirus E1A gene product. Cell (29):913-919,1982.
4. Nevins J, Winkler J: Regulation of early adenovirus transcription: a protein product of early region 2 specifically represses region 4 transcription. Proc Natl Acad Sci USA (77):1893-1897,1980.
5. Schrier PI, Bernards R, Vaessen RTMJ, Houweling A, Van der Eb AJ: Expression of class I major histocompatibility antigens swiched off by highly oncogenic adenovirus 12 in transformed rat cells. Nature (305):771-775,1983.
6. Borelli E, Hen R, Chambon P: Adenovirus-2 E1A products repress enhancer-induced stimulation of transcription. Nature (312):608-612,1984.
7. Klessig DF, Grodzicker T: Mutations that allow human Ad2 and Ad5 to express late genes in monkey cells map in the viral gene encoding the 72Kd DNA binding protein. Cell (17):957-963,1979.
8. Solnick D: An adenovirus mutant defective in splicing RNA from early region 1A. Nature (291):508-510,1981.
9. Van der Vliet PC, Sussenbach JS: An adenovirus type 5 gene function required for initation of viral DNA replication. Virology (67):415-427,1975.
10. Challberg MD, Desidero SV, Kelly TJ: Adenovirus DNA replication in vitro: characterization of a protein covalently linked to nascent DNA strands. Proc Natl Acad Sci USA (77):5105-5107,1980.
11. Stillman BW, Lewis JB, Chow L, Mathews NB, Smart JE: Identification of the gene and mRNA for the adenovirus terminal protein prcursor. Cell (23):497-508,1981.
12. Stillman BW, Tamanoi F: Adenovirus DNA replication: DNA sequences and enzymes required for initiation in vitro. Cold Spring Harbor Symp Quant Biol (47):741-752,1983.
13. Graham FL, Van der Eb AJ, Heyneker HL: Size and location of the transforming regions in human adenovirus type 5 DNA. Nature (251):687-691,1974.
14. Graham FL, Abrahams PJ, Mulder C, Heyneker HL, Warnaar SO, de Vrives FAJ, Fiers W, van der EbAJ: Studies on in vitro transformation by DNA and DNA fragments of human adenoviruses and simian virus 40. Cold Spring Harbor Symp Quant Biol (39): 637-650,1975.
15. van der Eb AJ, Mulder C, Graham FL, Houweling A: Transformation with specific fragments of adenovirus DNAs. 1.Isolation of specific fragments with transforming activity of adenovirus 2 and 5 DNA. Gene (2):115-132,1977.
16. van der Eb AJ, Van Ormondt H, Schrier PI, Lupker JH, Jochenisen H, Van der Elsen PJ, De Leys RJ, Kratt J,

Van Bevern CP, Dykema R, De Wood A: Structure and function of the transforming genes of human adenoviruses and SV40. Cold Spring Harbor Symp Quant Biol (44):383-399,1979.
17. Ginsberg HS, Ensinger MS, Kauffman RS, Mayer Aj, Lomdholm U: Cell transformation: a study of regulation with types 5 and 12 adenovirus temperature-sensitive mutants. Cold Spring Harbor Symp Quant Biol (39):412-426,1974.
18. Williams JF, Young CH, Austin P: Genetic analysis of human adenovirus type 5 in permissive and nonpermissive cells. Cold Spring Harbor Symp Quant Biol (39):427-438,1974.
19. Logan J, Nicolas JC, Topp WC, Girard M, Shenk T, Levine AJ: Transformation by adenovirus early region 2A temperature sensitive mutants and their revertants. Virology (115): 419-422,1981.
20. Lewis JB, Mathews MB: Control of adenovirus early gene expression: a class of immidiate early products. Cell (21): 303-313,1980.
21. Van der Vliet PC, Levine AJ: DNA binding proteins specific for cells infected by adenovirus. Nature (246):170-174,1973.
22. Jeng YH, Wold WSM, Sugawara K, Green M: Evidence for an adenovirus type 2 coded early glycoprotein. J Virol (28): 314-323,1978.
23. Rosenwirth B, Shiroki K, Levine AJ, Shimojo H: Isolation and characterisation of adenovirus type 12 DNA binding proteins. Virology (67):14-23,1975.
24. Tooze J: Molecular biology of tumor viruses, 2nd ede. Part 2, DNA tumor viruses. Cold Spring Harbor, New York, 1981.
25. Anderson CW, Baum PR, Gesteland RF: Processing of adenovirus 2-induced proteins. J Virol (12):241-252,1973.
26. Lewis JB, Atkins JF, Baum PR, Solem R, Gesteland RF, Anderson CW: Localisation and identification of the genes for adenovirus type 2 early peptides. Cell (7):141-151, 1976.
27. Esche H, Schilling R, Doerfler W: In vitro translation of adenovirus type 12-specific mRNA isolated from infected and transformed cells. J Virol (30):21-31,1979.
28. Russell WC, Hayashi K, Sanderson PJ, Pereira HG: Adenovirus antigens: a study of their properties and sequential development in infection. J Gen Virol (1):495-507,1967.
29. Levinson AD, Levine AJ: The group C adenovirus tumor antigens: infected and transformed cells and a peptide map analysis. Cell (11):871-879,1977.
30. Achten S, Doerfler W: Virus specific proteins in adenovirus type 12-transformed and tumor cells as detected by immunoprecipitation. J Gen Virol (59):357-366,1982.
31. Cepko CL, Changelian PS, Sharp PA: Immunoprecipitation with two-dimensional pools as a hybridoma sceening technique: production and characterisation of monoclonal antibodies against adenovirus 2 proteins Virology.(11): 385-401,1981.
32. Sarnow P, Sullivan CA, Levine AJ: A monoclonal antibody detecting the adenovirus 5 E1b-58Kd tumor antigen: characterization of the E1b-58Kd tumor antigen in adenovirus infected and transformed cells. Virology (34):650-657,1982.

33. Green M, Brackmann KH, Lucher LA, Symington SJ: Antibodies to synthetic peptides targeted to transforming genes of human adenoviruses: an approach to understand early viral gene functions. Curr Top Microbiol Immunol (109):167-192, 1984.
34. Schughart K, Bause E, Esche H: Structure and expression of adenovirus type 12 E1B 58K protein in infected and transformed cells: studies using antibodies directed against a synthetic peptide. Virus Research (3):41-56,1985.
35. Challberg M, Kelly TS: Adenovirus DNA replication in vitro. Proc Natl Acad Sci USA (76):655-659,1979.
36. Stillman BW: The replication of adenovirus DNA with purified proteins. Cell (35):7-9,1983.
37. Manley JL, Sharp P, Gefter M: RNA synthesis in isolated nuclei: in vitro mutation of adenovirus 2 major late m-RNA precursor. Proc Natl Acad Sci USA (76):160-164,1979.
38. Weil A, Luse DL, Segall J, Roeder RG: Selective and accurate initiation of transcription at the Ad2 major late promotor in a soluble system dependent upon purified RNA polymerase II. Cell (18):469-481,1979.
39. Weinmann R, Ackerman S, Bunick D, Concino M, Zandomeni R: In vitro transcription of adenovirus genes. Curr Top Microbiol Immunol (109):125-145,1984.
40. Harter M, Lewis J: Adenovirus type 2 early proteins synthesized in vitro and in vivo: identification in infected cells of the 38,000 to 50,000 molecular weight proteins encoded by the left end of the Ad2 genome. J Virol (26):736-744,1978.
41. Lewis JB, Esche H, Smart JE, Stillman BW, Harter ML, Mathews MB: Organisation and expression of the left third of the genome of adenovirus. Cold Spring Harbor Symp Quant Biol (44): 493-508,1979.
42. Esche H, Siegmann B: Expression of early viral gene products in adenovirus type 12-infected and transformed cells. J Gen Virol (60): 99-113,1982.
43. Chow LT, Roberts JM, Lewis JB, Broker TR: A map of cytoplasmic RNA transcripts from lytic adenovirus type 2 determined by electron microscopy of RNA-DNA hybrids. Cell (11): 819-836,1977.
44. Evans RM, Fraser N, Ziff E, Weber J, Wilson M, Darnell JE: The initiation sites for RNA transcription in Ad2 DNA. Cell (12):733-739,1977.
45. Berk AJ, Sharp PA: Structure of the adenovirus 2 early mRNAs. Cell (14):695-711,1978.
46. Harrison T, Graham F, Williams J: Host range mutants of adenovirus type 5 defective for growth in HeLa cells. Virology (77):319-329,1977.
47. Frost E, Williams J: Mapping temperature-sensitive and host-range mutants of adenovirus type 5 by marker rescue. Virology (91):39-50,1978.
48. Ross SR, Flint SJ, Levine AJ: Identification of the adenovirus early proteins and their genomic map positions. Virology (100):419-432,1980.
49. Graham FG, Harrison TJ, Williams JF: Defective transforming capacity of adenovirus type 5 host range mutants. Virology (86):10-21,1978.

50. Shenk T, Jones N, Colby W, Fowlkes D: Functional analysis of Ad5 host range deletion mutants defective for transformation of rat embryo cells. Cold Spring Harbor Symp Quant Biol (44): 367-375,1979.
51. Chow LT, Broker T, Lewis JB: The complex splicing patterns of RNA from the early regions of Ad2. J Mol Biol (134): 265-303,1979.
52. Maat J, Van Ormondt H: The nucleotide sequence of the transforming Hind III-G fragment of Ad5: the region between map positions 5.5 (HpaI site) and 8.0 (HindIII site). Gene (6): 75-90,1979.
53. Gingeras TR, Sciaky D, Gelinas RE, Bing-Dong J, Yen CE, Kelly MM, Bullock PA, Parsons BL, O'Neill KE, Roberts RJ: Nucleotide sequences from the adenovirus 2 genome. J Biol Chem (257):13475-13491,1982.
54. Smart JE, Lewis JB, Mathews MB, Harter ML, Anderson CW: Adenovirus type 2 early proteins: assignment of the early region 1A proteins synthesized in vivo and in vitro to specific mRNAs. Virology (112): 703-713,1981.
55. Yee SP, Rowe DT, Tremblay ML, McDermott M, Branton PE: Identification of human adenovirus early region 1 products using antisera against synthetic peptides corresponding to the predicted carboxy termini. J Virol (46):1003-1013,1983.
56. Esche H, Mathews MB, Lewis JB: Proteins and messenger RNAs of the transforming region of wild-type and mutant adenoviruses. J Mol Biol (142):399-417,1980.
57. Perricaudet M, Akusjarvi G, Virtanen A, Pettersson U: Structure of the two spliced mRNAs from the transforming region of human subgroup C adenovirus. Nature (281):694-696,1979.
58. Nevins J: Mechanism of activation of early viral transcription by the adenovirus E1A gene product. Cell (26):213-220,1981.
59. Ricciardi RL, Jones RL, Cepco CL, Sharp PA, Roberts BE: Expression of early adenovirus genes requires a viral encoded acidic polypeptide. Proc Natl Acad Sci USA (78):6121-6125,1981.
60. Ross S, Levine AJ, Galos R, Williams J, Shenk T: Early viral proteins in HeLa cells infected with adenovirus type 5 host range mutants. Virology (103):475-492,1980.
61. Gaynor RB, Berk AJ: Cis-acting induction of adenovirus transcription. Cell (33):683-693,1983.
62. Solnick D, Anderson MA: Transforming-deficient adenovirus mutant defective in expression of region E1a but not region E1b. J Virol (42):106-113,1982.
63. Persson H, Monstein HJ, Akusjarvi G, Philipson L: Adenovirus early gene products may control viral mRNA accumulation and translation in vivo. Cell (23):485-496,1981.
64. Shaw AR, Ziff EB: Selective inhibition of adenovirus type 2 early region II and III transcription by an anisomycin block of protein synthesis. Mol Cell Biol (2):789-799,1982.
65. Carlock LR, Jones NC: Transformation-defective mutant of adenovirus type 5 containing a single altered E1a mRNA species. J Virol (49):657-664.1981.
66. Montell C, Fisher EF, Caruthers MH, Berk AJ: Resolving the functions of overlapping viral genes by site-specific mutagenesis at a mRNA splice site. Nature (295):380-344,1982.

67. Logan JS, Shenk T: Transcriptional and translational control of adenovirus gene expression. Microbiol Rev (46):377-383, 1982.
68. Katze MG, Persson H, Philipson L: Control of adenovirus early gene expression: post-transcriptional control mediated by both viral and cellular gene products. Mol Cell Biol (1): 807-813,1981.
69. Gilead Z, Jeng YH, Wold WSM, Sugawara K, Rho HW, Harter ML, Green M: Immunological identification of two adenovirus induced early proteins possibly involved in cell transformation. Nature (264):263-266,1976.
70. Schrier PI, Van der Elsen PJ, Hertoghs JJL, Van der Eb AJ: Characterization of tumor antigens in cells transformed by fragments of adenovirus type 5 DNA. Virology (99): 372-385,1979.
71. Bos JL, Polder LJ, Bernards R, Schrier PI, Van den Elsen AJ, Van Ormondt H: The 2.2Kb E1B mRNA of human Ad12 and Ad5 codes for two tumor antigens starting at different AUG trplets. Cell (27):121-131,1981.
72. Matsuo T, Wold WSM, Hashimoto S, Rankin A, Symington J, Green M: Polypeptides encodet by the transforming region E1b of human adenovirus 2. Immunoprecipitation from transformed and infected cells and cell-free translation of E1b-specific mRNA. Virology (118):456-465,1982.
73. Persson H, Katze MG, Philipson L: Purification of a native membrane-associated tumor antigen. J Virol (42):905-917, 1982.
74. Lassam NJ, Bayley ST, Graham FL: Synthesis of DNA, late polypeptides and infectious virus by host-range mutants of adenovirus 5 in nonpermissive cells. Virology (87):463-467, 1978.
75. Jones N, Shenk T: Isolation of adenovirus type 5 host range delition mutants defective for translation of rat embryo cells. Cell (17):683-689,1979.
76. Sarnow P, Ho YS, Williams J, Levine AJ: Adenovirus E1b-58Kd tumor antigen and SV40 large tumor antigen are physically associated with the name 54Kd cellular protein in transformed cells. Cell (28):387-394,1982.
77. Linzer DIH, Levine AJ: Characterisation of a 54K dalton cellular SV40 tumor antigen present in SV40 transformed cells and uninfected embryonal carcinoma cells. Cell (17):43-52, 1979.
78. Lane DP, Crawford LV: T antigen is bound to a host protein in SV40-transformed cells. Nature (278):261-263,1979.
79. Sarnow P, Hearing P, Anderson C, Halbert DH, Shenk T, Levine AJ: Adenovirus E1b-58Kd tumor antigen is physically assiciated with an E4-25Kd protein in adenovirus productively infected cells. J Virol (49):642-700,1984.
80. De Leo AB, Jay G, Apella E, Dubois GC, Law LW, Old LJ: Detection of a transformation related antigen in chemically induced sarcomas and other transformed cells of the mouse. Proc Natl Acad Sci USA (76):2420-2424,1979.
81. Luka J, Jörnvall H, Klein G: Purification and biochemical charakterisation of the Epstein-Barr virus-determined nucleic antigen and an associated protein with a 53000-dalton subunit. J Virol (35):592-602,1980.

82. Thomas R, Kaplan L, Reich N, Lane DP, Levine AJ: Characterization of human p^{35} antigens employing primate-specific monoclonal antibodies. Virology (131):502-517,1983.
83. Oren M, Maltzman W, Levine AJ: Post-translational regulation of the 54K cellular tumor antigen in normal and transformed cells. Mol Cell Biol (1):101-110,1981.
84. Reich N, Levine AJ: Regulation of p^{53} gene expression during cell growth and division. Nature, in press.
85. Mercer WE, Nelson D, DeLeo AB, Old LJ, BasergaR: Microinjection of minoclonal antibody to protein p^{53} inhibits serum-induced DNA synthesis in 3T3 cells. Proc Natl Acad Sci USA (79):6309-6312,1982.
86. Tjian R, Fey G, Graessmann A: Biological activity of purified SV40 T-antigen proteins. Proc Natl Acad Sci USA (75): 1279-1283,1978.
87. Shenk T, Williams J: Genetic analysis of adenoviruses. Curr Top Microbiol Immunol (III)1-39,1984.
88. Houweling A, Van der Elsen PJ, Van der eb AJ: Partial transformation of primary rat cells by the leftmost 4,5% fragment of adenovirus 5 DNA. Virology (105):537-550,1980.
89. Shiroki K, Shimojo H, Sawada Y, Uemizu Y, Fujinaga K: Incomplete transformation of rat cells by a small fragment of adenovirus 12 DNA. Virology (95):127-136,1979.
90. Van der Eb AJ, Bernards R: Transformation and oncogenicity by adenoviruses. Curr Top Microbiol Immunol (110): 23-51, 1984.
91. Ho Y-S, Galos R, Williams J: Isolation of type 5 adenovirus mutants with a cold-sensitive host range phenotype: genetic evidence of an adenovirus transformation maintenance function. Virology (122):109-124,1982.
92. Chinnadurai G: Adenovirus 2 lp^{+} locus codes for a 19kd tumor antigen that plays an essential role in cell transformation. Cell (33):759-766,1983.
93. Bernards R, Schrier PI, Bos JL, Van der Eb AJ: Role of adenovirus types 5 and 12 early region 1b tumor antigens in oncogenic transformation. Virology (127):45-54,1983.
94. Bos JL, Jochemsen AG, Bernards R, Schrier PI, Van Ormondt H, Van der Eb AJ: Deletion mutants of region E1a of Ad12 E1 plasmids: effect on oncogenic transformation. Virology (129): 393-400,1983.
95. Van den Elsen P, Houweling A, Van der Eb AJ: Expression of region E1b of human adenoviruses in the absence of a functional E1A region is not sufficient for complete transformation. Virology (128):377-390,1983.
96. Rekosh DMK, Russell WC, Bellet AJD, Robinson AJ: Identification of a protein linked to the ends of adenovirus DNA. Cell (11):283-295,1977.
97. Galos RS, Williams J, Binger MH, Flint SJ: Location of additional early gene sequences in the adenoviral chromosome. Cell (17):945-956,1979.
98. Enomoto T, Lichy JH, Ikeda JE, Hurwitz J: Adenovirus DNA replication in vitro: purification of the terminal protein in a functional form. Proc Natl Acad Sci USA (78):6779-6783, 1981.
99. Lichy JH, Field J, Horwitz MS, Hurwitz J: Separation of the adenovirus terminal protein precursor from its associated

DNA polymerase: role of both proteins in the initiation of adenovirus DNA replication. Proc Natl Acad Sci USA (79): 5225-5229,1982.
100. Stillman BW, Tamanoi F, Mathews MB: Purification of an adenovirus-coded DNA polymerase that is required for initiation of DNA replication. Cell (31):613-623,1982.
101. Fowlkes DM, Lord ST, Linne T, Pettersson U, Philipson L: Interaction between the adenovirus DNA-binding protein and double-stranded DNA. J Mol Biol (132):163-180,1979.
102. Ensinger MJ, Ginsberg HS: Selection and prelimenary chacacterization of temperature-sensitive mutants of type 5 adenovirus. J Virol (10):328-339,1972.
103. Van der Vliet PC, Keegstra W, Jansz HS: Complex formation between the adenovirus type 5 DNA binding protein and single-stranded DNA. Eur J Biochem (86):389-398,1978.
104. Fowlkes DML, Land ST, Linne T, Pettersson U, Philipson L: Interaction between the adenovirus DNA binding protein and double stranded DNA. J Mol Biol (132):163-180,1979.
105. Levinson AD, Levine AJ: The isolation and identification of the adenovirus group C tumor antigens. Virology (76): 1-11,1977.
106. Kruijer W, van Schaik FMA, Sussenbach JS: Nucleotide sequence of the gene encoding adenovirus type 2 DNA binding protein. Nucleic Acids Res (10):4493-4500,1982.
107. Klein H, Maltzman W, Levine AJ: Structure - function relationship of the adenovirus DNA binding protein. J Biol Chem (254):11051-11060,1979.
108. Ariga H, Klein H, Levine AJ, Horwitz MS: A cleavage product of the adenovirus DNA binding protein is active in DNA replication in vitro. Virology (101):307-310,1980.
109. Van der Vliet PC, Levine AJ, Ensinger MJ, Ginsberg HS: Thermolabile DNA binding proteins from cells infected with a temperature sensitive mutant of adenovirus defective in DNA replication. J Virol (15):348-354,1975.
110. Van der Vliet PC, Landberg J, Jansz HS: Evidence for a function of the adenovirus DNA binding proteins in initiation of DNA synthesis as well as in elongation on nascent DNA chains. Virology (80):98-109,1977.
111. Challberg MD, Ostrove JM, Kelly TJ Jr: Initiation of adenovirus DNA replication: detection of covalent complexes between nucleotide and the 80-kilodalton terminal protein. J Virol (41):265-270,1982.
112. Kaplan LM, Ariga H, Hurwitz J, Horwitz MS: Complementation of the temperature-sensitive defect in H5ts125 adenovirus DNA replication in vitro. Proc Natl Acad Sci USA (76): 5534-5538,1979.
113. Horwitz MS: Temperature sensitive replication of H5ts125 adenovirus DNA in vitro. Proc Natl Acad Sci USA (75): 4291-4295,1978.
114. Lichy JH, Horwitz MS, Hurwitz J: Formation of a covalent complex between the 80,000 dalton adenovirus terminal protein and 5' CMP in vitro. Proc Natl Acad Sci USA (78): 2678-2682,1981.
115. Friefeld BR, Krevolin MD, Horwitz MS: Effects of the adenovirus H5ts125 and H5ts107 DNA binding proteins on DNA replication in vitro. Virology (125):380-389,1983.

116. Van Bergen GM, van der Vliet PC: Temperature-sensitive initiation and elongation of adenovirus DNA replication in vitro with nuclear extracts from H5ts149-, and H5ts125-infected HeLa cells. J Virol (46):642-648,1983.
117. Carter TH, Blanton RA: Autoregulation of adenovirus type 5 early gene expression. II. Effect of temperature-sensitive early mutations on virus RNA accumulation. J Virol (28): 450-456,1978.
118. Nicolas J-C, Ingrand D, Sarnow P, Levine AJ: A mutation in the adenovirus type 5 DNA binding protein that fails to autoregulate the production of the DNA binding protein. Virology (122):481-485,1982.
119. Babich A, Nevins JR: The stability of early adenovirus mRNA is controlled by the viral 72kd DNA-binding protein. Cell (26):371-379,1981.
120. Handa H, Kingston RE, Sharp PA: Inhibition of adenovirus early region IV transcription in vitro by a purified viral DNA binding protein. Nature (302):545-547,1983.
121. Eron L, Westphal H, Khoury G: Post transcriptional restriction of human adenovirus expression in monkey cells. J Virol (15):1256-1261,1975.
122. Klessig DF, Anderson CW: Block to multiplication of adenovirus serotype 2 in monkey cells. J Virol (16):1650-1668, 1975.
123. Klessig DF, Chow LT: Incomplete splicing and deficient accumulation of the fiber messenger RNA in monkey cells infected by human adenovirus type 2. J Mol Biol (139): 221-242,1980.
124. Lewis AM, Levine AJ, Crumpacker CS, Levin MJ, Samabra RJ, Henry PH: Studies of undefective adenovirus 2-SV40 hybrid viruses. V. Isolation of additional hybrids which differ in their SV40-specific biological properties. J Virol (11): 655,1973.
125. Grodzicker T, Anderson C, Sharp P, Sambrook J: Conditional lethal mutants of adenovirus 2-simian virus 40 hybrids. I. Host range mutants of Ad2+ND. J Virol (13):1237-1244,1974.
126. Grodzicker T, Lewis JB, Anderson CW: Conditional lethal mutants of adenovirus type 2 simian virus 40 hybrids. II. Ad2+ND1 host range mutants that synthesize fragments of the Ad2+ND1 30K protein. J Virol (19):559-568,1976.
127. Cole CN, Crawford LV, Berg P: Simian virus 40 mutants with deltions at the 3' end of the early region are defective in adenovirus helper function J Virol (30):683-691,1979.
128. Rabek JP, Zakian VA, Levine AJ: The SV40 A gene product suppresses the adenovirus H5ts125 defect in DNA replication. Virology (109):290-302,1981.
129. Goldman ND, Howley P, Khoury G: Functional interaction between the early viral proteins of simian virus 40 and adenovirus. Virology (109):303-313,1981.
130. Kitchingham GR, Westphal H: The structure of Ad2 early nuclear and cytoplasmic RNAs. J Mol Biol (137):23-48, 1980.
131. Herisse J, Coutois G, Galibert F: Nucleotide sequence of the Eco RI-D fragment of adenovirus 2 genome. Nucleic Acids Res (8):2173-2192,1980.
132. Herisse J, Rigalet M, Dupont R, de Dinectin C, Galibert F: Nucleotide sequence of adenovirus 2 DNA encoding for

carboxylic region of the fiber protein and the entire E4 region. Nucleic Acids Res (9):4023-4042,1981.

133. Persson H, Oberg B, Philipson L: Purification and characterization of an early protein (E14K) from adenovirus type 2 infected cells. J Virol (28):119-139,1978.

134. Persson H, Sigmus C, Philipson L: Purification and characterization of an early glycoprotein from adenovirus type 2-infected cells. J Virol (29):938-948,1979.

135. Kornfeld R, Wold WSM: Structures of the glycoprotein coded by early region E3 of adenovirus 2. J Virol (40):440-449, 1981.

136. Jones N, Shenk T: isolation of deletion and substitution mutants of adenovirus type 5. Cell (13):181-188,1978.

137. Persson H, Kvist S, Ostberg L, Peterson PA, Philipson L: Adenoviral early glycoprotein E3-19K and ts association with translation antigens. Cold Spring Harbor Symp Quant Biol (44):509-517,1979.

138. Tigges MA, Raskas HJ: Expression of adenovirus 2 early region E4: assigment of the early region 4 polypeptides to their respective mRNAs, using in vitro translation. J Virol (44):907-921,1982.

139. Sarnow P, Hearing P, Anderson CW, Reich N, Levine AJ: Identification and characterization of an immunologically conserved adenovirus early region 11,000 MW protein and its association with the nucear matrix. J Mol Biol (102): 565-583,1982.

140. Downey JF, Rowe DT, Bacchetti S, Graham FL, Bayley ST: Mapping of a 14,000 dalton antigen to early region 4 of the human adenovirus 5 genome. J Virol (45):514-523,1983.

141. Challberg SS, Ketner G: delition mutants of adenovirus 2: isolation and initial characterization of virus carrying mutations near the right end of the viral genome. Virology (114):196-209,1981.

142. Mosig G, Breschkin A, Dannenberg R, Bock S: Multiple interaction of a DNA-binding protein gene 32 protein of phage T4 during DNA replication and recombination. In: Molineux I, Koliyuma M (eds) DNA synthesis, present and future. Plenum, 1977, pp367-380.

8

ADENOVIRUS GENETICS

JIM WILLIAMS

Department of Biological Sciences, Carnegie-Mellon University,
Pittsburgh, PA 15213

INTRODUCTION

Over the past twelve years or so a large number of adenoviral mutants has been isolated, and many of these have been used to analyse the functions of viral gene products involved in the processes of infection and transformation. Within the past year two extensive reviews on the topic of adenovirus genetics have been published. One of these (1) surveys the physiological consequences of mutations in viral genes in cells infected or transformed by group C adenoviruses (types 2 and 5), while the other (2) deals largely with genetic methodology and the study of genetic interactions. Between them these reviews provide a comprehensive summary of group C adenoviral genetics until late 1983. Although some reiteration of this information is inevitable, my main purpose here is to provide an update on progress made during the past year. Most of this progress has been made in areas relating to adenoviral early genes, and accordingly, much of the discussion in this chapter will be confined to analysis of mutations in these genes. For more detailed discussion of late-function mutants, the reader is referred to the above reviews. Further, the mutants made to investigate transcriptional and other control regions are discussed elsewhere in this volume, and will not be dealt with in this chapter. Recently, considerable progress has been made in the genetic analysis of type 12 (group A) adenovirus, and since this material was not included in these two reviews, it will be dealt with here.

MAKING VIRAL MUTANTS

Adenoviral mutants have been generated by a wide range of mutagenic procedures, which operationally fall into two broad categories. In the first type of approach, virus is treated with chemical or physical mutagens which induce single base-pair changes in the DNA, and mutants with the desired phenotype, for example, temperature-sensitive (_ts_), are identified by an appropriate screening procedure. Many useful adenoviral mutants have been generated in this way, but since many of the mutagenic and screening methods used have been previously reviewed in detail elsewhere (3), they will not be discussed here. It is, however, worth pointing out that this approach does present a number of advantages. Detailed knowledge of the viral genome is not a prerequisite for mutant production, and theoretically, at least, mutations can be introduced into any of the viral genes. The disadvantage to this type of approach stems from the fact that while well-defined phenotypes can be identified, the genotypes are not immediately known, and must be established to allow meaningful functional analysis of the mutants. In the alternative approach, segments of the viral DNA are modified in vitro by one of a variety of specific mutagenesis procedures (4), and rebuilt into complete genome which can then be transfected into permissive cells to yield infectious virus. In principle, these procedures permit mutations to be introduced at specific sites in any region of the viral chromosome, and confer the obvious advantage that the genotype is defined from the outset. Of course, the phenotype must be determined, and there are no guarantees with regard to the phenotypic properties of the end product. In addition, it is not always a straight forward task to introduce the mutated gene back into the viral genome.

Three general procedures have been used to introduce mutated fragments back into complete genome. In the first method, the mutation is made in a restriction fragment generated by an enzyme which cuts the viral genome only once or twice, and the mutated fragment is then ligated back to the adjacent fragment(s). For example, the genomic DNA of mutant H5dl309 (5) is cleaved only once by Xba I at 3.7 map units (m.u.)to generate a small left end fragment of 0-3.7 m.u., and a large fragment spanning 3.7 - 100 m.u. Many mutations have been made in the E1A genes contained within the small fragment, and these have been rescued into complete genome by simple ligation of that fragment to the larger one (6, 7, 8, 9, 10). Mutations made in cloned fragments obtained from internal regions of the viral DNA have been recovered into complete genome by three-fragment ligation (11). Should it be feasible to introduce the mutation only into a small fragment generated by a multicut enzyme, then this fragment can be inserted first into a suitable larger fragment prior to the ligation designed to give complete genome.

In the second method, the process of overlap recombination (12) is used to rescue mutations into complete genome. Mutations made in the E1A and the E1B genes contained in left-end 0-7.7 m.u. fragments were recovered into complete genome when such fragments were co-transfected into permissive cells with the H5dl309 XbaIA (3.7-100 m.u.) fragment (13). Rescue is successful when recombination in the overlap region takes place to the right of the mutation in question. Mutations made in internal fragments can also be incorporated into infectious genome by overlap recombination of three fragments (14).

If neither of the above methods can be conveniently employed, a third method may be feasible in some cases. This involves rescue of the mutated fragment by recombination with complete viral genome in doubly-transfected cells. Recently, the method has been used successfully to generate deletion (dl) mutations in the E2A gene encoding the 72K single-stranded DNA-binding protein (S. Rice and D. Klessig, personal communication, see below). In this case, a strong genetic selection was applied to force rescue of the deletion. The rescuing genome was derived from the double ts mutant H5 ts 1+2, which carries mutations in both the hexon and the 100K genes flanking the 72K gene, and the deletion was built into a DNA fragment spanning all three genes, and carrying wild-type alleles for hexon and 100K. Crossover events replaced the ts mutations with wild-type sequence, and some of these carry the deletions into complete genome. This type of approach should be applicable to introduction of other mutated genes into infectious genomes.

It is worth mentioning that many viral mutants did not have to be manufactured, since the mutations were already pre-existent in the viral population, and had only to be screened for or selected. The first adenovirus mutants of this type to be isolated were the cytocidal (cyt) mutants of type 12, which were recognized by the fact that they gave rise to large, clear plaques on permissive cells, in contrast to the very small plaques made by wild-type (15). More recently, type 12 variants adapted to grow in Vero (monkey kidney) cells (16), and human tumor cell lines (17), have been shown to have terminal rearrangements of the genome. A pre-existing heat-stable (hs) variant of type 5 adenovirus was selected from a wild-type stock by application of alternate cycles of heat inactivation *in vitro*, and growth in permissive cells (18). By selecting for variants resistant to the restriction enzyme XbaI, type 5 variants lacking XbaI restriction sites have been produced (19, 5). As already mentioned above, such variants have been extremely valuable for genetic manipulation and analysis.

GROWING VIRAL MUTANTS

Various methods have been developed for growing adenoviral mutants. Some mutants show little or no growth defect, and grow with wild-type efficiency under standard conditions. This is true of the type 12 cyt mutants (15), the type 5 hs mutant (18), and mutants with lesions in genes which are non-essential for viral growth such as the E3 and VAII genes (11, 19, 20). A variant with a mutation in the VA1 gene is viable, but produces slightly lower yields than wild-type virus (11).

A variety of conditionally-defective viruses, including ts (21, 3) and host-range (hr) mutants (22, 23) have been isolated for serotypes 2 and 5. The ts mutants, of course, grow efficiently in standard cell lines at permissive temperatures. The hr mutants, on the other hand, give very low yields in non-transformed, standard cell lines such as HeLa or KB, but grow to wild-type levels in type 5 transformed 293 cells. These cells are human embryonic kidney cells established as a result of transfection with type 5 DNA fragments, and they carry and express the type 5 E1 genes (24, 25). The E1 products made in these cells complement the growth of the defective E1 mutants, although it should be mentioned that some of the E1B mutants produced in this system, besides growing in 293 cells, also grow well in the precursor, non-transformed primary kidney cells (see below). Recently, similar complementing cell lines have been created by transfection with type 12 E1 DNA (26), and these have been used to isolate type 12 hr mutants (27; D. Breiding and J. Williams, unpublished results). Additional cell lines, expressing either E1A or E1B genes and capable of complementing E1 mutants, have been made by cotransformation of human cells with the appropriate DNA fragment and a selectable marker (28). This method has also been used to make cell lines which complement defective E4 (29), and defective E2A (72K) dl mutants (30: S. Rice and D. Klessig, personal communication). Cell lines expressing other viral genes would also be very useful, and it is likely that more will emerge in the near future. Human cell lines carrying suppressor tRNA genes, similar to the mouse cell lines previously made (31) would likewise be extremely valuable.

In the absence of complementing cell lines, defective mutants can in some instances be propagated by co-infection with complementing helper viruses. This approach will really only be useful in situations in which little or no recombination between the two genomes occurs, and in which the mutant can be physically separated from the helper component. Such prerequisites are likely to preclude wide applicability of this method. However, type 5 ts mutants have been used successfully as helper virus for the growth of type 2 defective dl mutants, and the mutant with the largest deletion was separated from the helper by CsCl density gradient centrifugation (32). It has been shown that type 12 adenovirus efficiently complements a wide variety of type 5 ts and hr mutants, and that the serotypes do not recombine efficiently (33, 34). Recently, we have shown that the two serotypes can be separated on the basis of their different buoyant densities, and we are using type 12 mutants as helper to complement the growth of defective type 5 mutants (B. Miller, D. Breiding and J. Williams, unpublished results).

MAPPING MUTATIONS

In the case of mutations generated by random mutagenesis (or pre-existing mutants), it is important that the lesion is mapped as precisely as possible on the genome, if significant phenotypic analysis is to be made. Adenoviruses recombine efficiently during infection of permissive cells, and as a prelude to more precise mapping, a genetic map can be constructed, based upon the recombination frequencies obtained from a set of two-factor crosses between mutants grown under non-selective conditions (35). A variety of methods have been used to locate mutations physically on the genome. The intertypic recombination method (36, 37) depends upon the presence of distinctive restriction endonuclease cleavage sites on the genomes of two serotypes capable of efficient recombination. Fairly precise boundaries can be set by this procedure, but in some cases it may be limited by the availability of suitable, differential cleavage sites. In the marker rescue method (38, 39), intact genomic DNA of the mutant and pure restriction fragments from wild-type genome are cotransfected into cells kept under restrictive conditions, and if a single mutation is responsible for the mutant phenotype, then one of these sub-genomic fragments will rescue the mutant allele and generate wild-type virus. Very good resolution can be obtained in some cases by using sets of overlapping fragments generated by different enzymes. The method can be used to reveal second-site mutations elsewhere on the genome, if these give rise to an identifiable phenotype, but it might not do so if such pairs of mutations are very closely linked. The method can, of course, be used to introduce mutations contained in cloned fragments into complete genome (see above). In the case of overlap recombination, two terminal DNA fragments with overlapping regions, each of which is non-infectious, produces infectious virus when cotransfected into appropriate cells (12, 13). The method allows detection and ordering of the distances between mutations within the overlap sequences, and in some cases, closely-spaced pairs of mutations can be segregated (40). In addition to the above mapping methods, the deletion mapping procedure of Benzer (41) works very well with adenoviruses (42). In this case, point mutants are crossed under non-selective conditions with dl mutants whose deletions are precisely known; if the mutation lies beyond the deletion limits wild-type recombinants form, while if it lies within the deletion limits no recombinants arise. Furthermore, deleted, cloned adenoviral fragments can be used in marker rescue analyses to verify and refine mutational map limits (B. Karger, D. Breiding and J. Williams, unpublished results). It should be noted that in some cases, the location of the mutated gene can be predicted from the mutant phenotype (for example, hr mutants which grow on 293 cells but not on HeLa cells map to the E1 region of the genome). Overall, there is very good agreement between the physical maps produced by these methods, and earlier genetic maps (35, 37, 39).

Using one or more of these mapping procedures the mutation under study can usually be assigned to a small segment of the genome, and in some cases it may be located with certainty in the reading frame for a specific protein. However, in some cases this may not be so, and because overlapping reading frames are very common in adenoviruses, it may be necessary in many cases to determine the exact site of the mutation by nucleotide sequence analysis (43, 44). This has now been done for many mutants isolated by random mutagenesis (examples are given in the tables in this review). In the case of the adenoviral E1A gene, it has been possible to narrow the map limits of mutations within even finer limits prior to sequencing (G. Glenn and R. Ricciardi, personal communication). The larger of the two proteins encoded by this gene activates transcription of the other early adenoviral genes (see below), and rapid assay of this activity can be made by measuring the ability of the product made by adenoviral E1A plasmids to induce the expression of the bacterial chloramphenicol acetyltransferase (CAT) gene cloned in plasmid and under adenoviral E3 control. Cotransfection of HeLa cells with wild-type E1A and CAT plasmids results in induction of the CAT gene, while cotransfection with hr E1A and CAT fails to do so. By constructing a series of hybrid E1A clones containing reciprocally arranged *hr* and wild-type segments of the E1A region and testing these in the CAT assay, the mutations responsible for the functional defect of a number of mutants were localised to a narrow region of the gene.

While mapping is particularly important for mutants generated by random mutagenesis, it is also very important to map the mutations made by fragment-directed and site-directed mutagenesis. In the former case, multiple mutations may be induced in the fragment, while in the latter, the procedures used to build the mutation back into viral genome may introduce additional changes.

PHENOTYPIC ANALYSIS OF ADENOVIRUS MUTANTS

FUNCTIONS OF THE E1A GENE PRODUCTS IN PRODUCTIVE INFECTIONS

The organization of the E1 region in group C adenoviruses, located in the extreme left of the viral genome, has been examined in detail by nucleotide sequence analysis, RNA mapping, and translational analysis *in vitro* of the E1 messages (for reviews and many references on these topics, see other chapters in this volume and 45, 46, 47). The region is divided into two transcriptional units, E1A (1·3 to 4·5 map units) and E1B (4·6 to 11·2 map units), each with its own promoter. Rightward transcription from the E1A promoter gives rise to three mRNAs, with identical 5' and 3' ends and unequal size (13S, 12S and 9S) resulting from differential splicing. The 13S and 12S RNAs are made very early in infection, while the 9S species appears at late times only. In serotypes 2 and 5 the coding regions of the two early messages specify polypeptides of 32K- and 26K-daltons (289 and 243 amino acids) respectively. The coding region for the larger product extends from nucleotide position 560 to 1543, with the region from 1112 to 1229 spliced out. The coding region of the smaller protein also extends from 560 to 1543, but in this case the intron is from 974 to 1229. Thus, the proteins from this region are identical, and differ only in that one contains a unique stretch of 46 amino acids. Gel electrophoretic estimates of the apparent molecular weights of the *in vitro* translation products of the 13S and 12S mRNAs vary considerably and are larger than the predicted values, lying within the ranges 41-58K and 35-54K respectively. The techniques of 2-D gel electrophoresis and immunoprecipitation using tumor antisera and an antiserum made against a synthetic polypeptide corresponding to the C-terminus of the E1A proteins have shown that the 13S and 12S mRNAs in infected cells encode a family of 4 major and 2 minor polypeptides (3 from each message), ranging in size from 35 to 52K (48, 49, 50) and the proteins are located in both nucleus and cytoplasm. It is not yet known why the E1A proteins are heterogeneous and possess higher-than-predicted molecular weights, but these properties may be related to the fact that all of them are phosphorylated (49). Furthermore, it is not yet clear if the E1A protein heterogeneity is functionally significant. It is worth noting that there is an unassigned open reading frame (URF) in the 1-strand transcript within the E1A region, which in type 2 extends from nucleotide position 1713 to 1197, and in type 5 lies between positions 1715 and 1067 (see 45, 46). This could encode a 23K polypeptide, but as yet there is no evidence that it is used during infection.

Most of the variants with mutations in E1A have been isolated as hr mutants which grow well in 293 cells, but not in certain other non-transformed human cell types such as HeLa (see section on growing mutants, above). The first group of hr mutants isolated had point mutations, induced by treatment of type 5 virus with either nitrous acid or ultraviolet light (22). An example of this type of mutant is hr1, which was isolated on the basis of its hr phenotype (22, 51). The second group of mutants isolated (also serotype 5), had deletions and insertions around the XbaI cleavage site at 3·7 map units, and were selected on the basis of loss of that site and acquisition of resistance to cleavage by XbaI (23). The mutant, dl312 is a representative of this second group. These, and some other E1A mutants discussed below are described in Table 1.

Studies with these prototype E1A mutants revealed that this gene encodes a regulatory product which activates the expression of the other early transcriptional units on the adenovirus genome. HeLa cells infected by these mutants display very low steady-state levels of these early mRNAs (52, 53) and their corresponding proteins (54) at early times following infection. Requirement for the E1A product is not absolute, and the effect of the E1A defect is to delay the onset of early viral transcription. The block can be overcome by infecting with the mutant at high multiplicity (53), and even at low multiplicity, transcription rates from the early promoters gradually increase after an extended period of time (56, 57). Furthermore, not all early transcription units are affected to the same extent. For example, cells infected with hr440 (8, 58) display reduced levels of E2 and E3 mRNAs, but normal levels of E1B and E4 messages

TABLE 1. E1A mutants of serotypes 2 and 5

Mutant	Mutation (nucleotide 1 is the left terminal nucleotide of the 1 strand)	Polypeptide Altered (AA1 is the N-terminal met of each protein)	Reference
hr1	1 bp deleted at 1055. Frameshift.	289R protein truncated to 175R. No alteration to 243R protein.	13, 22, 51, 52, 54
hr3	transversion of T to A at n 1086. Missense.	289R. AA176 Met to Lys. No alteration to 243R.	13, 22, 52, G. Glenn and R. Ricciardi, personal comm.
hr4	transition of C to T at n 1076. Missense.	289R. AA173 Leu to Phe. No alteration to 243R.	13, 22, 52, G. Glenn and R. Ricciardi, personal comm.
hr5	transition of G to A at n 1229. Missense.	289R. AA185 Ser to Asn. 243AA Gly 139 to Asp. No alteration to 243R.	13, 22, 52, G. Glenn and R. Ricciardi, personal comm.
hr^{cs}11	transition of G to A at n 1098. Missense	289R. AA180 Gly to Asp. No alteration to 243AA.	13, G. Telling and J. Williams, unpublished results.
in500	8 nucleotide insert at bp1010 Frameshift.	289R truncated. No alteration to 243R.	61
hr440	2bp change. transition of G to A at n979 and transversion of G to T at n980.	289R truncated. 243R not made.	8, 58
dl101	deletion of bp 1008 to 1012 Frameshift.	289R truncated to 186R. 243R not altered.	110
dl105	deletion of bp1003-1071.	289R shortened by 23AA. 243R not altered.	110
in106	16bp insert at bp1009 (TAG at 1013).	289R truncated to 137R. 243R not altered.	110
dl522	deletion of bp 1001-1071. Frameshift.	289R truncated to 137R. 243R not altered.	129

Table 1. continued

Mutant	Mutation (nucleotide 1 is the left terminal nucleotide of the 1 strand)	Polypeptide Altered (AA1 is the N-terminal met of each protein)	Reference
dl1500	deletion bp1110 to 1119 (donor splice site 13S RNA).	289R not made. 243R not altered.	63
dl520	deletion bp1107 to 1117 (donor splice site 13S RNA)	289R not made. 243R not altered.	129
dl347	deletion of bp975 to 1228 (mutant carries cloned DNA segment corresponding to 12S mRNA).	Makes only 243R.	64
pm975	transversion of G to T at n975 (donor splice site of 12S RNA)	243R not made. 289R not altered.	62, 63
dl348	deletion of bp1113 to 1228 (mutant carries DNA segment corresponding to 13S mRNA).	Makes only 289R.	64
dl521	deletion of bp 920 to 1139 (donor splice sites of both 13S and 12S mRNA removed)	289R and 243R both absent (9S RNA is made)	129
dl312	deletion of bp 448 to 1349	No E1A proteins made.	23, 53, 54

Abbreviations used: n, nucleotide; bp, base pair; AA, amino acid; R, residue.

early in infection. A similar effect is observed in wild-type-infected cells treated with the protein synthesis inhibitor anisomycin (56, 59, 60). Various lines of evidence show that the 289 residue (R) protein encoded by the 13S mRNA is the regulatory element. Mutants hr1, 3, 4 (52) and in 500 (61), all of which are mutated in the 13S unique region, exhibit the defect. The pm975 mutant, which is mutated in the 12S mRNA donor splice site and can't make the 12S message, is not defective (62), while dl1500 (63), which cannot make the 13S message, is defective. The 243R protein is thus not needed for activation of transcription, and has little or no activity in this process. Results with another mutant, dl347 (64) which can only produce the 12S mRNA, suggest that the 243R polypeptide is also capable of activating early transcription, but does so much less efficiently than the 289R product. More recently, it has been directly demonstrated that the 289R protein activates expression from the adenovirus E3 promoter. This protein has been expressed in E. coli, purified, and injected along with E3-CAT plasmid into Xenopus oocytes, where it stimulates CAT expression (65). To date, studies with hr mutants have not conclusively pinpointed the level at which regulation occurs. One set of pulse-labelling experiments concludes that the E1A product acts at the level of transcriptional initiation (56), while another investigation argues for posttranscriptional regulation (66). Strong support for the first view comes from results obtained in a transient expression system (67). HeLa cells transfected with chimeric plasmids in which the 5' sequences of the E2A and E3 genes are fused to the coding sequence of the CAT gene show very low levels of CAT activity. Cotransfection of these cells with the chimeric plasmids and functional E1A gene results in elevated levels of CAT activity. As just mentioned above, the pure 289AA protein does the same thing in the Xenopus system (65). Thus, the protein appears to act as a positive regulator of early viral genes at the transcriptional level, and the sequences required for this effect reside in the 5' ends of the early viral genes.

The mechanism by which the E1A gene product stimulates transcription from early promoters is still conjectural. It is not yet clear if the protein acts directly or indirectly to bring about this activation. On the basis of the observation that pre-treatment of HeLa cells with the protein synthesis inhibitor cycloheximide apparently reverses the transcriptional defect of the E1A mutants (56), it has been suggested that the E1A product acts indirectly by inactivating a short-lived cellular protein (whose synthesis is prevented by the inhibitor), which normally represses transcription of early viral genes. Using a similar approach it was found that anisomycin pre-treatment led to an apparent increase in the steady-state levels of early mRNAs in the cytoplasm of mutant infected cells (66), and concluded that the E1A protein inhibits a cellular protein which destabilizes viral mRNAs. However, others have found that cycloheximide pre-treatment of cells does not reverse the transcription phenotype of dl 312 (57). Explanation of this disagreement is not yet forthcoming.

The 289R protein may, of course, act more directly to stimulate transcription, by binding to regulatory sites on the viral genome. In this position it may facilitate RNA polymerase binding, either by interacting with the enzyme, or by changing the chromatin arrangement around the site to make entry of the enzyme easier. Based on the results from an elegant set of superinfection experiments, it has been suggested that in the absence of E1A, early transcription is induced by a cis-acting modification of the template (57). It has been further suggested that this modification may be the assembly of stable transcriptional complexes, and that the E1A protein catalyzes this event. Presumably, to bring this about, the 289R protein has to interact with one or more of the components in this complex, and it seems most likely that the 46R unique region comprises the active site. Point mutations in this region (see hr 3, 4 and 5 in Table 1), which introduce single, missense amino-acid substitutions drastically reduce the capacity of the protein to activate early transcription (52). These substitutions either directly alter the active site of the protein, preventing effective interaction with DNA or protein in the complex, or they alter function by disrupting the secondary or tertiary structure of the protein. However, the predicted secondary structures of the mutant proteins generated by computer analysis were compared with that for wild-type, and no major changes were observed (G. Glenn and R. Ricciardi, personal communication). The unique region contains a set of four cysteine residues, which are apparently conserved in all sequenced adenoviral serotypes, and these may be involved in the formation of disulphide bridges within or between molecules. The point mutations lie adjacent or close to these residues and may perturb disulphide bond formation leading to alterations in the tertiary structure of the protein. Location of the protein domain(s) necessary for transcriptional activation will require further site-specific mutagenic analysis. Ultimately, a clear understanding of the mechanism of action of this protein will emerge from biochemical and genetical studies of the pure protein (65), and the development of an in vitro system which will allow the identity and function of each regulatory component to be defined.

In addition to these effects upon viral transcription, the E1A 289AA protein also appears to activate certain cellular genes. For example, there is evidence that it activates the 70K heat shock protein gene in HeLa cells (68, 69) and cellular genes cloned in plasmids (70). Further, as discussed below, the E1A products are required for transformation, and the E1A gene is sufficient for immortalization of rodent cells (71). These processes could involve activation of one or more genes required for cellular growth. Besides this apparent capacity to activate genes, the E1A proteins of at least certain serotypes are capable of switching off certain cellular genes. A case in point is the depression of class I major histocompatibility (MHC) antigens and their mRNAs by the E1A gene of type 12 (72). It has been observed recently that both E1A gene products repress the transcription-stimulating ability of SV40 and polyoma enhancers linked to β-globin gene in plasmids, and it was shown that increasing amounts of either SV40, polyoma or adenovirus 2 E1A enhancer sequences compete out the repressive effect (73). This suggests a possible mechanism by which the E1A transcription unit could be autoregulated during productive infection, although it must be pointed out that it is not yet known if this E1A protein function operates under such conditions. These authors also suggest that depression of class I MHC gene products may occur via repression of putative enhancer elements.

Until quite recently, it was considered that the E1A 243R protein was not required for growth in human cells, since pm 975 (62) and dl 348 (64), which make only the 289AA protein, grow normally in non-transformed HeLa cells. It has, however, recently been shown that replication of pm 975 is poor on growth-arrested diploid secondary lung fibroblasts, and viral DNA synthesis is delayed and decreased in these cells (63, 74). Thus, although the 243AA protein may not be required for viral replication in growing cells, it does seem to be needed in growth-arrested cells, which are likely to be the predominant cell types encountered in infections of the natural host. E1A products are required for induction of DNA synthesis by type 5 adenovirus in G_0-arrested rodent cells (75, 76). It turns out that in growth-arrested, non-permissive, rodent cells, both E1A proteins are capable of inducing cellular DNA synthesis, but only the 243AA protein induces wild-type levels (74). This property may be related to the requirement for the protein in the process of transformation (see below).

FUNCTIONS OF THE E1B GENE PRODUCTS IN PRODUCTIVE INFECTIONS

At early times in productive infection, rightward transcription from the type 2 E1B promoter at 4·6 map units gives rise to two major mRNAs with identical 5' and 3' ends and unequal size generated by differential splicing. The larger, 22S (2·6kb) mRNA extends from nucleotide 1699 to 4061, with the region from 3504 to 3589 spliced out, while the smaller, 13S (0·9kb) mRNA extends over the same limits but is spliced from 2249 to 3589. The larger mRNA encodes two proteins from two overlapping open reading frames. One of these, initiating at the 5' proximal ATG at position 1711 and terminating at 2236, encodes a 175R protein. The other initiates from the second ATG at position 2017 in a separate frame, terminates at 3501 and encodes a 495R protein. The 13S message encodes only the 175R protein. Translation in vitro of E1B specific mRNAs gives rise to proteins with apparent molecular weights of 55-58K and 15-19K (77, 78), and proteins of these sizes have been immunoprecipitated from infected cells (79, 80, 81). An additional protein of 155R (18K) has also been identified (82, 83). It is coterminal with the 495R species, containing 78 amino terminal and 77 carboxy terminal residues of the former, and is translated from a minor mRNA species (1·2kb) which is spliced between positions 2249 and 3270 in addition to the small splice from 3504 to 3589. In infected cells the 495R protein is found in both nucleus and cytoplasm (49), and it becomes phosphorylated in both infected and transformed cells (84). Immunoprecipitates of this protein from infected cells possess protein kinase activity (85), but it is not yet known if this activity resides in a virally-encoded product or is a cellular enzyme associating with the 495R protein. In adenovirus transformants, the 495R protein is found in association with the p53 cellular tumor antigen (86), which is expressed at elevated levels in a variety of other transformed and tumor cells (see section on transformation below). In addition, the 495R species also becomes physically complexed with the E4 25K protein in infected cells (87).

The E1B 175R species is a non-glycosylated protein found in

association with nuclear and cytoplasmic membranes of infected and transformed cells (88). Recent immunofluorescence and electron-microscopic studies confirm that it is found in both cytoplasmic and nuclear locations at intermediate times after infection, but that at later times it is found predominantly in the inner and outer membranes of the nuclear envelope (89). In at least some transformed lines, some of the protein is also found at the surface of the cells (90; B. Stillman, personal communication). Although the 175R protein appears to show a marked preference for cellular membranes, including the plasma membrane, the amino acid sequence predicted from the DNA sequence does not reveal a signal sequence. The cellular location of the E1B-155R species has not yet been determined.

At late times after infection another, smaller (9S) mRNA is transcribed from a separate E1B transcriptional unit mapping between 9·7 to 11·2 units. The mRNA is unspliced, and the coding region (bp 3609 to 4029) specifies a 14K polypeptide corresponding to the virion structural polypeptide IX (91). In type 2 (but not type 5) there is an URF extending from bp 2291 to 2006 in the l-strand transcript within E1B. A 22S, l-strand-specific mRNA encoding an 11K protein has been identified in type 2-infected cells, and it was suggested that this may correspond to the product of the E1B URF (92). As with the putative E1A 23K polypeptide mentioned above, there is as yet no evidence that it is required for successful infection or transformation.

Currently, very little is known about the functional roles of any of the E1B gene products in either productive infection or transformation, and as with the E1A products, attempts are being made to define these functions using mutants. Mutations of various kinds have been introduced into both the 495R and the 175R genes, and some of the mutants isolated and under study are listed in Table 2. The first E1B mutants obtained, hr6 and hr7, were isolated on the basis of their host-range phenotype (22). As with the E1A hr mutants, they grow well on permissive 293 cells, and poorly on HeLa cells. Initially, these mutants were mapped by deletion and marker rescue methods (42) to the region between 6·1 to 8·5 map units. More recently, hr6 and another E1B isolate, $hr^{cs}13$ (13) have been sequenced throughout that region (see Table 2), and it has been shown that the mutations lie exclusively within the gene encoding the 495R protein. hr7 maps between 6·1 and 7·7 map units, and most likely lies within that gene, also. None of these mutants appears to produce immunoprecipitable 495R protein under non-permissive conditions (93, 94, P. Sarnow, Y. Ho, A. Levine and J. Williams,

TABLE 2. E1B mutants of serotypes 2 and 5.

Mutant	Mutation (nucleotide 1 is the left terminal nucleotide of the 1 strand.	Polypeptide Altered (AA1 is the N-terminal met of each protein)	Reference
5 hr 6	1 bp deleted at 2347. Frameshift. Transversion of G to T at 2947.	495R protein truncated to 118R. AA310 cys to phe in 495R protein. 175R and 155R proteins not altered.	13, 22, 42, 54. B. Karger and J. Williams unpublished results.
5 hr^{cs}13	Transversion of A to T at n2469	AA151 lys to UAG termination codon in 495R. 175R and 155R not altered.	13, 22, 42. B. Karger and J. Williams unpublished results.
5 hr 7	Probably point, between bp2048 and 2804.	495R altered. Perhaps 175R and 155R also.	13, 42.
5 dl313	Deletion of bp 1334 to 3639.	495R, 175R, 155R and protein IX not made. (Also removes C termini of E1A 289R and 243R).	23, 53, 54.
5 dl338	Deletion of bp 2809 to 3328. Stop codon at deletion join.	Makes N-terminal 223AA of 495R, and first 77AA of 155R, 175R not altered.	95. T. Shenk, personal comm.
5 dl100	Deletion of bp 2333 to 2804 Frameshift, TAA at 2851.	Makes N-terminal 104AA of 495R, and 14 missense AA. 175R not altered.	111
5 in127	Insertion of 4bp at bp 2808. TAG codon in frame at insert.	495R truncated to 263AA. 175R and 155R not altered.	111
5 dl 118	Deletion of bp 2049 to 2591; reading frame of 495R not altered. (Donor splice sites of 1·2 and 0·9 kb mRNAs removed by dl).	495R deleted by 181AA in N-terminal half. Only N-terminal 111AA of 175R should be made.	111

Table 2. continued

Mutant	Mutation (nucleotide 1 is the left terminal nucleotide of the 1 strand)	Polypeptide Altered (AA1 is the N-terminal met of each protein)	Reference
5 dl337	Deletion of bp 1771 to 1915.	175R truncated to 47R (19AA from N-terminus plus 28 missense AA).	101
2 dl250	As for dl337.	As for dl337.	102
2 lp3	Transition of C to T at n1718.	AA2 Ala to Val in 175R protein.	99, 100, 102
2 lp5	Transversions of G to T at 1954 and G to T at 2237.	AA81 Asp to Tyr and TGA termination codon to leu in 175R. AA74 Met to Ile in 495R.	99, 100, 102
2 cyt 106	Transition of G to A at bp1769.	AA20 Ser to Asn in 175R.	100
2 cyt6	Transition of T to C at bp1840	AA44 Lys to Glu in 175R.	103
2 cyt22	Transition of G to A at bp1721.	AA4 Trp to UAG termination codon in 175R.	103

Abbreviations used: n, nucleotide; bp, base pair; AA, amino acid; R, residue.

and B. Karger, unpublished results). For hr6 and $hr^{cs}13$ this is not too surprising, since only the N-terminal quarter of the molecule is likely to be made by these mutants. In hr6, the first mutation presumably determines the phenotype, and the second, which replaces a cysteine with a phenylalanine has no effect. The growth of hr6 and $hr^{cs}13$ on HeLa cells is markedly cold-sensitive, and they attain essentially wild-type yields within the normal time span at permissive temperature (13). However, the mutants are not cold-sensitive, and grow to wild-type levels at all temperatures on normal human embryo kidney cells and certain other non-transformed human cell lines (22 and J. Williams, unpublished results). These results are consistent with the view that the 495R species is not absolutely required for growth in human cells, and that there may be a cellular protein present in some cell types which can perform the function of the viral protein. Of course, if the 495R protein functions via interaction with a cellular component, the cold-sensitivity may signify a change in the viral protein to do so, and the host-cell dependence may reflect differences in the ability of the host component from the various cell types to interact with the mutant protein. It should be kept in mind, however, that these mutations do not alter the 155R protein, which is coterminal with the 495R species and consists of the 78 N-terminal and 77 C-terminal residues of the latter. This protein may share some function(s) with the larger one, and may act more effectively under some conditions or in some cells. Of course, like the coterminal E1A 289 and 243R proteins, the 495R and 155R proteins may also possess certain independent functions.

In addition to the hr mutants, a set of E1B deletion mutants was isolated on the basis of the loss of the XbaI site at 3·7 map units (23). A prototype of this group is dl 313, which has a large deletion extending from bp1334 in the E1A gene to 3639 in the E1B region. This eliminates the C termini of the E1A proteins, the entire coding regions of the 175R and 495R proteins, and the N terminus of protein IX. However, since dl 313 complements dl 312, it does seem to possess at least some E1A function(s).

With the exception of hr7, most of the 495R-specific mutants examined have a similar and very interesting phenotype. Mutants hr6, hr^{cs}13 (B. Karger, Y.S. Ho, K. Castiglia, S.J. Flint, and J. Williams, manuscript in preparation), dl 338 (95 and T. Shenk, personal communication), dl110, dl118 (alters 175R, too), in127 and dl163 (96), all fail to switch normally from the early to the late phase of the viral growth cycle. Under non-permissive conditions all make normal amounts of viral DNA, normal amounts of all early mRNAs and proteins (the level of E2A product is elevated in some), but accumulate reduced levels of late mRNAs and proteins. Late transcription rates are normal, however, so the defect may be manifest either at the level of RNA processing, transport, or stability. There is, in these mutant-infected cells, incomplete shut-off of host cell synthesis. In the case of the cold-sensitive mutants, all of these processes take place normally at permissive temperature. Collectively these experimental results suggest that the 495R protein plays a vital role in the viral growth cycle, and perhaps also in the shut-off of host-cell functions which takes place late in adenoviral infections, but is not possible to distinguish between a direct or primary effect and an indirect or secondary one. In view of the results with hr6 and hr^{cs}13, however, it should be kept in mind that this role is perhaps not entirely indispensable and may be cell-dependent. The results also indicate that although viral DNA synthesis per se may be required, it is not sufficient for normal expression of late gene functions. In contrast to the mutants just described, hr7 is defective for DNA replication and shows delayed onset of this event in HeLa cells (97, 98). Explanation of this phenotype awaits definition of the genotype by nucleotide sequencing.

Recently, a large number of mutants have been mapped to the 175R gene (99, 100, 101, 102, 103), and some of these are being phenotypically characterized. Initially, such mutants were isolated as large plaque variants, which grow well on HeLa or KB cells, and do not have an overt host-range phenotype. Some mutants mapping to this gene (dl337, cyt106, dl250) have been shown to possess an enhanced cytopathic effect, and are designated cyt. HeLa cells infected by these mutants tend to round up and lyse more rapidly than cells infected by wild-type virus (100, 101), and show a marked degradation of viral and cellular DNA, the deg phenotype. This aspect of the phenotype was first demonstrated for dl313 (which completely lacks the 175R gene) and certain type 12 cyt mutants (104). Two of the lp mutants, lp3 and lp5 have been described as having a cyt phenotype on HeLa cells (100), but a cyt^{+} phenotype on KB cells (99, 102), which may reflect either a cell-specific response, or some variance in judging the degree of the cytopathic effect. The lp3 mutant was considered to be cyt in HeLa cells, but found to be deg^{+} (100), suggesting that DNA degradation does not simply result from the cell-killing effect. On the contrary it might well be that the enhanced cytopathic effect results from the increased breakdown of cellular DNA. It is not yet clear what induces this DNA degradation, but it seems rather unlikely that the presumed nuclease activity results from an aberrant form of the 175R protein per se, since mutants which do not produce the protein (dl 337, dl 250) possess a deg phenotype. It could be encoded by another viral gene, and if so, it is likely to be an early one, since it is expressed in the presence of hydroxyurea (100, 101) and in cells infected by dl 313 (104). Alternatively it could be of cellular origin, and if so it might be switched on by the E1A gene product (see above). Certainly, mutants defective for E1A transcriptional activation, such as *hr1* and *dl312*, make reduced levels of E1B proteins, and appear to be *cyt*$^+$, *deg*$^+$. As mentioned previously (89), the 175R protein is localised largely in the nuclear envelope of HeLa cells infected with wild-type virus. In cells infected with *lp3*, which is *cyt*, *deg*$^+$, the distribution of 175R is normal, while in cells infected with *cyt106* and *lp5*, the distribution is aberrant, and the protein is found in structures separate from the envelope (89). In cells infected with *dl337*, no 175R protein is found. There is thus a correlation between absence or altered location in the nuclear envelope in cells and the *deg* phenotype. These authors suggest that nuclear envelope structures play a role in determining the spatial organization and perhaps the function of chromatin, and that the 175R protein acts to prevent the breakdown of chromatin by nucleases in infected cells.

THE ROLES OF E1 GENE PRODUCTS IN CELLULAR TRANSFORMATION

The human adenoviruses transform rodent cells *in vitro*, and a multitude of molecular and genetic analyses has shown that the adenovirus gene products responsible for this process are encoded within the left terminal 11% of the genome. It has, for example, been demonstrated that adenovirus transformants contain and express the E1A and E1B regions, and some fully transformed lines contain only that sequence (105, 106, 107). Transfection experiments using DNA fragments have shown that left-end (E1) fragments alone are sufficient to induce complete transformation (108, and see below). The introduction of mutations into E1A or E1B genes gives rise to mutants with a transformation-defective phenotype (13, 22, 23, 58, 61, 63, 99, 109, 110, 111). At the present point in time, relatively little is known about the adenovirus-cell interactions involved in establishing transformation. The process is likely to be complex and multistep, and the various E1 proteins may determine different components of the transformed phenotype. The identification of the viral gene products required for initiation and perhaps for maintaining the transformed state is a prime prerequisite to detailed functional analysis.

Attempts to identify and define the roles of adenoviral gene products in transformation and oncogenicity have followed two main pathways. In the first, the capacity of DNA fragments to transform rat cells *in vitro* was measured (108, 112, 113, 114, 115). These investigations showed that fragments encompassing the E1A region (0-4.3%) can immortalize cells whose lifespan *in vitro* is normally limited. The cells are only partially transformed, and are incapable of growing to high cell density. Cells transformed by fragments containing E1A and part of E1B (7-8%) appear to be more fully transformed, but still do not grow to the high cell densities usually attained by virally-transformed cells. In the case of type 5 transformants it has been found that these cells apparently produce both the 175R protein and a truncated version of the 495R protein, so either one or both of these proteins is directly or indirectly responsible for bringing about the complete transformation of cells. However, the E1B region alone, even when expressed, is unable to transform rat cells (116). Frame-shift mutations have been introduced into type 5 or type 12 E1 sequences in plasmid, and the mutated forms tested for transformation of rat cells (117). Plasmids with mutations in the E1B 495R gene can still transform, but those in which the mutation is close to the N-terminal end of the gene do so at reduced efficiency and induce a partially transformed (E1A-like) phenotype. Since the 175R protein is still expressed in these cells, it would appear that E1A and E1B 175R is insufficient to induce complete transformation. Support for this view comes from findings that frame-shift mutations put into the 175R gene alone apparently do not alter the ability of the plasmids to induce complete transformation of rat cells (117). However, others have found that the transforming capacity of cloned E1 DNA is greatly reduced by insertional mutation of the 175R gene by the bacterial transposon *Tn*5 (118), and *lp* and *cyt* viral mutants with mutations in that gene are known to be transformation defective (99, 100, 101, and see below). It thus seems likely that both E1B proteins are required for complete transformation, and there is evidence that both play a role in the development of tumorigenicity in cells transformed by type 12 virus. Cells transformed by plasmid containing 0 to 6.6% of the genome, or plasmids in which either the 175R or the 495R gene is mutated, are non-tumorigenic in nude mice (115, 117). Others, however, have found that rat kidney cells transformed by plasmid containing 0-6.6% of the type 12 genome do produce tumors in syngeneic baby rats, albeit with longer latent periods, and that some of these also produce tumors in nude mice (27). This group has also found that a high proportion of rat kidney cell lines generated with type 12 E1A fragment (0-4.6%) are tumorigenic in syngeneic, newborn rats (again with long latent periods), but not in nude mice. The E1A plasmid does contain the sequence encoding the N-terminal 18 residues of the 175R protein, but only the E1A species were detected in the transformants, so it would seem that the E1A gene product also plays a role in conveying the tumorigenic phenotype. Two other pieces of evidence agree with this view. First, it has been found that the DNA of E1B 495R hr mutants (see table 2 and below), can transform rat and hamster cells, although the mutant viruses do not (119). The hamster cell transformants, which show no detectable levels of either E1B 175R or 495R proteins, are tumorigenic in hamsters. Second, it has been found that mutation of the E1A 289R gene in E1 plasmid destroys its ability to transform rat kidney cells, but that fully transformed cells arise when the SV40 enhancer sequence is placed in the plasmid (120). These cells are, however, non-tumorigenic in nude mice. Thus, it would appear that the fully tumorigenic phenotype depends on expression of at least the E1A 289R product, and both major E1B products.

Analysis of adenovirus oncogenicity has also been carried out using E1 recombinant plasmids containing reciprocal arrangements of type 5 and type 12 E1A and E1B sequences (121). Tests with a set of cell lines transformed by such hybrid plasmids revealed that the E1B region seems to determine the degree of tumorigenicity: cells containing Ad12 E1B formed tumors at high frequency in nude mice, while those with Ad5 E1B were much less tumorigenic (122). However, the cells which were tumorigenic in nude mice were unexpectedly unable to induce tumors in syngeneic newborn rats, while the less tumorigenic ones (Ad12 E1A and Ad5 E1B) did so at low level. An explanation for these apparently contradictory results may lie in the finding that the type 12 289R protein induces depression of the levels of the class I antigens of the major histocompatibility complex in the rat, while the corresponding type 5 E1A product does not (123). In this case, the cells transformed by the type 12 E1A - type 5 E1B recombinants would escape the onslaught of cytotoxic T lymphocytes in the host, since the latter only recognize foreign viral antigens in transformants when these are complexed with the class I transplantation antigens, which are of course missing from these transformants. Possibly, therefore, region E1B products are determinants of cellular tumorigenicity in nude mice, while the E1A product is an important determining factor for malignancy in immunocompetent hosts. This could perhaps explain why some transformants induced by type 12 E1A alone are tumorigenic in syngeneic rats, but not in nude mice (27), and why fully transformed cells in which the E1A 289R gene is mutated are non-tumorigenic (120). Others have recently found a similar depression of H2 antigens in type 12, but not in type 5 transformed mouse cells (124, 125), and it has been found that mouse interferon-γ stimulates the expression of the class I genes to high levels in the type 12 transformants (K. Eager, J. Williams, D. Breiding, S. Pan, B. Knowles, E. Appella and R. Ricciardi, manuscript submitted for publication). However, another report suggests that the phenomenon of class I antigen depression may not be universal and that tumorigenicity of type 12 transformants in immunocompetent animals may be controlled by factors other than cytotoxic T cell surveillance (126). In agreement with this view is the finding that, even when they display reduced levels of H-2 antigens, not all type 12 transformants are tumorigenic in immunocompetent syngeneic mice (124 and K. Eager et al., manuscript submitted for publication).

In an alternative and complementary approach, viral mutants have been employed to analyse the roles that adenoviral gene products play in shaping the transformed cell phenotype. Many of the mutants described above and in Tables 1 and 2 have been used to show that all of the major E1A and E1B proteins appear to play important functions in the process. Unraveling the functions of these various proteins in transformation, showing how the proteins interact with each other and with cellular components, determining if they are required in a transitory way, or if they must be continually expressed to maintain certain facets of the transformed cell phenotype, constitute the main goals of the genetic approach.

Strong genetic evidence for requirement of an E1A gene function in transformation first emerged from studies using the type 5 E1A point mutants hr1, hr2, hr3, hr4 and hr5 (109). None of these mutants transformed rat embryo cells at 37°C, but they did induce incomplete and abortive transformation of baby rat kidney cells at higher than wild-type frequencies. The mutant-transformed cells remained somewhat fibroblastic, and unlike the wild-type transformants could not be established easily in culture, and died rapidly upon passage. At face value this result could be taken to mean that these mutants can initiate transformation as manifest by the emergence of partially-transformed foci in kidney cell monolayers, but that the change is not complete or permanent because functional E1A gene product is directly required to maintain the transformed phenotype. In the light of results (discussed above) which showed that the E1A 289R product activates expression of other early genes, an alternative interpretation also had to be considered. In this context, the role of the E1A product is to switch on the E1B gene(s) which is in turn required to maintain certain aspects of the transformed cell phenotype. Probably, however, this latter explanation is not the case. Some kidney cell lines, incompletely transformed by hr1 have been established in continuous culture, where they remain fibroblastic and apparently express normal amounts of E1B 175R and 495R proteins (127). In addition it has been found that the two E1A 289R mutants in 500 (61) and hr 440 (58) exhibit normal E1B expression, but fail to transform rat kidney cells. Although these observations do not exclude the possibility that one (or both) of the E1B proteins is required to maintain transformed cell traits, they do suggest that expression of these proteins alone is insufficient to maintain the complete transformed phenotype, and that functional E1A product is also required for this.

Just as it has been very valuable to use temperature-sensitive (ts) mutants in analysis of other viral gene functions (see 1, 2), so it would be for analysis of transforming gene functions. Despite many attempts by various research groups, no one has yet succeeded in introducing mutations into the E1 genes which generate mutants with ts (heat-sensitive) phenotypes. Recently, however, E1A mutants which are cold-sensitive (cs) for transformation, and in some cases for lytic growth on HeLa cells have been isolated (13). These mutants are conditionally defective for transformation of rat embryo cells, inducing partial transformation at normal or higher frequency at 38.5^{o}C, and at extremely low frequency at 35.5^{o}C and 32.5^{o}C. For example, hr1, which is not cs for growth in HeLa cells, is strongly cs for transformation of rat embryo cells (13), and cells of the cloned rat line CREF (128), and the results of temperature-shift experiments with this and other cs mutants suggest that the E1A region encodes a transformation maintenance function. Support for this view stems from the finding that some phenotypic properties of mutant-transformed cell lines, such as anchorage-independence, are also cs. The single base pair deletion of the hr1 mutation (see Table 1) produces a frameshift which results in translation of a truncated product from the 13S mRNA, while the product of the 12S mRNA is not altered (51). The incompletely transformed and cs phenotype of these cells might result either from the action of truncated E1A 289A protein, or that of the 243R protein.

It turns out that both of the E1A proteins are required to produce the completely transformed phenotype, and the cold-sensitivity is conferred by the smaller protein. This information was derived from analyses of type 2 mutants in which mutations prevent the splicing of either the 13S or the 12S mRNAs encoding the 289R and 243R proteins respectively (62, 63). One of these mutants pm 975, makes only the 289R protein, while the other, dl 1500, makes only the 243R protein (see Table 1). Both are defective for transformation of rat embryo and CREF cells. The mutant pm 975 transforms both embryonic and CREF cells at reduced frequency, and transformants of the latter, unlike wild-type transformants, still show anchorage-dependent growth. The other mutant, dl 1500, is cold-sensitive for transformation, in exactly the same way as hr1. Others have constructed a mutant similar to dl 1500 in type 5 adenovirus (129), and have obtained similar results. In addition, they showed that the 243R protein (of mutant dl 520) is capable of immortalizing baby rat kidney cells, indicating that the larger protein is not essential for that process. It will be recalled that this protein is required for viral growth in growth-arrested cells and stimulates cellular DNA synthesis under these conditions (63, see above). More than likely, this is the main function of the 243R protein in immortalization and transformation. The 289R protein activates the expression of viral and cellular genes in the lytic cycle, and there is a very good chance that this is one of its functions in transformation, too.

It is of interest that not all of the 289R hr mutants display a cs transformation phenotype (13, 110). For example, hr3 and hr4, both of which bear missense mutations in the sequence unique to the large E1A protein (see Table 1), are not cs for transformation of rat embryo cells (13). Of course, the wild-type virus does not display a cs transformation phenotype, nor does it transform at the high frequencies shown by hr1 and some of the other 289R mutants. Quite possibly, when both intact E1A proteins are present in the same cell the two interact in some way, the activity of the smaller is regulated by the larger one, and in the case of some mutants, such as hr3 and hr4 there is an interfering effect of the 289R protein on the activity of the 243R species. The presence of secondary mutations elsewhere in the E1A sequence, which alter the 243R product also, cannot be completely ruled out. It is worth noting that mutant $hr^{cs}11$, which has a single missense mutation in the unique sequence of the 289R protein is still cs for transformation, but does not show the elevated frequency typical of hr1, dl1500 and dl520 (13). Both this mutation and that of hr3 alter sequence in the region also occupied by the URF in the l-strand transcript, but do not alter the putative 23K protein encoded by that RNA due to the degeneracy of the genetic code. Thus, while that putative product could be involved in transformation, it is not sufficient to bring about complete transformation.

As with the E1A gene products, genetic evidence of requirement for E1B products in transformation came from studies using type 5 hr mutants with point mutations in the 495R gene (22). Unlike the E1A hr mutants, however, these E1B mutants were unable to transform either rat embryo or baby rat kidney cells, and they were not temperature-dependent for transformation (13, 109). The mutants are apparently defective for synthesis of the 495R protein (79, 94) and $hr^{cs}13$ would only be capable of making the N-terminal 151 residues of that protein in any case, so it must be concluded that this protein is required for complete transformation. Since the E1A and E1B 175R genes are intact in these mutants, one might expect to obtain the incomplete or abortive type of transformation observed when E1A mutants or E1A fragments are used. This has not been found when embryonic or rat kidney cells are infected, so it appears that the presence of the E1A proteins and the E1B 175R protein in the absence of functional E1B 495R product is insufficient for the initiation or the maintenance of transformation. These results are in agreement with the mutant plasmid results discussed above (117). It should be noted that the small 155R protein which is co-terminal with the 495R species (83) would not be altered by the hr6 and hr13 mutations, since these lie within the region spliced out of the minor mRNA encoding it. Thus, although this protein may play a role in transformation, it too is not sufficient for establishment of the event. These observations strongly suggest that an initiation step is involved, but do not exclude a maintenance role for this protein, also. Conditional mutants might help to resolve this problem. Unfortunately, hr6 and $hr^{cs}13$, which are cs for growth in some human cell lines, are not cs for transformation. It has been reported that the E1B mutant dl 313, which lacks all of E1B and the C-terminal part of E1A (see Table 1) elicits partial transformation of 3Y1 cells (130), but this is not the case when embryonic cells are used (55, J. Williams, unpublished observations). The fact that 3Y1 is a permanent cell line which is already immortal may account for this apparent disagreement. Several other mutations, in the form of deletions and insertions, have been introduced into the 495R gene, and all have been shown to be defective for transformation of CREF cells (111, 95 and T. Shenk, personal communication). Deletions in the C-terminal part of the protein, such as dl 163 (111) and dl 338 (95 and T. Shenk, personal communication) appear to be less defective than those with more N-terminal alterations giving rise to shorter truncated product, such as dl 110 and in 127 (111). The transformants which did arise from these infections, however, were completely transformed in contrast to the partial transformants induced by the E1A mutants. It should be kept in mind that these studies were carried out on CREF cells, which again are already immortalized.

Collectively, the studies with the various 495R mutants show that this protein is required to effect complete transformation by adenovirus type 5. However, as pointed out above, when DNA is used, the left end 7.7% fragment of type 5 is sufficient for complete transformation of rat kidney cells (108). Such transformants do not contain immunoprecipitable 495R protein (131), so it would appear that either this protein is not required for transformation, or that at least its C-terminal half is not needed. It has been found, however, that the DNA of the E1B 495R mutants can fully transform rat and hamster kidney cells with wild-type frequency, and not surprisingly, all of the transformants lacked the 495R protein (119). Thus, it may be that this protein is required to initiate transformation, either directly or indirectly, when virus is used, but not when DNA is used. There is an apparent discrepancy between these DNA transfection results, and those obtained with mutant E1 plasmids in which frame-shift mutations were inserted in the 495R gene (117, discussed above). Mutants with lesions near the N-terminus of the gene transformed with reduced frequencies, and induced foci with a partially transformed (E1A-like) phenotype. The reason for these apparently conflicting results is not clear, but differences in culture conditions, cell types and gene dosage may be contributory factors.

There is now considerable genetic evidence that the E1B 175R gene product is also required for transformation. The various lp and cyt mutants which map to this gene are transformation-defective, whether virus or viral DNA is used (99, 100, 101, 102, 103, 111). These results agree with the finding that the transforming capacity of E1 DNA is reduced by insertional mutation of the 175R gene (118, see above), but disagree with other claims that frame-shift mutations in this gene do not reduce the transforming capacity of E1 plasmids (117). Further disagreement about the role of the 175R protein in transformation stems from separate studies which on the one hand suggest that some fully-transformed cells possess no detectable (immunoprecipitable) protein (119), and on the other hand find it in all of a large number of viral (82) and E1 DNA transformants (131). Of course, as with the E1B 495R mutants, none of these various investigations defines the role of the protein, and it is not yet clear if it is required in a transitory way for initiation, or is required for maintenance. Should it simply be required during the initial phases of the transformation process, then there would be no need for its continued presence in the cell. Obviously, the mere presence of the protein in a transformed cell does not mean that it is functionally required. Any interpretation of 175R mutant transformation results should take into account that most of the mutants used are cytocidal and cause extensive damage to the cellular genome. They may simply fail to transform certain cell types because they kill them first.

In summary, it appears that both E1A products are required for transformation of rodent cells. The E1A 243R protein may immortalize cells by stimulating cellular DNA synthesis and driving them out of a growth-arrested condition, and it may also be required to maintain certain facets of the transformed cell phenotype, such as anchorage-independence. The E1A 289R protein also plays a decisive role in transformation, and might do so by activating cellular genes whose products contribute to the complete phenotype. Both proteins might also exert influence on the phenotype by switching off, rather than on, certain cellular genes via repression of enhancer elements. The major E1B proteins also seem to be required for complete transformation. The 495R protein may be involved in initiation or establishment of the transformed state, but a role in the maintenance of certain aspects of the phenotype is by no means ruled out. It is of great interest, and perhaps more than mere coincidence that this protein interacts with the product of the p53 cellular oncogene in transformed cells (86), and seems to stabilize and increase the level of this protein in these cells. Recently, it has been shown that like the adenoviral E1A gene, the p53 gene is capable of immortalizing rat chondrocytes _in vitro_ (132), and complementing the ability of the _ras_ gene to elicit oncogenic transformation in primary rat cells (133, 134, 135). Thus, it is possible that one function of the 495R product might be to stabilize the immortalization proteins, and increase their effective concentrations during both the initial phases of transformation and later, in the fully transformed cell. However, there is at the present time no evidence that the 495R protein stabilizes the adenoviral E1A immortalization proteins. There is some evidence from studies using chimeric plasmids that the 495R protein influences the tumorigenicity of transformed cells (121), and recent evidence (see below) using type 12 hr mutants supports this view. Others, however, have suggested that this protein is not absolutely required for tumorigenicity (119). The role of the E1B 175R protein in transformation is not yet known, but it is possible, at least when virus is used, that it acts indirectly in the process by preventing genomic damage and cell death, and ensuring survival of the cell. As mentioned above, some of this protein is found at the surface of transformed cells where it could influence the behavior of the cell and act as a tumor-specific transplantation antigen (TSTA). There is some evidence, using E1B hr and dl mutants, to suggest that this latter possibility is the case (136). Further genetic analysis should help to resolve some of these problems and provide deeper insight into the mechanisms by which these viruses transform cells.

FUNCTIONS OF THE E2B GENE PRODUCTS: DNA REPLICATION PROTEINS

The E2 transcriptional unit, which gives rise to leftward-reading transcripts, maps between coordinates 11 and 75 on the adenoviral genome, and the proteins encoded by this unit are primarily involved in replication of the viral DNA. The first transcriptional mapping studies to be carried out located mRNAs lying between 62 to 67 map units, but did not show any others to the left of 62 units (137, 138). Revelation of additional early transcription from the l-strand came from physical studies which placed the mutations of early function group N ts mutants of type 5 virus between 18.0 and 22.6 units on the genome (139, 140). At early times after infection transcription was shown to occur at very low levels from l-strand sequence in this region, and also between 11 to 14.5 units (139, 140). Further studies have shown that these mRNAs are generated from large transcripts, which commence around position 75 (141), previously shown to be the promoter region for the E2 transcription unit (142, 143, 144), and terminate around position 11. The transcripts have leader sequences from around map positions 75, 68 and 39, spliced to main bodies at 30, 26 and 23 units. Translation of these messages *in vitro* generates products with apparent molecular weights of 75K, 87K and 105K (141); the 87K species corresponds to the 80K precursor of the 55K terminal protein (141), while the 105K is the 140K protein identified in DNA replication systems *in vitro* as a DNA polymerase (145, 146, 147, 148, 149).

Sequence analysis of the E2B region in serotype 2 showed the presence of two large open reading frames (150, 151, and reviewed in 45, 46). One of these extends between a start codon at nucleotide 10531 and a termination codon at 8572 (28.8 to 23.5 units) and encodes the precursor to the terminal protein (152). The second large open reading frame extends from an initial ATG at nucleotide 8354 to a terminator at 5189 (22.9 to 14.2 units), and specifies the adenovirus DNA polymerase just mentioned above. In addition to these two large reading frames, a number of small unidentified open frames come from l-strand sequence in E2B, but none of these is of sufficient size to encode the 75K *in vitro* translation product, referred to above (141). Thus, the origin and identity of that protein remain in doubt. The complementary r-strand in this region encodes the three major late and the i leader sequences, and the four unidentified reading frames in r-strand transcripts, capable of encoding small proteins ranging in size from 9K to 18K. The i leader (in contrast to the late leaders) contains an open reading frame sufficient to encode a protein of 16K. The *in vitro* translation of mRNA selected by a DNA fragment spanning the i leader, generates a 16K protein (153, 154, 155), now known as the adenovirus agnoprotein. This rather complex region of the viral genome is a good example of how the adenoviruses economise by putting a small stretch of base pairs to multiple and varied use. We know little or nothing of the functions of these potential products from this region other than the terminal protein and the polymerase, and the region has yet to be dissected genetically.

In fact, the only mutants known to map in this region are those in complementation group N, whose lesions lie between 18.0 and 22.6 map units in E2B (139, 140). As yet, none of these mutations have been sequenced, and potentially they may also alter one of the unidentified reading frames, the second late leader, or the i leader (containing the agnoprotein frame). The four N group mutants which have been examined, *ts*36, *ts*37, *ts*69 and *ts*149, fail to synthesize viral DNA at non-permissive temperature (156, 157), and the results of temperature-shift tests indicate that the initiation of synthesis is most likely the defective stage (157, 158).

In cells infected with these mutants at restrictive temperature, the pattern of early RNA expression is apparently normal (159), but no late mRNAs (159) or proteins are made (160). To date, the expression of E1B mRNAs has yet to be examined in cells infected with these mutants. The fact that these *ts* mutations were found to map in the reading frame encoding the 120K protein and to cause defective DNA replication provided genetic support to the conclusion that the 140K DNA polymerase is the product of that frame. Further support for this view stemmed from dominance interaction tests between these mutants and wild-type virus in mixed infection (139, 140). The results of these experiments were consistent with the view that the N group mutations reside in a gene encoding a catalytic product. With the development of a soluble, *in vitro* system for the replication of adenoviral DNA (161), it was shown that extracts of cells infected at restrictive temperature with *ts* 36 or *ts* 149 were defective for synthesis of viral DNA *in vitro*, and that the purified 140K protein, which has DNA polymerase activity, complements the defect (147, 148, 162, 163). This *in vitro* complementation of *ts* mutant DNA replication forms the basis of a useful method for purifying replication proteins. Collectively, these results show that adenovirus encodes its own DNA polymerase. There is strong biochemical evidence that the virus also encodes the precursor to the terminal protein, but since no mutants have been isolated for this gene, there is, as yet, no genetic support for this view.

Using the soluble, *in vitro* DNA synthesis system (161), considerable progress has been made in unraveling the complex process of adenoviral DNA synthesis. An extensive overview of this work has appeared in a series of recent reviews (164, 165, 166, 167, 168), and since an update of the topic appears in another chapter of this book, detailed discussion will not be made here. The initial step in DNA replication is the covalent linkage of the terminal protein precursor with the 5'-terminal nucleotide of the new DNA strand, dCMP, and this complex is the primer for nascent strand elongation by DNA polymerase. Besides bringing about elongation by adding nucleotides to the 3'-OH of the primer, the polymerase probably also catalyzes the linkage of dCMP and terminal protein. The details of these processes have yet to be worked out, and it is not known if these proteins fulfill other functions in the infected cell. Genetic analysis is likely to be important in understanding how these proteins function and interact with each other, with the genome, and with cellular proteins. Much of this genetic work can be carried out in the simplified in vitro replication system. Already, the sequence comprising the viral replication origin has been cloned and resolved by mutational analysis into two functional regions (169, 170, 171). The first comprises the first 18 base pairs of the genome and allows a low level of initiation while the second includes the next 49 base pairs and enhances the level of initiation. It should be relatively easy to clone and express the viral replication genes in expression vectors, and determine the effects of a series of defined mutations upon the functions of these proteins in the in vitro system, and after building them back into complete genome, within the infected cell. The in vitro replication system has also allowed the identification and characterization of host components required for viral DNA replication, and at least three such host factors have been isolated from extracts of uninfected human cells (see 167 and 171). One of these cellular proteins binds with very high affinity to the second functional region within the origin sequence, and it appears that this interaction enhances the initiation of viral DNA synthesis. Neither the mechanism of action of this protein in viral DNA replication, nor its function in the uninfected cell are understood at the present time, but it is of interest that the protein does seem to be conserved, and is found in active form in rodent cells (171). The group C adenoviruses replicate their DNA efficiently in mouse (172, D. Breiding and J. Williams, unpublished observations), and Chinese hamster ovary cells (173). These particular cell types have been used extensively for the production of genetic variants, and offer a rich supply of mutants in which various phases of the cell cycle, including DNA synthesis, are affected (for review, see 174). Thus, they would seem to offer a potentially valuable system for genetic and biochemical analyses of the cellular proteins involved in adenoviral DNA replication.

In addition to affecting viral DNA replication, mutations in the polymerase (N) gene also alter the ability of adenovirus to transform rat cells. The type 5 N gene mutants, ts 36, ts 37, ts 69 and ts 149 are all capable of transforming with wild-type frequency at permissive temperature, but do so with reduced frequency at non-permissive temperature (35, 140). This result is somewhat surprising in the light of the evidence presented above that the E1 region of the genome is sufficient for complete transformation of rodent cells. The available evidence more or less excludes the possibility that the transformation defect results from second-site mutations residing in the E1 transformation region, and suggests that the mutations in the polymerase gene cause it. For example, all four independently-isolated mutants are defective, while a *ts*$^+$ revertant of *ts* 36 transforms at wild-type frequency (J. Williams, unpublished results), as do left-end DNA fragments of *ts* 36 which eliminate the polymerase gene (J. Arrand, personal communication). It seems likely that an initial or preliminary step in the transformation process is involved, since cells held at permissive temperature for two days prior to shift-up to non-permissive temperature become transformed at wild-type frequencies, but it should be noted that a transformant which is *ts* for growth has been isolated following infection with *ts* 37 (35). The function of the viral DNA polymerase in transformation is not known, but almost certainly the process is not prevented by simply blocking viral DNA synthesis. Since DNA-negative E2A mutants (see below) are capable of transformation (some at elevated frequency) at restrictive temperature, it would appear that viral DNA synthesis per se is not required for the process. Possibly, the polymerase is involved either prior to or during the process of integration of the viral DNA into the cellular genome. If so, this event must be circumvented when DNA fragments are used for transformation, since E1 fragments which exclude the N gene can transform cells. It should, however, be pointed out that it has not been rigorously ruled out that E2B fragments also possess transforming capacity, or at least the ability to influence the transformed phenotype in some way. An alternative possibility, again when virus is used, is that the polymerase either directly or indirectly regulates the levels or activities of one or more of the E1 transforming gene products following infection of rat cells.

THE E2A GENE PRODUCT: A MULTIFUNCTIONAL DNA REPLICATION PROTEIN

The E2A region lies between 61.6 and 75.1 units, and the leftward-reading transcript gives rise to an mRNA which encodes the single-stranded DNA-binding protein (DBP), the only known translational product of the region. Nucleotide sequence determination of the DNA and the mRNA from this region (175, 176) revealed that the DBP mRNA of serotypes 2 and 5 has a main body extending from 66.5 to 61.6 map units. An open reading frame of 1587 nucleotides encodes a polypeptide of 529 amino acid residues with a molecular weight of 59, 042 daltons. The apparent molecular weight of type 2 and type 5 DBP is usually determined at 72,000 daltons, and it

TABLE 3. E2A mutants of serotypes 2 and 5

Mutant	Mutation (Nucleotide 1 is the A of the ATG on the r strand)	Change in protein (AA1 is the N-terminal met)	Reference
5*ts*107 and 5*ts*125	Transition of C to T at n1237	AA413 pro to ser	175, 181.
2*ts* 23	Transition of C to T at n844	AA282 leu to phe	176, 183, J. Williams and L. Fraser, unpublished results.
5r (*ts*107) 202	C to T at 1237, transition of G to A at n1055	AA413 pro to ser AA352 gly to asp	210, 213.
5r (*ts*125)13	C to T at 1237, transition of C to T at n1522	AA413 pro to ser AA508 his to tyr	210, 213.
5r (*ts*125)20	C to T at 1237, transversion of G to C at n1039	AA413 pro to ser AA347 ala to pro	210, 213.
5 *hr* 404	Transition of C to T at n388	AA130 his to tyr	175, 196.
2 *hr* 400	Transition of C to T at n388	AA130 his to tyr	196, 197, S. Rice and D. Klessig, personal comm.
2 *hr* 405	Transition of C to T at n388 and C to T at n443	AA130 his to tyr AA148 ala to val	209, S. Rice and D. Klessig, personal comm.
2 *ts* 400	Transition of C to T at n388 C to T at n433 C to T at n1237	AA130 his to tyr AA138 ala to val AA413 pro to ser	209, S. Rice and D. Klessig, personal comm.
5 *dl* 801	Deletion of bp 23 to 335	N-terminal 7AA fused to 15 missense AA after dl.	S. Rice and D. Klessig, personal comm.
5 *dl* 802	Deletion of bp 65 to 306	N-terminal 21AA fused to 32 missense AA	S. Rice and D. Klessig, personal comm.

Table 3. continued

Mutant	Mutation (Nucleotide 1 is the A of the ATG on the r strand)	Change in protein (AA1 is the N-terminal met)	Reference
5 dl 803	Deletion of bp217 to 1488	N-terminal 72AA fused to C-terminal 33AA	S. Rice and D. Klessig, personal comm.
5 dl 804	Deletion of bp 687 to 1488	N-terminal 228AA fused to 43 missense AA.	S. Rice and D. Klessig, personal comm.
5 dl 805	Deletion of bp 217-686	N-terminal 72AA fused to 119 missense AA.	S. Rice and D. Klessig, personal comm.

Abbreviations used: n, nucleotide; bp, base pair; AA, amino acid.

is believed that this discrepancy may result largely from the high proline content of the molecule, which generates a coiled, nonglobular structure, and perhaps in part from phosphorylation of the molecule. At early times in infection, transcription of the E2A unit commences at a promoter located at 75.1 units on the genomic l-strand (177), while at later times in infection an alternative promoter at 72.0 units is used. As mentioned above, the promoter at 75.1 units is also used for transcription of the E2B unit. The E2A mRNAs made at early times in infection possess short leader sequences from 75.1 and 68.8 (175, 177), while the species made later in infection have leaders from 72.0 and 68.8 units. In each case the same ATG and reading frame are used.

The DBP was first isolated on the basis of its capacity for binding to single-stranded DNA (178). It binds very efficiently, but not covalently, to such DNA (179), and to the termini but not the internal regions of double-stranded DNA (180). Extensive analyses of a large variety of DBP mutants (see table 3) have made this protein the functionally best-characterized of the adenoviral early proteins. To date, four classes of DBP mutants have been isolated and studied phenotypically.

The first class comprises the ts mutants, of which type 5 ts 125

(181) is the prototype. The DBP of this mutant, and that of the Ad2ND1 mutant, ts 23 (J. Williams and L. Fraser, unpublished results) are thermolabile (182, 183). In cells infected at nonpermissive temperature with these mutants, viral DNA replication is defective, and on the basis of temperature-shift tests and experiments using antibody to DBP (with infected nuclei), it was suggested that the protein is needed for both initiation and elongation steps (183, 184). However, it has been shown that in the in vitro system, the DBP is not required for the formation of the 80K preterminal protein - dCMP complex, which is the initial step in adenoviral DNA synthesis (see above). Extracts of cells infected with ts 125 make complex at restrictive temperature (185, 186), and complex forms in vitro in the absence of pure wild-type DNA (145, 187). In the mutant extracts incubated at restrictive temperature, the early part of the elongation phase is barely affected, so that either the DBP is not needed, or a cellular binding protein can substitute in this process (186). As the elongation process progresses, however, the DBP of the virus is obligatory (186). The 72K DBP can be cut by chymotrypsin to yield two fragments (188, 189), and it has been shown that the larger (45K) C-terminal fragment, which still binds to single-stranded DNA, complements the replication of ts 125 DNA at restrictive temperature in vitro (190). Further, a 34K subunit of the C-terminal piece is sufficient for efficient complementation of viral DNA synthesis in vitro (186). The N-terminal, phosphorylated, 25K fragment, on the other hand, fails to bind to single-stranded DNA, and is not necessary for DNA synthesis in vitro. Thus, the C-terminal half of the DBP molecule contains the functional domain required for viral DNA synthesis.

Defective DNA replication is not the only phenotypic trait displayed by ts 125. In cells infected with the mutant at restrictive temperature, the DBP is made, but does not accumulate to wild-type levels because it undergoes rapid degradation (182). At intermediate and late times after infection these cells do, however, contain elevated levels of the DBP and other early mRNAs (such as E1A and E1B), in contrast to cells infected with wild-type virus, which show a switch-off in early mRNA expression. On the basis of these results, it has been suggested that the DBP autoregulates the production of its own message (191), and it has been further suggested that overproduction at restrictive temperature results from enhanced stability of the mRNAs (192). Thus, in this context, wild-type

DBP would impose a negative regulation of its own and the E1A and E1B genes by altering the stability of these mRNAs in the cytoplasm of the infected cell. Based upon other studies with ts 125 it has been proposed that the DBP represses transcription from the E4 region (but not its own promoter), and that this repression may act at the level of transcriptional initiation (193). In neither case of apparent negative regulation has it been determined if the effect is mediated directly, or indirectly, via another viral or cellular gene product. In addition, it has also been suggested, on the basis of microinjection experiments using purified adenoviral genes or mRNAs, that the DBP also stimulates the expression of E1B and E4 genes (194, 195). Lastly, as will be discussed more fully below, Ad5ts125 and Ad2 ND1ts23 transform rat cells at elevated efficiency compared with wild-type virus.

The second class of DBP mutants is that group of variants which demonstrate altered host-range on monkey cells. The prototypes of this group are type 2 hr 400 and type 5 hr 404, which have been mapped to the N-terminal region of the gene (196, 197, 175). Wild-type human adenoviruses do not grow well in monkey cells, although early gene functions are normally expressed (198), and the level of viral DNA synthesis is similar to that found in human cells (199). However, the levels of some late mRNAs and late proteins are reduced by about 2-10 fold in such cells (200, 201, 202). In the case of fiber gene expression, however, the mRNA level is reduced about 5-10 fold (as for other late messages), but protein synthesis is reduced by 100-1000 fold (201, 203). This obstruction of fiber gene expression has been attributed in part to defective splicing of the RNA (204), and in part to defective translation of the message (203), with a strong possibility that the latter defect is related to the first. For a number of reasons it is unlikely that the fiber translation defect is imposed simply by the translational apparatus of the monkey cell. For example, the fiber message appears to associate normally with polyribosomes from abortively infected monkey cells (203), and fiber synthesis takes place normally in cell-free lysates from such cells programmed with viral mRNA from productively infected HeLa cells (197). Further, it has been shown that in monkey-human heterokaryons, the cytoplasm of the monkey cell can translate the fiber message made in the adenovirally-infected human cell nucleus (205). An alternative explanation for the fiber translational defect implicates the ancillary x and y leader sequences. These leader sequences are encoded by early region 3 (see below), and they are found only on the messages of the region 3 19K glyco-protein and the fiber (also a glycoprotein). A reasonably high proportion of the fiber mRNAs made in productive infections (in both human and simian cells) contain either the x or the y leader, while in the abortive infection of monkey cells, the fiber messages lack these ancillary leaders (206). These results are interpreted to mean that the presence (of one or other) of these leaders on the fiber mRNA is required for its translation, and that the primary defect in the abortive infection of monkey cells lies at the level of RNA processing. The authors point out that the fiber initiation codon lies immediately adjacent to the leader sequences, and that minor changes in the latter could lead to a major perturbation in translation of the message. They also suggest that the x and y leader sequences may constitute a signal (at the RNA level), which guides the fiber (and E3 and 19K) mRNAs to the appropriate cell compartments for translation and glycosylation of their products.

The DBP hr mutants, in contrast to wild-type virus, do not display any of these defects. They make normal amounts of late mRNAs, correctly spliced fiber mRNA, normal amounts of all late proteins, and grow efficiently in monkey cells (196, 204, 206). Although the mechanism by which this occurs is not yet understood, it seems likely that the DBP normally interacts with human cell factors involved in some event(s) in late viral gene expression (RNA splicing, for example), but cannot interact effectively with the equivalent components in monkey cells. The hr mutations alter the DBP so that it can interact more efficiently with the monkey cell factors, and thus permit normal expression of late genes in these cells. It is worth noting that the block to growth in monkey cells can also be overcome by co-infection with SV40, and through use of $Ad2^{+}SV40$ hybrids and SV40 deletion mutants, it has been shown that the helper activity resides in the C-terminal region of the SV40 large T antigen (20, 207, 208). This suggests, but doesn't prove that DBP and T antigen permit efficient expression of late adenoviral genes in monkey cells via the same mechanism. Whatever this mechanism might be, it is clear that these mutations lie in the sequence encoding the N-terminal part of the DBP, and all the evidence suggests that this region comprises a separate, late-functional domain of the protein. Phenotypic analysis of a ts variant of type 2 hr 400 supports this view, and provides evidence that the DNA replication function, altered by a mutation in the C-terminal region of the gene (and identical to that in ts125), acts independently of the function required for viral late gene expression, which is altered by mutations in the N-terminal region of the gene (209). These findings corroborate the biochemical evidence for separate functional domains in the DBP which was discussed above.

Studies with a third class of mutants, namely the temperature-independent revertants of mutants ts125 and ts107 (210, 211, 212, 213), provide additional evidence in support of the view that the DBP interacts with host cell components. Four genotypic groups, comprising two phenotypic classes of revertants have been isolated. The members of group I are true revertants, and have wild-type sequence across the entire DBP gene. Group II revertants (such as 5 or 13) have a second-site mutation in the DBP gene at nucleotide 1522, which results in substitution of the histidine at position 508 to a tyrosine. These mutants overproduce a stable DBP in HeLa cells, but make essentially wild-type levels in human 293 cells and monkey cells. This host-cell specific autoregulatory response could perhaps mean that the DBP interacts with one or more cellular factors to regulate its own synthesis, and that these factors differ in the various cell lines. Groups III and IV revertants (5r202 and 5r20) represent a second phenotypic class which displays a host-range temperature-sensitivity, in that the mutants are wild-type for growth in HeLa cells but ts in 293 cells. They produce more or less normal levels of early viral products, viral DNA (reduced about two fold) and late proteins, but do not make viral particles or infectious progeny at restrictive temperature in 293 cells (214). This result suggests that the DBP is involved, either directly or indirectly, in either viral late protein synthesis or virion assembly, and is again consistent with the hypothesis that the DBP interacts with cellular factors to fulfil some of its numerous functions.

The last group of DBP mutants comprises a set of recently isolated deletion mutants which either make truncated DBPs or fail to make detectable DBP (S. Rice and D. Klessig, personal communication). Isolation of these defectives was made possible by the development of human cell lines which contain and express integrated DBP gene (30). Previous attempts to obtain such cell lines suffered from the fact that the DBP is toxic to human cells, and to circumvent this problem, the cells were transformed by plasmids containing the Eco-gpt gene as a dominant selectable marker, and the DBP gene under control of a modulatable promoter, in this case the mouse mammary tumor virus promoter, which can be activated by glucocorticoids. Thus, in the absence of hormone, very low levels of DBP are produced, and the cells are able to grow. On adding dexamethasone, DBP production is switched on, and under these conditions the dl mutants are complemented and grow to moderately high levels. Unfortunately, these cells do not support plaque production by the dl mutants, and infectivity must be determined by titration on HeLa cells using an E1 hr helper virus. As expected, none of these dl mutants grow on HeLa cells, and they are defective for viral DNA replication. Mutants dl 801 and dl 802 have deletions which come close to the N-terminus of the DBP coding sequence, and bring about frame-shifts (see table 3). These mutants should make extremely small proteins with little or no identity to DBP. Unexpectedly, dl 801 produces two DBP-related proteins of 40 and 37K, but dl 802 makes no detectable DBP-related protein. The dl 802 mutant is being used to analyze the role that the DBP plays in regulation of adenoviral early gene expression. It will be recalled that previous microinjection experiments had suggested that the DBP stimulated the expression of E1B and E4 (194, 195), while studies with ts125 had shown that cells infected with this mutant built up very high levels of the early mRNAs late in infection, suggesting that the protein effects negative regulation of these genes (191, 192, 193). Recent experiments carried out by S. Rice and D. Klessig (personal communication) show rather clearly that the DBP does not positively or negatively regulate E1B or E4 mRNA levels in infected cells. Since dl 802 fails to make the DBP, one would expect to see a more profound effect (for example, greatly elevated levels of E1B and E4 mRNAs by comparison with the levels in wild-type infected cells), but this was not the case. It should also be pointed out that these investigators observed somewhat less dramatic accumulations of E1B and E4 mRNAs in cells infected at restrictive temperature with ts125 than previously reported (192, 193), which could be ascribed to such variables as differences in the cells and conditions used. The autoregulatory effect of the DBP on E2A could not be addressed in these experiments, since dl 802 does not normally express E2A, and only rather low steady-state levels of E2A mRNAs could be detected. Rice and Klessig suggest that the most likely explanation for this effect is that removal of sequence by deletion brings about structural changes in the E2A gene or its mRNA, resulting in a cis-acting defect. For example, transcription or RNA processing may be much less efficient in the structurally altered molecule, or the mRNA produced may be much less stable. Further study of these and other deletion mutants will undoubtedly add to our understanding of DBP functions; already, the results with one of these mutants require us to reappraise the role of this protein in the regulation of other early genes.

Temperature-sensitive mutations in the DBP gene influence the ability of adenoviruses to transform rat cells in vitro. However, in contrast to the temperature-dependent effect of E2B mutations (see above), mutants such as ts125 and ts23 transform with higher frequency than wild-type virus at both permissive and restrictive temperatures, with more marked effect at the latter (35, 157). It seems unlikely that the transformation phenotype of these mutants results from second-site mutations in E1. True ts^+ revertants (but not those with second-site mutations within the DBP gene) transform at wild-type frequency (212, 215), as do E1 DNA fragments from ts107 (212). As with the adenovirus polymerase, the role of the DBP in transformation is probably indirect. Explanation of the enhanced transformation frequency is still largely conjectural. The effect is probably not simply due to a less cytotoxic effect of the ts mutants for rat cells (216). A possible explanation of the result is that the DBP affects the integration of viral DNA into cellular DNA, and in this respect it is of interest that many ts125 transformants (initiated at restrictive temperature) carry sequence from most of the viral genome, in contrast to wild-type transformants which frequently contain only left-end sequence (217, 218). An alternative explanation of the enhanced transformation might be that rodent cells infected with these mutants contain higher levels of the E1 transformation products, or the E2B polymerase which is also required for transformation by virus (see above). It has been suggested that the DBP negatively regulates the expression of the E1A and E1B genes in human cells, perhaps by reducing the stability of their messages (192), but in the light of the results (discussed above) with dl 802, which exhibits normal expression of the E1A and E1B genes, this function of the protein is now in question. Possibly, the wild-type DBP does carry out this activity in rat cells, or perhaps it performs some other activity in these cells which represses transformation, and the altered DBPs of ts125 and ts23 fail to enact this activity, so that transformation is not suppressed. We have recently carried out experiments with the dl mutants, which make truncated or no DBP, and find that these transform rat cells at wild-type frequencies (J. Williams, S. Rice and D. Klessig, unpublished results). These results more or less exclude the suppression model. However, we have found that in rat cells coinfected with ts125 and wild-type virus, the high frequency typical of the mutant is obtained (the ts125 transformation phenotype is dominant), and this result supports the view that the mutated DBP does act in the rat cell to enhance the transformation frequency. As discussed above, it has been proposed that the DBP possesses two separate functional domains; the N-terminal one is thought to regulate only late viral functions, while the C-terminal one is involved in viral DNA synthesis, perhaps in regulation of early gene expression, and in transformation. We have also examined the transforming capacity of some type 2 hr mutants, which possess mutations in the N-terminal region of the gene, and find that they transform at wild-type (or slightly lower) frequencies. This result supports the concept that there are two separate functional domains for transformation, too. Hint of a possible modulatory effect of the N-terminal domain upon the transforming activity of the C-terminal region has emerged from a recent set of experiments (J. Williams, S. Rice and D. Klessig, unpublished results). The Ad2 mutant ts400 transforms with essentially wild-type frequency at both permissive and restrictive temperatures. This mutant carries a C-terminal mutation identical to that of ts125, and the N-terminal hr mutations of hr405 (see table 3). Whereas the enhanced transformation phenotype of ts125 is dominant to that of wild-type, cells coinfected with ts125 and hr405 display a wild-type transformation frequency, so that the N-terminal suppression phenotype appears to be dominant to that resulting from the C-terminal mutation. These N-terminal mutations may be responsible for suppressing the enhanced transformation expected of the C-terminal mutant. Unfortunately, because of serotypic differences, there are a number of other amino acid differences between the N-termini of ts125 and ts400, and reconstruction of the ts125 gene is now being carried out in an attempt to resolve this problem.

THE E3 GENE PRODUCTS: ARE THEY NON-ESSENTIAL FOR VIRAL GROWTH?

This transcriptional unit is read off the r-strand between 76 and 86 map units, and it encodes at least eight overlapping mRNAs, which arise from a single promoter and possess different splicing patterns (143, 144, 219, 220, 221, 222, 223). The nucleotide sequence has been determined for the E3 region of type 2 adenovirus (224, 225, and see also 45, 46), and eight open reading frames (seven in rightward transcripts), capable of encoding polypeptides ranging in size from around 95 to 160 amino acid residues have been located. Translation in vitro of E3 specific mRNAs gives rise to four proteins of 13K, 14K, 15.5K and 19K (81, 226, 227, 228, 229), but the identity and locations of only two of the E3 genes have been clearly established to date. The first of these is the gene encoding the 19K glycoprotein which is found to be associated with the membrane fraction of infected and transformed cells (81, 230, 231, 232, 233). An open reading frame sufficient to encode a polypeptide of 159 residues (18.4K) has been mapped between 79.9 and 81.2 units (n 3179 to n 3656) (224, 45), and the 19K glycoprotein has been mapped to this region by in vitro translation of E3 mRNAs and analyses of the proteins made in cells infected with E3 dl and AD2-ND mutants (which have deletions and substitutions in the E3 region) (228, 229, 232, 234, 235). The protein contains an N-terminal hydrophobic signal sequence of 18 residues (224), and it has been shown to associate with class I transplantation antigens in both infected and transformed human and mouse cells (231, 236). When present in transformed cells the 19K protein could perhaps act as a tumor-specific transplantation antigen (TSTA), although there is no evidence for this, and it is more likely that the TSTA(s) is encoded by the E1 region (136, 237). The second E3 gene of known identity and map position encodes an 11.6K protein (238), and is comprised of an open reading frame spanning the sequence n 3835 to n 4138 (45, given as n 1860 to n 2143 in ref. 238). This protein has been detected in extracts of infected cells and as an in vitro translation product of E3 mRNA by immunoprecipitation with an antiserum made against a peptide predicted by a sequence within that frame. As with the 19K protein, the 11.6K species has a 22 residue hydrophobic domain, in this case in the middle of the molecule, and it is thought that it might be a transmembrane protein (238). A third E3 protein, a 14K species, has been isolated from infected cells (239), and this is probably encoded by a reading frame which maps between n 4811 and n 5195 (45). It should be noted that the x and y leader sequences, which, it will be recalled, end up on the fiber mRNAs, and the z leader, all map in E3 (45, 46).

Very little is known about the functions that any of the E3 proteins play during viral infection and/or transformation, and the study of viral mutants has done little to clarify this situation so far. A variety of E3 deletion and substitution mutants has been isolated (11, 19), and all of these grow normally on human cell lines such as HeLa and KB. Even dl 324, a variant with a large E3 deletion of 2286 bp which effectively removes all of the E3 open coding regions is completely viable in human cells, so it would appear that these proteins are not essential for the growth of the virus in such cells. This does not exclude the possibility that at least some of these gene products play important roles during infection of the natural host. For example, they might be required for establishing infection in specific target cells (perhaps adenoidal or respiratory epithelial cells. This possibility could perhaps be tested by measuring the ability of the E3 dl mutants to grow in primary or low passage cultures derived from a variety of human tissues. It has also been suggested that the 19K protein may play a role in the establishment of persistent infection (240), which is a hallmark of natural human adenovirus infections (241). To test this hypothesis it will probably be necessary to first develop a suitable animal model for study of adenovirus infection.

FUNCTIONS OF THE E4 PRODUCTS: OPENING UP A NEGLECTED AREA.

The E4 transcriptional unit lies in the right end of the genome between 91.3 and 99 map units, is copied in a leftward direction and encodes a large number of overlapping, differentially-spliced mRNAs (143, 144, 242). The nucleotide sequence of this region has been determined for type 2 (150, 243), and seven open reading frames have been located on the 1 strand. Recently, the molecular structure of many of the E4 mRNAs has been analysed by sequence analysis of the corresponding cDNA clones (244, 245), and it has been shown that all have the same 5' and 3' ends (suggesting a common precursor), all carry a small 5' leader sequence, and all are formed by excision of one to three introns. It has also been shown that all except one of these seven open frames are translated from the mRNAs characterized to date. The region encodes at least seven polypeptides, ranging in size from around 3K up to 34K, but it is quite possible that more proteins may be generated by mRNAs in which two open reading frames are fused in phase by removal of an intron (245).

Translation *in vitro* of E4 DNA-selected mRNAs from this region has identified as many as ten polypeptides which range in size from around 9K to 24K (77, 246, 247), and three E4 proteins have been detected in infected cells and assigned to E4 reading frames. These are the 11K (248), 14K (249), and the 25K protein (87). In the infected cell the 11K species is found in close association with the nuclear matrix, and it appears to be strongly immunologically conserved among the serotypes of the different subgroups (248). The 25K protein corresponds to the product of the 34K reading frame, and in infected cells it is found to be complexed with the E1B 495R protein (87).

In order to define the roles that these E4 products play in viral replication, a set of deletion and insertion mutations in each of the seven open frames has been isolated and characterized (1, and T. Shenk, personal communication). Mutations in six of these frames had little or no effect upon the ability of virus to grow in HeLa cells. Likewise, it was previously found that a mutant bearing a frameshift mutation in the 11K gene did not make that protein, but was completely viable (248). Thus, either these proteins are not required for viral growth under the conditions used, the remaining N-terminal fragments may be functionally active, or the product of one of the other E4 reading frames might be functionally related and able to replace the damaged protein. Two E4 deletions show a marked effect upon the growth of the virus in HeLa cells. In one of these, *dl* 355, there is a 14 bp deletion removing bp 2332 to bp 2345 (n 1 is the right terminal nucleotide of the genome) which lies within the reading frame (n 1861 to n 2743) encoding the 34K (25K) polypeptide. In this case only the N-terminal half of the protein is made. The second deletion (*dl* 366), is a very large one (bp 576 to pb 2844) which essentially removes most of the E4 region. This mutant is almost totally incapable of growth on HeLa cells, and grows only on cells of the transformed W162 line (29) which contain and express the E4 region. The overall phenotypes of these mutants are quite similar and rather complicated. They show delayed (and in the case of *dl* 366), reduced viral DNA synthesis, and at least some of the late mRNAs do not accumulate to wild-type levels. There is a parallel reduction in late viral polypeptide levels and an overproduction of the E2A-DBP late in infection, accompanied by only partial shut-off of host cell protein synthesis. For each aspect of this complex phenotype, the change was more dramatic with *dl* 366, which possesses the larger deletion, than with dl 355. It is not yet clear if, in any of these phenotypic parameters, the E4 product acts directly or indirectly via another viral or cellular moiety. As mentioned above, the E4 25K protein is shown to exist in a physical complex with the E1B 495R protein (87). It is thus probably more than coincidental that the overall phenotype of E1B 495R mutants is generally similar (in lytic infection) to that of the E4 25K mutants (95, 96, B. Karger, Y.S. Ho, K. Castiglia, S.J. Flint and J. Williams, unpublished results, discussed above in section on E1B). Collectively, these results argue in favor of a functional interaction between these two proteins. Clearly, the E4 protein plays an important role in the development of the adenoviral lytic infection, but it may not be completely indispensable since the phenotype of dl 355 is less severe in certain cell types (T. Shenk, personal communication). This property is also reminiscent of the 495R hr mutants, which show a cell-dependent host-range (22, ed J. Williams, unpublished results). It is not yet clear if the E4 mutations influence the ability of the adenovirus to transform rodent cells, in the way that E1B mutations do, but it is worth noting that it has recently been claimed that E4 gene expression is required for establishment of anchorage - independence in cells transformed by type 12 E1 (250, see below). The possibility that most of the E4 genes are dispensable for growth in continuous cell lines such as HeLa, does not mean (as was argued for the E3 genes) that these are not required for development of the infection in the natural host. Again, analysis in an animal model is probably imperative.

GENETIC ANALYSIS OF ONCOGENIC SEROTYPE 12: A NEW BEGINNING

The human adenoviruses isolated to date have been arranged into groups on the basis of various taxonomic criteria, such as hemagglutination, the guanine-cytosine content of the DNA, the degree of DNA sequence homology among the various species, and oncogenicity in rodents (251, 252, 253). Perhaps the most interesting and biologically significant property used to classify these viruses is the striking differences which they display in their ability to induce cancer in rodents. For example, the members of group A (types 12, 18 and 31) rapidly induce sarcomas and fibrosarcomas at high frequency in newborn hamsters and in the newborns of at least certain strains of mice and rats. In contrast, viruses of group B (types 3, 7 and 11, for instance) are weakly oncogenic in hamsters only, while those of group C (types 1, 2 and 5) are found to be nononcogenic in any of these rodents (254, 255, 256, 257, 258, 259, 260, 261, 262, 263, J. Williams, unpublished results). The members of group D (types 9, 17, 22, etc.) are also considered to be nononcogenic, with the notable exception of serotype 9 which induces breast fibroadenomas in newborn, young and adult female Wistar/Furth rats (264). This differential oncogenicity is not reflected in a similar pattern of transformation in vitro, and it has been found that serotypes representative of these four groups are capable of transforming rodent cells with roughly equal frequencies (reviewed in 253). However, the tumorigenicity of transformants established after infection with the viruses of different subgroups does differ significantly. Rat cells transformed by type 12 (group A) and type 3 (group B) are tumorigenic in rats (265, 266), while cells transformed by types 1, 2 and 5 (group C) are nontumorigenic (267, 268, 269). The same responses are also found for type 12 and type 5 mouse transformants inoculated into syngeneic mice (124, and J. Williams, unpublished results). Hamster cells do not seem to conform to this rule, however, since both type 5 (270) and type 2 (271) transformants are tumorigenic in newborn hamsters.

The principles determining these striking differences in viral oncogenicity and cellular tumorigenicity are not understood, and the roles of adenoviral gene products in these processes have yet to be defined. As discussed above (in the section on the roles of E1 products in transformation), a considerable body of evidence indicates that the group C (type 2 and 5) genes required and sufficient for transformation of rodent cells in vitro reside in the E1 region of the viral chromosome. This evidence was obtained from experiments in which the transforming agents were fragments of viral genome cloned in plasmid, or transformation-defective mutants. Similarly, such plasmid experiments have shown that the transforming genes of oncogenic serotype 12 lie in E1, and they suggest that gene products of both E1A and E1B regions act to determine the tumorigenicity of transformed cells. The genetic approach has generated a wealth of valuable information concerning the requirement for E1 products in transformation by group C viruses, but this approach has yet to be fully exploited for the oncogenic viruses of group A.

In actual fact, genetic study of group A adenoviruses pre-dates that of the more commonly used non-oncogenic group C viruses by about three years, and as mentioned above (in the section on making viral mutants, the first mutants of any adenoviral type to be isolated were the cyt variants of type 12 (15, 272). These were isolated either as naturally occurring variants or mutants induced by ultraviolet irradiation of wild-type stock, and possess the distinctive and easily recognised phenotypic traits of causing enhanced cellular destruction and formation of large plaques on human embryonic kidney cells. In addition, these mutants were found to be incapable of transforming hamster embryo cells *in vitro*, and they induced tumors in hamsters with greatly reduced efficiency. Surprisingly, despite these findings, characterization and exploitation of these mutants was not carried further at that time, and the genotypes were not determined. By analogy with findings from the group C adenovirus system it was generally inferred that the mutational lesions lay in one of the genes located in the left 10% of the genome, and even now, to my knowledge, these mutations have yet to be physically mapped. Following a hiatus of some twelve years or so, phenotypic analysis of some of the original cyt isolates was carried a step further. Cells infected by the mutants (and type 5 dl313, in which the E1B region is entirely deleted), showed extensive degradation of both viral and cellular DNA, and it was concluded that the cyt mutations lay in a gene whose product normally prevents nuclease digestion of the DNA within the infected cell (104, 273). Degradation of DNA was prevented by coinfection with either an E1A mutant (dl312) or an E1B 495R mutant (hr 6), and it did not occur in 293 cells, which express all E1 gene functions except protein IX. The results of these complementation tests imply that the mutations lie in the gene encoding the E1B 19K protein. The fact that all subsequently isolated type 2 and type 5 cyt, deg or cyt, deg^+ mutations (discussed above, 99, 100, 101, 102, 103, 111) map to this gene adds support to this view.

Some four or five years after the isolation of the cyt mutants, there was a flurry of reports describing the isolation of ts mutants of group A serotypes 12 (274, 275, 276, 277) and 31 (278). Many of these ts mutants were classified functionally by complementation analysis, and representatives of most groups were tested for ability to produce viral DNA and major late proteins. Mutants of at least three complementation groups were found to be defective for viral DNA synthesis at restrictive temperature; three such DNA-negative groups were isolated in one laboratory (279), two in another (277), and one in a third laboratory (280). Since these have not been cross-checked by complementation tests, it is not known if this collection represents three or more than three gene functions required for DNA replication. Two of these early function mutants were found to be defective for transformation at restrictive temperatures. Mutant ts 401 was found to transform hamster brain and rat 3Y1 cells at 36° and 38.5°C (restrictive for growth of the mutant in human cells) at frequencies of between two and six fold greater than wild-type did at these temperatures, but failed to transform them at higher temperatures of 40° and 41.5°C (281). The block to transformation appeared to hold even after infected cells were held at 36°C for three days prior to switch up to 40° or 41.5°C , suggesting that a maintenance event rather than an initiation event was involved. Supportive evidence for this view came from the fact that certain aspects of the ts 401 transformed cell phenotype, such as anchorage-independence, were found to be thermosensitive (282). It will be recalled that the type 5 DBP mutants ts 125 and ts 23 transform rodent cells with elevated frequency at a temperature (38.5°C) which is restrictive for viral DNA replication and growth on human cells. The second mutant, ts 505, transforms 3Y1 cells at 37° but not at 40.5°C, and temperature-shift experiments suggest that the product of the mutated gene is required for initiation of transformation rather than maintenance (280). Again, however, in some ts-induced transformants, certain phenotypic traits (ability to grow in soft agar) are apparently thermosensitive, suggesting a maintenance role for the gene product. These are potentially very valuable mutants which could provide useful information concerning the roles of viral gene products in the transformation process. Unfortunately, these and other ts mutants have not been further characterised, and a major weakness of this work lies in the fact that none of these mutations has been mapped either genetically or physically on the viral chromosome. When the genotype is not known, it is virtually impossible to interpret any aspect of the phenotype in a meaningful way, or make comparisons with other adenovirus systems.

Recently, some fifteen years or so after the initial isolation of the type 12 cyt mutants by Takemori and his coworkers, genetic analysis of type 12 adenovirus began again, with most of the emphasis being put on analysis of the transforming genes. Of course, our knowledge of the type 12 genome has increased dramatically over that period, and a great deal can be inferred from the group C work concerning the functions of various products. The overall organization and expression of the type 12 genome is similar to that of the group C viruses (283). The entire nucleotide sequence of the type 12 E1 region has been determined, and the locations of the various mRNAs and coding regions mapped in detail (for review, see 45, 46). While there is considerable sequence heterology between the E1 regions of type 12 and the group C viruses, there are also many conserved stretches of similar sequence, and the general organization of the type 12 E1 region is broadly similar in the various serotypes. With type 12, as with types 2 and 5, rightward transcription from the E1A promoter generates three mRNAs with identical 5' and 3' ends, which differ in size (13S, 12S and 9S) due to differential splicing. The 13S and 12S messages are made early in infection and encode polypeptides of 266R (30K) and 235R (26K) respectively. Rightward transcription from the E1B promoter gives rise to a family of mRNAs with identical 5' and 3' ends, and unequal size due to splicing. The larger (22S of the two major species encodes two proteins, of 482R (54K) and 163R (19K) from two overlapping open reading frames initiating from separate start codons. The other major species (13S) encodes only the 163R protein. Proteins corresponding to these various species have been immunoprecipitated from cells infected and transformed by type 12 (27, 284). While these type A and type C E1 proteins may not cross-react immunologically, there is evidence that at least some are functionally related, since wild-type 12 virus complements the growth of type 5 hr mutants in non-permissive HeLa cells (285, 286, 287).

A variety of mutants with lesions mapping in the E1 region (see table 4) have been isolated as hr mutants which grow on type 12 transformed human cells such as Ad12 HER (which contain and express the type 12 E1 region), but not on non-transformed cells such as A549 or HeLa (27). These mutants have been classified functionally into two groups by intratypic complementation and intertypic complementation with prototype hr and dl mutants (D. Breiding and J. Williams, unpublished results). Many of them have been physically mapped to either the E1A or the E1B region by both marker rescue and nucleotide sequence analysis. A number of insertion mutations have been introduced to these genes, and currently a number of deletions and oligonucleotide-directed mutations are being made in cloned E1 sequence and built back into complete genome (D. Breiding, C. Edbauer, J. Williams, P. Byrd and P. Gallimore, work in progress). As expected, all of the mutants with lesions which alter the

TABLE 4. Host-range mutants of serotype 12.

Mutant	Mutation or map limits (nucleotide 1 is the left terminal nucleotide of the l strand.	Polypeptide Altered	Reference
hr 700	Mutation not known. Limits n 0-1455	Probably E1A 266R and/or 235R.	27, D.Breiding and J. Williams, unpublished results.
hr 707	Deletion of GC base pair at bp 690.	Both E1A 266R and 235R truncated to 69R.	D. Breiding and J. Williams, unpublished results.
in 600	Insertion of 8bp at n 1004 Frameshift.	E1A 266R truncated to 169R. 235R not altered.	27.
in 700	Insertion of 2bp at n 1594.	E1B 163R truncated to 37R. E1B 482R not altered.	C. Edbauer and J. Williams, unpublished results.
in205B	Insertion of 12 bp at n 1597 In frame.	Four missense AA inserted into E1B163R. 482R not altered.	288
in205C	Insertion of 16bp at n 1597. Frameshift.	E1B163R truncated to 26R. E1B482R not altered.	288
dl205	Deletion of bp 1598 to 1779. Frameshift.	E1B163R deleted and truncated to 22R. E1B 482R not altered.	288
hr 703	Transition of C to T at n 2234 on l strand.	E1B 482R AA134 Gln to termination codon. 163R not altered.	27, D. Breiding and J. Williams, unpublished results.
in 720	Insertion of 4bp at n 2318. Frameshift.	E1B 482R truncated to 162R. 163R not altered.	C. Edbauer and J. Williams, unpublished results.
hr 706	Mutation not known. Limits n 2318-3497.	E1B 163R not altered. 482R changed.	D. Breiding and J. Williams, unpublished results.

Abbreviations used: n, nucleotide; bp, base pair; AA, amino acid; R, residue.

E1A proteins and the E1B 482R protein are transformation-defective in both mouse and rat embryo and baby kidney cells, and are either weakly or nononcogenic in baby rats. Non-transformed human cells infected with the E1A 266R mutant in600 show evidence of truncated protein (235R still intact), and reduced levels of E1B 482R and 163R, reminiscent of the E1 protein phenotype displayed by type 5 hr 1 (which has a similar mutational lesion -- see table 1). Unlike hr 1, in600 does not appear to display a cs transformation phenotype, and to date none of these type 12 mutants show any cold-sensitivity, as do many of the type 5 hr mutants. The 482R E1B protein cannot be detected in A549 cells (or human embryonic kidney cells) infected with either of the E1B 482R mutants hr 703 or hr 704, while the E1B 19K protein levels are only slightly reduced. Viral DNA synthesis is somewhat reduced and delayed in A549 cells infected with these two mutants, and virtually non-detectable in cells infected with E1B 482R mutant hr 706, and with E1A mutant hr 707. Further phenotypic analysis of these and other E1 mutants is currently in progress. What is clear from our analysis to date is that both the E1A 266R and the E1B 482R products are required for transformation and oncogenicity. The latter protein does not seem to be required for induction of tumor-specific transplantation immunity (27), and we favor the E1B 163R protein for this role. Recently we have introduced a frameshift insertion into the coding region of the 163R protein, which should truncate it to 37R. The mutant is viable and seems to be slightly more cytocidal on A549 cells, makes larger plaques than wild-type, apparently does not cause excessive DNA degradation, and transforms mouse cells at twice the frequency of wild-type virus (C. Edbauer and J. Williams, unpublished results). Others have made somewhat similar mutations in the 163R gene, and find that the mutants fail to transform rat 3Y1 cells (288). These results suggest that this protein is also required for transformation, but that the N-terminal 37 residues are sufficient to induce the process. We anticipate that further studies of these and other type 12 mutants will provide valuable information concerning transformation. Clearly, conditional mutants would aid these investigations enormously.

LOOKING TO THE FUTURE

Adenovirus mutants have proved to be extremely valuable for defining the roles of viral proteins in a wide variety of processes essential to viral growth and cellular transformation. Their utility is limited to a large extent by the availability of suitable systems in which such processes can be manipulated and studied. As new systems are developed, this utility will expand. The study of molecular processes, such as DNA replication, can be investigated in systems *in vitro*; mutants are and will be invaluable in analyzing the steps of this process. The same mutants can be used to verify these events in the whole cell, and the use of cell mutants may help to define the roles of cellular functions in viral infection. At the other end of the complexity spectrum, mutants can be used to probe viral processes in the whole animal. The oncogenic process immediately springs to mind, but the role of viral genes in the development of other pathogenic conditions or in persistent infections may also demand study in an animal model. Some adenoviral genes (E3 and E4) appear to be non-essential for growth of virus in cultured cell lines, but they may be important for viral growth in the host. Genetic analysis of these genes may only be possible in a whole animal system.

Technically, mutations can be introduced at any specific location on the viral genome. In practice, however, the growth of defective mutants will remain a severe hurdle to their exploitation. That barrier can be overcome by the development of more cell lines, similar to the 293 line, which contain and express various viral genes. Lines expressing the E2A and E4 regions have been made, and more are likely to be produced in the future. Alternative means of growing defective variants, for example, by complementation with a non-homologous serotype, may also be successful.

It seems likely that the E1 transforming genes will continue to receive considerable attention over the next few years. However, it appears that other viral genes (outside E1) can also modulate the transforming capacity of the virus (E2A and E2B), and may influence the phenotype of the transformed cell (E4). In addition, it has been suggested that genes outside the E1 region may also influence the oncogenicity of adenoviruses (289). Further genetic analysis is likely to throw some much-needed light on these topics. One problem will undoubtedly continue to plague such investigations, and that is the dilemma of distinguishing between direct and indirect effects of mutations, in a system where many gene functions interact to regulate viral gene expression, and generate a complex phenotype. The best policy is to analyze the system thoroughly and be prepared for possible pitfalls, since there is no easy way around this perplexing situation; it is in fact the essence of the system.

Acknowledgements. I gratefully acknowledge the diligent secretarial assistance of Kathryn Galligan and Julie Marengo. Their patience almost matches that of Walter Doerfler and Merlyn, my wife. I wish to thank Beverly Miller for critically reading the manuscript and making sure that it came through with a minimum of errors. I am extremely grateful to my colleagues in the laboratory, David Breiding, Cheryl Edbauer, Brian Karger, Beverly Miller and Glenn Telling for many helpful discussions concerning their own and other work, and I thank my colleagues elsewhere, Phil Byrd, Jane Flint, Phil Gallimore, Gary Glenn, Roger Grand, Dan Klessig, Steve Rice, Rob Ricciardi, Tom Shenk and Bruce Stillman, for allowing me to quote some of their unpublished findings. Unpublished work quoted from my own laboratory was supported by grants from the National Institutes of Health (CA-21375 and CA-32940).

REFERENCES

1. Shenk, T., and Williams, J., In: current Topics in Microbiology and Immunology, Vol. 111: The Molecular Biology of Adenoviruses 3 (Ed. W. Doerfler), pp. 1-39. Springer-Verlag, Berlin, Heidelberg, 1984.
2. Young, C.S.H., Shenk, T., and Ginsberg, H.S., In: "The Viruses: Adenoviruses" (Ed. H.S. Ginsberg), pp. 125-172. Plenum, New York, 1984.
3. Ginsberg, H.S., and Young, C.S.H., In: Comprehensive Virology, Vol. 9 (Eds. H. Fraenkel-Conrat and R.R. Wagner), pp. 27-88. Plenum, New York, 1977.
4. Shortle, D., DiMaio, D., and Nathans, D., Ann. Rev. Genet. 15, 265-294, 1981.
5. Thimmapaya, B., Jones, N., and Shenk, T., Cell 18, 947-954, 1979.
6. Stow, N.D., J. Virol. 37, 171-180, 1981.
7. Carlock, L.R., and Jones, N.C., J. Virol. 40, 657-664, 1981.
8. Solnick, D., Nature 291, 508-510, 1981.
9. Osborne, T.F., Gaynor, R.B., and Berk, A.J., Cell 29, 139-148, 1982.
10. Hearing, P. and Shenk, T., J. Mol. Biol. 167, 809-822, 1983.
11. Thimmapaya, B., Weinberger, C., Schneider, R.J., and Shenk, T., Cell 31, 543-551, 1982.
12. Chinnadurai, G., Chinnadurai, S., and Brusca, J., J. Virol. 32, 623-628, 1979.
13. Ho, Y.S., Galos, R., and Williams, J., Virology 122, 109-124, 1982.
14. Kapoor, Q.S. and Chinnadurai, G., Proc. Natl. Acad. Sci. U.S.A. 78, 2184-2188, 1981.
15. Takemori, N., Riggs, J.L., and Aldrich, C., Virology 36, 575-586, 1968.
16. Werner, G., and zur Hausen, H., Virology 86, 66-77, 1978.
17. Schwarz, E., Reinke, C., Yamamoto, N., and zur Hausen, H., Virology 116, 284-296, 1982.
18. Young, C.S.H. and Williams, J., J. Virol. 15, 1168-1175, 1975.
19. Jones, N.C., and Shenk, T., Cell 13, 181-188, 1978.

20. Kelly, T.J. Jr., and Lewis, A.M. Jr., J. Virol. 12, 643-652, 1973.
21. Williams, J.F., Gharpure, M., Ustacelebi, S., and McDonald, S., J. Gen. Virol. 11, 95-101, 1971.
22. Harrison, T.J., Graham, F.L., and Williams, J., Virology 77, 319-329, 1977.
23. Jones, N.C., and Shenk, T., Cell 17, 683-689, 1979.
24. Graham, F.L., Smiley, J., Russell, W.C., and Nairn, R., J. Gen. Virol. 36, 59-72, 1977.
25. Aiello, L., Guilfoyle, R., Huebner, K. and Weinmann, R., Virology 94, 460-469, 1979.
26. Whittaker, J.L., Byrd, P.J., Grand, R.J.A., and Gallimore, P.H., Mol. Cell. Biol. 4, 110-116, 1984.
27. Gallimore, P., Byrd, P., Whittaker, J., Breiding, D., and Williams, J., Cancer Cells 2; Oncogenes and Viral Genes, pp. 519-526, Cold Spring Harbor Laboratory, 1984.
28. Babiss, L.E., Young, C.S.H., Fisher, P.B., and Ginsberg, H.S., J. Virol. 46, 454-465, 1983.
29. Weinberg, D.H., and Ketner, G., Proc. Natl. Acad. Sci. U.S.A. 80, 5383-5386, 1983.
30. Klessig, D.F., Brough, D.E., and Cleghorn, V., Mol. Cell. Biol. 4, 1354-1362, 1984.
31. Hudziak, R.M., Laski, F.A., Raj Bhandary, U.L., Sharp, P.A., and Capecchi, M.R., Cell 31, 137-146, 1982.
32. Challberg, S.S., and Ketner, G., Virology 114, 196-209, 1981.
33. Williams, J., Young, C.S.H., and Austin, P.E., J. Virol. 15, 675-678, 1975.
34. Williams, J., Ho, Y.S., and Galos, R., Virology 110, 208-212, 1981.
35. Williams, J., Young, C.S.H., and Austin, P.E., Cold Spring Harbor Symp. Quant. Biol., 39, 427-437, 1974.
36. Williams, J., Grodzicker, T., Sharp, P., and Sambrook, J., Cell 4, 113-119, 1975.
37. Sambrook, J., Williams, J., Sharp, P.A., and Grodzicker, T., J. Mol. Biol., 97, 369-390, 1975.
38. Arrand, J.E., J. Gen. Virol., 41, 573-586, 1978.
39. Frost, E., and Williams, J., Virology 91, 39-50, 1978.
40. Volkert, F.C., and Young, C.S.H., Virology, 125, 175-193, 1983.
41. Benzer, S., Proc. Natl. Acad. Sci. U.S.A. 47, 403-415, 1961.
42. Galos, R.S., Williams, J., Shenk, T., and Jones, N., Virology 104, 510-513, 1980.
43. Maxam, A.M., and Gilbert, W., Proc. Natl. Acad. Sci. U.S.A. 74, 785-788, 1977.
44. Sanger, F., Nicklen, S., and Coulson, A.R., Proc. Natl. Acad. Sci. U.S.A. 74, 5463-5467, 1977.
45. van Ormondt, H., and Galibert, F., In: Current Topics in Microbiology and Immunology, Vol. 110, The Molecular Biology of Adenoviruses 3 (Ed. W. Doerfler), pp. 73-142. Springer-Verlag, Berlin, Heidelberg, 1984.
46. Sussenbach, J.S., In: "The Viruses: Adenoviruses" (Ed. H.S. Ginsberg) pp. 35-124, Plenum, New York, 1984.
47. Pettersson, U., Virtanen, A., Perricaudet, M., and Akusjavri, G., In: Current Topics in Microbiology and Immunology, Vol. 109: The Molecular Biology of Adenoviruses 1 (Ed. W. Doerfler), pp. 107-123, 1983.
48. Smart, J.E., Lewis, J.B., Mathews, M.B., Harter, M.L., Harter, M.L., and Anderson, C.W. Virology 112, 703-713, 1981.
49. Yee, S.P., Rowe, D.T., Tremblay, M.L., McDermott, M., and Branton, P.E., J. Virol. 46, 1003-1013, 1983.

50. Feldman, L.T., and Nevins, J.R., Mol. Cell Biol. 3, 829-838, 1983.
51. Ricciardi, R.P., Jones, R.L., Cepko, C.L., Sharp, P.A., and Roberts, B.E., Proc. Natl. Acad. Sci. U.S.A. 78, 6121-6125, 1981.
52. Berk, A.J., Lee, F., Harrison, T., Williams, J., Sharp, P.A., Cell 17, 935-944, 1979.
53. Jones, N.C., and Shenk, T., Proc. Natl. Acad. Sci. U.S.A. 76, 3665-3669, 1979.
54. Ross, S.R., Levine, A.J., Galos, R.S., Williams, J., and Shenk, T., Virology 103, 475-492, 1980.
55. Shenk, T., Jones, N., Colby, W., and Fowlkes, D., Cold Spring Harbor Symp. Quant. Biol. 44, 367-375, 1979.
56. Nevins, J.R., Cell 26, 213-220, 1981.
57. Gaynor, R.B., and Berk, A.J., Cell 33, 683-693, 1983.
58. Solnick, D., and Anderson, M.A., J. Virol. 42, 106-113, 1982.
59. Persson, H., Monstein, H.J., Akusjarvi, G., and Philipson, L. Cell 23, 485-496, 1981.
60. Shaw, A.R., and Ziff, E.B., Mol. Cell. Biol. 2, 789-799, 1982.
61. Carlock, L.R., and Jones, N.C., J. Virol. 40, 657-664, 1981.
62. Montell, C., Fisher, E.F., Carruthers, M.H., and Berk, A.J., Nature 295, 380-384, 1982.
63. Montell, C., Courtois, G., Eng, C., and Berk, A.J., Cell 36, 951-961, 1984.
64. Winberg, G., and Shenk, T., EMBO Journal 3, pp. 1907-1912, 1984.
65. Ferguson, B., Jones, N., Richter, J., and Rosenberg, M., Science 224, 1343-1346, 1984.
66. Katze, M.G., Persson, H., and Philipson, L., Mol. Cell. Biol. 1, 807-813, 1981.
67. Weeks, D.L., and Jones, N.C., Mol. Cell. Biol. 3, 1222-1234, 1983.
68. Nevins, J.R., Cell 29, 913-919, 1982.
69. Kao, H.T., and Nevins, J.R., Mol. Cell. Biol. 3, 2058-2065, 1983.
70. Green, M., Treisman, R., and Maniatis, T., Cell 35, 137-148, 1983.
71. Houweling, A., van den Elsen, P.J., van der Eb, A.J., Virology 105, 537-550, 1980.
72. Schrier, P.I., Bernards, R., Vaessen, R.T.M.J., Houweling, A., and van der Eb, A.J. Nature 305, 771-775., 1983.
73. Borelli, E., Hen, R., and Chambon, P. Nature 312, 608-612, 1984.
74. Spindler, K.R., Eng, C.Y., and Berk, A.J., J. Virol. 53, 742-750, 1984.
75. Ho, Y.S., Ph.D. Thesis, Carnegie-Mellon University, 1981.
76. Braithwaite, A.W., Cheetham, B.F., Li, P., Parish, C.R., Waldron-Stevens, L.K., and Bellett, A.J.D., J. Virol. 45, 192-199, 1983.
77. Lewis, J.B., Atkins, J.F., Baum, P.R., Salom, R., Gesteland, R.F., and Anderson, C.W., Cell 7, 141-151, 1976.
78. Halbert, D.N., Spector, D.J., and Raskas, H.J., J. Virol. 31, 621-629, 1979.
79. Lassam, N.J., Bayley, S.T., and Graham, F.L., Cell 18, 781-791, 1979.
80. Schrier, P.I., van den Elsen, P.J., Hertoghs, J.J.L., and van der Eb, A.J., Virology, 99, 372-385, 1979.
81. Ross, S.J., Flint, S.J., and Levine, A.J., Virology 100, 419-432, 1980.
82. Matsuo, T., Wold, W.S.M., Hashimoto, S., Rankin, A., Symington, J., and Green, M., Virology 118, 456-465, 1982.
83. Anderson, C.W., Schmitt, R.C., Smart, J.E., and Lewis, J.B., J. Virol. 50, 387-396, 1984.
84. Malette, P., Yee, S.P., and Branton, P.E., J. Gen. Virol. 64, 1069-1078, 1983.

85. Branton, P.E., Lassam, N.J., Downey, J.F., Yee, S.P., Graham, F.L., Mak, S., and Bayley, S.T., J. Virol. 37, 601-608.
86. Sarnow, P., Ho., Y.S., Williams, J., and Levine, A.J., Cell 28, 287-294, 1982.
87. Sarnow, P., Hearing, P., Anderson, C.W., Halbert, D.N., Shenk, T., and Levine, A.J., J. Virol. 49, 692-700, 1984.
88. Persson, H., Katze, M.G., and Philipson, L., J. Virol. 42, 905-917, 1982.
89. White, E., Blose, S.H., and Stillman, B.W., Mol. Cell. Biol. 4, 2865-2875.
90. Fohring, B., Gallimore, P.H., Mellow, G.H., and Raska, K. Jr., Virology 131, 463-472.
91. Pettersson, U., and Mathews, M.B., Cell 12, 741-750, 1977.
92. Katze, M.G., Persson, H., and Philipson, L., EMBO J. 1, 783-789, 1982.
93. Lassam, N.J., Bayley, S.T., and Graham, F.L., Virology 87, 463-467, 1978.
94. Ross, S.R., Levine, A.J., Galos, R.S., Williams, J., and Shenk, T., Virology 103, 475-492, 1980.
95. Logan, J., Pilder, S., and Shenk, T., Cancer Cells 2; Oncogenes and Viral Genes. pp. 527-531, Cold Spring Harbor Laboratory, 1984.
96. Babiss, L.E., and Ginsberg, H.S., J. Virol. 50, 202-212, 1984.
97. Harrison, T.J., Ph.D. Thesis, University of Glasgow, 1976.
98. Stillman, B.W., UCLA Symp. Mol. Cell. Biol. 10, 381-393, 1983.
99. Chinnadurai, G., Cell 33, 759-766, 1983.
100. White, E., Grodzicker, T., and Stillman, B.W., J. Virol. 52, 410-419, 1984.
101. Pilder, S. Logan, J., and Shenk, T., J. Virol. 52, 664-671, 1984.
102. Subramanian, T., Kuppuswamy, M., Mak, S., and Chinnadurai, G., J. Virol., 52, 336-343, 1984.
103. Takemori, N., Cladaras, C., Bhat, B., Conley, A.J., and Wold, W.S.M., J. Virol. 52, 793-805, 1984.
104. Lai Fatt, R.B., and Mak, S., J. Virol. 42, 969-977, 1982.
105. Gallimore, P.H., Sharp, P.A., and Sambrook, J., J. Mol. Biol. 89, 49-72, 1974.
106. Sambrook, J., Botchan, M., Gallimore, P., Ozanne, B., Pettersson, U., Williams, J., and Sharp, P.A., Cold Spring Harbor Symp. Quant. Biol., 39, 615-632, 1974.
107. Flint, S.J., Sambrook, J., Williams, J., and Sharp, P.A., Virology 72, 456-470, 1976.
108. Graham, F.L., Abrahams, P.J., Mulder, C., Heijneker, H.L., Warnaar, S.O., deVries, F.A.J., Fiers, W., and van der Eb, A.J., Cold Spring Harbor Symp. Quant. Biol. 39, 637-650, 1974.
109. Graham, F.L., Harrison, T., and Williams, J., Virology 86, 10-21, 1978.
110. Babiss, L.E., Fisher, P.B., and Ginsberg, H.S., J. Virol. 49, 731-740, 1984.
111. Babiss, L.E., Fisher, P.B., and Ginsberg, H.S., J. Virol. 52, 389-395, 1984.
112. van der Eb, A.J., van Ormondt, H. Schrier, P.I., Lupker, J.H., Jochemsen, H., van den Elsen, H., DeLays, P.J., Maat, J., van Bevern, C.P., Dijkema, R., DeWard, A., Cold Spring Harbor Symp. Quant. Biol. 44, 383-399, 1979.
113. Shiroki, K., Shimojo, H., Sawada, Y., Uemizu, Y., and Fujinaga, K., Virology 95, 127-136, 1979.
114. Houweling, A., van den Elsen, P.J., and van der Eb, A.J., Virology 105, 537-550, 1980.

115. Jochemsen, H., Daniels, G.S.G., Hertoghs, J.J., Schrier, P.I., van den Elsen, P.J., and van der Eb, A.J., Virology 122, 15-28.
116. van den Elsen, P.H., Houweling, A., van der Eb, A.J., Virology 128, 377-390, 1983.
117. van der Eb, A.J. and Bernards, R., In: Current Topics in Microbiology and Immunology, Vol. 110: The Molecular Biolgoy of Adenoviruses 2 (Ed. W. Doerfler), pp. 23-51. Springer-Verlag, Berlin, Heidelberg, 1984.
118. McKinnon, R.D., Bacchetti, S., and Graham, F.L., Gene 19, 33-42, 1982.
119. Rowe, D.T., and Graham, F.L., J. Virol. 46, 1039-1044, 1983.
120. Bos, J.L., Jochemsen, A.G., Bernards, R., Schrier, P.I., van Ormondt, H., and van der Eb, A.J., Virology 129, 393-400, 1983.
121. Bernards, R. Houweling, A., Schrier, P.I., Bos, J.L., and van der Eb, A.J., Virology 120, 422-432, 1982.
122. Bernards, R., Schrier, P.I., Houweling, A., van der Eb, A.J., Zijlstra, M., and Melief, C.J.M., Nature 305, 776-779, 1983.
123. Schrier, P.I., Bernards, R., Vaessen, R.T.M.J., Houweling, A., and van der Eb, A.J., Nature 305, 771-775, 1983.
124. Eager, K.B., Williams, J., Breiding, D., Pau, S., Knowles, B., Appella, E., and Ricciardi, R.P., Proc. Natl. Acad. Sci., 1985, In press.
125. Tanaka, K., Isselbacher, K.J., Khoury, G., and Jay, G., Science, 1985, In press.
126. Mellow, G.H., Fohring, B., Dougherty, J., Gallimore, P.H. and Raska, K., Jr., Virology 134, 460-465, 1984.
127. Ruben, M., Bacchetti, S., and Graham, F.L., J. Virol. 41, 674-685, 1982.
128. Babiss, L.E., Ginsberg, H.S., and Fisher, P.B., Proc. Natl. Acad. Sci. U.S.A. 80, 1352-1356, 1983.
129. Haley, K.P., Overhauser, J., Babiss, L.E., Ginsberg, H.S. and Jones, N.C., Proc. Natl. Acad. Sci. U.S.A., 81, 5734-5738, 1984.
130. Shiroki, K., Maruyama, K., Saito, I., Fukui, Y., and Shimojo, H., J. Virol. 38, 1048-1054, 1981.
131. Schrier, P.I., van den Elson, P.J., Hertoghs, J.J., and van der Eb, A.J., Virology 99, 372-385, 1979.
132. Jenkins, J.R., Rudge, K., and Currie, G.A., Nature 312, 651-654, 1984.
133. Ruley, H.E., Nature 304, 602-606, 1983.
134. Eliyahu, D., Ras, A., Grass, P., Givol, D., and Oren, M., Nature 312, 646-649, 1984.
135. Parada, L.F., Land, H., Weinberg, R.A., Wolf, D., and Rotter, V., Nature 312, 649-651, 1984.
136. Gallimore, P.H., and Williams, J., Virology 120, 146-156, 1982.
137. Philipson, L., Pettersson, U., Lindberg, U., Tibbetts, C., Vennstrom, B., and Persson, T., Cold Spring Harbor Symp. Quant. Biol. 39, 447-456, 1974.
138. Sharp, P.A., Gallimore, P.H., and Flint, S.J., Cold Spring Harbor Symp. Quant. Biol. 39, 457-474, 1974.
139. Galos, R.S., Williams, J., Binger, M.H., and Flint, S.J., Cell 17, 945-956.
140. Williams, J., Galos, R.S., Binger, M.H., and Flint, S.J., Cold Spring Harbor Symp. quant. Biol. 44, 353-365.
141. Stillman, B.W., Lewis, J.B., Chow, L.T., Mathews, M.B., and Smart, J.E., Cell 23, 497-508, 1981.
142. Kitchingman, G.A., Lai, S.P., and Westphal, H., Proc. Natl. Acad. Sci. U.S.A. 74, 4392-4395.
143. Berk, A.J., and Sharp, P.A., Cell 14, 695-711, 1978.

144. Chow, L.T., Broker, T., and Lewis, J.B., J. Mol. Biol. 134, 265-303, 1979.
145. Enomoto, T., Lichy, J.H. Ikeda, J.E., and Hurwitz, J., Proc. Natl. Acad. Sci. U.S.A. 78, 6779-6783, 1981.
146. Lichy, J.H., Field, J., Horwitz, M.S., and Hurwitz, J., Proc. Natl. Acad. Sci. U.S.A. 79, 5225-5229, 1982.
147. Stillman, B.W., Tamanoi, F., and Mathews, M.B., Cell 31, 613-623, 1982.
148. Ostrove, J.M., Rosenfeld, P., Williams, J., and Kelly, T.J. Jr., Proc. Natl. Acad. Sci. U.S.A. 80, 935-939, 1983.
149. Stillman, B.W., and Tamanoi, F., Cold Spring Harbor Symp. Quant. Biol. 47, 741-752, 1983.
150. Gingeras, T.R., Sciaky, D., Gelinas, R.E., Bing-Dong, J., Yen, C.E., Kelly, M.M., Bullock, P.A., Parsons, B.L., O'Neill, K.E., and Roberts, R.J., J. Biol. Chem., 257, 13475-13491, 1982.
151. Alestrom, P., Akusjarvi, G., Perricaudet, M., and Pettersson, U., J. Biol. Chem. 257, 13492-13498, 1982.
152. Smart, J.E., and Stillman, B.W., J. Biol. Chem. 257, 13499-13506, 1982.
153. Lewis, J.B., Esche, H., Smart, J.E., Stillman, B.W., Harter, M.L., and Mathews, M.B., Cold Spring Harbor Symp. Quant. Bio. 44, 493-508, 1979.
154. Lewis, J.B., and Mathews, M.B., Cell 21, 303-313, 1980.
155. Virtanen, A., Alestrom, P., Persoon, H., Katze, M.G., and Pettersson, U., Nucl. Acids Res. 10, 2539-2548, 1982.
156. Wilkie, N.M., Ustacelebi, S., and Williams, J.F., Virology 51, 499-503, 1973.
157. Ginsberg, H.S., Ensinger, M.J., Kauffman, R.S., Mayer, A.J., and Lundholm, U., Cold Spring Harbor Symp. Quant. Biol. 39, 419-426, 1974.
158. Levine, A.J., van der Vliet, P.C., Rosenwirth, B., Rabek, J., Frenkel, G., and Ensinger, M., Cold Spring Harbor Symp. Quant. Biol. 39, 559-566, 1974.
159. Berget, S.M., Flint, S.J., Williams, J., and Sharp, P.A., J. Virol. 19, 879-889, 1976.
160. Russell, W.C., Skehel, J.J., and Williams, J.F., J. Gen. Virol. 24, 247-259, 1974.
161. Challberg, M.D., and Kelly, T.J., Jr., Proc. Natl. Acad. Sci. U.S.A. 76, 655-659.
162. Lichy, J.H., Nagata, K., Friefield, B.R., Enomoto, T., Field, J., Guggenheimer, R.A., Ikeda, J.E., Horwitz, M.S., and Hurwitz, J., Cold Spring Harbor Symph. Quant. Biol. 47, 731-740, 1982.
163. Friefield, B.R., Lichy, J.D., and Hurwitz, J., and Horwitz, M.S., Proc. Natl. Acad. Sci. U.S.A., 80, 1589-1593, 1983.
164. Kelly, T.J., Jr., In: "The Viruses-Adenoviruses" (Ed. H.S. Ginsberg), pp. 271-308, Plenum, New York, 1984.
165. Sussenbach, J.E., and van der Vliet, P.C., Current Topics in Microbiology and Immunology Vol 109: The Molecular Biology of Adenoviruses 1 (Ed. W. Doerfler) pp. 53-73, Springer-Verlag, Berlin, Heidelberg, 1983.
166. Tamanoi, F., and Stillman, B.W., In: Current Topics in Microbiology and Immunology, Vol 109: The Molecular Biology of the Adenoviruses 1 (Ed. W. Doerfler), pp. 75-87, Springer-Verlag, Berlin, Heidelberg, 1983.
167. Friefeld, B.R., Lichy, J.H., Field, J., Gronostajski, R.M., Guggenheimer, R.A., Krevolin, M.D., Nagata, K., Hurwitz, J., and Horwitz, M.S., In: Current Topics in Microbiology and Immunology, Vol. 110: The Molecular Biology of the Adenoviruses 2 (Ed. W. Doerfler), pp. 221-255, Springer-Verlag, Berlin, Heidelberg, 1984.

168. Futterer, J., and Winnacker, E.L., In: Current Topics in Microbiology and Immunology, Vol 111: The Molecular Biology of Adenoviruses 3 (Ed. W. Doerfler), pp. 41-64, Springer-Verlag, Berlin, Heidelberg, 1984.
169. Tamanoi, F., and Stillman, B.W., Proc. Natl. Acad. Sci. U.S.A. 80, 6446-6450, 1983.
170. Challberg, M.D., and Rawlins, D.R., Proc. Natl. Acad. Sci. U.S.A. 81, 100-104, 1984.
171. Rawlins, D.R., Rosenfeld, P.J., Wides, R.J., Challberg, M.D., and Kelly, T.J. Jr., Cell 37, 309-319, 1984.
172. Younghusband, H.B., Tyndall, C., and Bellett, A.J.D., J. Gen. Virol. 45, 455-467, 1979.
173. Longiaru, M., and Horwitz, M.S., Mol. Cell. Biol. 1, 208-215, 1981.
174. Hochstadt, J., Ozer, H.L., and Shopsis, C., In: Current Topics in Microbiology and Immunology 94/95, pp. 243-308, Springer-Verlag, Berlin, Heidelberg, 1981.
175. Kruijer, W., van Schaik, F.M.A., and Sussenbach, J.S., Nucl. Acids Res. 9, 4439-4457, 1981.
176. Kruijer, W., van Schaik, F.M.A., and Sussenbach, J.S., Nucl. Acids Res. 10, 4493-4500, 1982.
177. Baker, C.C., Herisse, J., Courtois, G., Galibert, F., and Ziff, E., Cell 8, 569-580, 1979.
178. van der Vliet, P.C., and Levine, A.J., Nature, New Biol. 239, 47-49, 1973.
179. van der Vliet, P.C., Keegstra, W., and Jansz, H.S., Env. J. Biochem. 86, 389-398, 1978.
180. Fowlkes, D.M., Lord, S.T., Linne, T., Pettersson, U., and Philipson, L., J. Mol. Biol. 132, 163-180, 1979.
181. Ensinger, M.J., and Ginsberg, H.S., J. Virol. 10, 328-339, 1972.
182. van der Vliet, P.C., Levine, A.J., Ensinger, M.J., and Ginsberg, H.S., J. Virol. 15, 348-354, 1975.
183. van der Vliet, P.C., Blanken, W.M., Zandberg, J., Jansz, H.S., In: DNA Synthesis, present and future (I. Molineux and M. Kolinyama, Eds.), Plenum, N.Y., pp. 869-885, 1977.
184. van der Vliet, P.C., and Sussenbach, J.S., Virology 67, 415-416, 1975.
185. Challberg, M.D., Ostrove, J.M., and Kelly, T.J. Jr., J. Virol. 41, 265-270, 1982.
186. Friefeld, B.R., Krevolin, M.D., and Horwitz, M.S., Virology 124, 380-389, 1983.
187. Ikeda, J., Enomoto, T., and Hurwitz, J., Proc. Natl. Acad. Sci. U.S.A., 78, 884-888, 1984.
188. Klein, H., Maltzman, W., and Levine, A.J., J. Biol. Chem. 254, 11051-11060, 1979.
189. Linne, T., Philipson, L., Eur. J. Biochem. 103, 259-270, 1980.
190. Ariga, H., Klein, H., Levine, A.J., and Horwitz, M.S. Virology 101, 307-310, 1980.
191. Carter, T.H., and Blanton, R.A., J. Virol. 25, 664-674, 1978.
192. Babich, A., and Nevins, J.R., Cell 26, 371-379, 1981.
193. Nevins, J.R., and Winkler, J.J., Proc. Natl. Acad. Sci. U.S.A. 77, 1893-1897, 1980.
194. Rossini, M., Virology 131, 49-58, 1983.
195. Richardson, W.D., and Westphal, H., Cell 17, 957-966, 1979.
196. Klessig, D.F., and Grodzicker, T. Cell 17, 957-966, 1979.
197. Klessig, D.F., and Quinlan, M.P., J. Mol. Appl. Genet. 1, 263-272, 1982.
198. Anderson, K.P., and Klessig, D.F., J. Virol. 42, 748-754, 1982.

199. Friedmann, M.P., Lyons, M.J. and Ginsberg, H.S., J. Virol. 5, 586-597, 1970.
200. Eron, L., Westphal, H., and Khoury, G., J. Virol. 15, 1256-1261, 1975.
201. Klessig, D.F., and Anderson, C.W., J. Virol. 16, 1650-1668, 1975.
202. Farber, M.S., and Baum, S.G., J. Virol. 27, 136-148, 1978.
203. Anderson, K.P., and Klessig, D.F., J. Mol. Appl. Genet. 2, 31-43, 1983.
204. Klessig, D.F., and Chow, L.T., J. Mol. Biol. 139, 221-242, 1980.
205. Zorn, G.A., and Anderson, C.W. J. Virol. 37, 759-769, 1981.
206. Anderson, K.P., and Klessig, D.F., Proc. Natl. Acad. Sci. U.S.A. 81, 4023-4027, 1984.
207. Lewis, A.M., Levine, A.J., Crumpacker, C.S., Levin, M.J., Samabra, R.J., and Henry, P.H., J. Virol. 11, 655-664, 1973.
208. Cole, C.N., Crawford, L.V., and Berg, P., J. Virol. 30, 683-691, 1979.
209. Rice, S.A., and Klessig, D.F. J. Virol. 49, 35-49, 1984.
210. Nicolas, J.C., Saurez, F., Levine, A.J., and Girard, M. Virology 108, 521-524, 1981.
211. Nicolas, J.C., Ingrand, D., Sarnow, P., and Levine, A.J., Virology 122, 481-485, 1982.
212. Logan, J., Nicolas, J.C., Topp, W.C., Girard, M., Shenk, T., and Levine, A.J., Virology 115, 419-422, 1981.
213. Kruijer, W., Nicolas, J.C., van Schaik, F.M.A., and Sussenbach, J.S., Virology 124, 425-433, 1983.
214. Nicolas, J.C., Sarnow, P., Girard, M., and Levine, A.J. Virology 126, 228-239, 1983.
215. Carter, T.H., Nicolas, J.C., Young, C.S.H., and Fisher, P.B. Virology 117, 519-521, 1982.
216. Fisher, P.B., Babiss, L.E., Weinstein, B., and Ginsberg, H.S., Proc. Natl. Acad. Sci. U.S.A. 79, 3527-3531, 1982.
217. Mayer, A.J., and Ginsberg, H.S. Proc. Natl. Acad. Sci. U.S.A. 74, 785-788, 1977.
218. Dorsch-Hasler, K., Fisher, P.B., Weinstein, I.B., and Ginsberg, H.S., J. Virol. 34, 305-314, 1980.
219. Flint, S.J., Berget, S.M., and Sharp, P.A., Virology 72, 443-455, 1976.
220. Kitchingman, G.R., and Westphal, H., J. Mol. Biol. 137, 23-48, 1980.
221. Baker, C.C., and Ziff, E.B., J. Mol. Biol. 149, 189-221, 1981.
222. Ahmed, C.M.I., Chanda, R.S., Stow, N.D., and Zain, B.S., Gene 20, 337-344, 1982.
223. Stalhandske, P., Persson, H., Perricaudet, M., Philipson, L., and Pettersson, U., Gene 22, 157-165, 1983.
224. Herisse, J., Courtois, G., and Galibert, F. Nucl. Acids Res. 8, 2173-2192, 1980.
225. Herisse, J. and Galibert, F., Nucl. Acids Res. 9, 1229-1240, 1981.
226. Lewis, J.B., Atkins, J.F., Baum, P.R., Solem, R., Gesteland, R.F., and Anderson, C.W., Cell 7, 141-151, 1976.
227. Green, M., Wold, W.S.M., Brackmann, K., Carlas, M.A., Cold Spring Harbor Symp. Quant. Biol. 44, 457-470, 1979.
228. Persson, H., Jansson, M., and Philipson, L., J. Mol. Biol. 136, 375-394, 1980.
229. Persson, H., Jornvall, H., and Zabielski, J., Proc. Natl. Acad. Sci. U.S.A. 77, 6349-6353, 1980.
230. Persson, H., Signas, C., and Philipson, L., J. Virol. 29, 938-948, 1979.
231. Persson, H., Kvist, S., Ostberg, L., Pettersson, P.A., and Philipson, L., Cold Spring Harbor Symp. Quant. Biol. 44, 509-517, 1979.
232. Ross, S.R., and Levine, A.J. Virology 99, 427-430, 1979.
233. Wold, W.S.M., and Green, M., J. Virol. 30, 297-310, 1979.

234. Storch, T.G., and Maizel, J.V. Jr., Virology 103, 54-67, 1980.
235. Kapoor, Q.S., Wold, W.S.M., and Chinnadurai, G., Virology 112, 780-784, 1981.
236. Signas, C., Katze, M.G., Persson, H., and Philipson, L., Nature 299, 175-178, 1982.
237. Gallimore, P.H., and Paraskeva, C., Cold Spring Harbor Symp. Quant. Biol. 44, 703-714, 1979.
238. Wold, W.S.M., Cladaras, C., Magie, S.C., and Yacoub, N., J. Virol. 52, 307-313, 1984.
239. Persson, H., Oberg, B., and Philipson, L., J. Virol. 28, 119-139, 1978.
240. Levine, A.J., In: Current Topics in Microbiology and Immunology Vol. 110: The Molecular Biology of the Adenoviruses 2 (Ed. W. Doerfler), pp. 143-167, Springer-Verlag, Berlin, Heidelberg, 1984.
241. Rowe, W.P., Huebner, R.J., Gilmore, L.K., Parrott, R.H., Ward, T.G., Proc. Soc. Expt. Biol. Med. 84, 570-573, 1953.
242. Tigges, M., and Raskas, H., J. Virol. 50, 106-117, 1984.
243. Herisse, J., Rigolet, M., DuPont de Dinechin, S., and Galibert, F., Nucl. Acids Res. 9, 4023-4042, 1981.
244. Freyer, G.A., Katoh, Y., and Roberts, R.J., Nucl. Acids Res. 12, 3503-3519, 1984.
245. Virtanen, A., Gilardi, P., Naslund, A., LeMoullec, J.M., Pettersson, U., and Perricaudet, M., J. Virol. 51, 822-831, 1984.
246. Matsuo, T., Hashimoto, S., Wold, W.S.M., Symington, J., Rankin, A., and Green, M., J. Virol. 41, 334-339, 1982.
247. Tigges, M.A., and Raskas, H.J., J. Virol. 44, 907-921, 1982.
248. Sarnow, P., Hearing, P., Anderson, C.W., Reich, N., and Levine, A.J. J. Mol. Biol. 162, 565-583, 1982.
249. Downey, J.F., Rowe, D.T., Bacchetti, S., Graham, F.L., and Bayley, S.T., J. Virol. 45, 514-523, 1983.
250. Shiroki, K., Hashimoto, S., Saito, I., Fukui, Y., Fukui, Y., Kato, H., and Shimojo, H., J. Virol. 50, 854-863, 1984.
251. Schlesinger, R.W., Adv. in Virus Research 14, 1-61, 1969.
252. Tooze, J., Molecular Biology of Tumor Viruses, 2nd Ed., Part 2, DNA Tumor Viruses, Cold Spring Harbor, N.Y. 1980.
253. Straus, S.E., In: "The Viruses: Adenoviruses (Ed. H.S. Ginsberg) pp 451-496. Plenum, New York, 1984.
254. Huebner, R.J., Rowe, W.P., and Lane, W.T., Proc. Natl. Acad. Sci. U.S.A. 48, 2051-2058, 1962.
255. Trentin, J.J., Yabe, Y., and Taylor, G., Science 137, 835-841, 1962.
256. Pereira, M.S., Pereira, H.G., and Clarke, S.K.R., Lancet 1, 21-23, 1965.
257. Huebner, R.J., Rowe, W.P., Turner, H.C., and Lane, W.T., Proc. Natl. Acad. Sci. U.S.A. 50, 379-389, 1963.
258. Rabson, A.S., Kirchstein, R.L., and Paul, F.J., J. Natl. Cancer Inst. 32, 77-87, 1964.
259. Yabe, Y., Samper, L., Bryan, E., Taylor, Co., and Trentin, J.J., Science, 143, 46-47, 1964.
260. Allison, A.C., Berman, L.D., and Levey, R.H., Nature 215, 185-187, 1967.
261. Girardi, A.J., Hilleman, M.R., and Zwickey, R.E. Proc. Soc. Expt. Biol. Med. 115, 1141-1150, 1964.
262. Huebner, R.J., Casey, M.J., Chanock, R.M. and Schell, K., Proc. Natl. Acad. Sci., U.S.A. 54, 381-388, 1965.
263. Trentin, J.J., van Hoosier, G.L., and Samper, L., Proc. Soc. Expl. Biol. Med. 127, 683-689, 1968.

264. Jonsson, N., and Ankerst, J., Cancer 39, 2513-2519, 1977.
265. Freeman, A.E., Black, P.H., Wolford, R., and Huebner, R.J., J. Virol. 1, 362-367, 1967.
266. Freeman, A.E., Vanderpool, E.A., Black, P.H., Turner, H.C., and Huebner, R.J., Nature 216, 171-173, 1967.
267. Freeman, A.E., Black, P.H., Vanderpool, E.A., Henry, P.H., Austin, J.B., and Huebner, R.J., Proc. Natl. Acad. Sci. U.S.A. 58, 1205-1212, 1967.
268. McAllister, R.M., Nicolson, M.O., Lewis, A.M. Jr., Macpherson, I. and Huebner, R.J., J. Gen. Virol. 4, 29-36, 1969.
269. Gallimore, P.H., J. Gen. Virol. 16, 99-102.
270. Williams, J.F., Nature 243, 162-163, 1973.
271. Lewis, A.M., Breeden, J.H., Wewerka, Y.L., Schnipper, L.E., and Levine, A.S., Cold Spring Harbor Symp. Quant. Biol. 39, 651-656, 1974.
272. Takemori, N., Riggs, J.L., and Aldrich, C.D., Virology 38, 8-15, 1969.
273. Ezoe, H., Lai Fatt, R.B., and Mak, S., J. Virol., 40, 20-27, 1981.
274. Lundholm, U., and Doerfler, W., Virology 45, 827-829, 1971.
275. Shiroki, K., Irisawa, J., and Shimojo, H., Virology 49, 1-11, 1972.
276. Hama, S., and Kimura, G., Japan. J. Microbiol. 16, 337-338, 1972.
277. Ledinko, N., J. Virol., 14, 457-468, 1974.
278. Suzuki, E., and Shimojo, H., J. Virol. 13, 538-540, 1974.
279. Shiroki, K., and Shimojo, H., Virology 61, 474-485, 1974.
280. Hama, S., and Kimura, G., Microbiol., Immunol. 24, 629-647, 1980.
281. Ledinko, N., J. Virol. 26, 828-831, 1978.
282. Ledinko, N., Schaeufele, J., and Soorma, O., J. Virol. 29, 250-260, 1979.
283. Smiley, J.R., and Mak, S., J. Virol. 28, 227-239, 1978.
284. Grand, R.J.A., and Gallimore, P.H., J. Gen. Virol. 65, 2149-2166, 1984.
285. Williams, J., Ho, Y.S., and Galos, R., Virology 110, 208-212, 1981.
286. Rowe, D.T., and Graham, F.L., J. Virol. 38, 191-197, 1981.
287. Brusca, J.S., and Chinnadurai, G., J. Virol. 39, 300-305, 1981.
288. Fukui, Y., Saito, I., Shiroki, K., Shimojo, H., J. Virol., 49, 154-161, 1984.
289. Bernards, R., deLeeuw, M.G.W., Vaessen, M.J., Houweling, A., and van der Eb, A.J., J. Virol. 50, 847-853, 1984.

9

ALTERATIONS IN CELLULAR FUNCTIONS IN ADENOVIRUS-INFECTED AND TRANSFORMED CELLS

J. FLINT
Department of Molecular Biology
Princeton University
Princeton, NJ 08544

INTRODUCTION

Infection of mammalian cells by human adenoviruses induces dramatic alterations in the cells' abilities to continue their normal business. Indeed, adenoviruses were first recognised by their ability to kill cells which supported the efficient replication of the virus (1, 2). Such cell death is the culmination of several processes which redirect the host cells' biosynthetic apparatus. Thus, when adenoviruses infect rapidly-dividing, fully permissive cells, they induce rapid and complete inhibition of synthesis of both cellular DNA and cellular proteins. At first sight, such redirection of cellular functions appears of obvious advantage to the virus: the relevant cellular machinery becomes devoted to the production of the components which will form progeny virions, the raison d'etre of the infectious cycle. The complete cessation of cellular DNA or protein synthesis and the vast quantities of viral macromolecules produced attest to the efficiency of the molecular mechanisms by which the virus carries out such usurption. It is, therefore, quite surprising that as little as 10% of the newly made viral material is assembled into virions (see ref. 3 for a review).

Although it is the adenovirus-induced inhibition of cellular biosynthetic processes in rapidly growing, human cells which is best understood, it is important to recall that the natural host cells in which adenoviruses replicate are quite different, for example, near-terminally differentiated, quiescent cells of the upper respiratory tract lining. In this context, it is of considerable interest that quiescent human or rodent cells in culture do not respond to adenovirus infection by cessation of biosynthetic processes, but rather are induced to enter the cell cycle and resume cellular DNA synthesis. Although the cell cycle progression induced in rodent cells, a potentially important, early step in transformation, has been examined in some detail and shown to be aberrant in several respects (see ref. 4, for a review), its molecular basis has not been established. As relatively little work has been performed with quiescent, normal epithelial cells as hosts for adenoviruses, this review will discuss the alterations in cellular gene expression which characterise infection of rapidly dividing, immortalised cells, such as HeLa and their potential relationships to adenovirus transformation functions.

The dramatic inhibition of cellular gene expression induced during the late phase of infection under such circumstances was recognised more than fifteen years ago (5), but seems to have been regarded as a convenience, rather than a topic of intrinsic interest until a recent renaissance: the molecular mechanisms mediating inhibition of cellular gene expression in particular have been given considerable attention. As we shall see, these mechanisms have proved to be quite novel, providing insights into translational regulation and an unusual example of post-transcriptional regulation of mRNA production.

INHIBITION OF CELLULAR GENE EXPRESSION IN PRODUCTIVELY INFECTED CELLS.

When permissive, growing human cells in culture are infected by a subgroup C adenovirus (Ad2 or Ad5) under typical laboratory conditions, that is, infection at a sufficiently high multiplicity to ensure that all cells in the culture are infected more or less simultaneously, synthesis of viral DNA commences at 6 to 8 hours after infection. This event defines the onset of the late phase of infection and from this point onwards viral, structural proteins are made in large, and increasing, quantities. Concommitantly, moreover, synthesis of cellular proteins decrease (5-8), as illustrated in Figure 1. Although the inhibition of cellular protein synthesis is eventually complete, it is not instantaneous, but takes place over several hours (for example, 8, 9). All proteins made in uninfected HeLa cells appear to be equally affected during the late phase of adenovirus infection. It is now clear (see next section) that adenoviruses, like certain other types of virus, notably picornaviruses (10), possess the ability to redirect the host

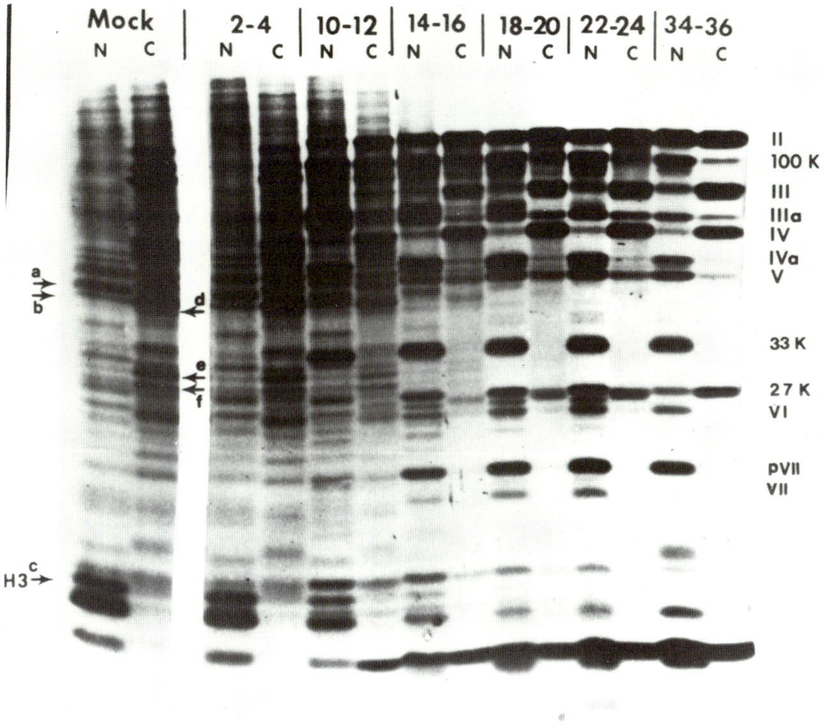

Figure 1: Inhibition of Synthesis of Cellular Proteins in adenovirus-infected cells.

HeLa cells were infected with 40pfu/cell Ad2 and labelled with ^{3}H-leucine for 2 hr. periods at the times indicated. Mock-infected cells were labelled in the same way. Cells were harvested at the end of the labelling period and nuclear and cytoplasmic fractions separated in 0.65% NP40. Aliquiots of each fraction were then analysed in a 15%-polyacrylamide gel. Viral proteins are indicated at the right. The amount of radioactivity present in the six cellular protein bands labelled a-f was determined after excision from a similar gel to confirm that the labelling of each cellular polypeptide band was completely inhibited by 18-20 hrs. after infection.

cells' translational machinery to their own ends. In addition, adenovirus infection disrupts the biogenesis of cellular RNA. Such a phenomenon is not unique to adenovirus infections, a fact of little surprise in view of its obvious advantage to the virus: rhabdoviruses, for example, induce inhibition of transcription by RNA polymerase II (11), apparently mediated by the leader RNA (12, 13). Nevertheless, the mechanism whereby adenoviruses induce inhibition of cellular RNA biogenesis is presently the only example of its kind and, as discussed in a subsequent section, provides a rare opportunity to investigate nuclear to cytoplasmic transport in mammalian cells.

Selective Translation of Viral mRNA Species in Adenovirus-infected Cells

It has been apparent for some time that adenovirus-infected cells contain cellular mRNA species which fail to be translated during the late phase of infection: cytoplasmic, poly(A)-containing RNA prepared from infected cells harvested after the onset of viral DNA synthesis directs the synthesis not only of large quantities of viral, late polypeptides but also of cellular proteins, when added to heterologous, cell-free translation systems, despite the fact that the corresponding cellular proteins are no longer made *in vivo* (for example, 14-16). The implication of such observations, that cellular mRNA species persist in the cytoplasm throughout the late phase of adenovirus infection, has now been directly demonstrated in several instances, for example mRNA species encoding DHFR (17), β-tubulin (S.J.F., unpublished observations) and actin (18). The concentrations of these cellular mRNA species, and of several other abundant HeLa mRNA species, do not appear to be much reduced during the late phase of Ad2 or Ad5 infection, compared to their concentrations in uninfected cells (17, 18; S. J. Flint, unpublished observations). Moreover, Babich and colleagues have shown that the actin mRNA present, but not translated, in Ad2-infected cells at 15 hours after infection can be translated perfectly well in an *in vitro* system. Thus, there can be little doubt that adenovirus infection does not inactivate pre-existing cellular mRNA species, but rather must modify the hosts cells' translational machinery to establish the pattern of virus-specific protein synthesis which is so characteristic of the late phase of infection.

What was perhaps an even more dramatic demonstration of the translational changes which typify the late phase of adenovirus infection came with the isolation of mutants of Ad5 which cannot direct the synthesis of VA-RNAI (19). Such mutants as H5dl330 and H5dl331, which carry lesions within the transcriptional control region of the

VA-RNAI gene, direct the synthesis of normal quantities of viral, late mRNA species, which are structurally indistinguishable from their wild-type counterparts and can be translated efficiently in *in vitro* translation systems (18, 19). Nevertheless, mutant-infected cells show greatly impaired synthesis of viral late proteins. Thus, VA-RNAI must be required for the efficient translation of viral mRNA species (19).

Whether the second small RNA species encoded by the adenovirus genome, VA-RNAII, (20), which displays considerable sequence and structural homology to VA-RNAI (21, 22), also facilitates translation in adenovirus infected cells is not clearly established. Cells infected by the VA-RNAI-negative mutants H5dl330 and H5dl331 produce wild type VA-RNAII, but nevertheless fail to synthesise proteins efficiently during the late phase of infection (19), indicating that VA-RNAII cannot be a functional homologue of the more abundant (20) VA-RNAI. Mutants of Ad5 which cannot support the transcription of VA-RNAII have been constructed, but display very similar growth properties and viability to the Ad5 parent (19). These phenotypes, or, rather, the lack of them, suggest that VA-RNAII does not play so crucial a role during productive infection as VA-RNAI. On the other hand, the quantities of VA-RNAII made in cells infected by H5dl330 do not approach the levels of VA-RNAI accumulating in wild-type Ad5 infections, although they are somewhat increased (19). Thus it could be argued that H5dl330-infected cells simply do not contain enough VA-RNAII to compensate for the lack of VA-RNAI. A mutant in which both VA-RNA genes are deleted has been constructed and grows extremely poorly, yielding titers 1/60 to 1/80 of those of wild-type viruses (23, 24). The decreased viability, compared even to H5dl330 or H5dl331, of the double mutant indicates that VA-RNAII must fulfill some important role. Further studies with such double mutants and elucidation of the properties of mutants which direct the synthesis of amounts of VA-RNAII approaching wild-type quantities of VA-RNAI, if these can be constructed, would help address the question of the function of VA-RNAII.

Far less than the normal numbers of polyribosomes containing viral mRNA are observed in cells infected by H5dl330 or H5dl331 (18, 19, 25). Such observations and the results of measurement of the mean times required for ribosomes to traverse viral mRNA in mutant compared to wild-type infected cells, approximately two minutes in each case, suggested that VA-RNAI facilitates initiation of translation of viral mRNA species in adenovirus infected cells (25). These properties of mutant-infected cells therefore appeared to provide clues about the mechanisms by which this viral small RNA species might mediate, selective translation of viral mRNA species during the late phase of adenovirus infection. However, further insight into the molecular function of VA-RNAI gained from the use of cell-free translation systems suggests that the role of VA-RNAI is, in fact, quite different.

Translational systems prepared from H5dl331-infected cells show much less activity than those made from wild-type, Ad5-infected cells, whether translating either endogenous or exogenous viral mRNA (25, 26). Addition of VA-RNAI to defective H5dl331-infected cell extracts failed to restore translational activity to wild-type levels (26), an observation implying that VA-RNAI does not act directly to mediate initiation of translation of viral mRNA species. The routine observation that viral mRNA species purified from H5dl331-infected cells harvested during the late phase of infection are perfectly translatable in reticulocyte lysates or in extracts made from mock-or Ad2-infected HeLa cells, as is the corresponding mRNP (18, 19, 26), strongly suggests that it is the translational machinery itself which is defective in adenovirus infected cells in which VA-RNAI cannot be made.

Indeed, addition of purified eukaryotic initiation factors eIF2 or eIF2B (GTP recycling factor or guanosine nucleotide exchange factor) to extracts of H5dl331-infected cells stimulated translation some 10 to 20-fold, to restore protein synthesis to levels within 50 to 100% of those displayed by wild type advenovirus-infected cell extracts (26, 27). Such stimulation does not represent a non-specific effect, for addition of these same factors to extracts made from mock- or wild-type adenovirus-infected cells did not stimulate translation (26, 27) and none of six other initiation factors or two elongation factors tested by Reichel and colleagues (26) restored the translational activity of H5dl331-infected cell extracts. Thus, it can be concluded that initiation factor eIF2 becomes inactivated in adenovirus infected cells in the absence of VA-RNAI (26, 27), a block which certainly explains the failure of translation in H5dl331-infected cell extracts at an early step in the initiation process. Moreover, the ability of eIF2B to restore translational activity of H5dl331-infected cell extracts argues that the recycling of eIF2 is defective in mutant-infected cells: during formation of the initiation complex comprising the 40S ribosomal subunit and met-tRNA, eIF2.GTP becomes hydrolysed to eIF2.GDP and must be recharged if the initiation factor is to act catalytically, rather than stoichiometrically, a step which requires eIF2B (see 28, 29 for reviews).

Inhibition of initiation of translation as a result of loss of eIF2 function has been described in several systems, including reticulocytes subjected to haem deficiency or double-stranded RNA and cultured cells exposed to certain viruses or double-stranded RNA and thus induced to synthesis interferon (28, 30-34). In such situations, specific protein kinases become activated and phosphorylate the α-subunit of eIF2, a modification which prevents the recycling of eIF2-GDP to eIF2-GTP discussed in the previous paragraph, probably by sequestration of eIF2B (see 29). It might therefore be expected that VA-RNAI protects eIF2 from phosphorylation, normally induced by adenovirus infection (26). This explanation predicts that a specific kinase analogous or indentical to the ribosome associated, double-stranded RNA-dependent protein kinase induced by interferon in cultured cells phosphorylates this initiation factor in adenovirus infected cells. Indeed, Schneider and colleagues (27) have recently demonstrated both the presence of increased levels of phosphorylated eIF2 in human cells infected with H5dl331, compared to mock or wild-type Ad5-infected control cells, and the three to five fold more efficient phosphorylation of exogenously added purified eIF2 by extracts of mutant-infected compared to mock- or Ad5-infected cell extracts. Moreover, H5dl331 grows nearly as efficiently as wild-type Ad5 in a line of human fibroblasts which cannot synthesise the specific kinase (P1/eIF2α kinase) in response to interferon treatment (27). These observations, together with the increased quantities of phosphorylated ribosomal protein P1 also present in mutant-infected cells (27) provide compelling evidence that it is the P1/eIF2α kinase which is activated in adenovirus infected cells in the absence of VA-RNAI (27). Of course, the corollary of this conclusion must be that adenovirus infection induces the synthesis of, or activates, this kinase. As certain human adenoviruses have been reported to induce the synthesis of interferon, at least in chick embryo fibroblasts (35-37), and large quantities of complementary RNA molecules are produced when the major late and E2 transcriptional units are simultaneously expressed during the late phase of infection (see Chapter 2) appropriate mechanisms for kinase activation are not difficult to find in adenovirus-infected cells.

The mechanism whereby VA-RNAI protects eIF2 activity in adenovirus-infected cells has not been elucidated, but this small, viral RNA might directly inhibit P1/eIF2α kinase, alter the activity of the enzyme which dephosphorylates eIF2 (38) or act in some indirect fashion (27). The first of these alternatives appears particularly attractive in view of the ability of high concentrations of double-stranded RNA to prevent activation of P1/eIF2α kinase (39-41) and the highly double-stranded character of VA-RNAI (22, 42, 43).

This model makes several interesting predictions, for example, that the function of VA-RNAI is not determined by specific primary sequences (27) and that any small RNA with strong double-stranded character, such as a tRNA, or indeed VA-RNAII, should be able to perform the VA-RNAI function, if produced in sufficient quantities in infected cells. The sequences encoding two small EBV RNA species transcribed by RNA polymerase III have been introduced into the Ad5 genome, replacing only VA-RNAI (H5sub730) or both VA-RNAI and VA-RNAII (H5sub731). The former mutant exhibited growth properties which are very similar to those of wild-type virus, whereas the latter produces some 50-fold lower than normal yields of virus (23). The EBV RNA species can adopt extensively base-paired structures, which are strikingly similar to those proposed for VA-RNAI and VA-RNAII (44). Thus, the failure of H5sub731 to grow well might suggest that VA-RNAI function does not depend solely on its double-stranded character. However, the concentrations of EBV small RNA species produced in H5sub731-infected cells were well below those attained by VA-RNAI in wild-type infected cells (23), so it can be argued that insufficient quantities of EBV RNA species are produced to compensate for the absence of VA-RNAII.

Thimmappaya and colleagues (24) have constructed, in a VA-RNAII negative background, a large set of mutants of Ad5 bearing deletions or insertions at different points throughout the VA-RNAI coding sequence and examined their growth properties and ability to produce cytoplasmic VA-RNAI. All mutants grew significantly better than the VA-RNAI and VA-RNAII-negative double mutant mentioned previously. Moreover, the majority showed little impairment of growth, compared to the VA-RNAII negative parent, but removal of the region between +43 and +53 or between +107 and the 3' end did reduce virus yields more substantially. While these results point to the greater importance of the latter two regions, their significance in terms of the model of VA-RNAI function discussed previously is not easy to assess. At first sight, the finding that many different parts of the VA-RNAI molecule can be altered without dramatic effects on virus growth seems to suggest that no one primary sequence is essential to its function and thus to favor the conclusion that it is secondary structure which is important. Indeed, many of the mutant VA-RNAI species created by Bhat et al (24) are predicted to adopt structures in which most of the bases are hydrogen-bonded. Nevertheless, these authors argue against the notion that it is the double-stranded nature of VA-RNAI which is functionally important on two counts, the failure of EBV small RNA species which, as we have seen, can adopt similar hairpin structures, to impart wild-type growth properties upon viruses which direct the synthesis of neither VA-RNAI nor VA-RNAII (23) and the lack of any correlation between the cytoplasmic concentration of mutant VA-RNAT species and the growth. properties of the mutant viruses (24). The first argument is undermined by the low concentrations of EBV-RNA produced, but the second is difficult to dismiss in the absence of any experimental data to establish whether the mutant RNA species do indeed adopt an extensively hydrogen-bonded configuration, like that collected for wild-type VA-RNAI (43). Clearly, less ambiguous experiments will be required to establish the molecular details of the mechanism whereby VA-RNAI maintains translational activity in adenovirus-infected cells.

While these observations have rapidly advanced our understanding of the function of at least VA-RNAI, they do not provide any explanation of the selective translation of viral mRNA species during the late phase of adenovirus infection: the inactivation of eIF2 via its phosphorylation by the kinase induced or activated as a consequence of adenovirus infection must inhibit translation of both cellular and viral mRNA species. Similarly, any antagonistic effect of VA-RNAI would be expected to restore translation of all mRNA species. Indeed, the in vitro translation systems prepared from wild-type or mutant infected cells display just these properties: extracts made from H5dl331-infected cells not only fail to translate viral mRNA, as discussed previously, but also translate neither total HeLa cell mRNA nor several purified mRNA species, unless they are supplemented with eIF2 or the S100 fraction from rabbit reticulocytes (26). Thus, it appears that we must look elsewhere for the molecular basis of the selective synthesis of viral proteins which characterises the late phase of adenovirus infection. It should, nevertheless be noted that a role for VA-RNAI, in addition to protection against eIF2 inactivation, cannot be completely ruled out at this time. The main argument against such a possibility is the failure of purified VA-RNAI to restore translation activity to extracts of H5dl331-infected cells (26). However, any specific effect on translation of viral mRNA could be detected in in vitro systems derived from H5dl331-infected cells only once the activity of eIF2 had been restored and the appropriate experiments, for example, addition of both eIF2 (or eIF2B) and VA-RNAI, or better, VA-RNAI in association with the proteins to which it appears to be bound in the cell (45-47), have not been performed. Indeed, the use of deproteinized RNA might have precluded certain VA-RNA functions in vitro. Good precedent for such a situation is provided by the 7SL RNA component of signal recognition particle, which displays no ability to arrest translation of proteins destined for membrane translocation or insertion unless it is complexed with the protein components of the signal recognition particle (48).

While little evidence presently favors a direct role for VA-RNAI in regulation of translation during the late phase of infection, it does seem clear that certain viral RNA sequences are indeed important, namely those which comprise the tripartite leader forming the 5'-terminal segment of all mRNA species processed from transcripts of the major late transcription unit (49, 50). In an elegant series of experiments, Logan and Shenk (51) constructed a set of viruses in which the transcriptional control and 5' untranslated regions of the Ad5 E1A region were replaced by the control region of the major late (mL) transcriptional unit and various combinations of the three segments which comprise the tripartite leader. The addition of the mL control region and all three leader segments led to increased production of E1A polypeptides, during the late phase of infection. Measurement of the concentrations of E1A proteins produced under these circumstances and in cells infected by viruses which carry the wild-type E1A region or E1A expressed from the mL control region in the absence of any leader segment established that the presence of the tripartite leader sequences was responsible for about a five-fold increase in E1A protein production. The addition of just the l1 segment or each of the possible combinations of pairs of leader segments induced modest increases in the production of E1A proteins, but the presence of all three leader segments was required for maximal stimulation (51). Such increased production of E1A proteins occured in the absence of any detectable increase in the quantities of E1A mRNA produced and was restricted to the late phase of infection (51). The influence of tripartite leader segments upon translation during the late phase of infection has also been seen with chimeric genes containing SV40 (52) or dihydrofolate reductase (53) coding sequences.

The E1A mRNA species carrying all three leader segments were found to be associated with 3 to 6 ribosomes during the late phase of infection, by contrast to the E1A mRNA containing no tripartite leader sequences which was associated with only one or two ribosomes (51). These observations provide strong evidence in support of the view that sequences of the tripartite leader facilitate efficient translation during the late phase of adenovirus infection. The stimulation of ribosome association, and the observation that leader sequences exert no effect upon translation of E1A mRNA in infected cells when placed at the 3' end of the mRNA (51) imply that the leader influences initiation of translation, but this has yet to be demonstrated directly.

In the experiments of Logan and Shenk (51) the chimeric E1A mRNA carrying sequences from all three leader segments was still translated substantially less efficiently than hexon mRNA. This difference could reflect the fact that the variant mRNA carried its normal 3' untranslated sequence, or the absence from the chimeric E1A mRNA of the last 30 or so nucleotides of the 202 nucleotide late leader (51). The latter interpretation seems most reasonable at present, for it is clear that sequences derived from all three leader segments contribute to increased translational efficiency.

Viral mRNA species which are not derived from the major late transcriptional unit and therefore do not carry the tripartite leader, such as IVA2 and E2A mRNA species (see Chapter 2), are also translated efficiently during the late phase of infection However, neither their 5' nor 3' untranslated regions share a sequence homology with the tripartite leader (U-C. Yang, personal communication). Perhaps similarities amongst all viral mRNA species translated during the late phase of infection will be discerned when potential secondary structures are considered. Indeed, the observation that all segments of the tripartite leader must be present to confer maximal stimulation of translation (51) certainly suggests that it is a unique secondary or tertiary structure of the tripartite leader, rather than a simple linear nucleotide sequence, which facilitates translation.

Despite many uncertainties about the precise role of tripartite leader sequences and the mechanism whereby they influence translational efficiency, the demonstration that sequences which contribute to viral mRNA species translated during the late phase of infection determine translational efficiency might suggest that such sequences are all that is required to achieve selective translation of viral mRNA. In other words, it might not seem unreasonable that signals within the tripartite leader permit preferential translation of viral mRNA species in the absence of any other virus-induced changes, perhaps because such signals permit viral mRNA species to compete very successfully with cellular mRNAs during some early step in the process of translation. Attractive as such an idea might seem, a considerable body of evidence demonstrating that the inhibition of translation of cellular mRNA species is a much more active process, requiring specific viral proteins, has now been collected. One of the first indications that viral functions other than the RNA sequences discussed in this section influence protein synthesis during the late phase of infection was, in fact, the failure of cells infected by the VA-RNAI-negative mutants discussed in this section to resume cellular protein synthesis. Thus, for exammple, HeLa cells infected by the Ad5 mutants H5dl330 or H5dl331, produce neither viral nor cellular proteins efficiently during the late phase of infection (18, 19). This property clearly implicates additional viral functions in inhibition of translation of cellular mRNA species, which, as we have seen, persist in adenovirus-infected cells.

Analysis of the phenotypes displayed by several mutants whose lesions lie within sequences encoding viral, early proteins has implicated two such proteins in regulation of protein synthesis during the late phase of infection. The first mutant clearly exhibiting altered patterns of proteins made at this time, H2dl807, was described by Challberg and Ketner (54): cells infected by this mutant virus fail to produce viral late proteins in normal quantities and continue to synthesise cellular proteins. The deletion introduced into the Ad2 genome to create H2dl807 is large, spanning the region 82.5 to 95.0 map units and removes part of the E3 transcriptional unit, fiber coding sequences and the 3' portion of the E4 transcriptional unit (54; see Chapter 2 for a map). As the E3 region is not essential for growth of adenoviruses in tissue culture cells (55, 56) and viruses mutated in the fiber gene are not defective for inhibition of cellular protein synthesis (57, 58) the unique phenotypes exhibited by H2dl807 can be ascribed to deletion of E4 sequences. However, the E4 transcriptional unit potentially encodes a large number of proteins, perhaps as many as sixteen (59-61), several of whose coding sequences would be affected by the H2dl807 deletion (see Chapter 2). However, a mutant in which a very small deletion was introduced into the E4 reading frame encoding a polypeptide of predicted molecular weight 33,000 daltons has been constructed more recently and displays a phenotype very similar to that of H2dl807 (J. Cutt and T. Shenk, personal communication). The effects of these mutations do not reflect defects in earlier steps in the infectious cycle, upon which entry into the late phase depends, for synthesis of viral early products and viral DNA are not impaired in cells infected by them.

Aberrant patterns of protein synthesis during the late phase of infection have also been observed in cells infected by several viruses carrying deletions or point mutations within the sequences encoding the larger E1B polypeptide, predicted molecular weight 55kd. The structures of these mutant E1B regions, as well as the predicted mutant proteins

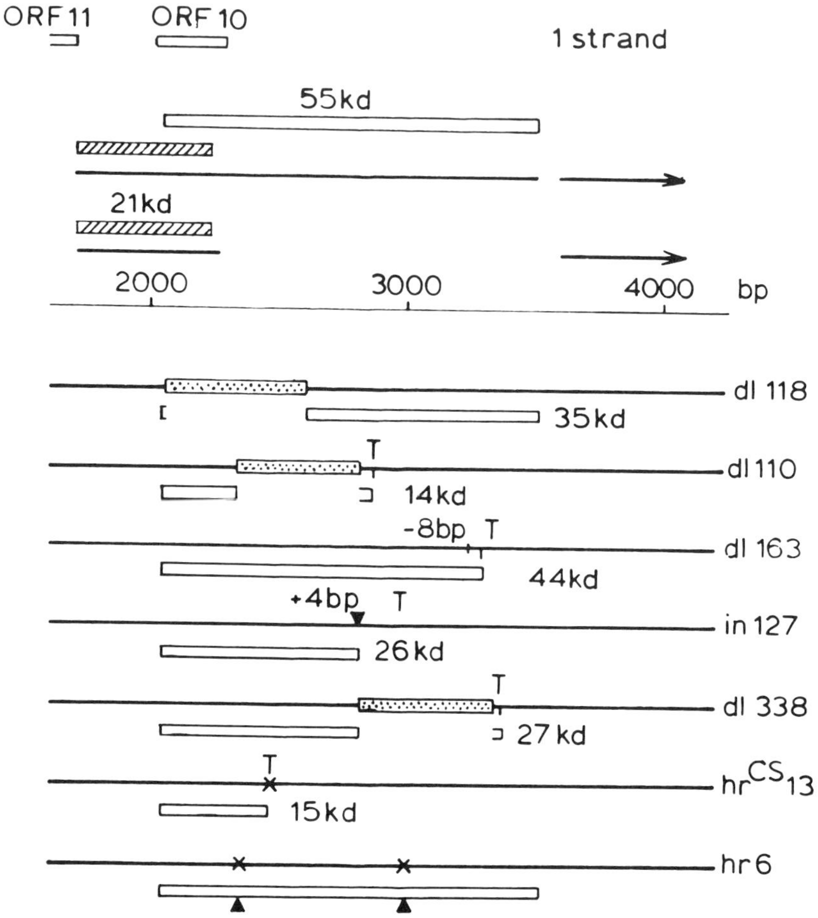

Figure 2: Deletion, insertion and point mutations in the Ad5 E1B Region
The region of the Ad5 genome between 2000 and 4000 base pairs is shown by the solid horizontal line. The major E1B mRNA species are drawn, in the direction of transcription, above the genome and sequences removed during splicing and poly(A) are shown as gaps, and arrow heads, respectively. The E1B 21kd and 55kd protein reading frames, which are shifted by two nucleotides (130) are represented by [hatched box] and [open box], respectively. The position of two open frames, OR10 and ORF11 (131) in the l-strand are also indicated. The structures of the E1B regions of the various E1B mutants discussed in the text are illustrated below the wild-type genome, where [stippled box], ▼, and X represent deletions, insertions and single base changes, respectively. The positions of termination codons introduced by these mutations are indicated by T. Below each mutant genome is shown the E1B 55kd-related protein it could encode and the approximate molecular weight of that protein. The two point mutations carried by H5hr6 substitute phenylalanine for cysteine, ▲ residues. The sources of these data are listed in the text.

they encode, are illustrated in Figure 2. The deletion mutants engineered by Babiss and Ginsberg (62) and Logan and colleagues (63) are not completely defective, but do exhibit longer growth cycles and 10 to 100-fold reduced yield of virus. The point mutants, such as H5hr6 and H5hr^{cs}13 are host range mutants, severely impaired for growth in the normal host cell, HeLa, compared to 293 cells, but are also cold-sensitive for growth in HeLa cells (64, 65). All mutants examined showed little impairment of viral DNA synthesis in infected cells, suggesting that a defect, or defects, in the late phase of infection is responsible for the reduced yields of virus. This is true not only of the point mutations, and deletions or insertions confined to the E1B 55kd protein coding sequences, such as H5dl338 (63) or H5dl110, H5dl163 and H5in127 (62), but also of H5dl118, the mutant examined in most detail by Babiss and Ginsberg (62), whose deletion also remove the C-terminal half of the E1B 21kd protein coding sequence (see Figure 2). This property of H5dl118 is perhaps a little surprising, for a relatively large number of E1B 21kd. protein mutants, bearing deletions or point mutations, have now been constructed and all show increased degradation of intracellular DNA (66 - 68). Be that as it may, the important property of these E1B 55kd protein mutants for the purposes of this discussion is the deranged pattern of protein synthesis observed in cells infected by them: during the late phase, synthesis of viral proteins is reduced compared to wild type or permissive infections and production of cellular proteins continues unabated (62, 63; S. Pilder and T. Shenk, personal communication; B. Karger, S. Ho, J. F. Williams, C. L. Castiglia, T. Mann and S. J. Flint. in preparation).

The failure of cells infected by these mutants to synthesise normal levels of viral, late proteins might be the result of defects in either the production of viral, late mRNA species or their translation. Cells infected by H5dl118 produce wild-type amonts of newly-synthesised nuclear RNA hybridising to L3 or L5 DNA sequences until 40 hrs after infection (62). Similarly, transcription of sequences which comprise the major late transcription unit is not impaired in HeLa cells infected by H5hr^{cs}13 at a non-permissive temperature of 33°C (Karger et al, in preparation). By contrast, some decrease in transcription of late RNA sequences has been observed in H5dl338-infectd cells (S. Pilder and T.

Shenk, personal communication), although not until very late times after infection. All three groups have employed very similar assays, hybridisation of pulse-labelled RNA to viral DNA fragments, so this difference is likely to reflect the exact nature of the mutations, rather than trivial, experimental differences. Be that as it may, normal transcription of late RNA sequences has been observed in cells infected by all mutants when synthesis of viral, late proteins is impaired. Despite normal transcription, cells infected by the E1B 55kd protein mutants exhibit steady state levels of viral, late mRNA which are reduced compared to wild-type or permissive infections (62; S. Pilder and T. Shenk, personal communication; Karger et al, in preparation). The most substantial reductions, up to ten-fold, are observed in cells infected by H5hr6 or H5hr^{cs}13 at a non-permissive temperature. Although these results might suggest that the primary defect in cells infected by mutants which cannot express the normal E1B 55kd polypeptide lies in processing or transport of viral, late mRNA (62), this is not the case, at least for H5hr6 or H5hr^{cs}13: spliced, viral late mRNA species are produced efficiently in the nucleus of mutant-infected cells maintained at a non-permissive temperature and newly-synthesised viral late RNA sequences enter the cytoplasm of infected cells as efficiently at 33°C as at 39°C (T. Mann, S. J. Flint, B. Karger and J. F. Williams, in preparation). Thus, it appears that viral late mRNA sequences which reach the cytoplasm normally in mutant-infected cells under non-permissive conditions are less stable than at a permissive temperature or than those produced in wild-type Ad5-infected cells at either temperature. (Mann et al, in preparation).

Whether such instability is the primary result of the absence of functional E1B 55kd protein or a secondary consequence of the failure of viral, late mRNA species to be translated efficiently is not yet known. However, the latter alternative, which postulates a critical role for the E1B 55kd protein in efficient translation of viral mRNA species during the late phase of productive adenovirus infection, appears the more attractive at present: cells infected by these E1B 55kd protein mutants clearly display one translational defect, the failure to inhibit translation of cellular mRNA species (62; S. Pilder and T. Shenk, personal communication; Karger et al, in preparation). It might be argued that the continued synthesis of cellular proteins in mutant-infected cells is merely a result of the absence of the normal, very large quantities of viral, late mRNA species, but this interpretation can be eliminated on the basis of the phenotypes exhibited by cells infected with H5hr6 or H5hr^{cs}13: at a permissive temperature, viral, late mRNA species are present in the cytoplasm of infected cells at close to wild-type concentrations but no inhibition of cellular protein synthesis takes place (Karger et al, in preparation).

At this juncture, much remains to be learnt about the mechanisms whereby the adenoviral E1B 55kd and E4 33kd proteins inhibit translation of cellular mRNA species and facilitate expression of viral, late genes. However, the finding that these two proteins are involved in regulation of gene expression during the late phase of infection is provactive, for they have been shown to exist as a complex in infected cells (69, 70). A monoclonal antibody raised against the E1B 55kd protein was found to immunoprecipitate specifically not only this viral polypeptide, but also one of apparent molecular weight 25kd (69). The immunoprecipitated 25kd protein was subsequently shown, by partial amino acid sequence analysis, to be the product of E4 subregion 6, that which encodes the E4 protein of predicted molecular weight 33kd discussed in a previous paragraph. Compelling evidence for the physical association of these two proteins in infected cells, rather than fortuitous recognition of the E4 protein by the antibody, was obtained when 293 cells, which express only the E1A and E1B transcriptional units (71, 72) or HeLa cells were infected with a mutant which cannot produce the E1B 55kd protein: only in the former circumstance was the E4 33kd protein precipitated by the anti-E1B 55kd protein monoclonal antibody (70).

It is therefore tempting to speculate that a complex of these adenoviral E1B and E4 proteins plays an important role in regulation of protein synthesis during the late phase of infection. No direct evidence to support this view is presently available, nor, indeed are the molecular functions performed by either protein understood. Furthermore, most of the E1B 55kd-E4 33kd protein complex appears, as illustrated in Figure 3, to be located in the nucleus. Quite substantial quantities of the E1B 55kd protein can be found in both the nucleus and cytoplasm during the late phase of Ad2 or Ad5 infection, although the protein is clearly concentrated in the nucleus (Figure 3), as expected on the basis of many immunoflouresence studies (see ref. 73 for review) and the work of Sarnow <u>et al</u>, (69,70). Despite the presence of some fraction of the E1B 55kd protein in the cytoplasm, no E4 33kd protein can be immunoprecipitated from the cytoplasm in association with this E1B protein, as it can from nuclear extracts (Figure 3). This finding implies either that participation of the complex comprising

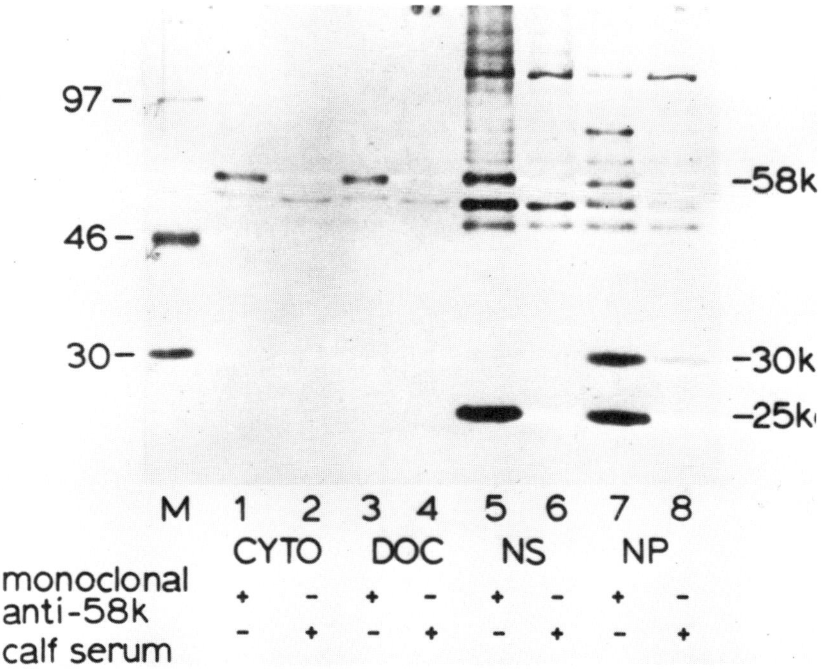

Figure 3: Distribution of E1B55kd and E433kd Proteins in Adenovirus Infected Cells

HeLa cells were infected with 40 pfu/cell Ad5 and labelled with ^{35}S-methionine at 20 to 22 hrs. after infection. The cells were harvested and exposed to 0.5% NP40 in RSB containing 5mMgCl2 (CYTO) followed by 1% Tween - 40 and 0.5% deoxycholate in the same buffer (DOC). The resulting nuclei were lysed in 0.5MNaCl and digested with DNA ase I. Nuclear soluble (NS) and insoluble (NP) material was separated by centrifugation. Aliquiots of each fraction were precipitated with the 2A6 anti-55kd monoclonal antibody of Sarnow <u>et al</u>, (69) or calf serum, as indicated in the figure. The positions of the E1B 58(55)kd, and the E433(25)kd proteins are indicated. ^{14}C-labelled molecular weight markers were applied to track 7.

these two proteins in regulation of translation might be indirect or, less likely, that such a complex exists only transiently in the cytoplasm, but is quite stable in the nucleus. Despite such caveats, the identification of viral, early proteins which influence translation during the late phase of infection represents an important advance, one which should lead to an improved understanding of the molecular details of the mechanisms mediating selective expression of viral genetic information in adenovirus infected cells.

Inhibition of Production of Cellular RNA Species

It is now clear that adenovirus infection disrupts not only translation of cellular mRNA, as discussed in the previous section, but also production of certain cellular RNA species. Early hybridisation studies suggested that most of the newly-synthesised, poly(A)-containing RNA entering the cytoplasm during the late phase of subgroup C adenovirus infection was viral in origin (74), a result subsequently confirmed and extended to several, specific mRNA species (9, 17, 18, 75; J. J. Plunkett and S. J. Flint, unpublished observations). Surprisingly, it has been established that transcription and processing of the products of cellular class II genes are not significantly inhibited during a similar period of the infectious cycle (9, 17, 18, 75, 76; see ref. 4 for a detailed review). These results have led to the conclusion that adenovirus infection of permissive, human cells induces selective transport of viral mRNA from the nucleus to the cytoplasm, concommitant with the onset of the late phase of infection. Direct evidence that would distinguish this mechanism from discrimination among viral and cellular, processed class II transcripts at some earlier step, for example displacement of cellular transcripts form their normal site(s) of synthesis or processing, is not available and, will, indeed, be hard to come by, for such essential steps in the production of functional mRNA as translocation to the cytoplasm are poorly understood. In the case of adenovirus-infected cells, however, some additional information supporting the conclusion that adenovirus infection alters nuclear to cytoplasmic transport of RNA has come from analysis of ribosomal RNA synthesis in infected cells.

The appearance in the cytoplasm of newly-synthesised 28S ribosomal RNA is also inhibited following adenovirus infection (77-79), with a time course of inhibition parallel to that of inhibition of appearance in the cytoplasm of newly-synthesised cellular mRNA (79). Thus, inhibition of cytoplasmic appearance of labelled 28S rRNA takes place during the period 8 to 12 hours after infection under typical conditions, in the absence of any corresponding change in the production of new cytoplasmic 18S rRNA or of nucleolar 45S and 32S pre-rRNA species (79). Precursors to cellular rRNA and mRNA species are transcribed by different RNA polymerases at different sites within the nucleus and are processed via pathways which are not known to share any processing enzymes or RNA packaging proteins. In fact, the only step known to be common to the biogenesis of these two classes of RNA is the transport to the cytoplasm of mature species, although the implication of these observations, that transport of cellular mRNA and transport of 28S, but not 18S, rRNA share some common feature(s) is surprising.

Neither the mechanism(s) whereby these changes in the production of cellular RNA species are mediated, the viral functions which might induce such altered transport nor the relation of such changes to the altered patterns of translation which typify the late phase of infection, if any, have been elucidated. However it must be the case that some signal exists which permits discrimination of viral mRNA species destined for translocation to the cytoplasm from cellular RNA species, transported in normal, uninfected HeLa cell, but retained in the nucleus from about 8 hours after infection onward in adenovirus infected cells. In the final analysis, such a signal must reside in the viral RNA molecules themselves, but no likely candidates have yet been identified. It is to be hoped that further study of this phenomenon, particularly identification of viral RNA sequences and perhaps proteins which are responsible for selective appearance in the cytoplasm of newly-synthesised, viral mRNA, will provide answers to the questions raised at the beginning of this paragraph. Particularly exciting is the possibility that this system will also yield important information about the normal process of nuclear to cytoplasmic transport, an essential step in the production of many functional RNA species which has not been amenable to rigorous analysis (for example, by genetic methods) in cells of higher eukaryotes.

Later during the adenovirus infectious cycle, the appearance in the cytoplasm of 18S rRNA is also inhibited, but this response appears to reflect a different mechanism, for synthesis of 45S pre-rRNA is also inhibited (79). Indeed, such later inhibition of rRNA gene expression has been reported to be the result of a deficiency in a factor essential to transcription of rRNA genes (80; see ref. 4 for a detailed discussion), presumably one effect of the inhibition of cellular protein synthesis, which is severe by the time that transcriptional inhibition of 45S pre-rRNA is observed (79).

Although transcription of the majority of human class II genes whose expression in adenovirus-infected cells has been examined is not impaired, a few whose transcription is inhibited concomitant with the onset of the late phase have been identified. Transcription of histone genes is, for example, inhibited following adenovirus infection, in parallel with the virus induced inhibition of cellular DNA synthesis (81). It therefore appears that the regulation of histone gene expression in adenovirus-infected cells resembles the normal state of affairs, when transcription of histone genes is strictly coupled to DNA replication (see ref. 82, for a review). Transcription of a human heat shock protein gene which is stimulated during the early phase of infection (see next section) is also inhibited during the period corresponding to entry of infection into late phase (83), but whether expression of this gene might normally be regulated during the cell cycle is not known.

In view of the profound effects of adenovirus infection upon expression of cellular genes transcribed by RNA polymerases I and II, a natural question is whether class III gene expression is also altered or disrupted. Indeed, the vast amounts of VA-RNAI transcribed by RNA polymerase III (20, 21, 84) might suggest that the capacity of the RNA polymerase III transcriptional machinery becomes overwhelmed. In fact, no significant alterations in the production of such class III transcripts as 5S RNA, tRNAs or cytoplasmic 7S RNA species have been observed in the 24 hour period following Ad2 or Ad5 infection of HeLa cells (C.L. Castiglia and Flint unpublished observations).

INDUCTION TO CELLULAR GENE EXPRESSION IN PRODUCTIVELY INFECTED CELLS

It has long been known that adenovirus infection of quiescent, by contrast to actively growing, human cells stimulates rather than inhibits cellular DNA synthesis (for example, 85-87). Not surprisingly, it has also been observed that adenovirus infection induces the synthesis of such enzymes as thymidine kinase (85, 88-90) and of dihdrofolate reductase (DHFR) mRNA (17). The molecular mechanism responsible for stimulation of expression of thymidine kinase gene has not been investigated, but more recent work has shown that the expression of several human genes is stimulated in HeLa cells infected with subgroup C adenoviruses under the typical laboratory conditions that lead to inhibition of cellular DNA synthesis.

Ad5 infected HeLa cells were, for example, found to synthesise increased quantities of a protein of apparent molecular weight 70,000 daltons, identified as a human heat shock protein (91). This response, seen at 5 hour after infection, appears to be the result of transcriptional activation of the human hsp70 gene, leading to an increase of some 100-fold in the concentration of hsp70 mRNA accumulating in the cytoplasm of infected cells (83). Cells infected by high multiplicities of the E1A deletion mutant H5dl312 showed no induction of expression of hsp70 mRNA or protein (83). The larger E1A product whose coding sequences are deleted from the genome of H5dl312 is essential to efficient expression of other, viral, early genes (92-96; see Chapter 4). However, such conditions of high multiplicity infection circumvent the requirement for the E1A protein and permit delayed expression of other, viral, early transcriptional units (94-97). Thus, the failure of H5dl312 to stimulate expression of the host hsp70 gene can be ascribed to the absence of products (presumably the larger, 289R protein), encoded by the E1A transcriptional unit.

More recent studies have established that adenovirus infection also stimulates expression of human β-tubulin and DHFR genes, responses which also depend on viral E1A protein(s) (17, 98; S. S. Yoder and S. M. Berget, personal communication). Although one can imagine why infected cells, which produce extremely large quantities of viral DNA, might benefit from increased concentrations of DHFR (and thymidine kinase), there is no obvious rationale for the stimulation of production of β-tubulin or the hsp70 protein. Of course, the normal function of the latter is poorly understood. Nevertheless, these observations raise the question of whether adenovirus infection stimulates transcription of many cellular genes during the early phase of infection, perhaps as a secondary consequence of the production of the E1A protein which is essential to efficient transcription from viral transcriptional units other than E1A itself. This does not, however, seem to be the case for little evidence for increased synthesis of the majority of cellular proteins during the early phase of infection has been collected (9, 91). Moreover, adenovirus infection induces no alterations in transcription of at least the human genes encoding actin, globin (98) an abundant HeLa mRNA species identified by isolation of a cDNA clone (83). The mechanism whereby adenovirus infection stimulates cellular gene expression at the transcriptional level, at least in the case of the hsp70 and β-tubulin genes (83, 91), has not been established, but is assumed to be analogous to that whereby the 289R protein in particular stimulates transcription from adenoviral transcriptional units. While such an analogy seems reasonable, it is not very helpful at present, for the mechanism responsible for stimulation of transcription of viral sequences is not yet understood (see, for example 94, 97, 99, 100, and Chapter 4). It does seem clear that viral and cellular genes whose transcription responds to an adenovirus E1A protein are not distinguished by a specific sequence element (101-103), but little else about the mechanism of action of the E1A protein can be stated with certainty.

CHANGES IN CELLULAR FUNCTIONS IN ADENOVIRUS TRANSFORMED CELLS

It is a striking fact that the changes in cellular gene expression and cellular DNA synthesis wich accompany adenovirus infection of fully-permissive human cells are apparently mediated by the products of the viral E1A and E1B transcriptional units, just those essential for transformation of non- or semi-permissive rodent cells (see Chapter 10). Thus, the E1A 289R protein stimulates transcription not only of other viral, early genes, but also of certain cellular genes; the 243R E1A protein does not exhibit such activity under most circumstances (but see 104), but does seem to be required for efficient replication of the virus when strictly growth-arrested human fibroblasts are infected (105; A. J. Berk, personal communication). This latter property is of particular interest in relation to the stimulation of cellular DNA synthesis and production of certain enzymes which occurs under these conditions (see previous section) and upon infection of rodent cells, for the relevant adenovirus function(s) has been mapped to the E1A region (106, 107; see ref. 4 for a detailed review). As we have seen, the E1B 55kd protein appears to be an important governor of translation during the late phase of productive infection, whereas the unrelated 21kd protein also specified by this transcriptional unit (see Chapter 2) protects both cellular and viral DNA from degradation in adenovirus infectd cells and may be required for replication of viral DNA (66-68).

Although the DNA comprising adenovirus E1A and E1B transcriptional units (see Chapter 10) is all that is required to mediate transformatin of rodent cells, these viral transcriptional units have not evolved under selection for transformation activity, a property of no value during the natural life history of the virus. Thus a strong case can be made that the molecular functions of the E1A and E1B gene products during transformation are manifestations, albeit perhaps unusual ones, of their normal activities in permissive cells. Indeed, these activities can be readily translated into attractive postulates describing the molecular roles of E1A or E1B proteins in the transformation process. An obvious example is provided by the ability of the 289R E1A protein to stimulate transcription of certain cellular class II genes: such an activity could be critical during adenovirus transformation, for example, to stimulate transcription of such cellular oncogenes as c-fos or c-myc. The expression of both these genes is known to be very rapidly induced in normal cells in response to mitogenic stimuli (108-113). The re-entry of quiescent cells into the cell cycle and of cellular DNA synthesis induced by adenovirus infection, mentioned in previous sections, might be the result of such activity of E1A proteins and thus represent the net result of an unusual mechanism of activation of the cascade of molecular events which regulates cell cycle progression in normal cells. Of course, any such activation of transcription of cellular genes would be permanent in adenovirus transformed cells, for they invariably retain and express the E1A transcriptional unit (see 73, 114; Chapter 10, for reviews). It is therefore clear that the activities in which E1A proteins engage in permissive cells could provide the basis for reasonable explanations of the loss of growth control which is an important characteristic of transformed cells. In similar fashion, it is not difficult to imagine how the adenoviral E1B 55kd protein might alter the expression of key cellular gene products which normally mediate or regulate mitogenic responses, by a mechanism analogous to that whereby the protein facilitates translation of, or stabilises, viral, late mRNA.

Reasonable as such speculations might seem, there is at present, no way to judge whether they hold a grain of truth or are wildly off the mark: next to nothing is known about the molecular changes which distinguish adenovirus transformed rodent cells from their normal parents. Cells transformed by adenoviruses show such typical transformation phenotypes as altered morphology, increased growth rates, reduced dependence on growth factors present in serum and increased capacity for anchorage-independent growth and are tumorigenic both in nude mice and, in many cases, in syngeneic animals (see 115). Molecular correlates of such altered phenotypes have not, for the most part, been identified. This state of affairs is disappointing, for the few molecular or biochemical properties of adenovirus transformed cells that have been described are intriguing.

It has, for example, recently been found that adenovirus transformation of rat cells can alter expression of major histocompatability antigen genes. Such alterations in production of these cellular proteins are dependent upon E1A gene products and appear to be important determinants of the tumorigenicity displayed by adenovirus transformed cells in immunocompetent animals (115-117; see chapter 11 for a detailed discussion).

A second exciting observation concerns the E1B 55kd protein. As we have seen, some fraction of this polypeptide is found in productively infected cells in a complex with the 33kd protein encoded within the E4 transcriptional unit. In adenovirus transformed cells, by contrast, the E1B 55kd protein is associated with a cellular protein termed p53, on the basis of its apparent molecular weight: the two proteins are co-precipitated by monoclonal antibodies which recognise either the adenoviral E1B 55kd protein or the cellular p53 protein (69). The p53 protein was first identified by virtue of its association, in SV40 transformed cells, with SV40 large T-antigen and thus precipitation by anti-T antigen antibodies (118, 119). Subsequently, the p53 protein has been shown to be present in increased quantities in many types of transformed cell (120). Its expression in normal cells is regulated during the cell cycle (110, 121, 122) and induced soon after resting cells are exposed to mitogenic stimuli (108, 110, 123, 124). Indeed expression of the p53 protein appears to be an essential prerequisite for entry into the cell cycle, for microinjection of monoclonal antibodies to this protein prevents serum-induced stimulation of cellular DNA synthesis (124, 125). It has recently been reported, moreover, the p53 protein will cooperate with Ha c-ras oncogenes to transform secondary rat embyro fibroblasts (126, 127) and can independently immortalise adult rat xiphisternum chondrocytes (128). In these experiments, p53 coding sequences were linked to strong transcriptional control regions and the p53 protein therefore overproduced as a result of efficient transcription of its coding sequences. In SV40 transformed cells, binding of the p53 protein to large T antigen increases the half life of the cellular protein from less than 60 minutes to about 24 hours (121, 129). It therefore seems likely that any mechanism which leads to overproduction of the cellular p53 protein can contribute to transformation. That by which the adenoviral E1B protein might increase the intracellular concentrations of p53 protein has not been elucidated. Further work is obviously required before the consequences and significance of association of the p53 protein with the E1B 55kd protein in adenovirus transformed cells is understood. Nevertheless, the value of this kind of information is already apparent.

It is to be hoped that more attention will in future be paid to the biochemical and molecular characterisation of changes in cellular functions induced in adenovirus transformed cells, for without such information little progress towards any real understanding of adenovirus transformation can be made.

Acknowledgements

I thank my colleagues who have communicated data prior to publication and Laine Gordon for her patient assistance in preparation of this manuscript. Unpublished work of the author's cited was supported by a grant, AI 17265, from the National Institutes of Health.

REFERENCES

1. Hilleman, M. R., and Werner, J. H. Proc. Soc. Exp. Biol. Med. 85: 183-188, 1954
2. Rowe, W. P., Huebner, R. J., Gillmore, L. K., Parrott, R. H., and Ward, T. G. Proc. Soc. Exp. Biol. Med. 84: 570-573, 1953
3. Flint, S. J. and Broker, T. R. In Molecular Biology of Tumor Viruses .2 DNA Tumor Viruses (ed. J. Tooze) Cold Spring Harbor, New York, 1980 pp 443-546.
4. Flint, S. J. In Comprehensive Virology (eds. H. Frankel-Conrat and R. R. Wagner) Plenum Press, New York and London, Vol. 19, 1984, pp 297-358.
5. Ginsberg, H. S., Bello, L. J. and Levine, A. J. In The Molecular Biology of Viruses (eds. J. S. Colter and W. Paranchych) Academic Press, New York, 1967, pp 547-572.
6. White, D. O. , Scharff, M. D. and Maizel, J. V. Virology 38: 395-406, 1969.
7. Russell, W. C. and Skehel, J. J. J. Gen. Virol. 15: 45-57, 1972.
8. Anderson, C. W., Baum, P. R. and Gesteland, R. F. J. Virol. 12: 241-252, 1973.
9. Beltz, G. A. and Flint, S. J. J. Mol. Biol. 79: 371-373, 1979
10. Ehrenfield, E. Cell 28: 431-436, 1982.
11. Wagner, R. R., Thomas, J. R. and McGowan, J. J. In Comprehensive Virology (eds. H. Frankel-Conrat and R. R. Wagner), Plenum Press, New York and London, Vol. 19, 1984, pp 223-296.
12. Weck, P. K., Carroll, A. R., Shattuck, D. M. and Wagner, R. R. J. Virol. 30: 746-753, 1979.
13. Ginnell, B. W. and Wagner, R. R. Cell 36: 533-543, 1984.
14. Anderson, C. W., Lewis, J. B., Atkins, J. F. and Gesteland, R. F. Proc. Natl. Acad. Sci. USA 71: 2756-2760, 1974.
15. Lewis, J. B., Atkins, J. F., Anderson, C. W., Baum, P. R. and Gesteland, R. F. Proc. Natl. Acad. Sci., USA 72: 1344-1348, 1975.
16. Paterson, B. M., Roberts, B. and Kuff, E. L. Proc. Natl. Acad. Sci. USA 74: 4370-4374, 1977.
17. Yoder, S. S., Robberson, B. L., Leys, E. J., Hook, A. G., Al-Ubaidi, R., Yeung, C-Y., Kellems, R. E. and Berget, S. M. Mol. Cell. Biol. 3: 819-828, 1983.
18. Babich, A., Feldman, L. T., Nevins, J. R., Darnell, J. E. and Weinberger, C. Mol. Cell. Biol. 3: 1212-1221, 1983.
19. Thimmappaya, B., Weinberger, C., Scheider, R. J. and Shenk, T. Cell 31: 543-551, 1982.
20. Söderlund, H., Petersson, U., Vennströmm, B., Philipson, L. and Mathews, M. B. Cell 7: 585-593, 1976.
21. Mathews, M. B. and Pettersson, U. J. Mol. Biol. 119: 293-328, 1978.
22. Akusjärvi, G., Mathews, M. B., Andersson, P., Vennstromm, B. and Pettersson, U. Proc. Natl. Acad. Sci. USA 71: 2424-2428, 1980.
23. Bhat, R. A. and Thimmappaya, B. Proc. Natl. Acad. Sci. USA 80: 4789-4793, 1983.
24. Bhat, R. A., Domer, P. H. and Thimmapaya, B. Mol. Cell. Biol. 5: 187-196, 1985.
25. Schneider, R. J., Weinberger, C. and Shenk, T. Cell 37: 291-298, 1984.
26. Reichel, P. A., Merrick, W. C., Siekierka, J. and Mathews, M. B. Nautre 313: 196-200, 1985.
27. Schneider, R. J., Safer, B., Munemitsu, S. M., Samuel, C. E. and Shenk, T. Proc. Natl. Acad. Sci. USA, 82, in press.

28. Jagus, R., Anderson, W. P. and Safer, B. Prog. Nuc. Acid Res. Mol. Biol. 25: 127-185, 1981.
29. Safer, B. Cell 33: 7-8, 1983.
30. LeBleu, B., Sen, G. C., Shaila, S. Cabrer, B. and Lengrel, P. Proc. Natl. Acad. Sci. USA 73: 3107-3111, 1976.
31. Roberts, W. K., Hovanessian, A., Brown, R. E., Clemens, M. J. and Kerr, I. M. Nature 264: 477-480, 1976.
32. Zilberstein, A., Federman, P., Schulman, L. and Revel, M. FEBS LeHers 68: 119-124, 1976.
33. Samuel, C. E., Parris, D. A. and Epstein, D. A. Virology 38: 56-71, 1977.
34. Jackson, R. J. In Protein Biosynthesis in Eukaryotes (ed. R. Perez-Bercoff). Plenum Press, New York and London, 1980, pp 363-408.
35. Beladi, I. and Pusztai, R. Z. Naturforsch 22b: 165-168, 1967.
36. Ho, M. and Kohler, K. Archivfur die Gesamte Virusforsch. 22: 69-74, 1967.
37. Ustacelabi, S. and William, J. F. Nature 235: 52-54, 1972.
38. Crouch, D. and Safer, B. J. Biol. Chem. 259: 10363-10368, 1984.
39. Hunter, T., Hunt, T., Jackson, R. J. and Robertson, K. D. J. Biol. Chem. 250: 409-417, 1975.
40. DeBenditti, A. and Baglioni, C. Nature 311: 79-81, 1984.
41. Miyamoto, N. G., Jacobs, B. L. and Samuel, C. E. J. Biol. Chem. 258: 15232-15237, 1984
42. Zain, S., Gingeras, T. R., Bullock, P., Wong, G. and Gelinas, R. E. J. Mol. Biol. 135: 413-433, 1979.
43. Monstein, J-H. and Philipson, L. Nuc. Acids. Res. 9: 4239-4250, 1981.
44. Rosa, M. D., Gottlieb, E., Lerner, M. R. and Steitz, J. A. Mol. Cell Biol. 1: 785-796, 1981.
45. Lerner, M. R. , Boyle, J. A., Hardin, J. A. and Steitz, J. A. Science 211: 400-402, 1981,
46. van Eekelen, C., Buitels, H., Linne, T., Ohlsson, R., Philipson, L. and van Venrooij, W. Nuc. Acids Res. 10: 3039-3052, 1982.
47. Francouer, A. M. and Mathews, M. B. Proc. Nalt. Acad. Sci. USA 79: 6772-6776, 1982.
48. Walter, P. and Blobel, G. Cell 34: 525-533, 1983.
49. Berget, S. M., Moore, C. and Sharp, P. A. Proc. Natl. Acad. Sci. USA 74: 3171-3175, 1977.
50. Chow, L. T., Gelinan, R. E., Broker, T. R. and Roberts, R. J. Cell 12: 1-8, 1977.
51. Logan, J. and Shenk, T. Proc. Natl. Acad. Sci. USA 81: 3655-3659, 1984.
52. Thummel, C., Tjian, R., Hu, S. L. and Grodzicker, T. Cell 33: 455-464, 1983.
53. Berkner, K. L. and Sharp, P. A. Nucleic Acids Res. 12: 1925-1941, 1984.
54. Challberg, S. S. and Ketner, G. Virology 114: 196-209, 1981
55. Flint, S. J., Werwerka-Lutz, Y., Levine, A. S., Sambrook, J. and Sharp, P. A. J. Virol 16: 662-671, 1975.
56. Jones, N. and Shenk, T. Cell 13: 181-188, 1978.
57. Ensinger, M. J. and Ginsberg, H. S. J. Virol 10: 328-339, 1972.
58. Russell, W. C., Skehel, J. J. and Williams, J. F. J. Gen. Virol 24: 247-259, 1974.
59. Chow, L. T., Broker, T. R. and Lewis, J. B. J. Mol. Biol. 134: 265-303, 1979.

60. Tigges and Raskas, H. J. J. Virol. 50: 106-117, 1984.
61. Virtanen, A., Gilardi, P., Näslund, A., LeMoullec, J. M., Pettersson, U. and Perricaudet, M. J. Virol 51: 822-831, 1984
62. Babiss, L. E. and Ginsberg, H. S. J. Virol 50: 202-212, 1984.
63. Logan, J., Pilder, S. and Shenk, T. Cancer Cells 2: 527-535, 1984.
64. Harrison, T. J., Graham, F. and Williams, J. F. Virology 77: 319-329,
65. Ho., Y. S., Galos, R. and Williams, J. F. Virology 122: 109-124, 1982.
66. White, E., Grodzicker, T. and Stillman, B. W. J. Virol. 52: 410-419, 1984.
67. Pilder, S., Logan, J. and Shenk, T. J. Virol 52: 664-671, 1984.
68. Subramanian, T. Kuppuswamy, M., Gysberg, J., Mak, S. and Chinnadurai, G. J. Biol. Chem. 259: 11777-11783.
69. Sarnow, P., Ho, Y. S., Williams, J. F. and Levine, A. J. Cell 28: 381-394, 1982.
70. Sarnow, P., Hearing, P., Anderson, C. W., Halbert, D. N., Shenk, T. and Levine, A. J. J. Virol 49: 692-700.
71. Lassam, N. T., Bayley, S. T. and Graham, F. L. Cell 18: 781-791, 1979.
72. Aiello, L., Guilfoyle, R., Huebner, K. and Weinmann, R. Virology 94: 460-469, 1979.
73. Flint, S. J., Pharmac. Ther. 26: 59-88, 1984.
74. Lindberg, U., Persson, T. and Philipson, L. J. Virol. 10: 909-919, 1972.
75. Flint, S. J., Beltz, G. A. and Linzer, D. I. H. J. Mol. Biol. 96: 47-68, 1983.
76. Yoder, S. and Berget, S. M. J. Virol, 1985, in press.
77. Raskas, H. J., Thomas, D. C. and Green, M. Virology 40: 893-902, 1970.
78. Ledinko, N. Virology 49: 79-89, 1972.
79. Castiglia, C. L. and Flint, S. J. Mol. Cell. Biol. 3: 662-671.
80. Learned, R. M., Smale, S. T., Haltiner, M. M. and Tjian, R. Proc. Natl. Acad. Sci. USA 80: 3558-3562, 1983.
81. Flint, S. J., Plumb, M. A., Yang, U. C., Stein, G. S. and Stein, J. L. Mol. Cell Biol. 4: 1363-1371, 1984.
82. Stein, G. S., Sierra, F., Stein, J. F., Plumb, M., Marashi, F., Carozzi, N., Prokopp, K. and Baumbach, L. In Histone Genes and histone gene expression (ed. G. S. Stein, J. L. Stein and W. F. Marzluff), J. Wiley and Sons Inc., New York, 1984, pp 397-455.
83. Kao, H. T. and Nevins, J. R. Mol. Cell Biol. 3: 2058-2065, 1983.
84. Price, R. and Penman, S. J. Mol. Biol. 70: 431 - 450, 1972
85. Takahashi, M., Ogino, T. Baba, K. and Onaka, M. Virology 37: 513-520, 1969.
86. Ledinko, N. Cancer Res. 27: 1459-1469, 1967.
87. Yamashita, T. and Shimojo, H. Virology 38, 351-355, 1969.
88. Kit, S., Dubbs, D. R., deTorres, R. A. and Melnick, J. L. Virology 27: 453-457, 1965.
89. Zimmerman, J., Raska, K. and Strohl, W. K. Virology 42: 1147-1149, 1970.
90. Takahashi, M., Veda, S. and Ogino, T. Virology 30: 742-743, 1966.
91. Nevins, J. R. Cell 29: 913-919, 1982.
92. Jones, N. and Shenk, T. Proc. Natl. Acad. Sci. USA 76: 3665-3669, 1979.
93. Berk, A. J., Lee, F., Harrison, T., Williams, J. F. and Sharp, P. A. Cell 17: 935-944, 1979.

94. Nevins, J. R. Cell 26: 213-220, 1981.
95. Ricciardi, R. P., Jones, R. L., Cepko, C. T., Sharp, P. A. and Robwerts, B. E. Proc. Natl. Acad. Sci. USA 78: 6121-6125, 1981.
96. Montell, C., Fisher, E. F., Carruthers, M. H. and Berk, A. J. Nature 295: 380-384, 1982.
97. Gaynor, R. B. and Berk, A. J. Cell 33: 683-693, 1983.
98. Stein, R. and Ziff, E. Mol. Cell. Biol. 4: 2792-2801, 1984.
99. Katze, M. G., Persson, H. and Philipson, L. Mol. Cell Biol. 1: 807-813, 1981.
100. Katze, M. G., Persson, H., Johansson, B. M. and Philipson, L. J. Virol. 46: 50-59, 1983.
101. Green, M. R., Treisman, R. and Maniatis, T. Cell 35: 137-148, 1983.
102. Bos, J. L. and ten Wolde-Kraamwinkel, H. C. The EMBO J. 2: 73-76, 1983.
103. Kingston, R. E., Kaufman, R. J. and Sharp, P. A. Mol. Cell Biol. 4: 1970-1977, 1984.
104. Leff, T., Elkaim, R., Goding, C. R., Jalinot, P., Sassone-Corsi, P., Perricaudet, M., Kedinger, C. and Chambon, P. Proc. Natl. Acad. Sci., USA 81, 4381-4385, 1984.
105. Montell, C., Courtois, G., Eng., C. and Berk, A. J. Cell 36: 951-561, 1984.
106. Rossini, M., Jonak, G. J. and Baserga, R. J. Virol. 38: 982-986, 1981.
107. Braithewaite, A. W., Cheetham, B. F. , Li, P. , Parish, C. R., Waldron-Stevens, L. K. and Bellett, A. J. D., J. Virol, 45: 192-199, 1983.
108. Kelly, K., Cochran, B. H., Stiles, C. D. and Leder, P. Cell 34: 789-798, 1983.
109. Greenberg, M. E. and Ziff, E. B. Nature 311: 433-437, 1984.
110. Campisi, J., Gray H. S., Pardee, A. B., Dean, M. and Sonnenshein, G. E. Cell 36: 241-247, 1984.
111. Makino, R., Hayashi, K. and Sugimura, T. Nature 310: 697-698, 1984.
112. Muller, R., Bravo, R. , Burckhardt, J. and Curran, T. Nature 312:, 716-720, 1984.
113. Kruijer, W., Cooper, J. A., Hunter, T. and Verma, I. M. Nature 312: 711-716, 1984.
114. Flint, S. J. In Internal Review of Cytology (eds. G. H. Bourne and J. F. Danielli) Academic Press, New York, Vol. 76, 1982, pp 47-66.
115. Bernards, R., Schrier, P. I. Houweling, A., Bos, J. L., van der Eb, A. J., Zijlstra, M. and Malef, C. J. M. Nature 305: 776-779, 1984
116. Schrier, P. I., Bernards, R., Vaessen, R. T. M., Houweling, A. and van der Eb, A. J. Nature 305: 771-775, 1983.
117. Mellow, G. H., Führing, B., Dougherty, J., Gallimore, P. H. and Raska, K. Virology 134: 460-465, 1984.
118. Lane, D. P. and Crawford, L. V. Nature 278: 261-263, 1979.
119. Linzer, D. I. and Levine, A. J. Cell 17: 43-52, 1979.
120. Crawfold, L. V. Int. Rev. Exp. Path. 25: 1-5-, 1983.
121. Reich, N. C., Owen, M. and Levine, A. J. Mol. Cell Biol. 3: 2143-2150, 1983.
122. Reich, N. C. and Levine, A. J. Nature 308: 199-201, 1984.
123. Milner, J. and Milner, S. Virology 112: 785-788, 1981.
124. Mercer, W. E., Nelson, D., DeLeo, A. B., Old L. J. and Baserga, R. Proc. Natl. Acad. Sci. USA 79: 6309-6312, 1982.

125. Mercer, W. E., Auignoto, C. and Baserga, R. Mol. Cell. Biol. 4: 276-281, 1984.
126. Eliyahau, D., Raz, A., Gruss, P., Givol, D. and Owen, M. Nature 312: 646-649, 1984.
127. Parada, L. F., Land, H., Weinberg, R. A., Wolf, D. and Rotter, V. Nature 312: 649-651, 1984.
128. Jenkins, J. R., Rudge, K. and Currie, G. A. Nature 312: 651-654, 1984.
129. Owen, M., Malzman, W. and Levine, A. J. Mol. Cell Biol. 1: 101-110, 1981.
130. Bos, J. L., Poler, L. J., Bernards, R., Schrier, P. I., van den Elsen, P. J., van der Eb, A. J., and van Ormondt, H. Cell 27: 121-131, 1981.
131. Gingeras, T. R., Sciaky, D., Gelinas, R. E., Bing-Dong, J., Yen, C. E., Kelly, M. M. Bullock, P. A., Parsons, B. L., O'Neill, K. E. and Roberts, R. J. J. Biol. Chem. 257: 13475-13491, 1982.

10

EXPRESSION OF CLASS I MAJOR HISTOCOMPATIBILITY ANTIGENS IN ADENOVIRUS-TRANSFORMED CELLS

A.J. van der Eb, B.A. Oostra, Sylvius Laboratories, Univerity of Leiden, Leiden, The Netherlands.

1. INTRODUCTION

Human adenoviruses have been used extensively as a model for studies on the mechanism of oncogenic transformation. The interest in this group of viruses is due to the fact that they combine a number of unique features, that are not found in other viruses: (a) Oncogenic transformation by adenoviruses is the result of the activity of 2 or 3 distinct viral gene products, rather than by the action of a single viral protein, as in the case of SV40. This seemingly greater complexity of the transforming region of adenovirus is an advantage rather than a disadvantage. It should considerably simplify genetic and functional analysis of transformation as compared to the situation in SV40, where the transforming functions are all compressed into a single multifunctional protein (1). (b) Oncogenic transformation by adenoviruses can be considered to be a "multistep" process in which 2 or 3 genes are involved (2). This resembles the situation in spontaneous cancer in animals, which also requires the concerted activity of two, or possibly more, cellular oncogenes. Interestingly, a functional similarity has been established between the adenovirus E1A and E1B regions on one hand, which together constitute the viral transforming region, and the two cooperating cellular oncogenes myc and ras on the other (3,4). (c) The human adenoviruses can be divided into oncogenic and non-oncogenic serotypes (e.g. Ad12, an oncogenic virus, and Ad2 or Ad5, both non-oncogenic viruses). Despite this difference in biological behavior, all human adenoviruses tested so far are capable of morphologically transforming cultured cells. The structural and organizational relatedness of the Ad5 and Ad12 transforming regions implies that the difference in oncogenicity must be brought about by one or more subtle differences in gene function, and may enable us to dissociate oncogenicity, i.e. the ability to grow as a tumor in normal animals, from morphological transformation in vitro.

In an attempt to establish which adenovirus gene products are responsible for the difference in oncogenic potential between Ad5 and Ad12, Bernards et

W. Doerfler (ed.), ADENOVIRUS DNA.

Table 1. Cellular Proteins Present in a Series of Adenovirus-transformed Cell Lines with Different Oncogenic Potential

transformed by[a]	Expression of Ad5 or Ad12 region E1A	E1B	Proteins detected in transformed cells 32kD	(45kD)	(12kD)	Oncogenicity[b] mice	rats
BRK (untransformed)	-	-	+	+	+	-	-
pAd5XhoC	5	5	+	+	+	+/-	-
pAd5HindIIIG	5	5 (only 19kD)[d]	+	+	+	-	n.d.
pAd5dlSac	5	5 (only 58kD)[e]	+	+	+	-	n.d.
pAd5HpaIE	5	-	+	+	+	-	n.d.
pAd12RIC	12	12	-	-	+	+	+
pAd12HindIIIG	12	12 (only 19kD)[f]	-	-	+	-	-
pAd12dlAcc	12	12 (only 54kD)[e]	-	-	+	-	-
pAd512	5	12	+	+	+	+	-
pAd125	12	5	-	-	+	+/-	+/-
pAd51212	5 +12	12	+	+	+	+	-
pST12	5 +12	Ad12 19kD+Ad5 58kD	+	+	+	+/-	-
pLT12	5 +12	Ad5 19kD+Ad12 54kD	+	+	+	+	-
pAd512pm975	12 + 5 (only 13S product)[c]	12	+	+	+	n.d.	n.d
pAd512HL1007	12 + 5 (only 12S product)[c]	12	-	-	+	n.d.	n.d.
pSVR7	12 (only 12S product)[c]	12	+	+	+	+	n.d.
pSVR11	12 (no 12S product)[c] (no 13S product)	12	+	+	+	+	n.d.
pAd512E1a + pAd12E1b	5/12	12		+		n.d.	n.d.
pAd125E1a + pAd12E1b	5/12	12		-		n.d.	n.d.

a) The composition of the various recombinant pasmids is given in 14,19
b) +, Highly oncogenic; +/-, weakly oncogenic; -, nononcogenic; n.d.,not done
c) Truncated products from mutated genes can still be present
d) Truncated product from mutated gene <34kD
e) Truncated product from mutated gene < 2kD
f) Truncated product from mutated gene <18kD

al.(5) have constructed recombinant plasmids consisting of region E1A of nononcogenic Ad5 and E1B of oncogenic Ad12, and vice versa. By transforming primary cultures of baby rat kidney (BRK) cells with these constructs and testing the oncogenic potential of the resulting transformed cells in athymic nude mice and in normal syngeneic rats, the following results were obtained (Table 1): (1) The capacity of the transformed cells to induce tumors in nude mice is determined by the identity of the E1B region, i.e. when E1B is derived from oncogenic Ad12, the transformed cells are highly oncogenic and induce tumors in 100% of the animals inoculated. However, when E1B is derived from nononcogenic Ad5, the cells are only weakly oncogenic in nude mice, causing tumors in 10-50% of injected animals (6). (2). The oncogenicity of the transformed cells in syngeneic immunocompetent rats (5-6 days old), however, is not only determined by region E1B but, in addition, by the identity of region E1A. Transformed BRK cells that contain region E1A of oncogenic Ad12 were found to be as oncogenic in normal rats as in nude mice. When region E1A originated from non-oncogenic Ad5, however, the cells invariably were non-oncogenic in rats, irrespective of their oncogenicity in nude mice. A possible explanation for these results is that cells expressing region E1A from Ad12 can escape the immune defence of the host whereas cells expressing region E1A from Ad5 are sensitive. Since recognition and subsequent destruction of tumor cells by the immune system is mediated, amongst others, through virus- or cell-encoded neoantigens on the cell membrane, the results would suggest that Ad5-transformed cells contain strong transplantation antigens (TSTAs) while Ad12-transformed cells lack such antigens. The following observations, however, suggested that this interpretation may not be correct: Immunization of hamsters and mice with Ad12 virus was shown to induce immunity against subsequent challenge with Ad12-transformed cells (7,8). The presence of TSTAs on Ad12-induced mouse tumor cells was subsequently demonstrated by Sjögren et al. (9) while Shiroki et al. (10) provided evidence that proteins encoded by the E1B region of Ad12 are responsible for induction of viral TSTAs. Furthermore, Raska et al. (11,12) showed that the transforming region of both oncogenic Ad12 and non-oncogenic Ad2 (Ad2 is almost identical to Ad5) encodes polypeptides that elicit complement-dependent cytotoxic antibodies in rats, indicating that both types of viruses induce tumor-specific antigens in transformed cells.

Thus, it appears unlikely that differences in expression of virus-specific tumor antigens can explain the results discussed above.

2. EXPRESSION OF CLASS I MAJOR HISTOCOMPATIBILITY COMPLEX ANTIGENS IS REDUCED IN Ad12-TRANSFORMED CELLS

In an attempt to explain the phenomenon that Ad12-transformed cells exhibit the same degree of oncogenicity in nude mice as in immunocompetent animals, Schrier et al. (12) have tried to detect antigenic differences between Ad12-transformed cells on one hand and untransformed or Ad5-transformed cells on the other. By using sera from Balb/c mice, immunized with untransformed primary baby rat kidney cells (derived from kidneys of 6-day old Wag-Rij rats), a number of cellular proteins were identified that were present in both untransformed and Ad5-transformed cells but absent in Ad12-transformed cells. The clearest effects were observed for a 38kD and a 32kD protein. Analysis by immunoprecipitation of a panel of BRK cell lines transformed by mutant Ad5 or Ad12 E1 regions or Ad5/Ad12 hybrid E1 regions showed that the absence of these proteins in transformed cells correlates with the presence of

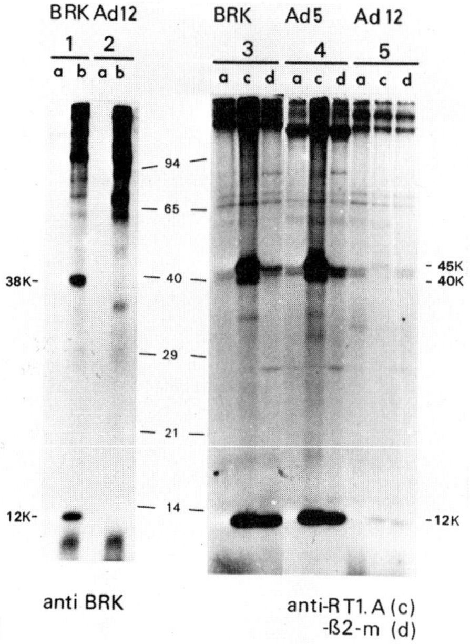

Figure 1. Proteins immunoprecipitated from extracts of ^{125}I labeled rat cells with anti-BRK mouse serum, anti-Wag/Rij alloantiserum, and an antiserum against human β_2-m. Extracts of the following cells were used for immunoprecipiation: (lanes 1 and 3) primary BRK cells; (lanes 2 and 5) pAd12RIC-transformed BRK cells; (lane 4) pAd5XhoC-transformed BRK cells. The following sera were used: normal mouse serum (lanes a); mouse anti-BRK serum (lanes b); Lewis anti-Wag/Rij hyperimmune alloantiserum (lanes c); and anti-human β_2-m serum (lanes d). (Reprinted with permission from Nature vol.305, pp.771-775 Copyright (c) 1983, MacMillan Journals limited).

an intact Ad12 E1A region and is independent of E1B functions. A strict correlation was found between absence of the 38kD and 32kD protein and oncogenicity in immunocompetent rats (6,13,14). Since recognition of transformed cells by the immune system occurs via an interaction with plasma membrane components, the question was asked whether these proteins are exposed on the outside of the cellular membrane. Immunoprecipitation of ^{125}I-labeled transformed or normal BRK cells with the anti-BRK serum showed that the 32kD protein was not labeled and hence presumably not exposed on the outer surface of the membrane. (Fig.1). However, a labeled 38kD protein was precipitated from untransformed BRK cells, but not from Ad12-transformed cells. This 38kD protein was probably identical to the 38kD species detected in ^{35}S-methionine-labeled extracts. In addition to the 38kD protein, a polypeptide of 12kD was also precipitated from BRK cells which was absent in Ad12-transformed cells. (This 12kD protein had remained undetected in the initial studies due to the electrophoresis conditions). The coprecipitation of the 38kD protein with a 12kD component suggested that the two proteins might represent the heavy and the light chain of the class I Major Histocompatibility Complex (MHC) antigens, which are known to be components of the plasma membrane. The 12kD light chain would then represent β_2-microglobulin (β_2-m). Immunoprecipitation with specific rat alloantisera directed against the class I MHC antigens of Wag/Rij rats ($RT1^u$ haplotype) or antiserum against human β_2-m showed that untransformed and Ad5-transformed cells contain 45kD and 12kD polypeptides, representing the class I heavy chain and β_2-m, respectively, but that very little, if any, 45kD protein was present in Ad12-transformed cells (Fig.1). Small amounts of the 12kD protein could be precipitated from ^{125}I-labeled Ad12-transformed cells with the human anti β_2-m serum. Further work showed, however, that expression of the β_2-m gene was hardly suppressed in Ad12-transformed cells, but that its concentration in the plasma membrane was reduced as a result of the absence of the heavy chain molecules, to which β_2-m is complexed. Thus, Ad12-transformed BRK cells generally express greatly reduced amounts of proteins with molecular weights of 32kD, 38kD and 45kD and this property correlates with expression of region E1A of Ad12. The latter two polypeptides are exposed on the outer surface of the plasma membrane. The 45kD protein is complexed with β_2-microglobulin and belongs to the class I MHC products. The 38kD protein is probably also complexed to β_2-m and may similarly represent a class I MHC protein but its precise identity remains to be determined. Attempts to establish the nature of the 32kD protein have been unsuccessful so far.

3. CHARACTERIZATION OF INHIBITION OF CLASS I MHC GENE EXPRESSION IN AD12-TRANSFORMED CELLS

As discussed in the previous section, the reduced expression of the class I MHC proteins correlates with the presence of a functional E1A region of oncogenic Ad12 and is not influenced by the identity of the E1B region, nor by mutations in one of the two E1B proteins (13,14). Further work on the mechanism of suppression of class I MHC products in Ad12-transformed cells has led to the following conclusions.

3.1. Inhibition of class I gene expression occurs at the level of mRNA accumulation

The strong reduction of the concentration of class I MHC heavy chains in Ad12-transformed BRK cells can be caused by: (a) a reduced concentration of translatable mRNA as a result of either a decreased rate of mRNA synthesis or a decreased stability of mRNA; (b) a reduced rate of translation of class I MHC mRNA; (c) a reduced stability of the MHC protein; (d) absence of β_2-m light chains, to which the 45kD heavy chains are complexed.

Immunoprecipitation studies with anti- β_2-m serum have shown that expression of this protein is not greatly affected by Ad12 E1A (13, Table I) To distinguish between the other three alternatives, Northern blot analyses were carried out, using a cDNA clone of human HLA-B7 as a probe. This cDNA clone contains sequences of the conserved constant (third domain) of the heavy chain genes (15) and, therefore, should cross-react with class I MHC genes from other mammalian species. As expected, the radioactively-labeled HLA-B7 probe hybridized to a 1.7 kb RNA species in untransformed and Ad5-transformed cells (Fig.2). However, very little hybridization was obtained with RNA isolated from Ad12-transformed cells or from cells transformed by the E1A region of Ad12 and the E1B region of Ad5. The average reduction of class I specific RNA in transformed BRK cells expressing Ad12 region E1A varied between 4 to 10 times. Thus, it appears that suppression of class I MHC gene expression is caused by a reduced level of mRNA. Preliminary results have shown that the concentration of class I-specific RNA is also reduced in nuclear RNA from Ad12-transformed cells but not in that of Ad5-transformed cells (R. Offringa and R.T.M.J. Vaessen, personal communication).

3.2. Inhibition of class I MHC is a function of the product of the first exon of 13S E1A mRNA

In transformed cells the Ad5 DNA-region Ela codes for two coterminal mRNAs of 12S and 13S. A third E1A mRNA species of 9S is found only in lytically infected cells and not in transformed cells. The 12S and 13S mRNAs share the same 3' splice acceptor site but have different 5' splice donor sites (reviews in 1,16,17). The 13S mRNA product contains a stretch of 46 (Ad5) or 31 (Ad12) amino acid residues that do not occur in the respective 12S mRNA products. Studies with an Ad12 region E1 mutant plasmid (R7) containing a deletion in the sequence affecting the 13S mRNA product only have shown that this plasmid is defective for transformation of primary rat cells, unless the Ad12 E1A promoter is replaced by the SV40 promoter + enhancer (18) . Cells transformed by this mutant plasmid, SVR7, showed normal expression of class I MHC, indicating that the 12S mRNA product alone is not sufficient for turning off expression of class I genes and that the activity is probably associated with the 13S mRNA product (6,14, Table I).

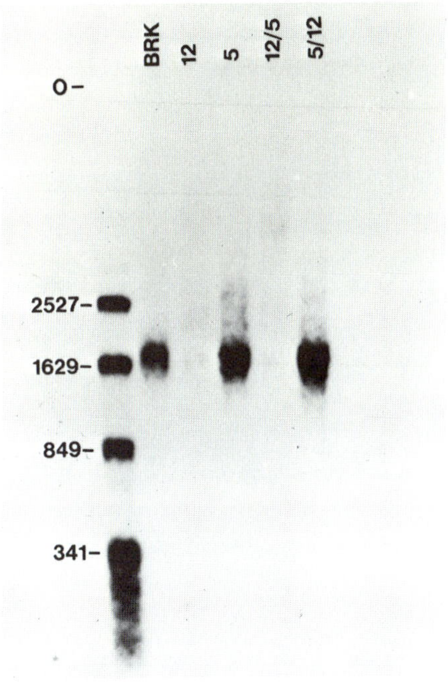

Figure 2. Northern blot analysis of RNA extracted from primary and adenovirus-transformed BRK cells, using a nick-translated human HLA B7 cDNA clone as described by Schrier et al. (1983). (Lane 1) HaeIII-digested M13mp8 marker DNA; (lane 2) RNA isolated from primary BRK cells; (lane 3) RNA from cells transformed by pAd12RIC; (lane 4) RNA from cells transformed by Ad5XhoC; (lane 5) RNA from cells transformed by pAd125; (lane 6) RNA from cells transformed by pAd512. (Reprinted with permission from Nature vol.305, pp. 771-775 Copyright (c) 1983, MacMillan Journals Limited).

Jochemsen et al.(19) investigated which exon of the Ad12 E1A mRNAs is responsible for the MHC-suppressing activity. To that end, two chimaeric E1A regions were constructed in which the DNA segment coding for the first E1A exon of Ad12 was linked to the segment coding for the second exon of Ad5, and vice versa. Cells transformed by these hybrid plasmids, in the presence of region E1B of Ad12, did not express class I MHC when the first E1A exon was from Ad12, but did express class I MHC when the first exon was derived from Ad5 (Table 1). Thus, the suppression of class I MHC is a function of the polypeptide encoded by the first exon of the Ad12 E1A mRNAs. Since the activity is associated with the 13S mRNA product, it follows that the segment which is unique to the 13S mRNA must have an important function in this phenomenon.

3.3 Reduction of class I gene expression is a general property of Ad12-transformed cells derived from primary cell cultures

Inhibition of class I MHC gene activity is not restricted to Ad12-transformed BRK cells, but it also occurs in Ad12-transformed baby mouse kidney cells, hamster embryo cells and human embryonic retinoblast cells. The effect was also found in Ad12 virion-transformed cells. The corresponding Ad5-transformed cells showed normal, or even somewhat enhanced, expression of class I genes as compared to untransformed primary or secondary cultures of the 3 cell types (20). Suppression of class I gene activity was also observed in BRK cells transformed by another highly oncogenic human adenovirus, Ad31, but not in cells transformed by weakly oncogenic Ad7 (R.T.M.J. Vaessen and B.M.M. Dekker, personal communication), suggesting that the phenomenon is characteristic for cells transformed by highly oncogenic adenoviruses. In all cases, expression of class I MHC genes was estimated by Northern blot analysis, using the HLA-B7 cDNA clone as a probe. In the case of transformed rat cells, class I gene expression was also measured by immunoprecipitation with alloantisera specific for the haplotype of the rat cells used. Since the two methods yielded basically the same results, it can be concluded that inhibition of translation of class I mRNA cannot be a major mechanism for reduction of class I proteins.

It should be noted, however, that reduction of class I MHC gene expression was not observed in all types of Ad12-transformed or Ad12 E1-expressing cells. When the E1 region of Ad12 was introduced into established cell lines, e.g. rat 3Y1 or BRLtk$^-$ cells or in mouse Ltk$^-$ cells, expression of class I genes was not reduced (20). Apparently, the inhibition of class I gene expression does not occur in cells that are already immortalized at the time of introduction of the transforming genes. (cf. Table 1).

3.4. Ad5 Ela prevents Ad12 from suppressing the activity of class I MHC genes

Since Ad12 E1A turns off expression of class I MHC, while Ad5 E1A apparently does not inhibit these genes, it appeared of interest to investigate the effect of the simultaneous presence of both E1A regions. BRK cells transformed by the E1A regions of Ad5 and Ad12, and the E1B region of Ad12, expressed both E1A regions with equal efficiency (6,21). Analysis by immunoprecipitation showed that these cells produced normal levels of class I MHC (6, Table I). This indicated that Ad5 E1A can prevent Ad12 E1A from turning off the expression of class I genes. Further experiments by Vaessen et al.(20) demonstrated that when Ad12-transformed cells, which showed a strongly reduced expression of class I genes, were supertransfected with the E1 region of Ad5 class I gene activity was restored in the cell clones that expressed both Ad12 and Ad5 transforming genes. This result shows that Ad5 not only prevents Ad12 from turning off MHC expression but that Ad5 is also dominant over Ad12 with respect to the interaction with class I genes, and, in addition, that the Ad12-transformed MHC-negative cells are intrinsically capable of expressing normal levels of class I MHC antigens.

By transforming cells with plasmids expressing either Ad12 E1 plus Ad5 13S E1A RNA or Ad12 E1 plus Ad5 12S E1A RNA, it was found that the dominant effect of Ad5 is caused by the product of the 13S mRNA (14, Table I). A still more precise localization of the dominant activity of Ad5 E1A was obtained by using the hybrid E1A region described above, in which the first exon-domain of Ad5 region E1A was coupled with the second exon domain of Ad12 E1A, or vice versa (19). Co-transfection of BRK cells with Ad12 region E1 and either of these two hybrid E1A regions resulted in transformed cells which exhibited reduced class I MHC expression when the first exon of the hybrid E1A region was derived from Ad12, but normal class I MHC expression when the first E1A exon was derived from Ad5. Thus, the dominant activity of Ad5 E1A is a function of the first exon of the 13S mRNA.

The results can be summarized as follows: (a) the Ad12-induced suppression of class I MHC activity is a function of the product of the 13S E1A mRNA, more specifically of the sequence encoded by the first E1A exon; (b) the dominant effect of Ad5 in preventing the Ad12-induced inhibition of class I MHC expression is also a function of the product of the 13S mRNA, again specifically of the first E1A exon. This suggests that the suppressing activity by Ad12 E1A and the dominant activity by Ad5 E1A may be based on a similar interaction of the E1A products with a common receptor in the cell. This interaction can result either in an inhibition of class I MHC expression in Ad12-transformed cells, or in the prevention of this inhibition when Ad5 E1A is present in addition to Ad12 E1A. Whether these effects take place at the level of mRNA transcription or mRNA accumulation can not be concluded from these results.

3.5. Evidence that Ad12 can actively suppress class I expression

Evidence that Ad12 E1A can actively suppress class I MHC activity was provided by the following observation. The mutant Ad12 E1 plasmid pR11 has lost its transforming activity due to a deletion in region E1A (18). Replacement of the E1A promoter by the SV40 early promoter + enhancer resulted in restoration of transforming activity in BRK cells. The transformed cells expressed region E1B but showed no reduction of class I gene expression, presumably as a result of the E1A mutation. When these pSVR11-transformed cells were supertransformed with wild-type Ad12 region E1, two transformed lines were obtained, one of which still expressed normal levels of class I MHC mRNA whereas the other contained strongly reduced levels. S1 nuclease protection analysis showed that the concentration of wild-type E1A mRNA was very low in the former line, and very high in the latter (R.T.M.J. Vaessen, unpublished results). This suggests that there is a certain threshold level of Ad12 wild-type E1A mRNA and that only levels exceeding this value are capable of turning off class I gene expression.

3.6. Suppression by Ad12 E1A is not complete

The reduction of class I MHC expression which appears to be a general phenomenon of cells transformed by region E1 of Ad12 is by no means complete. A residual class I antigen expression was found in all Ad12-transformed cell lines tested, the reduction varying from about 4-fold to more than 10-fold (in the majority of the lines) as compared to the expression in untransformed or Ad5-transformed cells (13, R.T.M.J. Vaessen and B.A. Oostra, personal communication). By means of indirect immunofluorescence with antisera directed against the individual K, D and L loci of the H-2 complex of the mouse and analysis by flow cytometry it has recently been shown that expression of all three loci was strongly reduced in Ad12-transformed baby mouse kidney cells. Essentially similar results were obtained by means of S1 protection analysis, using probes specific for the K, D and L loci of the same mouse cells (A. Israel, P. Kourilsky and R.T.M.J. Vaessen, unpublished results). Further confirmation that Ad12 region E1-transformed rat cells (derived from inbred

LIS rats) have reduced levels of class I MHC was provided by Mellow et al. (22). By use of immunofluorescence and flow cytometry they showed that, although MHC expression was reduced in Ad12-transformed cells, the difference with Ad5-transformed and untransformed cells was much smaller than in the previously mentioned studies. The reason for this discrepancy is not clear.

A surprising result was recently reported by Grosveld et al. (23). These authors showed that infection of mouse embryo cultures with Ad5 or Ad12 virus resulted in a strong stimulation of expression of the H-2K^b gene with either virus. Infection with adenovirus mutants affecting the E1A or the E1B genes indicated that products of both region are required for H-2 gene activation. These results are in contrast to those reported by Schrier et al. (13). It has been suggested that the opposing results may be explained by the fact that the stimulation of MHC expression may be an early effect of infection with virus, eliciting a stronger immune response against virus-infected cells, while the inhibition of class I MHC expression by Ad12 region E1 may only become apparent after stable transformation. The possibility that other viral genes than those encoded by region E1 have a role in the stimulation of class I gene expression cannot be ruled out, either.

4. BIOLOGICAL CONSEQUENCES OF REDUCED CLASS I MHC EXPRESSION IN ADENOVIRUS-TRANSFORMED CELLS

4.1. Oncogenicity depends on the nature of the transforming virus and the immunocompetence of the host

The tumor-inducing capacity of adenoviruses or Ad-transformed cells in rodent hosts is dependent on several factors, determined both by the virus or virus-transformed cell, and the host animal. Ad12 is highly oncogenic when inoculated into newborn rodents, but weakly oncogenic, if at all, in immunocompetent young or adult rodents, unless the animals are immunosuppressed by anti-thymocyte serum or steroid hormones, or by thymectomy (24-28). In contrast, Ad2 even when inoculated into newborn rodents does not induce tumors, the exception being when the newborn animals (rats) are immunosuppressed by anti-thymocyte serum (29). This indicates that tumor formation depends both on the Ad serotype and on the degree of immune competence of the host.

The tumor-inducing capacity of rodent cells transformed in vitro basically parallels that of the virus. Rodent cells transformed in vitro by Ad12 are highly oncogenic in athymic nude mice (6,21,30,31), newborn or 5-6 days old rats and hamsters and in syngeneic adult hamsters (6,32,33,34). In older syngeneic rats (3 weeks and older) Ad12-transformed cells are only weakly oncogenic, about 10% of the animals developing tumors (R. Bernards, personal communication). In contrast, rodent cells transformed by Ad2 or Ad5 are oncogenic only when inoculated into newborn normal hamsters or athymic nude mice (6,21,31-35), but nononcogenic in newborn rats or adult hamsters (34) unless the animals are immunosuppressed (31,36).

Thus, the tumor-inducing capacity of Ad viruses or Ad-transformed cells is dependent on: (a) the nature of the virus: Ad2 and Ad5 being non-oncogenic, Ad12 oncogenic; (b) the immune status of the prospective rodent host: tumors are produced more efficiently in newborn (immunologically immature) or immuno-suppressed animals than in immunomature and non-suppressed animals; (c) the genetic constitution of the rodent host: inbred strains of the same animal species may differ in their susceptibility to Ad tumor induction (28,37). Furthermore, hamsters are generally more susceptible to tumor induction than rats or mice (32-34).

These findings demonstrate the importance of the host immune system in determining whether or not cells transformed by adenoviruses will induce tumors. Although in experimental animals Ad-transformed cells can elicit production of complement-dependent antibodies which are cytotoxic for Ad-transformed cells in vitro (11,12), it is not clear what role this line of defence plays in tumor rejection. Furthermore, no correlation was found to exist between the degree of immunogenicity of a series of Ad2-transformed rat and hamster cells (as measured by their ability to confer transplantationresis-tance), and the oncogenic potential of the cells (33,36). These findings, and the observations that immune suppression by anti-thymocyte serum or thymectomy increase tumor induction by Ad-transformed cells suggest that the cellular immune system acts as the major defence mechanism against tumor induction.

4.2. Evasion of cellular immunity by Ad12-transformed cells

One of the types of effector lymphocytes that play an important role in recognition and destruction of tumor cells expressing foreign antigens are the cytotoxic thymus-derived lymphocytes (CTLs). CTLs can specifically recognize foreign antigens on transformed cells and are capable of lysing these cells. Recognition and cytolysis can only occur, however, when the tumor cells also express their class I MHC antigens (dual recognition). When the cells do not express class I MHC proteins they will not be recognized by CTLs and escape from CTL-mediated killing (38,39).

Since Ad12-transformed cells express reduced amounts of class I MHC antigens, Bernards et al. (6) have investigated whether Ad5- and Ad12-transformed cells differ in their sensitivity to in vitro lysis by allogeneic CTLs. Allogeneic CTLs, against the Wag-Rij rat cells from which the Ad-transformed cells were derived (class I MHC haplotype $RT1^u$), were raised in primary mixed lymphocyte culture by incubating spleen cells of ACI rats (haplotype $RT1^a$) in vitro with X-irradiated spleen cells of Wag-Rij rats. Thus, the resulting CTLs were only sensitized against the $RT1^u$ haplotype of the Ad-transformed Wag-Rij cells and were not directed against the viral tumor antigens. It was found that BRK cells transformed by Ad12EI, or Ad12E1A + Ad5E1B, were significantly more resistant to in vitro cytolysis than cells transformed by Ad5EI or Ad5E1A + Ad12E1B. A residual CTL activity was observed with all Ad12 E1A-expressing cells tested but this cytolytic activity was 4 to 6 fold lower than with Ad5 E1A-expressing cells. This finding suggested that tumor induction by Ad12-transformed cells in immunocompetent rats may be explained, at least in part, by evasion of the T-cell-mediated immunity. This view was further supported by the observation that transformed BRK cells expressing the E1A regions of both Ad5 and Ad12 and the E1B region of Ad12 showed full expression of class I MHC proteins (due to the dominant effect of Ad5 E1A), and were completely non-oncogenic in syngeneic rats, but still highly oncogenic in nude mice (6). The low remaining CTL activity observed with Ad12-transformed cells was attributed to the residual expression of class I MHC antigens. It is not clear which viral T antigen is recognized by CTLs as the transplantation antigen. Föhring et al. (40) have presented evidence that the surface antigen conferring sensitivity to CTLs in Ad12-transformed cells must be encoded by region E1B (probably the 19kD protein).

Confirmation that expression of class I MHC antigens is reduced in Ad12-transformed rat or mouse cells was provided by Mellow et al. (22), and Tanaka et al. (41). The reduction found by Mellow et al. (22) was substantially less pronounced than that reported by Schrier et al. (13) and Tanaka et al. (41). Probably as a result of the high residual class I antigen expression in the first mentioned study, the sensitivity of the Ad12-transformed cells to allogeneic CTLs did not significantly differ from that of Ad5-transformed cells. In spite of the high sensitivity to sensitized CTLs, the Ad12-transformed cells were all oncogenic when inoculated into newborn syngeneic rats. Confirmation that oncogenicity of Ad12-transformed rat cells is not necessarily correlated with reduced cytolysis by sensitized CTLs was provided earlier by

Raska and Gallimore (42). These results indicate that in these cases tumor formation cannot be ascribed to a resistance to cytotoxic T-lymphocyte. Thus, other mechanisms must exist which determine whether or not transformed cells will be capable of initiating a tumor in vivo. Considerable evidence has been accumulated in recent years that oncogenicity by adenoviruses in young animals (rat, hamster) is determined to a large extent by a resistance to lysis by natural killer (NK) cells (spontaneously cytolytic lymphoid cells) and activated macrophages (42-45). NK cells and activated macrophages are not class I MHC-restricted. Since they do not require prior sensitization by antigens, as opposed to cytotoxic T lymphocytes, they can act faster and may constitute the first line of defence against tumor cells.

It is not yet clear which factors determine whether Ad-transformed cells are resistant or sensitive to lysis by NK cells. Cook et al. (46) showed that expression of viral T antigens is required for susceptibility of Ad2-transformed cells to NK cells. Recent experiments with cells transformed by Ad5/Ad12 recombinant viruses, consisting of an Ad5 genome in which either region E1, E1A or E1B was replaced by the corresponding region of Ad12, suggested, however, that resistance to NK cell killing correlates with expression of region E1A of Ad12. On the other hand, susceptibility to NK cells was found to require expression of E1A of Ad5 (K.Raska, Y. Sawada and T.E. Shenk, personal communication). The observation that resistance to lysis by NK cells correlates with expression of Ad12 E1A is not in agreement with results of Bernards et al. (5,6). These authors found that BRK cells transformed by Ad5 E1A + Ad5 E1B were oncogenic in 50% of the nude mice tested, whereas rat cells transformed by Ad12 E1A + Ad5 E1B are only oncogenic in 10% of the nuce mice. Since the Ad5 E1B regions were expressed with approximately equal efficiencies, the difference in oncogenicity must be ascribed to the nature of the E1A regions. As nude mice lack CTL activity but do contain high NK cell activities (47), the results of Bernards et al. (6) do not support the conclusion that Ad12 E1A confers resistance to NK cells. It is clear, however, that Ad2- and Ad5-transformed cells in general are susceptible to cytolysis by CTLs, NK cells and macrophages, whereas Ad12-transformed cells are comparatively resistant to these effector cells.

Evidence that class I MHC antigens indeed may play a role in tumor immunity was provided by the following observations. (a) Serial passage of SV40-transformed C3H mouse fibroblasts resulted in the isolation of highly oncogenic variants which differed from the parental cells by the loss of expression of the H-$2K^k$ allele of the murine class I MHC. Thus acquisition of the oncogenic phenotype correlated with the loss of expression of a class I MHC gene (48). (b) Even more convincing evidence was provided by Hui et al.(49) who reported that introduction of the H-$2K^k$ gene by DNA transfection into the AKR tumor cell line K36.126, which lacked detectable expression of endogeneous H-$2K^k$, resulted in a loss of oncogenicity of these cells if the introduced H-2 gene was expressed. (c) Further confirmation that class I MHC antigens are involved in Ad oncogenesis was recently provided by Tanaka et al. (41). These authors transfected cloned H-$2L^d$ genes into mouse cells transformed by Ad12 virus. The parental transformed cells were highly oncogenic and expressed little or no endogenous class I MHC antigens. Interestingly, one transfectant showed high expression of the newly introduced H-2 gene and had lost its oncogenicity. However, the other transfectants showed low expression of the H-$2L^d$ gene and this suggests that the loss of oncogenicity is caused by the reappearance of class I MHC molecules. The observation by Tanaka et al. (41) that the newly introduced H-2 gene was expressed to a low level in several of the Ad12-transformed cells studied is in agreement with preliminary results obtained by A.G. Jochemsen (personal communication). He showed that introduction of cloned H-2 genes into BMK cells transformed by Ad5 or Ad12, results in strong expression of the transfected H-2 gene in the Ad5-transformants but very weak expression in the Ad12-transformants. This suggests that genes stably introduced by DNA transfection can also be regulated by Ad region E1A.

In summary, it has become clear that adenovirus malignancy is determined by a variety of factors. Apart from differences in intrinsic growth potential of transformed cells in vivo and the hypothetical possibility that oncogenic cells may exert immunosuppressive effects on their hosts, the major factor appears to be the ability of tumor cells to evade the cellular immune defence. This latter property is dependent on (a) resistance of the transformed cells to the action of natural killer cells and activated macrophages, immune systems that are not class I MHC-restricted. (b) resistance to the cytolytic action of class I MHC-restricted cytotoxic T lymphocytes. The contribution of each of these factors may vary depending on the genetic constitution of the host animal, the immunocompetence of the host at the time of tumor transplantation or virus infection, and the host species. Resistance to NK cells may constitute the first line of defence, because no prior antigen-sensitization is required. The more slowly developing CTL resistance will become effective at later stages. Still difficult to explain is why Ad12-transformed rat cells are highly oncogenic when injected into newborn rats but very weakly oncogenic when inoculated into older animals (3 weeks and older) (6). The tumors arising in rats exposed to transformed cells within the first 5 or 6 days of life become visible at the age of about 6 weeks when the animals have become fully immunologically mature. It seems likely, however, that tumor development already initiates within the first 1-2 weeks after inoculation of tumor cells when the animals are still immunologically immature. Tumor initiation was possibly preceeded by a process of adaptation or selection, to yield cells with increased in vivo growth potential and/or increased resistance to the host's immune defence. Inoculation with tumor cells at the age of 3 weeks or older, when the animals have become immunocompetent, results in an almost complete loss of oncogenicity. Apparently, the cellular immune defence is then capable of eliminating the tumor cells in most cases, in spite of their relative resistance to CTLs and NK cells.

5. REGULATION OF GENE EXPRESSION BY ADENOVIRUS REGION E1A

5.1. Activities of region E1A

Region E1A is transcribed in Ad5-transformed cells into two major mRNAs of 12S and 13S (see 3.2.). The products of the 12S and 13S mRNAs have been shown to be involved in several different activities, including immortalization and morphological transformation (30,50,51), cooperation with activated c-ras oncogenes in oncogenic transformation of primary rat cells (4) and activation of gene expression. With the help of mutants of Ad5 containing lesions in region E1A, it has been shown that the 13S E1A mRNA is required for transcriptional activation of region E1B and the other early regions of the viral genome (52-57). The dependence of transcription of the viral early regions on E1A function is not absolute, since HeLa cells infected with Ad5 E1A mutants, including dl312 which lacks the entire E1A coding sequence, eventually show a delayed expression of early viral genes followed by production of progeny virus. Infection with dl312 at high multiplicity of infection even results in a normal virus replication and production of near-wild-type levels of progeny virus (54,58). This indicates that the E1A product is required for efficient transcription of early viral genes but is not strictly indispensable. The activation by region E1A of early viral transcription was initially attributed to inactivation by the 13S E1A product of labile repressor-like molecules that normally prevent expression of the Ad early regions (54,59,60). More recent evidence indicated, however, that E1A may have a direct, trans-acting stimulatory activity on early gene transcription (61). The delayed onset of early gene transcription after infection of HeLa cells with E1A mutants such as the deletion mutant dl312, may be mediated by a different mechanism than the activation of early gene transcription by functional E1A products. The former process may be the result of a slow spontaneous assembly of early viral templates into stable transcription complexes, whereas the latter may represent a rapid and specific formation of early transcription complexes in which E1A products are directly involved, either stoichiometrically or catalytically.

Region E1A proteins not only activate transcription of early viral genes but also of certain cellular genes, such as the gene coding for the 70kD heat shock protein (62,63) and β-tubulin (64). Infection with adenoviruses does not, however, lead to a generalized increase of cellular gene expression (65). This indicates that the transcription-enhancing effect of Ad5 E1A is specific for viral early genes and certain cellular genes only, suggesting that E1A products may recognize specific base sequences in these genes. This was supported by the observation of Bos and Ten Wolde-Kraamwinkel (66) who presented evidence that E1A products of Ad5 and Ad12 might recognize specific DNA sequences in the promoter area of Ad12 region E1B. These sequences appear to be located within a 135 bp stretch 5' from the E1B cap site. It is noteworthy that this region contains a conserved 10 bp sequence, located about 50 residues upstream from the cap site, which is also found at the same relative position in other early adenovirus transcription units that are positively regulated by E1A. Whether the stimulation of transcription of the cellular genes is fortuitous and caused by accidental homology of sequences in their promoter region with Ad early promoters, or has a specific function in viral replication, is unknown.

Region E1A not only stimulates the activity of RNA polymerase II genes but also of RNA polymerase III genes, as demonstrated by R.G. Roeder and colleagues (personal communication). The significance of this observation is not clear.

Region E1A products have also been shown to stimulate transient expression of cellular genes introduced into cultured cells by means of calcium phosphate-mediated DNA transfection. This has been demonstrated for the human and rabbit β-globin gene (67,68) and the rat preproinsulin gene (69). The corresponding endogenous genes remained inactive, however. The fact that E1A can activate newly introduced genes from their own promoters, but not the corresponding endogenous genes, may be explained by the circumstance that the former are not yet incorporated in an ordered chromatin structure, as opposed to the endogenous genes. It is possible, therefore, that region ElA will stimulate transcription of any newly introduced gene in an aspecific manner. That the ElA-activated transcription of introduced genes may be aspecific is supported by the observation that the transcription-initiation of such genes is rather imprecise, as compared to that of early viral genes (69-71). Thus, the Ad ElA products appear to act as trans-activators of gene expression, both from RNA polymerase II and III promoters, possibly by interacting with the promoter region, or facilitating interaction of transcription factors with this region.

In addition to the stimulatory influence on gene expression, region ElA products have also been found to be capable of inhibiting gene activity. It was recently shown that Ad2 and Ad5 ElA products can repress the activity of the enhancer of SV40, polyoma virus and Ad2 region ElA (72). The repressing activity is directed to the enhancer sequences themselves and not to the transcription promoter signals. The 12S and 13S ElA products were equally effective in suppressing the SV40 and polyoma enhancers. A suppressing activity has also been observed for expression of fibronectin genes. Cells transformed by EI region of either Ad5 or Ad12 contain considerably lower levels of fibronectin mRNA than untransformed cells (H. van Kranen and J.L. Bos, unpublished results). The phenomenon is dependent on the activity of region ElA, but it is not clear whether it is caused by a direct interaction of ElA products with fibronectin gene expression or is an indirect consequence of the transformed state of the cell.

A final example of ElA-mediated repression of cellular gene activity is the inhibition of expression of class I MHC genes (and a number of other, as yet unidentified, cellular genes; see above) by Ad12 region ElA, but not by Ad5 region ElA. Ad5 ElA not only lacks this repressing activity but even may stimulate expression of class I genes in transformed rat cells (B.A. Oostra and R.T.M.J. Vaessen; unpublished results).

5.2. A putative mechanism of inhibition of MHC expression

Very little is known about the mechanism by which adenovirus region ElA influences the expression of class I MHC genes. The observation that the inhibition of class I gene activity by Ad12 region ElA, and the prevention of this inhibition by Ad5 region ElA, are both functions of the first exon of the respective 13S ElA mRNA products, suggests that the two phenomena are based on a similar type of reaction. This reaction could involve, for example, the participation of E1A products in the formation of transcription complexes, as discussed above. If this interpretation is correct, the question may be asked how the opposite effects on class I gene expression arise. A possible explanation would be that the observed phenomena are based on differences in the potential of the E1A products to stimulate or modify gene expression. Bos and Ten Wolde-Kraamwinkel (66) have shown that Ad5 E1A is capable of activating the Ad12 E1B promoter 5-6 times more efficiently than Ad12 E1A. Similarly, the Ad5 E1 region is 10- to 50-fold more active in inducing morphological transformation in primary cultures of BRK cells than the Ad12 E1 region, a property which has been attributed to the respective E1A regions (5) If we assume that transformation is dependent on altered cellular gene expression, and, in the case of (adeno)viral transformation, on viral control of the expression of certain cellular genes, it can be tentatively concluded that Ad12 E1A products are weaker modulators of gene expression than Ad5 E1A products. Since the E1A effects on gene expression appear to be dependent on an interaction with specific base sequences, it is conceivable that genes exist with recognition sequences that cause inappropriate interactions with E1A products. If E1A products indeed participate in the formation of transcription complexes, this would mean that Ad5 E1A proteins might be capable of forming active complexes with class I MHC genes, but that Ad12 E1A proteins would form inactive complexes which could even block endogenous expression of the genes. The Ad5 E1A product apparently has higher affinity for the reaction complex and can compete out the Ad12 E1A product. This explanation would be valid only of inhibition of class I gene activity occurs at the level of gene transcription. It cannot yet be excluded, however, that the effect is caused by destabilization of class I MHC mRNA. In the latter case, we would have to assume that Ad12 E1A products are capable of decreasing the stability of class I mRNAs.

The results summarized in the preceeding sections indicate that reduction of class I MHC gene expression in Ad12-transformed cells results in reduced sensitivity to CTLs and in this way may contribute to oncogenicity in immunocompetent animals. The exact contribution of reduced sensitivity for CTLs, relative to that for NK cells or macrophages, however, is not well known. The question whether reduction of class I MHC gene expression by oncogenic adenoviruses has biological significance for the virus can also not yet be answered. It is conceivable that his activity can contribute to the survival of the virus in certain hosts, but this is speculative. Whatever its natural role, the phenomenon appears to be a valuable tool for studies on E1A functions and the significance of class I MHC genes in oncogenesis.

ACKNOWLEDGEMENTS

The authors thank Dr. H. van Ormondt for critically reading the manuscript and our colleagues for communicating their results prior to publication. They are also indebted to Mrs. Rita Veeren for the patience and secretarial help.

REFERENCES

1. Tooze J: DNA Tumor Viruses. 2nd edition. Cold Spring Harbor Laboratory. 1981.
2. Van der Eb AJ, Van Ormondt H, Schrier PI, Lupker JH, Jochemsen H, DeLeys RJ, Maat J, Van Beveren, CP, Dijkema R, de Waard A: Structure and function of the transformed genes of human adenoviruses and SV40. Cold Spring Harbor Symp Quant Biol (44): 383-399, 1979.
3. Land H, Parada LF, Weinberg RA: Tumorigenic conversion of primary embryo fibroblasts requires at least two cooperating oncogenes. Nature (304):596-602, 1979.
4. Ruley HE: Adenovirus region 1A enables viral and cellular transforming genes to transform primary cells in cultures. Nature (304): 602-606, 1983.
5. Bernards R, Houweling A, Schrier PI, Bos JL, Van der Eb AJ: Characterization of cells transformed by Ad5/Ad12 hybrid early region 1 plasmids. Virology (120): 422-432, 1982.
6. Bernards R, Schrier PI, Houweling A, Bos JL, Van der Eb AJ, Zijlstra M, Melief CJM: Tumorigenicity of cells transformed by adenovirus type 12 by evasion of T-cell immunity. Nature (305): 776-779, 1983.
7. Eddy BE, Grubbs GE, Young RD: Tumor immunity in hamsters infected with adenovirus 12 or simian virus 40. Proc Soc Exp Biol Med (117): 575-579, 1964.
8. Trentin JJ, Bryan E: Virus-induced transplantation immunity to human adenovirus type 12 tumors of the hamster and mouse. Proc Soc Exp Biol Med (121) 1216-1219, 1966.
9. Sjögren HO, Minowada J, Ankerst J: Specific transplantation antigens of mouse sarcomas induced by adenovirus type 12. J Exp Med (25): 689-701, 1967.
10. Shiroki K, Shimojo H, Maeta Y, Hamada C: Tumor-specific transplantation and surface antigen in cellc transformed by the adenovirus 12 DNA fragments. Virology (99): 188-191, 1979.
11. Raska K, Morrongiello MP, Föhring B: Adenovirus type 12 tumor antigen. III. Tumorigenicity and immune response to syngeneic rat cells transformed with virions and isolated transforming fragment of adenovirus 12 DNA. Int J Cancer (26): 79-86, 1980.
12. Raska K, Dougherty J, Gallimore PH: Product of adenovirus type 2 early gene block E1 in transformed cells elicits cytolytic response in syngeneic rats. Virology (117): 530-535, 1982.

13. Schrier PI, Bernards R, Vaessen RTMJ, Houweling A, Van der Eb AJ: Expression of class I major histocompatibility antigens switched off by highly oncogenic adenovirus 12 in transformed rat cells. Nature (305): 771-775, 1983.
14. Van der Eb AJ, Bernards R, Schrier PI, Bos JL, Vaessen RTMJ, Jochemsen AG, Melief CJM: Altered expression of cellular gene in adenovirus-transformed cells. in: Van de Woude GF et al. (eds). Cancer Cells 2, Cold Spring Harbor Laboratory, 1984, pp 501-510.
15. Sood AK, Pereira D, Weissman SM: Isolation and partial nucleotide sequence of a cDNA clone for human histocompatibility antigen HLA-B by use of an oligodeoxynucleotide primer. Proc Natl Acad Sci USA (78): 616-620, 1981.
16. Van Ormondt H, Galibert F: Nucleotide sequences of adenovirus DNAs. Curr Topics Microbiol Immunol (110): 73-142, 1984.
17. Pettersson U, Virtanen A, Perricaudet M, Akusjärvi G: The messenger RNAs from the transforming region of human adenoviruses. Curr Topics Microbiol Immunol (109): 107-123, 1983.
18. Bos JL, Jochemsen AG, Schrier PI, Van Ormondt H, Van der Eb AJ: Deletion mutants of region E1A of Ad12 E1 plasmids: effect of oncogenic transformation. Virology (129): 393-400, 1983.
19. Jochemsen AG, Bos JL, Van der Eb AJ: The first exon of region E1A genes of adenoviruses 5 and 12 encodes a separate functional protein domain. EMBO J (3): 2923-2927, 1984.
20. Vaessen RTMJ, Bernards R, Oostra BA, Houweling A, Jochemsen AG, Bos JL, Van der Eb AJ: Expression of MHC genes in adenovirus:transformed cells. In: Bishop JM, Rowley JD and Greaves M (eds) Genes and Cancer. Alan R.Liss, Inc., New York 1984, pp 303-314.
21. Bernards R, Schrier PI, Bos JL, Van der Eb AJ: Role of adenovirus type 5 and 12 early region 1b tumor antigens in oncogenic transformation. Virology (127): 45-53, 1983.
22. Mellow G, Föhring B, Dougherty T, Gallimore PH, Raska K: Tumorigenicity of adenovirus-transformed rat cells and expression of class I major histocompatibility antigens. Virology (134): 460-465, 1984.
23. Grosveld F, personal communication.
24. Yabe Y, Trentin JJ, Grant T: Cancer induction in hamsters by human type 12 adenovirus. Effect of age and of virus dose. Proc Soc Exp Biol Med (111): 343-344, 1982.
25. Kirschstein RL, Rabson AS, Peters EA: Oncogenic activity of adenovirus 12 in thymectonized BALB/c and C3H/HeN mice. Proc Soc Exp Biol Med (117): 198-200, 1964.
26. Yohn DS, Funk CA, Kalninus VI, Grace JT: Sex-related resistance in hamsters to adenovirus 12 oncogenesis. Influence of thymectomy at three weeks of age. J Natl Cancer Inst (35): 617-624, 1965.
27. Yohn DS, Funk CA, Grace JT*f* Sex-related resistance in hamsters to adenovirus 12 oncogenesis. III. Influence of immunologic impairment by thymectomy or cortisone. J Immunol (100): 771-780, 1968.
28. Allison AC, Berman LD, Levey RH: Increased tumor induction by adenovirus type 12 in thymectonized mice and mice treated with anti-lymphocyte serum. Nature (215): 185-187, 1967.
29. Harwood, LMJ, Gallimore PH: A study on the oncogenicity of adenovirus type 2-transformed rat embryo cells. Int J Cancer (16): 498-508, 1975.
30. Van den Elsen P, De Pater S, Houweling A, Van der Eb AJ: The relationship between region E1a and E1b of human adenoviruses in cell transformation. Gene (18): 175-185, 1982.

31. Gallimore PH, McDougall JK, Chen LB: In vitro traits of adenovirus-transformed cell lines and their relevance to tumorigenicity in nude mice. Cell (10): 669-678, 1977.
32. Cook JL, Lewis AM: Host response to adenovirus 2-transformed hamster embryo cells. Cancer Res (39): 1455-1461, 1979.
33. Lewis AM, Cook JL: Spectrum of tumorigenic phenotypes among adenovirus 2, adenovirus 12 and simian virus 40-transformed Syrian hamster cells defined by host cellular immune-tumor cell interactions. Cancer Res (42): 939-944, 1982.
34. Lewis AM, Cook JL: The interphase between adenovirus-transformed cells and cellular immune response in the challanged host . Curr Toppics Microbiol Immunol (110): 1-22, 1984.
35. Graham FL, Abrahams PJ, Mulder C, Heyneker HL, Warnaar SO, De Vries FAJ, Fiers W, Van der Eb AJ: Studies on in vitro transformation by DNA and DNA fragments of human adenoviruses and simian virus 40. Cold Spring Harbor Symp Quant Biol (39): 637-650, 1974.
36. Gallimore PH, Paraskeva C: A study to determine the reasons for difference in the tumorigenicity of rat cell lines transformed by adenovirus 2 and adenovirus 12. Cold Spring Harbor Symp Quant Biol (44):703-713, 1980.
37. Yabe Y, Samper I, Bryan E, Taylor G, Trentin JJ: Oncogenic effect of human adenovirus type 12 in mice. Science (143): 46-47, 1964.
38. Zinkernagel RM, Doherty PC. Restriction of in vitro T cell-mediated cytotoxity in lymphocytic choriomeningitis within a syngeneic or semi-allogeneic system. Nature (248) 701-702, 1974.
39. Zinkernagel RM, Oldstone MB: Cells that express viral antigens but lack H-2 determinants are not lysed by immune thymus-derived lymphocytes but are lysed by other antiviral immune attack mechanisms. Proc Natl Acad Sci USA (73): 3666-3670, 1976.
40. Föhring B, Gallimore PH, Mellow GH, Raska K: Adenovirus type 12 specific cell surface antigen in transformed cells is a product of the Elb early region Virology (131): 463-472, 1983.
41. Tanaka K, Isselbacker KJ, Khoury G, Jay G: Reversal of oncogenesis by the expression of a MHC class I gene. Science in press, 1985.
42. Raska K, Gallimore PH: An inverse relation of the oncogenic potential of adenovirus-transformed cells and their sensitivity to killing by syngeneic natural killer cells. Virology (123): 8-18, 1982.
43. Cook JL, Hibbs JB, Lewis AM: DNA virus-transformed hamster cell-host effector cell interactions: level of resistance to cytolysis correlated with tumorigenicity. Int J Cancer (30): 795-803, 1982.
44. Cook JL, Lewis AM: Differential NK cell and macrophage killing of hamster cells infected with non-oncogenic and oncogenic adenoviruses. Science (224): 612-614, 1984.
45. Sheil JM, Gallimore PH, Zimmer SG, Sopori ML. Susceptibility of adenovirus transformed rat cell lines to natural killer (NK) cells: direct correlation between NK resistance and in vivo tumourigenesis. J.Immunol (132): 1578-1582, 1984.
46. Cook JL, Hauser J, Patch C, Lewis AM, Levine AS: Adenovirus 2 early gene expression promoter susceptibility to effector cell lysis of hybrids formed between adenovirus 2 and simian virus 40-transformed hamster cells. Proc Natl Acad Sci USA (80): 5995-5999, 1983.
47. Herberman RB, Holden HT: Natural cell-mediated immunity. Adv Cancer Res (27) 305-377, 1978.
48. Gooding LR: Characterization of a progressive tumor from CSH fibroblast transformed in vitro with SV40 virus. Immunoresistance in vivo correlates with phenotypic loss of H-2K^{k} J Immunol (129): 1306-1312, 1982.

49. Hui K, Grosveld F, Festenstein H: Rejection of transplantable AKR leukemia cells following MHC DNA-mediated cell transformation. Nature (311): 750-752, 1984.
50. Houweling A, Van den Elsen PJ, Van der Eb AJ: Partial transformation of primary rat cells by the leftmost 4.5% fragment of adenovirus 5 DNA. Virology (105) 537-550, 1980.
51. Van den Elsen PJ, Houweling A, Van der Eb AJ: Morphological transformation of human adenoviruses is determined to a large extent by gene products of region Ela. Virology (131): 242-246, 1983
52. Jones, N, Shenk T: Isolation of adenovirus type 5 hrt host range deletion mutants defective for transformation of rat embryo cells. Cell (17): 683-689, 1979.
53. Berk AJ, Lee F, Harrison T, Williams J, Sharp PA: Pre-early adenovirus 5 gene product regulate synthesis of early viral messenger RNAs. Cell (17): 935-944, 1979.
54. Nevins JR: Mechanism of activation of early viral transcription by the adenovirus E1A gene products. Cell (26): 213-220, 1981.
55. Ricciardi RP, Jones RL, Cepko CL, Sharp PA, Roberts BE: Expression of early adenovirus genes requires a viral-encoded acidic polypeptide. Proc Natl Acad Sci USA (78): 6121-6125, 1981.
56. Montell C, Fisher EF, Caruthers MH, Berk AJ: Resolving the functions of overlapping viral genes by site-specific mutagenesis at mRNA splice site. Nature (295): 380-384, 1982.
57. Carlock LR, Jones NC: Transformation-defective mutant of adenovirus type 5 containing a single altered Ela mRNA species. J Virol (40): 657-664,1981.
58. Shenk T, Jones N, Colby W, Fowlkes D: Functional analysis of adenovirus type 5 host range deletion mutants defective for transformation of rat embryo cells. Cold Spring Harbor Symop Quant Biol (44): 367-375, 1979.
59. Katze MG, Persson H, Philipson L: Control of adenovirus early gene expression: a post-transcriptional control mediated by region Ela products. Mol Cell Biol (1): 807-813, 1981.
60. Katze MG, Persson H, Johansson BM, Philipson L: Control of adenovirus gene expression: cellular gene products restrict expression of adenovirus host-range mutants in non-permissive cells. J Virol (46): 50-59, 1983.
61. Gaynor RB, Berk AJ: Cis-acting induction of adenovirus transcription Cell (33): 683-693, 1983.
62. Nevins JR: Induction of the synthesis of a 70.000 dalton mammalian heat shock protein by the adenovirus E1A gene product. Cell (29): 913-919, 1982.
63. Kao HT, Nevins JR: Transcriptional activation and subsequent control of the human heat shock gene during adenovirus infection. Mol Cell Biol (3): 2058-2065, 1983.
64. Stein R, Ziff EB: HeLa cell -tubilin gene transcription by adenovirus 5 in paralel with viral early genes by an E1A dependent mechanism. Mol Cell Biol (4): 2792-2801, 1984.
65. Beltz GA, Flint SJ: Inhibition of HeLa cell protein synthesis during adenovirus infection: restriction of cellular messenger RNA sequences to the nucleus. J Mol Biol (131): 353-373, 1979.
66. Bos JL, Ten Wolde-Kraamwinkel HC: The Elb promoter of Ad12 in mouse L tk^- cells is activated by adenovirus region Ela. EMBO J (2): 73-76,1983.
67. Green MR, Treisman R, Maniatis T: Transcriptional activation of cloned human -globin genes by viral intermediate-early gene products. Cell (35) 137-148, 1983.

68. Svenson C, Akusjärvi G: Adenovirus 2 early region E1A stimulates expression of both viral and cellular genes. EMBO J (3): 789-794, 1984.
69. Gaynor RB, Hillman D, Berk AJ: Adenovirus early region 1A protein activates transcription of a nonviral gene introduced into mammalian cells by infection or transfection. Proc Natl Acad Sci USA (81): 1193-1197, 1984.
70. Curtois G, Berk A: Adenovirus E1A protein activation of an integrated viral gene. Embo J (3): 1145-1149, 1984.
71. Baker CC, Ziff EB: Promoters and heterogeneous 5' termini of the messenger RNAs of adenovirus serotype 2. J Mol Biol (149) 189-221, 1983.
72. Borelli E, Hen R, Chambon P: Adenovirus 2 E1A products repress enhancer-induced stimulation of transcription. Nature (312): 608-612, 1984.

11

ONCOGENIC TRANSFORMATION BY HUMAN ADENOVIRUSES AND THEIR DNA

STANLEY MAK AND IRENE MAK

Department of Biology, McMaster University,
Hamilton, Ontario L8S 4K1, CANADA

SUMMARY

Some adenoviruses such as Ad12 can induce tumors in newborn animals while others such as Ad2 and Ad5 cannot. However, all of them can cause morphological transformation of rodent cells in vitro. Hamster and rat cells transformed by the highly oncogenic viruses, for example Ad12, are tumorigenic in syngeneic immunocompetent hosts. Although hamster cells transformed by the non-oncogenic group (Ad2 and Ad5) are tumorigenic, rat cells transformed by this group of viruses are not. This difference in their biological properties affords an opportunity to dissect the viral genes responsible for cellular transformation and for tumorigenesis in vivo. The left 11% of the adenovirus genome (the E1 region) is sufficient to transform cells and for tumor induction. This region is further subdivided into E1A (0-4.5%) and E1B (4.5-11%). E1A encodes several polypeptides some of which have the capacity to regulate the activity of other viral and cellular genes during lytic infection, and is responsible for immortalization of primary cells. E1B, which encodes three to five polypeptides, confers other properties to the transformed cells, such as anchorage independent growth and cell morphology in some cases. It appears that both E1A and E1B are required for tumorigenicity of Ad12-transformed cells in immunocompetent animals. The E1B 58K polypeptide or its N-terminal portion is responsible for the "fully" transformed or tumorigenic phenotype. Available experimental data indicate that the E1B 19K polypeptide plays an important role in the efficiency of transformation of primary cells, but is not required for tumor induction by the transformed cells. Much of what is known about the biological functions of various viral genes has come from studies using cloned viral DNA fragments or virus containing mutations in the E1 region. It is anticipated that the increasing use of recombinant DNA technology to construct viral genomes containing specific mutations (site-directed mutagenesis) will further our understanding of the functions of transforming proteins.

INTRODUCTION

It was more than 20 years ago that human adenovirus type 12 was found to be capable of tumor induction in newborn hamsters (1). However, not all adenoviruses are oncogenic. Based on their ability to induce tumors in newborn hamsters, human adenoviruses can be classified into three subgroups: the highly oncogenic group A (e.g. serotypes 12, 18, and 31), the weakly oncogenic group B (e.g. serotypes 3, 7, 11, 21) and the non-oncogenic group C (2, 3, 4). The group A viruses induce tumors with high frequency and the latent period is very short while those belonging to group B cause tumors only in a small fraction of the animals and tumors become evident only after a relatively long latent period. Using DNA-DNA hybridization analysis, Green and his colleagues (5) have subdivided the non-oncogenic group into C (e.g. serotypes 2 and 5), D (e.g. serotypes 3, 13, and 15) and E (serotype 4) based on their DNA homology. Although common DNA sequences are lacking among different groups, their functional organizations are similar, with only minor differences in the size and the number of mRNA species synthesized by different serotypes. For example, E1A of Ad12 encodes four mRNAs while the same region of Ad2 encodes only two mRNA species (6). Similar differences also have been suggested for the mRNAs encoded in region E1B from Ad2 and Ad12 (7). However, the pattern of expression of various viral genes during lytic infection are essentially the same (8, 9, 10, 11).

Although some serotypes are unable to induce tumors in vivo, all the human adenoviruses can induce morphological transformation of rat or hamster cells in culture (12-16). The transformed cells can be established as cell lines with unlimited ability to divide. Many of the transformed hamster cell lines can induce tumors when injected into newborn hamsters whether they are transformed by the non-oncogenic group C or the highly oncogenic group A (17, 18). However, rat cells transformed by Ad2 or Ad5 fail to induce tumors while those transformed by Ad12 are highly oncogenic even in immunocompetent animals, reflecting their respective oncogenic potential in hamsters injected with infectious virus (19, 20). These studies indicate that although cells may be morphologically transformed and acquire the ability to divide indefinitely, they may remain non-tumorigenic when tested in syngeneic hosts. Thus, studies of morphological transformation of mammalian cells, in particular of rat cells, by various adenoviruses can contribute to the understanding of the mechanism of conversion of normal cells to transformed cells and how these cells may overcome the immune surveillance of the host animals to give rise to tumors.

VIRAL GENES INVOLVED IN THE MAINTENANCE OF THE TRANSFORMED PHENOTYPE

As stated earlier, cells transformed by adenoviruses can have very different properties, ranging from the ability to divide indefinitely in culture (immortalization), to the acquisition of tumor formation in immunocompetent animals. This wide range of phenotypes requires a definition of a transformed cell. In this article, transformed cells are defined as mammalian cells that have an altered phenotype resulting from infection with virus or from transfection with viral DNA. These cells generally have lower serum requirement for growth than "normal" cells, and usually form dense foci when cultured on plastic.

It is now widely accepted that transformation by adenoviruses requires the insertion of viral DNA into the cellular chromosome. Although not the entire viral genome is present in the majority of the transformed cells, the left 14% of the viral sequence is always found in cells transformed by all serotypes (21-24). These observations suggest that viral genes located at the left extremity are required and sufficient to maintain the transformed phenotype. Further supporting evidence is provided by transfection experiments. When different viral fragments generated by restriction endonucleases were used for transfection, it was found that only the left hand end of the adenovirus genome has transforming activity. This has been shown for members from the highly oncogenic group (20,25), the weakly oncogenic group (26,27) as well as the non-oncogenic group (28-30). The transforming E1 region is located between 0 to approximately 11% and is further subdivided into E1A and E1B. E1 is one of five early regions expressed before viral DNA replication during lytic infection (31). Cells transformed by the E1 region of the viral DNA have the same morphology, growth properties and tumorigenicity as those transformed by virus (20, 25, 29, 32).

Viral DNA fragments smaller than E1 have also been shown to be able to transform cells. However, these cells appear to have properties different from those transformed by virus (see later section). Viral information is generally expressed in the transformed cells since both virus-specific RNA (33,34) and virus-specific antigens (35) can be detected. These studies indicate that continued expression of viral genes is necessary for the maintenance of the transformed phenotype.

Cells transformed by adenoviruses usually contain more than the E1 region of the viral DNA. The size of the viral DNA segment integrated varies greatly from cell line to cell line. As mentioned previously, the majority of the cell lines transformed by the non-oncogenic group contains only part of the viral genome. Exceptions are found in some cell lines transformed by certain host range mutants of Ad5 or Ad5-ts125, where it was found that the entire genome was integrated (36,37). Furthermore, in some transformed cells, the left and right hand end segments of the viral DNA are joined, sometimes in head-to-tail fashion (38). Circles of Ad5 DNA found in infected cells (39) have been postulated as possible intermediates in the integration of viral DNA during transformation (40). Ad12 tumor and transformed cells frequently contain the whole viral genome, though in some tumor cells, less than half of the Ad12 genome was found (41-44). Analysis of the integration pattern of the viral DNA sequences seems to indicate that there is no specificity with respect to the integration sites in the cellular DNA (45-49). These studies indicate that the integration sites as well as the viral genes other than those in E1 play no important role in the maintenance of the transformed phenotype.

POLYPEPTIDES ENCODED IN REGION E1

The polypeptides encoded in region E1 have been intensively studied for members from the non-oncogenic group (serotypes 2 and 5) and to a lesser extent for serotype 12, a member of the highly oncogenic group. Two approaches are most commonly used to identify these proteins. One is to label either the infected or transformed cells with radioactive amino acid or ^{32}P (for phosphoproteins) and the cell extracts are immunoprecipitated with sera from animals bearing adenovirus induced tumors or monoclonal antibodies. Antibodies directed against synthetic peptides which have the same amino acid sequence as a specific region of the virus-specific protein can also be used for the immunoprecipitation studies. The precipitated polypeptides are then analyzed by polyacrylamide gel electrophoresis. In some cases, the polypeptides are further analyzed by two dimensional gel electrophoresis. Another approach is to select viral messenger RNA from specific regions of the viral genome by DNA-RNA hybridization and the eluted mRNA is then translated in vitro using either wheat germ or the rabbit reticulocyte system. The polypeptides thus obtained are then analyzed as mentioned before.

Three species of mRNA, 13S, 12S and 9S, are encoded in region E1A of Ad2 or Ad5. All three mRNAs have common 5' and 3' ends and differ only in the size of the introns (50-54). This region of the genome of Ad2 and Ad5 has been sequenced (55,56) and the predicted molecular weights of the 12S and 13S mRNAs are 26K and 32K respectively. These polypeptides would have the same N- and C-terminal amino acid sequence and differ by only 46 amino acids. Cell free translation of the 12S mRNA resulted in two polypeptides, 35-47K while translation of the 13S mRNA gave another two polypeptides, 41-52K (57-59). In addition, a 28K was detected from a 9S mRNA (60). More recently, Branton and his colleagues studied E1A proteins by in vivo labeling with ^{32}P of KB cells infected with Ad5 followed by immunoprecipitation with antiserum against synthetic peptides corresponding to either the C- or N-termini. The precipitated polypeptides were analyzed by two dimensional gel electrophoresis. They have found that E1A region of Ad5 induced four major species (45-52K) and four minor ones (25-37.5K) (61-64). Studies using Ad5 mutants that fail to make either the 13S or the 12S messenger RNA provided evidence that two major (52 and 48.5K) and two minor (37.5 and 29K) polypeptides were specified by the 13S mRNA. The other two major (50K and 45K) and two minor polypeptides (35K and 25K) are the translation products of the 12S mRNA (62,63). Using a monoclonal antibody against E1A proteins, Winberg and Shenk detected a polypeptide of about 25K, presumably that encoded by the 9S mRNA (65).

The E1B region of Ad2 (and Ad5) is transcribed into two major mRNAs, 22S and 13S, in addition to a 9S mRNA which encodes the structural protein IX. The larger species can be translated from two different reading frames, giving rise to a 58K or a 19K polypeptide. The 13S mRNA encodes only the 19K (55,56). Using antibodies to precipitate infected cell extracts or by in vitro translation of mRNA, various investigators have confirmed that these are the E1B encoded proteins (57,61,62,66-69). A third 18K (20K) polypeptide which has the same N-terminal and C-terminal portions as those of the 58K polypeptide has been detected (69,70).

Polypeptides encoded by the E1 region of the highly oncogenic Ad12 have not been studied as intensively as those from the non-oncogenic group (71-75). Consequently, there is a great deal of uncertainty with respect to the size and species of the proteins. We have studied the polypeptides encoded in E1A of Ad12 by in vitro translation of E1A specific mRNA. Total cytoplasmic RNA was extracted from Ad12-infected

KB cells at 24 hours after infection and was hybridized to a plasmid containing the left 4% of the Ad12 genome. The RNA was eluted and translated in a rabbit reticulocyte system, with ^{35}S-methionine as label. The polypeptides were immunoprecipitated with antiserum C1, obtained from rats bearing tumors induced by cells transformed by the left 16% of the Ad12 genome, and were then analyzed by polyacrylamide gel electrophoresis. The autoradiogram in Figure 1 shows only two polypeptides, a major one at about 55K and a minor species at about 23K. These polypeptides were further analyzed, without prior immunoprecipitation, by two-dimensional gel electrophoresis. The larger polypeptide can be resolved into two closely related species, with pI value of about 5, and the pI value of the minor species is 5.1 (Figure 1).

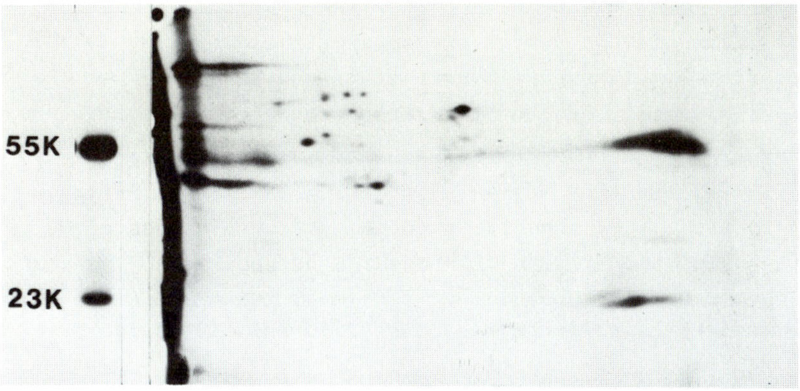

FIGURE 1. Polypeptides encoded in E1A of Ad12. Cytoplasmic RNA was extracted from infected cells at 24 hours after infection, hybridized to a cloned fragment (the left 4% of Ad12), eluted and translated in vitro, with ^{35}S-methionine as label. The polypeptides were analyzed by polyacrylamide gel electrophoresis after immunoprecipitation with antiserum C1 or by two-dimensional gel electrophoresis without immunoprecipitation.

According to the DNA sequence data region E1A of Ad12 encodes two related polypeptides with molecular weights 26K and 29.7K. It seems likely that the polypeptide with an apparent molecular weight of 55K represents the two predicted polypeptides which were resolved by the 2-D gel. The small amount of 23K polypeptide may represent the translation product of a minor mRNA species. Apparent molecular weights of 38K and 42K, and 37K and 39K found in infected cells have been assigned to this region (71,72,76). More recently, Green and his colleagues, by in vivo labeling and immunoprecipitation with antiserum against synthetic peptide targeted to the common C-terminus of Ad12 antigens encoded by the E1A 13S and 12S mRNA, found only two polypeptides, 45K and 47K (77). They further demonstrated that these are phosphoproteins with pI values of 5.4 and 5.2, respectively. These results are similar to those shown in Figure 1. However, Esche and his colleagues have reported that in vitro translation of Ad12 E1A-specific mRNA produce at least eight polypeptides ranging from 22K to 38K (78). The differences in the apparent molecular weights reported by different investigators may be due to different gel systems used, but the reason for the differences in the number of polypeptides detected is not clear. It should be noted that the apparent molecular weights reported by all investigators are larger than the predicted values, a phenomenon which is the same as that seen with the E1A polypeptides from the non-oncogenic group.

To determine the polypeptides encoded in the E1B region of Ad12 we again employed the technique of hybrid-selection of mRNA and in vitro translation. Plasmids containing Ad12 DNA sequences spanning 4.5 to 6.7% or 6.7 to 10.5% were used to select E1B specific mRNA. The translation products after immunoprecipitation with antiserum C1 were analyzed by polyacrylamide gel electrophoresis. The autoradiogram in Figure 2 shows that three major species (55K, 19K, and 17K) and two minor species (15K and 14K) were encoded by mRNAs from the E1B region of Ad12. The autoradiogram was underexposed so as to resolve the smaller polypeptides. Consequently,

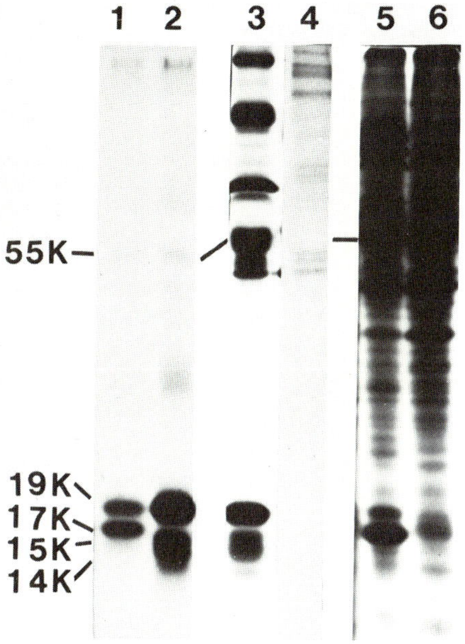

FIGURE 2. Polypeptides encoded in E1B of Ad12. Cytoplasmic RNA was extracted from cells at 24 hours after infection and hybrid-selected for translation in vitro, with ^{35}S-methionine as label or the infected cells were labeled in vivo with ^{35}S-methionine. The polypeptides were analyzed by polyacrylamide gel electrophoresis. Lanes 1 and 2 show the polypeptides resulting from in vitro translation of mRNA selected by cloned fragments of Ad12 DNA, 6.7-10.5% (lane 1) and 4.5-6.7% (lane 2), immunoprecipitated by antiserum C1 before analysis. Lanes 3 and 5 are data from in vivo labeling of infected cells, immunoprecipitated with antiserum C1 (lane 3) or without immunoprecipitation before analysis (lane 5). Lanes 4 and 6 show polypeptides from non-infected KB cells, with immunoprecipitation with antiserum C1 (lane 4) and without immunoprecipitation (lane 6).

the 55K polypeptide is not apparent in the Figure. The 55K polypeptide is believed to be the translation product of the 22S E1B mRNA and the 19K resulted from the translation of both the 22S and 13S mRNA (55,71). It is not clear what mRNA is responsible for the 17K polypeptide. However, it cannot be translated from either one of the two major species of mRNA since the mRNA selected by the DNA fragment between 4.5 to 6.7% did not produce the 17K polypeptide (lane 2). It is unlikely to be the structural protein IX because we could detect from translation of E1B messenger RNA a 10K polypeptide, which comigrated with a virion protein (unpublished data). Furthermore, mRNA selected by a plasmid containing the left 8% of the Ad12 genome produced the 17K polypeptide (unpublished data) and this region of the DNA is outside the coding sequence for polypeptide IX (55). This 17K polypeptide may be initiated with the same ATG as the 19K polypeptide but lacks the majority of the coding sequence of the latter. Instead, it may share amino acid sequence of the C-terminus of the 55K polypeptide. The two minor polypeptides originate from the region between 4.5-6.7% of the genome. All five polypeptides can also be detected by in vivo labeling of Ad12 infected cells with ^{35}S-methionine followed by polyacrylamide gel electrophoresis analysis. The 58K, 19K, 15K and the 14K are clearly visible in lane 3 while the 17K is best demonstrated in lane 5. These results provide evidence that the E1B region of Ad12 encodes 5 proteins.

Katz et al. have reported that an 11K polypeptide is the product of an l-strand transcript, which spans the border of E1A and E1B of Ad2 (79). We carried out experiments to determine if any of the 5 polypeptides from this region of Ad12 may be originated from l-strand transcripts. Either l- or r-strand DNA spanning the region from 4.5-6.7% and from 6.7-10.5% of Ad12 was used to hybrid select viral mRNA and then translated in vitro. Figure 3 illustrates the strategy for obtaining either the l- or r-strand of viral DNA spanning the viral sequence between 6.7 to 10.5%. This

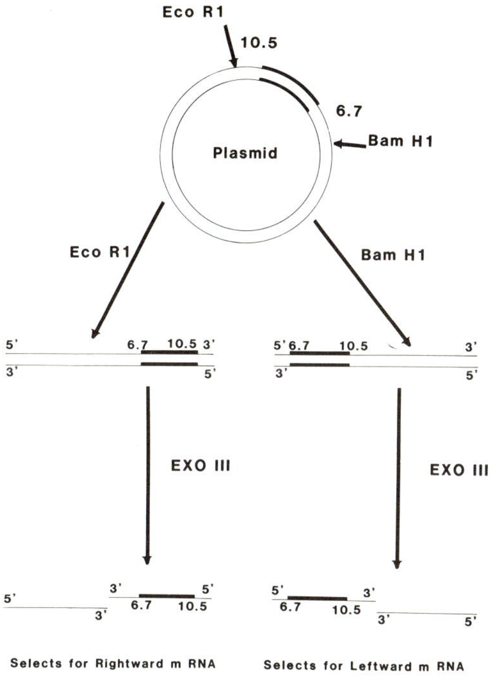

FIGURE 3. Strategy for the separation of strands for selection of rightward and leftward transcripts encoded in region between 6.7-10.5% of the Ad12 genome. See text for explanation.

region of the viral DNA was cloned into the Hind III site of PBR322 in the orientation as shown. The plasmid was then restricted with either EcoR1 or BamH1. The linearized DNA was then digested with Exo III, and the resulting DNA was used for selection of mRNA for translation. Similar strategy was used to isolate r- and l-strand transcripts from 4.5-6.7% of the viral genome for isolation of mRNA.

Results as shown in Figure 4 clearly demonstrate that all 5 polypeptides are the products of r-strand transcripts. Thus, 14K and 15K proteins may be related to the 11.6K

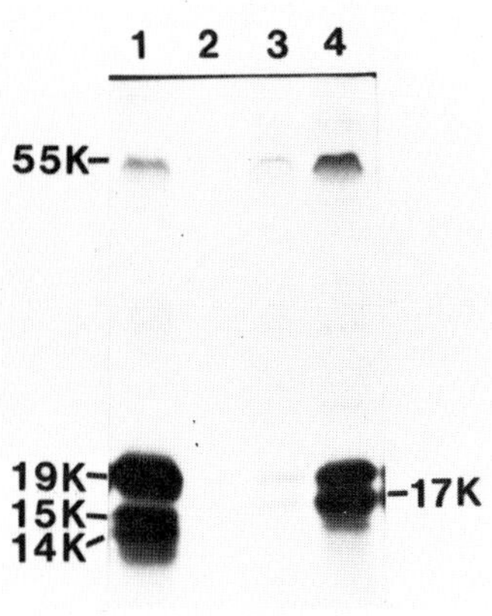

FIGURE 4. Polypeptides encoded by different strands of E1B DNA of Ad12. Cytoplasmic RNA from cells at 24 hours after infection was hybridized to separated strands of DNA from E1B region of Ad12. The eluted RNA was translated in vitro using ^{35}S-methionine as label, and the products were analyzed by polyacrylamide gel electrophoresis, after immunoprecipitated with antiserum C1. The RNA was selected by l-strand of DNA from 4.5-6.7% (lane 1), by r-strand of DNA from 4.5-6.7% (lane 2), by r-strand of DNA from 6.7-10.5% (lane 3) and by l-strand DNA from 6.7-10.5% (lane 4).

polypeptide that can theoretically be translated from the 13S mRNA (80).

Comparison of proteins encoded by E1A of Ad2 (Ad5) and those of Ad12 suggest that all of them are phosphorylated. However, there appears a genuine difference in the number of protein species produced. Each of the mRNAs from Ad5 can give rise to 4 polypeptides, while those from Ad12 encode one polypeptide each with the exception of the results obtained by Esche (78). If post-translational modifications are responsible for the generation of multiple species of Ad5 E1A polypeptides, as suggested, the kind of modifications may differ for the proteins from the two serotypes. The biological significance of these molecular differences between the two serotypes is not known.

The E1B region of both Ad2 (Ad5) and Ad12 codes for the same major polypeptides, namely, the 58K (55K for Ad12) and the 19K. However, differences exist between the two serotypes. Ad2 encodes two other polypeptides (16K and 18K) which have the same N-terminal sequence as the 58K polypeptide. Although the Ad12 E1B region encodes a polypeptide of similar size (17K), it does not appear to contain the N-terminus of the 55K polypeptide (see above). This conclusion was confirmed by our observation that the antibody directed against synthetic peptide, which has the same amino acid sequence as the N-terminus of the E1B 58K protein, did not specifically precipitate the 17K (unpublished results). It is not clear whether the differences in these E1B polypeptides can explain in part the difference in the tumorigenic potential of the two different serotypes.

FUNCTIONS OF E1 PROTEINS

Much information concerning the number, size, and coding regions of the E1 proteins of adenoviruses has been accumulated (see previous section); however, the specific biological and biochemical functions of the various polypeptides are still unclear. The majority of information has been derived from cells infected by E1 mutants of the non-oncogenic group. Ad5 dl312 which lacks essentially all of E1A is replication defective in human HeLa and KB cells (81). An Ad2 mutant, pm975, that fails to produce the 12S mRNA, has very little effect on its replicative cycle except that it shows a reduction in virus yield in non-growing human WI-38 cells (82). This virus also fails to stimulate DNA synthesis in growth arrested rodent cells. On the other hand, mutants having defects in the 13S mRNA region are generally defective in virus replication (83). However, none of the E1A proteins is absolutely required for the lytic cycle since Ad5 dl312 could give comparable virus yield as wild type when HeLa cells were infected at high multiplicities (84).

Much evidence has been accumulated demonstrating that the translation products of the E1A 13S mRNA and not those from the 12S mRNA, can regulate the expression of other early viral genes, the E1B, E2, E3, and E4. However, these early regions are not all affected to the same extent (85,86). For example, the synthesis of mRNA from E1B and E4 is less severely affected than those from E2 or from E3 (87). It has been proposed that the E1A proteins act in trans to stimulate transcription (88) or to stabilize the mRNA from the non E1 early regions (89). Inhibition of protein synthesis by cycloheximide allows the expression of early genes in cells infected with mutants with defects in the 13S mRNA. These studies led to the suggestion that E1A polypeptides can inactivate a cellular protein which inhibits transcription of early viral genes (88). It has also been suggested that the E1A proteins affect the structure of the DNA or the transcription complex to increase the transcriptional activities of early viral genes (90). In addition, E1A gene products can stimulate cellular gene expression, notably the 70K heat shock and the b-tubuline proteins (91,92). Ad12 E1A protein from the 13S mRNA, but not those from the non-oncogenic adenoviruses, can repress the expression of the class I transplantation antigen (93). Thus, the function of the E1 proteins can alter the activity of both viral and cellular genes, perhaps, through different mechanisms.

The E1B proteins, like the E1A polypeptides, are not absolutely required for productive infection since Ad5 dl313, lacking most of E1B region, can produce virus yield comparable to that produced by wild type virus, when the infection was carried out at high multiplicities (84). Not all the E1B polypeptides are required to complete the infectious cycle, even at low multiplicities of infection. This is shown by a number of E1B mutants such as the cyt mutants of Ad12 (94), cyt106 of Ad2 (95) and the large plaque mutants of Ad2 (96, 97). However, the cyt mutants in general produce reduced amounts of virus (96,98, unpublished data).

In order to provide possible explanations for the low virus yield, we studied DNA degradation in cells infected by the cyt mutants (98,99). KB cells were infected with either cyt70, wild type virus, or both, and were labeled with ^{3}H-thymidine between 13 and 40 hours after infection. The labeled DNA was then analyzed by rate zonal sedimentation in sucrose gradients. Results presented in Figure 5 show that the labeled DNA from mutant infected cells is extensively degraded, in contrast to that from the cells

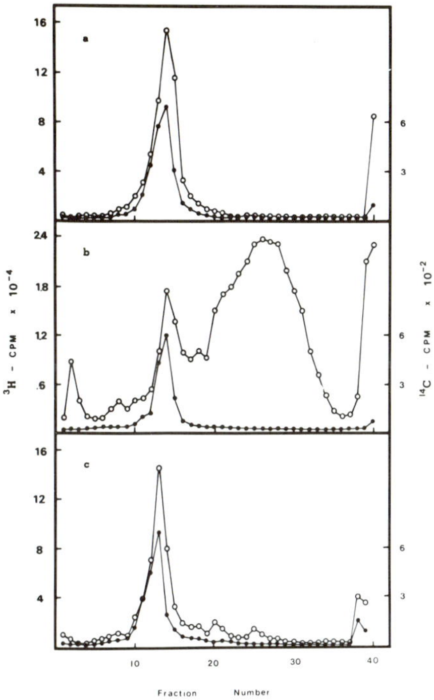

FIGURE 5. Sedimentation profiles of intracellular DNA in KB cells infected with Ad12 parental strain, the cyt mutant (H12cyt70), and a mixture of the parental strain and the cyt mutant in a 5 to 20% alkaline sucrose gradient. Sedimentation is from right to left. (a) Cells infected with the parental strain (o); (^{14}C) Ad2 marker DNA (•). (b) Cells infected with the cyt mutant (o); (^{14}C) Ad2 marker DNA (•). (c) Cells co-infected with the parental strain and the cyt mutant (o); (^{14}C) Ad2 marker DNA (•).

infected by wild type virus. Co-infection of cyt mutant with wild type virus prevents DNA degradation (Panel C) suggesting that a diffusible substance can control the production of or inhibit the action of a nuclease (100). Nucleic acid hybridization studies have shown that both cellular and viral DNA were degraded (99), thus providing an explanation for the low yield of virus in cells infected by cyt mutants. Complementation studies with Ad5 mutants, that have known molecular defects, such as dl312, hrl, and hr6, led us to propose that the ElB 19K polypeptide is probably responsible for the cyt phenotype (100). Subsequently, we showed that a cyt mutant synthesizes no detectable 19K protein (101). Many mutants since then have been isolated from the non-oncogenic group, and studies on these mutants show indeed that the 19K polypeptide is required for the protection of DNA from degradation (95,96, 102). The 19K polypeptide may also have regulatory capacity since KB cells infected with 19K minus mutants such as Ad2 dl250 and cyt68 of Ad12 overproduce ElA RNA at late times after infection (Caussy and Mak, unpublished data). Thus, it is possible that this 19K protein may regulate the expression of an endonuclease gene. It is of interest to find that another mutant (cyt61) is dominant in inducing DNA degradation. The dominant characteristic indicates that the endonuclease or its inducer is virus coded. It appears that the mutation in cyt61 probably lies outside the ElB region and prevents the normal ElB factor (19K?) from controlling the expression or activity of this endonuclease or its inducer (100). In addition to the function(s) in lytically infected cells, the 19K protein is also involved in transformation (103,104, also see later section).

The ElB 58K protein appears to control the transcription of late viral mRNA and is involved in shutting off host protein synthesis (102,105,106). It is of interest to note that it is not required for viral DNA synthesis during lytic infection by the non-oncogenic group while the corresponding protein from the highly oncogenic Ad12 is involved in viral DNA synthesis (106,107). No function has been assigned to other ElB polypeptides, such as the Ad12 17K, 15K and 14K and the Ad2 20K.

Protein kinase activity was detected when extracts from KB cells infected with either Ad5 or Ad12 were immunoprecipitated with antisera from tumor-bearing animals (108-110). Subsequent experiments showed that the kinase activity was found to be associated with immunoprecipitates when sera specific for either E1A or E1B polypeptides were used (111). Raska et al. (112) also found protein kinase activity in Ad12 infected cells. It is not clear from these experiments whether activity is an intrinsic property of the viral proteins or due to the binding of cellular enzymes to the viral antigens.

MUTANTS DEFECTIVE IN TRANSFORMATION

The E1 region of adenoviruses has been shown to be important in transformation and regulation of both viral and cellular genes. Consequently, this region has been intensively studied and many mutants have been constructed and isolated. Most of the E1 mutants belong to the non-oncogenic group Ad5 and Ad2. This perhaps is due to the availability of a very useful cell line, the 293 human embryonic kidney cells which contain and express the E1 region of Ad5 (113). This cell line supports the replication of mutants that are defective in E1 functions and unable to replicate in human HeLa or KB cells. The recent advancement in recombinant DNA technology also allows one to construct mutants with specific changes in the DNA (site-directed mutagenesis). The altered DNA sequence is then rescued back to the virus and their biological functions studied.

E1A mutants

A series of deletion mutants of Ad5 was isolated by Jones and Shenk (81). One of these, the Ad5 dl312, with most of E1A region deleted is unable to transform rat embryo or rat embryo brain cells or primary baby rat kidney cells even with high virion input (81, see also Figure 6). These results unequivocally established that E1A is absolutely required for immortalization or for the maintenance of transformed phenotype.

Since E1A region of the adenoviruses encodes two early mRNAs, the 13S and the 12S, mutants have been constructed

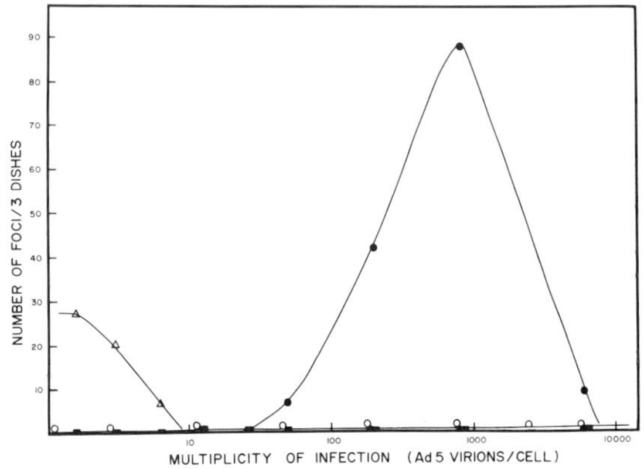

FIGURE 6. Transformation of rat kidney cells by an Ad5 host range group II mutant and deletion mutants at different multiplicities of infection. Symbols: Δ , Ad5 wild type; ● Ad5 dl313, ▬ , Ad5 host range mutant (hr6); ○ , Ad5 dl312.

so that only one of the messages is affected. One of such mutants is the Ad2 pm975 (114). It was elegantly constructed by introducing a change from a T to G at nucleotide position 975 in ElA. This change prevents the splicing of the 12S mRNA without affecting the amino acid sequence specified by the 13S mRNA because of the degeneracy of the genetic code. This mutant was later found to transform primary rat kidney cells as well as an established rat cell line (CREF) with 4 to 5 fold reduction compared to wild type virus. The transformed foci are generally smaller and are defective in anchorage independent growth. These results indicate that the proteins specified by the ElA 12S message may play a role in the establishment of transformation and are required for the "fully transformed" phenotype. (115).

The functions of the proteins encoded by the ElA 13S mRNA were studied by the construction of a series of mutants that have defects in this mRNA. H5in500 (87), which has an insertion in the region unique to the 13S mRNA, resulted in a truncated polypeptide while the translational products of the 12S mRNA are not affected. Mutant hr440 constructed by Solonick and Anderson has a mutation which prevents the proper splicing of the 12S mRNA (116). Both of these mutants were shown to be defective in transformation of primary cells in culture, and E1B expression was not affected in the lytically infected cells (87,116). These studies indicate that it is the E1A defect that is responsible for the deficiency in transformation. Furthermore, the fact that H5in500 is defective only in the 13S mRNA suggests that a functional product from this message plays a very important role in immortalization or maintenance of the transformed state.

Ad5 host range mutants, which grow in 293 cells but not in HeLa cells, were first isolated by Graham et al. (83). The group I mutants (hr1 and hr2) were shown to be transformation negative for rat embryo or rat embryo brain cells, but these mutants transformed baby rat kidney cells more efficiently than wild type virus. However, cell lines cannot be established with single foci unless mixed populations from many transformants were used (83). One mutant, hr1, was shown subsequently to have a single base deletion affecting the 13S mRNA. Consequently, the mutant produces a truncated polypeptide (28K) instead of the normal 51K protein (117). Because the mutation is located in the intron of the 13S message, it is expected that the 12S mRNA is not affected. The fact that hr1 transforms cells more efficiently than wild type at 37°C implies that not the entire protein of the 13S product is required for immortalization. However, when transformation was carried out at 32.5°C and 35.5°C both hr1 and hr2 were unable to transform. Two additional mutants from E1 region (hr11 and hr12), isolated by Ho et al., show the same cold sensitivity in transformation (118). Since the host range mutants were isolated by chemical mutagenesis of virus stock, the possibility that the phenotype of hr1 is due to mutation in the unsequenced region cannot be ruled out. A group of better defined mutants are H5dl101 and dl106, which were constructed to have mutations in the introns of the

12S and 9S mRNA and leaving their products unaltered (119). The predicted product of altered 13S mRNA of dl101 and dl106 are 30K and 24K, respectively. Both of these mutants were found to be cold sensitive in transformation using cell line CREF (clone rat embryo fibroblast). A similar mutant, dl105, with a predicted polypeptide of 47K instead of the normal 51K was found to be transformation negative at all temperatures (119). Although the cells transformed by these cold sensitive mutants are fibroblastic instead of the usual epitheloid, nevertheless, these studies suggested that the carboxyl portion of the 13S product up to about 27K was not required for conditional transformation functions.

Montell et al. constructed an Ad5 deletion mutant dl1500 that synthesizes only the 12S mRNA and found that this mutant was cold sensitive in transformation (115). This mutant, like the hr1, transformed both primary rat embryo cells and CREF more efficiently than the wild type virus. This study indicates that the process of transformation by the protein specified by the 12S message is cold sensitive. It was suggested that the absence or an altered form of the 51K protein decreases the expression of other viral functions toxic to rodent cells, thus resulting in higher transformation frequency.

The number of E1A mutants belonging to the highly oncogenic group is very limited. However, the results from such studies are similar to those obtained with the non-oncogenic Ad2 and Ad5 (120).

E1B mutants

Group II host range mutants of Ad5 (hr6 and hr7) with defects mapped in region E1B are unable to transform primary rat embryo brain or baby rat kidney cells (83,121). However, DNA extracted from the group II mutant virions can transform cells as efficiently as wild type viral DNA, and these cells express the "fully transformed" phenotype (122). The 58K protein encoded by the 22S message was not detected in HeLa cells infected by the mutants, but the 19K polypeptide encoded by the 13S message appear not to be affected by the mutation (63,66). DNA sequencing data of hr6 reveal that the mutation is located in the region unique to the 22S message, predicting a normal 19K polypeptide as suggested in an earlier study (123). Although additional changes in other parts of the E1B region cannot be ruled out, nevertheless, these studies led to the suggestion that the 58K polypeptide was involved in integration of viral DNA when viruses were used to transform, but not during transfection when relatively large amounts of viral DNA was used (124). Recently, Babiss et al. constructed a series of Ad5 mutants with deletions of varying sizes within the coding region of the 58K polypeptide (125), and found that the transformation frequency decreased as the size of the deletion increased. Ad5 dl313, a mutant with nearly all of E1B deleted, does not transform rat embryo or rat embryo brain cells under normal conditions (81). However, dl313 can transform primary baby rat kidney cells at high input multiplicity and the data are shown in Figure 6. Transformation was also observed by this virus with a rat cell line 3Y1 (126). These results then indicate that transformation can occur without E1B (103, Figure 6). In contrast, hr6, an E1B mutant, is transformation defective for primary baby rat kidney cells even at high multiplicities (Figure 6). One possible explanation for the difference in the transformation by the two viruses is that when sufficient amounts of viral DNA are available at high multiplicity of infection (dl313), the function of the 58K (integration?) is no longer required, just as in the case of transfection with viral DNA. However, the additional information in region E1B of hr6 may induce toxic substances in the infected cells either directly or indirectly, thereby killing the transformants. Consistent with this possibility is the observation that hr6 can replicate DNA in non-permissive human HeLa cells while Ad5 dl313 cannot unless cells are infected at high multiplicities (81,106).

Another very interesting group of E1B mutants (cyt mutants) was originally isolated by Takemori from highly oncogenic Ad12 (94). These mutants are non-tumorigenic in hamsters and produce large plaques on human embryonic kidney cells, in contrast to the fuzzy-edged plaques produced by the wild type virus. Furthermore, this group of mutants induces extensive DNA degradation in infected cells (see previous section). We have studied transformation of primary baby rat kidney cells by five of these mutants, two of which, cyt52 and cyt70, are spontaneous mutants, and cyt61, cyt62 and cyt68 are UV induced mutants. The transformation data presented in Figure 7 show that all of the mutants except cyt61 are extremely inefficient in transformation; however, foci can be obtained at high input multiplicities and permanent cell lines can be established. Cyt61 possesses a normal transformation capacity in spite of its cytocidal nature (99). Thus, defective transforming activity of the cyt mutants is not entirely related to the enhanced cell killing by these mutants.

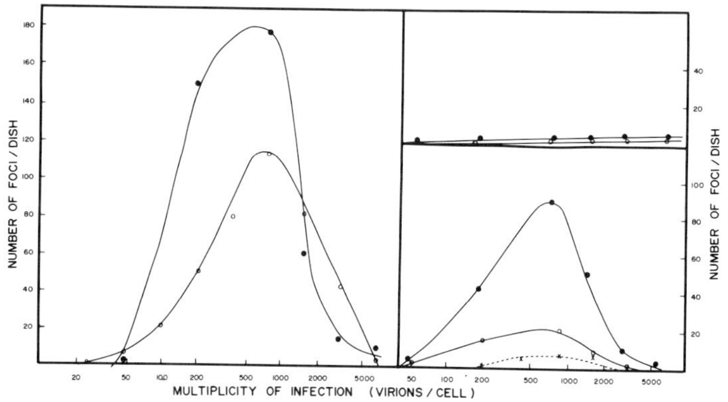

FIGURE 7. Transformation of rat kidney cells by Ad12, cyt mutants, and a revertant at different multiplicities of infection. (a) ○, wild type (strain 1131); ●, a revertant of H12 cyt70. (b) Spontaneous cyt mutants. ○, H12 cyt52; ●, H12 cyt70. (c) UV-induced cyt mutants. ●, H12 cyt61; ×, H12 cyt68; ○, H12 cyt62.

Since E1B functions can affect the phenotype of the transformants, we studied the tumorigenicity of the rare transformants induced by the majority of the cyt mutants. Usually 10^6 cells were injected subcutaneously into 3- to 4-week old immunocompetent rats.

TABLE 1. Tumorigenicity of transformed cell lines (a)

Cell line	Transformed by:	Fractions of rats with tumors	Latent period[b] (wk)
P-C1	Wild type	11/11	3.5
P-C2	Wild type	11/11	4
P-C3	Wild type	7/8	3
P-C4	Wild type	9/15	4
61-C1	cyt61	16/21	4
61-C2	cyt61	9/10	4
62-C1	cyt62	1/20	
62-C2	cyt62	3/11	11
62-C4	cyt62	1/13	9
62-C5	cyt62	0/13	
62-C4	cyt62 with $5x10^6$ cells	4/19	6 months
68-C1	cyt68	20/22	7
68-C3	cyt68	9/15	10
68-C4	cyt68	11/15	9
70-C1	cyt70	15/20	
70-C3	cyt70	10/19	7-9
70-C4	cyt70	15/20	

[a]Cells (10^6 or as indicated) were injected subcutaneously into weanling rats (age 3 to 4 weeks) and observed weekly.

[b]The latent period is defined as the time required for the appearance of the tumor in 50% of the maximum number of tumor-bearing animals.

Data in Table 1 show that most of the transformants are highly oncogenic, with the exception of those transformed by cyt62. The simplest interpretation is that the cyt mutants are not defective in the maintenance of the tumorigenic state but they transform primary cells with very low efficiency. Consistent with this idea is the observation that coinfection of a cyt mutant (cyt70) with the weakly oncogenic Ad7 increased the number of cyt70-transformants by 15 to 30 times more than expected if cyt70 and Ad7 were used to transform cells independently (103). These transformants possess the characteristic morphology of Ad12-transformed cells (different from Ad7-transformed cells), express Ad12 mRNA and antigens, and induce tumors in immunocompetent rats (103). It seems that Ad7 can supply a factor, which is missing in the cyt mutant, for the establishment of transformants.

To determine the polypeptides responsible for the defect in the cyt mutants, cytoplasmic RNA of infected KB cells was hybridized to cloned fragments of either E1A or E1B of Ad12 DNA, and the hybrid-selected RNA was translated in vitro using a reticulocyte system. The translation products were analyzed by polyacrylamide gel electrophoresis. Results show that the polypeptides encoded in the E1A region of the cyt mutants have the same apparent molecular weights as those by the wild type virus (Figure 8). This suggests that the cyt mutants may not have defects in E1A. Figure 9 shows the polypeptides of the E1B proteins from cells infected by different mutants and those from the wild type virus-infected cells. Translation of RNA from cells infected by cyt mutants or by wild type virus produced about the same amounts of 55K and 17K polypeptides. However, the amount of 19K protein induced by the cyt mutants varied from none by cyt68 to small amounts by cyt70 to the same amount by cyt61 as that induced by the wild type virus. Preliminary DNA sequencing data on cyt68 indicate that there is a deletion of about 100 bp in the coding region of the E1B 19K near the N-terminus (Schaller and Mak, unpublished data), predicting the absence of the 19K polypeptide. Thus, the E1B 19K polypeptide of Ad12 may be involved in the

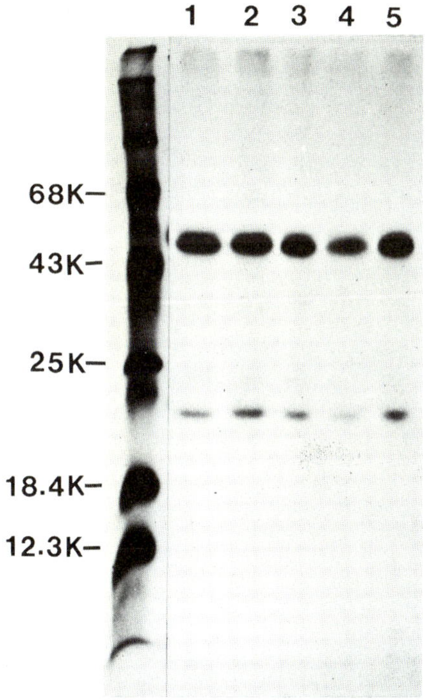

FIGURE 8. Polypeptides encoded by E1A of Ad12 and cyt mutants. Cytoplasmic RNA was extracted from infected cells at 24 h after infection, hybridized to a cloned fragment of Ad12 (the left 4%), eluted, and translated in vitro, with ^{35}S-methionine as label. The polypeptides were immunoprecipitated by antiserum C1 and analyzed by SDS-polyacrylamide gel electrophoresis. RNA was from cells infected with Ad12(lane 1), cyt61(lane 2), cyt62(lane 3), cyt68(lane 4), and cyt70(lane 5).

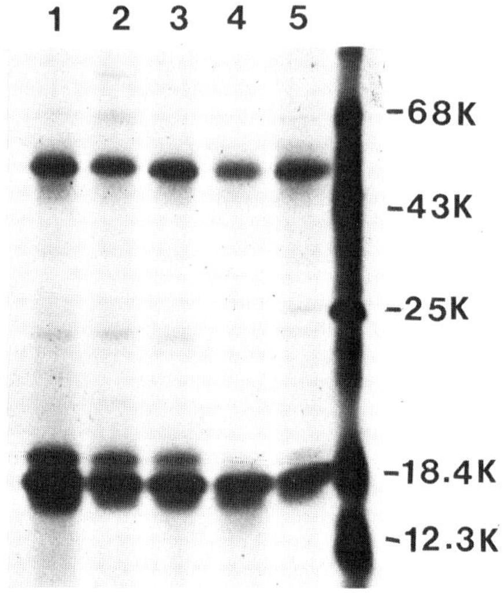

FIGURE 9. Autoradiograms of ^{35}S-labeled polypeptides synthesized in a cell-free protein synthesizing system and analyzed by polyacrylamide gel electrophoresis. Total cytoplasmic RNA was extracted from infected KB cells at 24 h after infection, hybridized to a cloned viral DNA fragment (6.7 to 17.8%), eluted, and translated in vitro. The polypeptides were immunoprecipitated with antiserum C1 and analyzed. Cells were infected with Ad12(lane 1), cyt61 (lane 2), cyt62(lane 3), cyt68(lane 4), and cyt70(lane 5).

efficiency of transformation and protection of DNA degradation, but not in the maintenance of the tumorigenic state since cyt68 transformed cell lines are highly tumorigenic (101). Mutations introduced into the left hand portion of the E1B region of Ad12 also resulted in defective transformation function (127).

Recently, a number of mutants from the non-oncogenic Ad2 and Ad5 have been constructed. Ad2 dl250 (104), Ad5 dl111 and Ad5 dl112 (125) are mutants with mutations affecting only the E1B 19K and leaving the 58K undisturbed. Like the cyt mutants of Ad12, they are severely defective in transformation of either established rat cell lines or primary rat kidney cells. Large plaque E1B mutants of Ad2, the lp3 and lp5, isolated by Chinnadurai are also transformation defective in rat cell line 3Y1 (97), and primary baby kidney cells (unpublished data). Ad2 lp3 has a single base substitution at the N-terminal portion of the 19K while Ad2 lp5 has a substitution at the N-terminal portion of the 19K in addition to a change in the termination codon of this polypeptide, giving rise to a polypeptide of larger size. These data indicate that the N-terminal portion of the 19K is important for the efficiency of transformation. Transfection by plasmid containing transposon Tn5 in Ad5 DNA also showed that the N-terminal portion of the 19K is important for transformation (128).

Up to now, the number of E1 mutants from the highly oncogenic group has been limited, partly due to lack of a human cell line that expresses the E1 genes of Ad12. We have recently established three human embryonic kidney cell lines transformed by the left 16% of the Ad12 genome. These cell lines do synthesize the E1 antigens and can complement the Ad12 E1B cyt mutants in virus yield and prevention of DNA degradation (unpublished results). The availability of these cell lines together with other transformed human cell lines (129,130) should facilitate the isolation of E1 mutants of the highly oncogenic group.

Mutants mapped outside of E1 region

Several mutants which are mapped outside of E1 region are also defective in transformation. The mutation of group N mutants of Ad5, ts36, 69 and 149, is located in the E2B region between 18.5 and 22 mu (131). These mutants are temperature sensitive for initiation of transformation, but not for the maintenance of the transformed phenotype (132), and the virus-coded DNA polymerase is likely to be temperature-sensitive (133). The reason for the requirement for the polymerase for transformation is not known; however, it has been suggested that the complex formed by the viral DNA, the precursor for the terminal protein and the polymerase may play a role in integration of the viral genome (134). Another group of mutants, such as Ad5 ts125, which is mapped in the E2A between 61 and 75 mu produces a non-functional 72K DNA binding protein. These mutants transform cells more efficiently than wild type at both permissive and non-permissive temperatures (135). The reason for this enhanced transformation is not clear, but these observations indicate that cellular transformation by DNA-containing viruses is a complex process and may be influenced by viral genes other than those absolutely necessary and may also depend on the host cells used for such experiments.

TRANSFORMATION BY VIRAL DNA FRAGMENTS

It is generally accepted that the region E1 of human adenoviruses is sufficient to cause neoplastic transformation of rodent cells in vitro. Temperature-sensitive mutants with defects in E2A and E2B can influence the transformation frequency, as mentioned in the previous section. In order to gain an understanding of the functions of E1 genes in the process of transformation without the influence of other viral genes, transformation with DNA fragments originated from region E1 may prove to be informative. This kind of study is made possible by the calcium phosphate technique developed by F. Graham (136,137).

We have compared the transformation efficiency of primary baby rat kidney cells by plasmids containing different amounts of E1 information of Ad12. Results indicate that complete E1 region of Ad12 is required for efficient transformation of these cells (Table 2). Inclusion of the complete E1B viral sequence in the transfecting plasmid can increase the efficiency 400 times as compared to DNA containing only the E1A sequence. It was also found that treatment of transfected cultures with a protein inhibitor, cycloheximide, increases the transformation frequency quite significantly, especially when the plasmids containing less than the E1 region were used. Although the mechanism of enhancement of transformation by cycloheximide is not known, the enhancement may be partly due to increased transcription of the transfecting DNA (unpublished data), and/or stabilization of the E1 proteins (62). These results clearly indicate that E1A of Ad12 DNA could effect morphological transformation even though the selection medium contained low calcium and horse serum. Normally, no foci can be obtained when primary cells are transfected with E1A of Ad5 if the same stringent selection medium is used (McKinnon and Graham, personal communications). However, morphological transformation of primary rat cells has also been achieved with the E1A fragment of the non-oncogenic Ad5 and with that from the weakly oncogenic Ad7, when the selection media contain normal concentrations of calcium and fetal bovine serum (27,138, McKinnon and Graham, personal communication). These transformants tend to have fibroblastic morphology. On the other hand, cells transformed by E1A of Ad12 have the same morphology as those transformed by complete E1 DNA (epitheloid). This is not surprising since the morphology of adenovirus transformants is controlled by region E1A (139).

It was noted that DNA fragments smaller than the E1 region of Ad12, up to 8%, have a very low transformation efficiency (Table 2, and unpublished data). However, the equivalent region of Ad5 had transforming efficiency similar to that of the complete E1 (30). Since these plasmids can theoretically express the E1B 19K and a truncated 58K (33K for Ad5 and 32K for Ad12), the difference in the transforming activity may be reflected in the function of the E1B 58K protein. It is possible that the functional domains required for transformation in the 58K protein may be different for the two types of viruses. Plasmids containing Ad5 E1 region with mutation at the TthI site (6.5%) had a drastic reduction in transformation frequency (140), consistent with the idea

TABLE 2. Transformation of primary rat kidney cells by plasmids containing different Ad12 sequences in the absence or presence of cycloheximide.

	No. of Foci (a)					
Plasmid DNA	PHA7 (0-4.5%)		PHAB8R (0-6.7%)		PHAB6 (0-16%)	
μg/dish	Without CH	With CH (b)	Without CH	With CH (b)	Without CH	With CH (b)
0.5					41	180
1	2.0	18	0	11	97	303
2	0	43	0	4	98	335
4	2.0	60	0	5	220	384
6	1.0	72	0	2	445	864
8	0	56	1	6	201	492

(a) The number represents the total number from 4 dishes

(b) CH (cycloheximide) was added at 20 μg/ml between 4-6 hours and removed at 17 hours after transfection

No foci were obtained in cells transfected with 8 μg/dish of PBR322 with or without cycloheximide treatment

TABLE 3. Tumorigenicity of cells transformed by the left PHA7 (4.5%) or PHAB8R (6.7%) of Ad12 DNA

Cell lines	Transformed by	Tumor incidence
HA7-c33	pHA7	0/10
HA7-c32	pHA7	0/10
HA7a-4a	pHA7	0/12
HA7a-c2	pHA7	0/6
HA7-c11	pHA7	0/7
HABa-c1	pHAB8R	0/11
HAB8a-c2	pHAB8R	8/10
HAB8a-c3	pHAB8R	2/10
HAB8a-20a	pHAB8R	0/9

that the region of the left 8% of Ad5 affects the efficiency of transformation. Although the mechanism is not known, it is possible that E1B or the left half of E1B increases the expression of E1A, facilitating the establishment of transformed phenotype (Caussy and Mak, unpublished data).

Although the inclusion of E1B region of adenoviruses can increase the efficiency of transformation, E1B itself has no transforming activity in primary rat cells (141). In order to overcome the requirement of E1A for E1B expression, plasmids were constructed in which the E1B was ligated to the SV40 promoter region and the transformation capacity was tested. No transformation activity was observed by these plasmids in a rat cell line, 3Y1, even though the E1B is known to be expressed in the transfected cells. It is clear from these experiments, E1A alone can cause morphological transformation. However, E1B contributes to both the efficiency of transformation and the phenotype of the transformed cells. The conclusion drawn from studies using DNA fragments is in agreement with the transformation studies using various viral mutants. It is of interest to note that the efficiency of transformation by E1A of Ad2 and of Ad12 can be increased when cotransfected with human ras gene, suggesting that some of the functions of E1B of adenoviruses can be substituted by other oncogenes (142,143, unpublished data).

ONCOGENICITY OF TRANSFORMED CELLS

The classification of adenoviruses into non-oncogenic, weakly oncogenic and highly oncogenic groups was based on tumor induction by injection of infectious virus into newborn hamsters. However, it was soon found that the non-oncogenic group C viruses (Ad2 and Ad5) can induce morphological transformation of hamster cells in vitro under certain conditions, such as with temperature sensitive mutants at the non-permissive temperature, and the transformed cells are tumorigenic in newborn hamsters. The discrepancy between tumor induction by virus and by transformed cells has been attributed to the fact that hamster cells are permissive to adenovirus type 2 and 5 replication. Thus, all the potential transformants would be killed when hamster cells are infected with virus from this group. On the other hand, rat cells transformed by group C viruses are non-tumorigenic in newborn or weanling rats, perhaps due to the difference in the immune response from the two species of animals.

Since it is well established now that only region E1 is required for transformation of both hamster and rat cells, much effort has been devoted to ascertain the importance of various E1 proteins in tumorigenicity. Baby hamster kidney cells transformed by Ad5 XhoI-C fragment (0-16%) or by Hind III-G fragment (0-8%) could induce tumors in newborn hamsters, although the tumor incidence varied from cell line to cell line (144). Cell lines transformed by the Xho-C fragment synthesize the 58K and the 19K polypeptides encoded in the E1B region, as expected. Although the 19K is made in all the cell lines transformed by the Hind III-G fragment, none of them produces the E1B 58K polypeptide. This is not surprising since the Hind III-G fragment contains information for only the N-terminal portion of this protein. However, sera from animals bearing tumors induced by some of the Hind III-G transformed cell lines can immunoprecipitate a 58K protein from Ad5 infected cells (144). A reasonable explanation for this observation is that these transformed cells synthesize a truncated form of the 58K protein or a hybrid species. These results indicate that the Ad5 E1B 58K protein in its entirety is not required for tumorigenicity in newborn hamsters. Rowe and Graham reported that hamster cells transformed by DNA isolated from host range group II mutants of Ad5 which do not synthesize detectable amounts of E1B 58K protein can induce tumors in newborn hamsters (122). This observation suggests not even the N-terminal portion of the 58K is required for tumorigenicity in hamsters unless it is synthesized and then rapidly degraded. Alternatively, the tumorigenic state could be maintained by the newly described 19-20K E1B protein which shares both the N- and C-terminus of the 58K protein (69,70). It is also possible that "immortalized" hamster cells can easily progress to the highly oncogenic state. The observation that Ad12 transformed hamster cells which have lost all viral sequence but are still tumorigenic lends support to the idea that the "immortalized" cells can progress to highly tumorigenic state (43).

It is generally accepted that primary rat cells or a rat cell line, 3Y1, transformed by the left 16% of Ad12 are tumorigenic in newborn or weanling rats (3-4 weeks old). Tumorigenicity in nude mice appears to correlate with the expression of the E1B 55K protein (74). Some studies have shown that rat cells transformed by the Hind III-G fragment of Ad12 (0-6.7%) are non-tumorigenic in either nude mice or new born syngeneic rats (74). Yet, we found that some rat cell lines transformed by the same fragment of Ad12 DNA can induce tumors even in immunocompetent animals (see Table 3). These results indicate that neither the carboxyl portion of the E1B 55K nor the 17K are necessary for tumor induction, since most of the coding sequence of the 17K protein is not included in the Hind III-G fragment of Ad12 (see above). Rat cells transformed by a E1B 19K-minus mutant (cyt68) synthesize no 19K protein (101) and are highly tumorigenic (see Table 2). The available data tend to indicate that only the N-terminal portion of the E1B 58K or a hybrid protein in combination with E1A is sufficient for the maintenance of the tumorigenic state. Alternatively, some other E1B proteins, such as the 14K and 15K detected both by in vivo labeling of infected cells and by in vitro translation of E1B mRNA, may play a role in tumor induction.

Although region E1A of Ad12 has very low transformation efficiency of primary rat kidney cells, cell lines were nonetheless established from transformed foci. Data presented in Table 3 show that all five cell lines tested fail to induce tumors even after one year of observation. However, Gallimore et al. reported that four out of seven cell lines transformed by E1A are tumorigenic in newborn rats after unusually long latent periods (120). The difference between these studies may be due to the age of the animals used, 3-4 weeks old vs. newborn.

Studies on the oncogenicity of rat cells transformed by plasmids containing E1A of Ad12 and E1B of Ad5 and vice versa have indicated that transformed cells that harbor the E1A region of Ad12 are not rejected by immunocompetent animals (145). It has been postulated that Ad12 transformed cells evade the immunological surveillance of the host by inhibiting the expression of the class I major histocompatability antigens (145). However, other studies have shown that there is no strict correlation of tumorigenicity and the inhibition of expression of this class of genes (146). Thus, the mechanism of induction of tumors by the highly oncogenic adenoviruses remains unclear.

Morphological transformation and tumor induction by DNA-containing viruses occur only under non-permissive conditions for virus replication. The non-permissivity may be due to the cells, such as hamster or rat cells infected with Ad12, or due to infection of a normally permissive host by defective virus or temperature sensitive mutant virus at non-permissive temperature. Preparations of Ad12 virions often contain a significant fraction of virions with lighter density and these are replication defective in human KB cells, but can induce tumors in hamsters as effectively as the infectious virus (147,148). Furthermore, human embryonic kidney cells have been transformed by Ad5 DNA fragments (the 293 cells) and shown to be capable of tumor induction in nude mice (113, Graham, F.L., personal communication). Transformation of human retinoblasts and human embryonic kidney cells by Ad12 DNA fragments has also been reported (129,130, unpublished data). These findings raise the possibility that some human tumors may be caused by human adenoviruses or their DNA. Green and his colleagues have examined many human tumors for the presence of adenovirus DNA and found none (149,150). More recently, we have examined six human retinoblastoma cell lines for the presence of Ad12 specific DNA and RNA and also found no detectable viral sequence (151). The assay has sufficient sensitivity to detect less than one copy of the transforming region of Ad12 per cell. These studies indicate that adenoviruses are probably not responsible for human cancer. However, the possibility that adenovirus induces tumors by hit-and-run mechanism cannot be completely ruled out in view of the finding that Ad-transformed hamster cells retain tumorigenic potential even though they have lost all viral sequences (43).

ACKNOWLEDGEMENTS

We thank our many colleagues for providing reprints and information. We also want to thank Mrs. H. Lindemann for assistance in the preparation of the article and Mrs. P. Hayward for typing the manuscript. Work from our laboratory is supported by the National Science and Engineering Research Council and the National Cancer Institute of Canada.

REFERENCES

1. Trentin, J.J., Yabe, Y., and Taylor, G. Science 137: 835-841, 1962.
2. Green, M. Annu. Rev. Biochem. 39: 701-756, 1970.
3. Huebner, R.J. In: Perspectives in Virology, Vol. 5 (Ed. Pollard, M.), Academic Press, New York, pp. 147-166, 1967.
4. Trentin, J.J., Van Hoosier, G.L. and Samper, L. Proc. Soc. Exp. Biol. Med. 127: 683-689, 1968.
5. Green, M., Mackey, J.K., Wold, W.S.M., and Råden, P. Virology 93: 481-492, 1979.
6. Sawada, Y. and Fujinaga, F. J. Virol. 36: 639-651, 1980.
7. Virtanen, A., Pettersson, U., le Moullec, J-M., Tiollais, P., and Perricaudet, M. Nature 295: 705-707, 1982.
8. Pettersson, U., Tibbetts, C., and Philipson, L. J. Mol. Biol. 101: 479-501, 1976.
9. Sharp, P. A., Gallimore, P.H. and Flint, J.S. Cold Spring Harb. Symp. Quant. Biol. 39: 457-474, 1974.
10. Scheidtmann, K.H., and Doerfler, W. J. Virol. 22: 585-590, 1977.
11. Smiley, J.R. and Mak, S. J. Virol. 28: 227-239, 1978.
12. McBride, W.D. and Weiner, A. Proc. Soc. Exp. Biol. Med. 115: 870-874, 1964.
13. Freeman, A.E., Black, P.H., Vanderpool, E.A., Henry, P.H., Austin, J.B., and Huebner, R.J. Proc. Natl. Acad. Sci. USA 58: 1205-1212, 1967.
14. Casto, B. J. Virol. 2: 376-383, 1968.
15. McAllister, R.M., and Macpherson, I. J. Gen. Virol. 2 99-106, 1968.
16. Gilden,R.V., Kern, J., Freeman, A.E., Martin, C.E., McAllister, R.C., Turner, H.C., and Huebner, R.J. Nature 219: 517-518, 1968.
17. Cook. J.L. and Lewis, A.M., Jr. Cancer Res. 39: 1455-1461, 1979.
18. Lewis, A.M., Jr., and Cook, J. L. Cancer Res. 42: 939-944, 1982.
19. Gallimore, P.H. and Paraskeva, C. Cold Spring Harbor Symp. Quant. Biol. 44: 703-713, 1979.
20. Mak, S., Mak, I., Smiley, J.R. and Graham, F.L. Virology 98: 456-460, 1979.
21. Sharp. P.A., Pettersson, U. and Sambrook, J. J. Mol. Biol. 86: 709-726, 1974.
22. Gallimore, P.H., Sharp, P.A. and Sambrook, J. J. Mol. Biol. 89: 49-72, 1974.
23. Flint, S.J., Sambrook, J., Williams, J. and Sharp, P.A. Virology 72: 456-470, 1976.
24. Sambrook, J., Botchan, M.B., Gallimore, P.H., Ozanne. B., Pettersson, U., Williams, J.F. and Sharp, P.A. Cold Spring Harbor Symp. Quant. Biol. 39: 615-632, 1974.
25. Shiroki, K., Handa, H., Shimojo, H., Yano, S., Ojima, S., and Fujinaga, K. Virology 82: 462-471, 1977.

26. Sekikawa, K., Shiroki, K., Shimojo, H., Ojima, S. and Fujinaga, K. Virology 88: 1-7, 1978.
27. Dijkema, R., Dekker, B.M.M., van der Feltz, M.J.M. and van der Eb., A.J. J. Virol. 32: 943-950, 1979.
28. Graham, F.L., van der Eb, A.J., and Heijneker, H.L. Nature 251: 687-691, 1974.
29. Graham, F.L., Abrahams, P.J., Mulder, C., Heijneker, H. L., Warnaar, S.O., de Vries, F.A.J., Fiers, W. and van der Eb, A.J. Cold Spring Harbor Symp. Quant. Biol. 39: 637-650, 1974.
30. van der Eb, A. J., Mulder, C., Graham, F.L., and Houweling, A. Gene 2: 115-132, 1977.
31. Pettersson, U., Virtanen, A., Perricaudet. M., and Akusjarvi, G. In: The Molecular Biology of adenoviruses 1. Current Topics in Microbiology and Immunology Vol. 109. (Ed. W. Doerfler) Springer Berlin Heidelberg, New York, 1983, pp. 107-123.
32. Yano, S., Ojima, S., Fujinaga, K. and Shimojo, H. Virology 82: 214-220, 1977.
33. Fujinaga, K. and Green, M. Proc. Natl. Acad. Sci. USA 55: 1567-1574, 1966.
34. Fujinaga, K. and Green, M. J. Virol. 1: 576-582, 1967.
35. Huebner, R.J., Rowe,W.P., Turner,H.C., and Lane, W.T. Proc. Natl. Acad. Sci. USA 50: 379-389, 1963.
36. Ruben, M., Bacchetti, S., and Graham, F.L. J. Virol. 41: 674-685, 1982.
37. Dorsch-Hasler, K., Fisher, P., Weinstein, B and Ginsberg, H. J. Virol. 34: 305-134, 1980.
38. Visser, L., Reemst, A.C.M.B., van Mansfeld, A.D.M. and Rozijn, T.H. Nucleic Acids Res. 10: 2189-2198, 1982.
39. Ruben, M., Bacchetti, S. and Graham. F.L. Nature 301: 172-174, 1983.
40. Graham, F.L. EMBO J. 3: 2917-2922, 1984.
41. Green, M., Chinnadurai, G., Mackey, J. and Green, M. Cell 7: 419-428, 1976.
42. Ibelgaufts, H., Doerfler, W., Scheidtmann, K.H. and Wechsler, W. J. Virol. 33: 423-437, 1980.
43. Kuhlmann. I.. Achten. S., Rudolph. R. and Doerfler, W. EMBO J. 1: 79-86, 1982.
44. Lee, K.C. and Mak, S. J. Virol. 24: 408-411, 1977.
45. Sambrook, J., Greene, R., Stringer, J., Mitchison, T., Hu, S.-L. and Botchan, M. Cold Spring Harbor Symp. Quant. Biol. 44: 569-584, 1979.
46. Doerfler, W., Stabel, S., Ibelgaufts, H., Sutter, D., Neumann, R., Gronenberg, J., Scheidtmann, K. H., Deuring, R. and Winterhoff, U. Cold Spring Harbor Symp. Quant. Biol. 44: 551-564, 1979.
47. Stabel, S., Doerfler, W. and Friis, R. R. J. Virol. 36: 22-40, 1980.
48. Vardimon, L. and Doerfler, W. J. Mol. Biol. 147: 227-246, 1981.
49. Green, M., Wold, W.S.M. and Buttner, W. J. Mol. Biol. 151: 337-366, 1981.
50. Berk, A.J. and Sharp, P.A. Cell 14: 695-711, 1978.

51. Kitchingman, G.R. and Westphal, H. J. Mol. Biol. 137: 23-48, 1980.
52. Chow, L. T., Broker, T.R. and Lewis, J.B. J. Mol. Biol. 134: 265-303, 1979.
53. Virtanen, A. and Pettersson, U. J. Mol. Biol. 165: 496-499, 1983.
54. Svensson, C., Pettersson, U. and Akusjarvi, G. J. Mol. Biol. 165: 475-479, 1983.
55. Bos, J.L., Polder, L.J., Bernards, R., Schrier, P.I., van den Elsen, P.J., van der Eb, A.J. and van Ormondt, H. Cell 27: 121-131, 1981.
56. Gingeras, T.R., Sciaky, D., Gelinas, R.E., Bing-Dang, J., Yen, C., Kelly, M., Bullock, P., Parsons, B., O'Neill, K. and Roberts, R. J. J. Biol. Chem. 257: 13475-13491, 1982.
57. Halbert, D. N., Spector, D.J. and Raskas, H.J. J. Virol. 31: 621-629, 1979.
58. Green. M., Wold, W.S.M., Brackmann, K.H. and Cartas, M.A. Virology, 97: 275-286, 1979.
59. Lewis, J.B., Esche. H., Smart, J.E., Stillman, B., Harter, M. L. and Mathews, M.B. Cold Spring Harbor Symp. Quant. Biol. 44: 493-508, 1979.
60. Spector, D.J., Crossland, L.D., Halbert, D.N. and Raskas, H.J. Virology 102: 218-221, 1980.
61. Yee, S-P.,Rowe, D.T., Tremblay, M?L., McDermott, M. and Branton, P.E. J. Virol. 46: 1003-1013, 1983.
62. Branton, P.E., Rowe, D.Y., Yee, S-P., Tremblay, M.L. and Graham, F.L. In: Genes and Cancer-UCLA Symposia on Molecular and Cellular Biology, New Series, Vol. 17 (Eds. Bishop, J.M., Greaves, M. and Rowley, J.D.) Alan R. Liss Inc., New York, 1984, pp. 339-350.
63. Rowe, D.T., Yee, S-P., Otis, J., Graham, F. L. and Branton, P.E. Virology 127: 253-271, 1983.
64. Downey, J.F., Rowe, D.T., Bacchetti, S., Graham, F.L. and Bayley, S. T. J. Virol. 45: 514-523, 1983.
65. Winberg, G. and Shenk, T. Eur. Mol. Biol. Org. J. 3: 1907-1912, 1984.
66. Lassam, N.J., Bayley, S. T. and Graham, F.L. Cell 18: 781-791, 1979.
67. Jochemsen, J., Hertoghs, J.L., Lupker, J.H., Davis, A. and van der Eb, A.J. J. Virol. 37: 530-534, 1981.
68. Green, M., Brackman, K.H., Lucher, L.A. and Symington, J.S. In: The molecular Biology of Adenoviruses 1. Current Topics in Microbiology and Immunology Vol. 109, (Ed. W. Doerfler) Springer Berlin Heidelberg, New York, 1983, pp. 167-192.
69. Green, M., Brackmann, K.H., Cartas, M.A. and Matsuo, T. J. Virol. 42: 30-41, 1982.
70. Anderson, C. W., Schmitt, R. C., Smart, J. E. and Lewis, J. B. J. Virol. 50: 387-396, 1984.
71. Saito, I., Shiroki, K. and Shimojo, H. Virology 127: 272-289, 1983.
72. Segawa, K., Saito, I., Shiroki, K., and Shimojo, H. Virology, 107: 61-70, 1980.
73. Jochemsen, H., Daniels, G.S.G., Lupker, J.H. and van der Eb, A.J. Virology 105: 551-563, 1980

74. Jochemsen, J., Daniels, G.S.G., Hertoghs, J.J.L., Schrier, P.I. and van der Eb, A.J. Virology 122: 15-28, 1982.
75. Achten, S. and Doerfler, W. J. Gen Virol. 59: 357-366, 1982.
76. Scott, M. O., Kimselman, D., Norris, D. and Ricciardi, R.P. J. Virol. 50: 895-903, 1984.
77. Lucher, L. A., Kimelman, D., Symington, J.S., Brackmann, K.H., Cartas, M.A., Thorton, H. and Green, M. J. Virol. 52: 136-144, 1984.
78. Esche, H., Reuther, M. and Schughart, K. In: The Molecular biology of adenoviruses 3. Current Topics in Microbiology and Immunology Vol. 111 (Ed. Doerfler, W.) Springer-Berlag, Berlin Heidelberg, 1984, pp. 91-106.
79. Katz, M.G., Perrsson, H. and Philipson, L. Eur. Mol. Biol. Org. J. 1: 783-789, 1982.
80. Fujinaga, K., Yoshida, K., Yamashita, T. and Shimizu, Y. In: The Molecular Biology of Adenoviruses 2, Current Topics in Microbiology and Immunology Vol. 110 (Ed. Doerfler, W.) Springer-Verlag, Berlin Heidelberg, 1984, pp. 53-72.
81. Jones, N. and Shenk, T. Cell 17: 683-689, 1979.
82. Spindler, K. R., Eng, C.Y. and Berk, A. J. Virol. 53: 742-750, 1985.
83. Harrison, T., Graham, F.L. and Williams, J. Virology 77: 319-329, 1977.
84. Shenk, T., Jones, N., Colby, W. and Fowlkes, D. Cold Spring Harbor Symp. Quant. Biol. 44: 367-375, 1979.
85. Berk, A. J., Lee, F., Harrison, T., Williams, J. and Sharp, P. A. Cell 17: 935-944, 1979.
86. Jones, N. and Shenk, T. Proc. Natl. Acad. Sci. USA 76: 3665-3669, 1979.
87. Carlock, L. R. and Jones, N. C. J. Virol. 40: 657-664, 1981.
88. Nevins, J. R. Cell 26: 213-220, 1981.
89. Katz, M.G., Persson. H. and Philipson, L. Mol. Cell Biol. 1: 807-813, 1981.
90. Gaynor, R. B. and Berk, A.J. Cell 33: 683-693, 1983.
91. Nevins, J. R. Cell 29: 913-919, 1982.
92. Stein, R. and Ziff, E. B. Mol. Cell Biol. 4: 2792-2800 1984.
93. Schrier, P. I., Bernards, R., Vaessen, R.T.N.J., Houweling, A. and van der Eb, A.J. Nature 305, 771-775, 1983.
94. Takemori, N., Riggs, J.L. and Aldrich, C. Virology 36: 575-586, 1968.
95. White, E., Grodzicker, T. and Stillman, B.W. J. Virol. 52: 410-419, 1984.
96. Subramanian, T., Kuppuswamy, M., Gysbers, J., Mak, S. and Chinnadurai, G. J. Biol. Chem. 259: 11777-11783, 1984.
97. Chinnadurai, G. Cell 33: 759-766, 1983.
98. Ezoe, H. and Mak, S. J. Virol. 14: 733-739, 1974.
99. Ezoe, H., Lai Fatt, R.B. and Mak, S. J. Virol. 40: 20-27, 1981.
100. Lai Fatt, R.B. and Mak, S. J. Virol. 42: 969-977, 1982.

101. Mak, I., Galet, H. and Mak, S. J. Virol. 52: 687-690, 1984.
102. Pilder, S., Logan, J. and Shenk, T. J. Virol. 52: 664-671, 1984.
103. Mak, I. and Mak, S. J. Virol. 45: 1107-1117, 1983.
104. Subramanian, T., Kuppuswamy, M., Mak, S. and Chinnadurai, G. J. Virol. 52: 336-343, 1984.
105. Babiss, L. E. and Ginsberg, H.S. J. Virol. 50: 202-212, 1984.
106. Lassam. N. J., Bayley, S.T. and Graham, F.L. Virology 87: 464-467, 1978.
107. Fukui, Y., Shiroki, K., Saito, I. and Shimojo, H. J. Virol. 50: 132-136, 1984.
108. Lassam, N. J., Bayley, S.T., Graham, F. L. and Branton P. E. Nature 277: 241-243, 1979.
109. Branton, P. E., Lassam, N. J., Graham, F. L., Mak, S. and Bayley, S. T. Cold Spring Harbor Symp. Quant. Biol. 44: 487-491, 1979.
110. Branton, P. E., Lassam, N. J., Downey, J.F., Yee, S-P., Graham, F.L., Mak, S. and Bayley, S.T. J. Virol. 37: 601-608, 1981.
111. Yee, S-P. and Branton, P. E. J. Gen. Virol. 64: 2305-2309, 1983.
112. Raska, K., Jr., Geis, A. and Fohring, B. Virology 99: 174-178, 1979.
113. Graham, F. L., Smiley, J., Russell, W. C. and Nairn, R. J. Gen. Virol. 36: 59-72, 1977.
114. Montel, C., Fisher, E. F., Caruthers, M. H. and Berk, A. J. Nature 295: 380-384, 1982.
115. Montell, C., Courtois, G., Eng, C. and Berk, A. Cell 36: 951-961., 1984.
116. Solnick, D. and Anderson, M. A. J. Virol. 42: 106-113, 1982.
117. Ricciardi, R. L., Jones, R. L., Cepko, C. L., Sharp, P. A. and Roberts, B. E. Proc. Natl. Acad. Sci. USA. 78: 6121-6125, 1981.
118. Ho, Y.-S., Galos, R. and Williams, J. Virology 122: 109-124, 1982.
119. Babiss, P., Fisher, P. and Ginsberg, H.S. J. Virol. 49: 731-740, 1984.
120. Gallimore, P., Byrd, P., Grand, R., Whittaker, J., Breiding, D. and Williams, J. In: Cancer Cells 2: Oncogenes and Viral Genes (Eds. Vande Woude, Levine, A. J., Topp, W. C. and Watson, J. D.) Cold Spring Harbor Press, N. Y. pp. 519-526, 1984.
121. Graham, F.L., Harrison, T. and Williams, J. Virology 86: 10-21, 1978.
122. Rowe, D.T. and Graham, F.L. J. Virol. 46: 1039-1044, 1983.
123. Branton, P. E., Bayley, S. T. and Graham, F. L. BBA Reviews on Cancer (In Press), 1984.
124. Graham, F.L., McKinnon, R., Ruben, M., Rowe, D.T. and Bacchetti, S. In: Proceedings of the 13th International Cancer Congress, Vol. 132, Alan Liss, New York, 1983, pp. 313-326.
125. Babiss, L.E., Fisher, P.B. and Ginsberg,H.S. J.Virol. 52: 389-395,1984.

126. Shiroki, K., Maruyama, K., Saito, I., Fukui, Y. and Shimojo,H. J.Virol. 38: 1048-1054, 1981.
127. Fukui, Y., Saito, I., Shiroki, K. and Shimojo, H. J. Virol. 49: 154-161, 1984.
128. McKinnon, R.D., Bacchetti, S. and Graham, F.L. Gene 19: 33-42, 1982.
129. Byrd, P., Brown, K. W. and Gallimore, P. H. Nature 298: 69-71, 1982.
130. Whittaker, J. L., Byrd, P. J., Grand, R.J.A. and Gallimore, P.H. J. Mol. Cell Biol. 4: 110-116, 1984.
131. Galos, R.S., Williams, J., Binger, M. H. and Flint, S.J. Cell 17: 945-956, 1979.
132. Williams, J. F., Galos., R. S., Binder, M. H. and Flint, S. J. Cold Spring Harbor Symp. Quant. Biol. 44: 353-366, 1979.
133. Stillman, B. W., Tamanoi, F. and Mathews, M. B. Cell 31: 613-623, 1982.
134. Graham, F.L. In: The Adenoviruses, 1984, Plenum Press N.Y. pp. 339-398.
135. Logan, J., Nicolas, J.C., Topp, W.C., Girard, M., Shenk, T. and Levine, A.J. Virology 115: 419-422, 1981.
136. Graham, F. L. and van der Eb, A.J. Virology 52: 456-467, 1973.
137. Graham, F. L. and van der Eb, A. J. Virology 54: 536-539, 1973.
138. Houweling, A., van den Elsen, P.J. and van der Eb, A.J. Virology 105: 537-550., 1980
139. Van den Elsen, P.J., Houweling, A. and van der Eb, A.J. Virology 131: 242-246, 1983.
140. Van der Eb, A.J. and Bernards, N. In: The molecular Biology of adenoviruses 2. Current Topics in Microbiology and Immunology, Vol. 110 (Ed. Doerfler, W.) Springer-Verlag, Berlin Heidelberg, 1984, pp. 23-51.
141. Van den Elsen, P.J., Houweling, A. and van der Eb, A.J. Virology, 128: 377-390, 1983.
142. Land, H., Parada, L. F. and Weinberg, R. A. Nature 304: 596-602, 1983.
143. Ruley, H. E. Nature 304: 602-606, 1983.
144. Rowe, D.T., Branton, P.E., Yee, S-P., Bacchetti, S. and Graham, F.L. J. Virol. 49: 162-170, 1984.
145. Bernards, R., Schrier, P.E., Houweling, A., Bos. J.L., van der Eb. A.J., Zijlstra, M. and Melief, C.J.M. Nature 305: 776-779, 1983.
146. Mellow, G.H., Föhring, B., Dougherty, J., Gallimore, P. H. and Raška, Jr., K. Virology 134: 460-465, 1984.
147. Mak, S. J. Virol. 7: 426-433, 1971.
148. Mak, I., Ezoe, H. and Mak, S. J. Virol. 32: 240-250, 1979.
149. Green, M., Wold, W.S.M., Mackey, J. K. and Rigden, P. Proc. Natl. Acad. Sci. USA 76: 6606-6610, 1979.
150. Green, M., Wold, W.S.M., Brackman, K.H., Cartas, M.A., Sanders, P. R., Olson, K., Lee, T.K., Young, L., Matsuo, T. and Kapoor, Q. In: Viruses in Naturally Occurring Cancers: Cold Spring Harbor Conferences on Cell Proliferation, Vol.7 (Eds. Essex, M., Todaro, G. and zur Hausen, H.),

Cold Spring Harbor Press, New York, pp. 373-397, 1980.
151. Mak, S., Mak, I., Gallie, B.L., Godbout, R. and Phillips, R. A. Int. J. Cancer 30: 697-700, 1982.

12

CLASSIFICATION AND EPIDEMIOLOGY OF ADENOVIRUSES

R. WIGAND and Th. ADRIAN

Institut für Medizinische Mikrobiologie und Hygiene, Abteilung für Virologie, Universitätskliniken, D-6650 Homburg (Saar)

1. INTRODUCTION

Adenoviruses (AV) were discovered by Rowe et al. (1), who recovered them from surgically removed adenoids of children. Viruses of the same group were independently isolated from throat swabs of patients with acute respiratory diseases by Hilleman and Werner (2). The group name *Adenoviruses*, indicating the occurrence in adenoid tissue, was created in 1956 by Enders et al. (3). The AV group was given a family status *Adenoviridae* in 1976 (4), with two genera *Mastadenovirus* and *Aviadenovirus*. They largely encompass AV occurring in mammals or birds respectively. The viruses are widely spread in nature in many mammalian and avian species; an AV has also been recovered from the frog. In man, 41 types or serotypes (5-8) have been described up to now. A great number of types has also been found in monkeys, cattle, and chickens (see Table 1). In view of this diversity, a classification scheme is essential, both to help correctly identifying viruses and to establish phenotypic and genomic relationships between them.

2. GENERAL PROPERTIES OF ADENOVIRUSES

AV are medium-sized, non-enveloped particles with a double-stranded DNA, mol. wt of 20 to 30 MegaD, corresponding to 30 to 45.5 kilobase pairs, and with at least 10 structural polypeptides ranging from 5 to 120 kD. The isometric particles, 70 to 90 nm in diameter, have an icosahedral symmetry with 252 capsomers. The 12 vertex capsomers (or penton bases) carry one or two filamentous projections (fibers)(genus mastadenovirus or aviadenovirus respectively).

Table 1. Adenovirus types in man and animals[a]

Genus	Host	Number of serotypes definite	candidates	References[b]
Mastade-novirus	man	41		6, 7, 8
	nonhuman primates	24		
	cattle	9		
	sheep	6	1	9, 10
	pig	4		
	horse	2		11
	dog	2		12
	goat		1	
	mouse	1	1	
	peromyscus		1	
	opossum		1	
	tree shrew (tupaia)	1		
	rabbit		1	13
	guinea pig		1	14
Aviade-novirus	fowl	9	3	
	turkey	3		15
	goose	3		16
	pheasant		1	
	duck		1	
	frog		1	

a from Wigand et al. (5)
b References are given for new types or new definite types

240 nonvertex capsomers (or hexons) are different both from penton bases and fibers. As AV are unenveloped viruses, they are stable against lipid solvents. They are also stable on storage at different temperatures.

"Essentials" to identify a virus *as* AV are its particle size and shape, the absence of a membrane, 252 capsomers, and the presence of a (double-stranded) DNA. Many AV types in the past have been identified solely by their cytopathic effect, by formation of intranuclear inclusion bodies, and by the production of a genus-specific antigen (hexon). These properties, though useful and mostly demonstrable, may not be essential for all AV. If new AV types are found, the above physicochemical properties should be demonstrated.

Most basic investigations have been carried out in human adenovirus 2 (HAV2, Ad2) and Ad5, which are closely related to one another, fewer with Ad12 and Ad7, and only in recent years some information became available on other AV.

The organization of the AV genome and its replicative strategy appear to have many conserved features:

1. The transcription maps of Ad3, 7 (17) and Ad12 (18) have been found to be largely similar to those of Ad2 and Ad5 (19), although these viruses belong to three different subgenera (see below). Apparently various genome sections of distantly related AV possess a functional analogy. For example, the hexon gene is localized between map units (approx.) 50 to 60 for Ad2, 12 (19), 3 (17), simian AV 8 and 16 (20), and bovine AV 3 (21). The transforming genes of different HAV also appear to have similar replicative functions (22, 23).

2. An inverted terminal repetition (ITR) of about 100 b.p. length has been found in all AV studied from human, simian, other mammalian and avian origin (24). Although the length of the ITR sequences varies (24-27), *all* AV appear to have an ATAATA sequence close to the 5'end (27, 28).

3. When nucleotide sequences of other short genome regions were compared for Ad5, 7, and 12 (29, 30), various homologous parts of the genome were revealed.

4. A protein covalently linked to the 5' terminus of each DNA strand (55K for HAV) has been found in all AV studied. Although its mol.wt may vary (31, 32), the primary sequence of this protein appears to be highly conserved (33).

3. DESCRIPTION OF GENERA

While the original definition of the two genera (4) was mainly based on the absence of immunological cross-reactivity - each genus has its own genus-specific hexon antigen -, other structural and biochemical properties also distinguish the two genera (Table 2). Only two short homologous sections

Table 2. Characteristics of adenovirus genera

	Mastadenovirus	Aviadenovirus	EDS
DNA: mol.wt (x 10^6)	20-25	28-30	22.5
G+C content[a] (per cent)	44-61	51-57	
Fiber	one	two	one
Fiber length (nm)	10-33[b]	see text	42[c]
Genus-specific antigen	yes[d]	yes	own

a Data from (5, 15, 34)
b Only HAV considered
c Data from (35)
d Except for bovine AV4 to 8 (36)

of the DNAs were found between fowl AV1 and Ad2 by hybridization under stringent conditions (37), namely at 18.1 to 19.3 and 57 to 58 map units. There appear to be more differences between the two genera than those shown in Table 2. Thus, the ITR sequences are much shorter in fowl AV1 than in all mammalian AV studied (24). Also the distribution of virion polypeptides, particularly the mol.wt of the core-associated polypeptides, is very different for fowl AV1 and Ad5 (38).

Bovine AV types 4 to 8 are unique in that they show little or no genus-specific reactivity in their hexons by precipitation or complement-fixation (36). With more sensitive methods like immunofluorescence, however, a low-titered cross-reactivity was demonstrated (39). When the DNA of bovine AV7 was compared with bovine AV3 and Ad2 (40), only slight relationship was found. If these results are confirmed and extended, a separate position of bovine AV4 to 7 must be considered.

A further group of AV from avian origin, mainly from ducks, are related to the EDS-76 prototype virus (41). The name stands for "egg-drop syndrome", as the affected birds tend to produce less eggs and ones with defective shells. The virus is very different from the genus aviadenovirus and is a candidate for a third genus. A detailed comparison, however, with mammalian AV has not been reported as yet.

4. ADENOVIRUS TYPES

Historically, different types of AV, like in other virus families, have been differentiated immunologically by neutralization (SN) tests. It is not just the ease to identify strains in the laboratory which up to now has led to give SN a priority in typing. The fundamental immune response of the host is important which, as far as it is type-specific, indicates past infection and immunity against reinfection and disease. The prevalence of neutralizing antibodies is a decisive parameter in seroepidemiologic studies and gives indications of necessity and success of active immunization. It is true, however, that only a very small part of the genome is responsible for the antigen determinants active in SN.

In 1981 the International Committee for the Taxonomy of Viruses (ICTV) had provisionally accepted the following definition of an AV *species* (5, 42):

"A species (formerly type) is defined on the basis of its immunological distinctiveness, as determined by quantitative *neutralization* with animal antisera. A species has either no cross-reaction with others or shows a homologous-to-heterologous titer ratio of $>$ 16 in both directions. If neutralization shows a certain degree of cross-reaction between two viruses in either or both directions (homologous-to-heterologous titer ratio of 8 to 16), distinctiveness of species is assumed if:

(i) the hemagglutinins are unrelated, as shown by lack of cross-reaction on hemagglutination-inhibition; or

(ii) substantial biophysical/biochemical differences of the DNAs exist."

A species naming has also been provisionally accepted (5, 42) However, at the last meeting of the ICTV in Sendai, Japan, in September 1984, these two proposals have not been definitely approved, mainly because the above concept fails to take into account the genomic relatedness. Hence at present it seems that the terms *type* or *serotype* have to be reintroduced. Also the proposed species names have lost their legitimacy and should not be used. Even the term *subgenera* of HAV (see below and 5) may not be definite. They are well-established entities, but it is possible that these groups once become the final AV *species*.

The above definition of species should be applied for *serotypes;* it takes into account that some types show a certain degree of cross-SN. Only in these cases results from hemagglutination-inhibition (HI) and/or DNA restriction analysis are to be considered to define new types. Identification problems are discussed below.

The number of known AV types are listed in Table 1. Ideally each AV type should show no cross-reactivity in SN with antisera in both directions not only against the AV from the same host, but also from different hosts. Adair and McFerran (47) have studied SN of HAV antisera against various nonprimate animal AV; they also performed cross-SN with bovine, ovine and porcine AV types. Willimzik et al. (44) made SN tests with 25 simian and 31 AV from nonprimate mammals and birds, almost all types known at that time, against HAV 1 to 31. Antisera to HAV were also tested against 25 simian prototype AV. With the exception of some cross-reactions between human and chimpanzee AV, all these tests were negative.

With antisera of AV from the same host, most of the cross-SN tests were done by two or more laboratories with corresponding results. In types listed as candidates in Table 1, this has either not been done or the results of two laboratories did not agree (like in mouse AV 2), or else, the reported data were not sufficient to establish firmly the type in question.

5. SUBGENERIC CLASSIFICATION

A division into subgeneric entities (subgenera, species) should help to understand the relationship between certain groups of AV. Furthermore, it may help to identify viruses by comparison with known data, and also to test hypotheses on evolution of viruses.

HAV have been subdivided into subgenera A to E (5) and either F or F/G (45), the so-called enteric adenoviruses, based on several of the criteria listed below. Criteria (in arbitrary order) are biological, morphological, immunological or biochemical ones. After they have been proved valuable for HAV, these criteria should also help to classify nonhuman AV into subgeneric entities.

5.1 Host specificity

The natural occurrence of AV is generally confined to a certain host species, with some exceptions. Likewise the propagation of AV in the laboratory is performed best (or only) in cells from the same species. HAV have not been recovered from animals, neither has an "animal" AV ever been recovered from man. Most HAV replicate in cells of simian origin only abortively (46). An exception are Ad40 and 41, as they replicate in cynomolgus kidney cell cultures, while most human cells are refractory (8). HAV of subgenus C (but not others) have a limited capacity to multiply in rodent cells (47). Some AV from chimpanzees are immunologically and biochemically related to HAV (48, 49). Several simian AV appear to be infective for different host species or even genera (S. S. Kalter, personal communication). Several bovine and ovine AV are infective for other animal species even under natural conditions (50). Canine AV 1, the causative agent of hepatitis contagiosa, is also infective for foxes. In several avian AV, the host specificity is not absolute (51). Despite these qualifications it is apparent that AV do not spread freely from one host species to another. The important question, whether this host specificity of AV also indicates special relationships in other, notably in biochemical properties, is largely unanswered and should be studied, in order to attain a subgeneric classification which involves not only HAV. Until more is known, the host range subdivision (see Table 1) is pragmatic and should not be considered as a step to final subgeneric classification of AV.

5.2 Hemagglutination (HA)

Differences in susceptibility of erythrocyte species and in the kind of the HA pattern have been observed by Rosen (52, 53); he therefore classified HAV into three groups or subgroups. These findings, with some modifications and additions, are still useful criteria to classify HAV and are helpful to identify HAV from subgenera B and D (Table 3), which comprise the greatest number of types. A similar division has been proposed for simian AV (54). HA has thus proved to be a useful criterion to show AV relationships. Some limitations exist, however, as Ad18, 25, and 28 usually show no HA. Furthermore, variant strains of other subgenus D types without HA may be found (unpublished). Also, not all simian or even all rhesus blood cell lots are suitable for all subgenus B viruses (55), and only certain rat strains are suitable to demonstrate HA with subgenus A and C viruses (56).

5.3 Oncogenicity

Some AV types, derived from man and animals, are oncogenic in newborn hamsters or other rodents (57) and most if not all other species can transform rodent cells in vitro. The degree of oncogenicity has been used for AV classification (45, 58, 59). However, as many types are not oncogenic and as the term "weak oncogenicity" is impractical, this interesting property of AV appears to be of limited value for classification.

5.4 Fiber length

HAV of 4 of the 6 subgenera are endowed with fibers of different length (Table 3). Surprisingly little is known on the fiber length in other mammalian AV, data which may be particularly helpful in the classification of simian AV.

Table 3. Properties of human adenovirus subgenera[a]

Subgenus	Types	Length of fibers (nm)	DNA homology[b] %	G+C %	number of *Sma*I fragments after cleavage at CCC/GGG	Molecular weight of internal polypeptides (kD) V	VI	Hemagglutination
A	12,18,31	28-31	48-69 (8-20)	47-49	4-5	51.0-51.5 46.5-48.5	25.5-26	rat[c]
B	3,7,11,14,16, 21,34,35	9-11	89-94 (9-20)	50-52	8-10	53.5-54.5	24	monkey
C	1,2,5,6	23-31	99-100 (10-16)	57-59	10-12	48.5	24	rat[c]
D	8,9,10,13,15, 17,19,22,23,24, 26,27,29,30,32, 33,36,37,38,39, 20,25,28	12-13	95-99 (4-17)	57-60	14-18	50.0-50.5	23.2	rat
E	4	17	(4-23)	57	19	48	24.5	atypical rat[c]
(F)	40, 41	28-33			9-11	46;48.5	25.5	

a Adapted from Wadell et al. (48) and Wigand et al. (5)
b Percent homology within the subgenus; figures in parentheses: homology with other types (60)
c Incomplete HA pattern

While fibers are often difficult to visualize on the virion surface, they are easier seen on isolated pentons. Eleven AV species from chicken had all *two* fiber projections (35). The length of the fibers differed somewhat between the types, but was drastically different for fowl AV1, as compared with all others (35).

5.5 Capsid proteins

Besides some cross-reaction in SN and/or HI between individual types, no obvious cross-reactivity between prototypes of any subgeneric setting is apparent. More subtle relations between related AV, however, exist in all three capsic proteins, particularly in the hexons (61, 62), although this kind of relationship is not easily demonstrable. To stress both the importance and the complexity of the hexon, it may be mentioned that the hexon, in addition to its genus-specific determinant α , has a type-specific determinant ε which is responsible for SN, and besides those a subgenus-specific determinant.

Johansson and Wadell (63) were able to prepare antisera with subgenus-specific reactivity by affinity chromatography. It is likely that monoclonal antibodies with this kind of specificity will soon become available.

5.6 Early antigens

T antigens, demonstrated by sera from tumor-bearing animals, are cross-reacting among types of one subgenus (58). Reactions with early antigens could also be demonstrated by immunoperoxidase in AV inoculated cell cultures, inhibited by cytosinearabinoside (64). A largely subgenus-specific reaction was noted in this system, except for Ad4 (subgenus E) which showed a cross-reaction with all other HAV.

5.7 Size of viral polypeptides

Wadell et al. (48, 65) compared the apparent mol.wts of viral polypeptides from virions in the SDS-PAGE. Considerable variations of capsid polypeptides II, III, and IV from one type to another were observed, whereas the internal polypeptides V,

VI, and VII showed more uniform properties. The mol.wt of the core polypeptide V and, to a lesser extent, of polypeptide VI was found to be largely constant for AV from the same subgenus and different from viruses outside the subgenus (Table 3). Attempts to use polypeptide size also for classification of simian AV (59, 66) were less successful; they were on the whole similar in size to those of HAV. When data from different laboratories are compared, however, the reported mol.wts showed considerable divergence (48, 59, 65, 67).

5.8 DNA size

Differences in the mol.wt of DNA exist among mammalian AV. HAV have mol.wts between 22 and 24 MegaD, whereas the mol.wt of DNA from mouse AV1 (34), bovine AV2 (68) and canine AV1 (69) is about 20 MegaD. Hence, differences in the overall size of DNA may be useful for classification. In addition, the length of the ITR sequences may also be considered (24).

5.9 GC content of DNA

This criterion was used to subdivide human (Table 3) and simian (70) AV. The upper and lower limits of GC content for AV of the two genera are listed in Table 2. The relation of GC content to the oncogenicity of the respective types, once suspected to have some implications, is only fortuitous.

5.10 DNA restriction analysis

It is a powerful method to screen viral genomes by dissection with base-specific endonucleases. As restriction analysis can be made from the DNA of infected cells without virus purification, the techniques can be performed in a large number of strains with relative ease. SmaI was one of the first enzymes, used by Wadell et al. (48, 71), since its recognition site (CCC/GGG) has some connection to the overall GC content of the respective DNAs: viruses with a high GC content may have more restriction sites than those with less GC (Table 3). Restriction patterns with some enzymes have been or are being published for the majority of HAV prototypes and for several animal AV.

To determine the relationship of AV, the number of restriction sites and the distribution of fragments can be helpful, as was found by Wadell et al. (48) for the subgenera of HAV. This similarity can be quantitated by pairwise analysis of comigrating fragments. A statistical study of genome relatedness can be made, whose accuracy depends on the number of restriction fragments to be compared. Preferably more than one enzyme should be used and at least 25 restriction fragments should be compared. In related types, comigrating fragments are supposed to be homologous and hence to have analogous functions, although exceptions may occur. On the other hand, in distantly related AV, observed comigrating fragments may be random, i.e. not homologous and without functional significance.

The potential as well as the limitations of comigration analysis as a method to study the relatedness of AV and as a criterion to construct subgeneric settings is exemplified by results with HAV prototype strains.

Subgenus A: These three viruses, although serologically related (56), are relatively different in their genomes from one another (72, see Table 3 for hybridization). It is therefore not surprising that only about 20 % comigrating fragments were found.

Subgenus B: The eight prototypes can be divided into two clusters: (1): Ad3, 7, 16, 21; (2): Ad11, 14, 34, 35 on the basis of DNA restriction analysis, as proposed by Wadell et al. (48) and confirmed with more restriction fragments recently (73). The rate of comigrating fragments within the cluster is at least 55 %, but no more than 32 % by inter-cluster comparison (Table 4, see next page).

Subgenus C: The four viruses showed a close relationship with 60 to 80 % comigrating fragments.

Subgenus D: The 23 prototype viruses showed no clustering in their genetic relatedness (Table 5); mean percentages of comigrating fragments were between 37 and 80 %. Serologically related types did not show any closer relationship with one another than others.

Table 4. Comigration analysis of subgenus B viruses, clusters 1 and 2

	3	7	16	21	11	14	34	35	Totals
3		39	33	30	13	12	14	14	47
7	81		31	27	10	10	11	12	49
16	69	63		31	10	12	12	11	49
21	63	56	64		13	10	15	13	48
11	30	22	22	30		28	26	31	40
14	27	22	26	22	68		31	24	42
34	30	23	26	32	61	71		34	45
35	31	26	24	29	76	57	78		42

Upper right: number of comigrating fragments, summarized for the restriction endonucleases BamHI, BglII, BstEII, HindIII, and SmaI
Lower left: Mean percentage of comigrating fragments

When the DNA restriction patterns of HAV from different subgenera (A, C, and E) were compared, 5 to 15 % comigrating fragments were counted.

From these data it may be concluded that a high rate of comigrating fragments would indicate a relatedness of the respective AV, whereas a low comigration rate per se would not exclude a relationship.

Two studies on avian AV have been reported. 17 strains of 11 fowl AV serotypes could be divided on the basis of fragment comigration into 5 groups which largely corresponded to their serological relations (74). All of seven strains from geese (16) had more than 25 % comigrating fragments, whereas only few fragments were common with fowl AV1 and none with EDS-76 DNA.

Restriction site maps or physical mapping of the genomes have been reported for relatively few AV types as yet (Table 6 and 7). Genome maps of closely related viruses are rather similar, as expected: Ad1, 2, 5, and 6 (19, 75), also Ad3 and 7 (17). Although restriction site maps reflect similarities of differences between DNAs in more detail than

Table 5. Comigration analysis of subgenus D viruses

	9	10	13	15	17	19	20	22	23	24	25	26	27	28	29	30	32	33	36	37	38	39	Totals
8	21	21	19	22	19	17	23	21	24	19	19	21	22	21	24	23	20	20	19	15	20	22	42
9		32	24	26	24	27	28	22	27	25	27	29	29	28	32	27	22	23	23	25	22	22	38
10			25	27	23	26	28	25	27	27	27	30	28	26	31	28	21	23	23	24	22	24	42
13				25	19	20	28	25	26	24	22	23	25	24	24	24	23	22	21	19	21	21	41
15					26	21	32	29	29	25	25	25	27	27	30	34	26	27	24	24	29	31	48
17						22	25	24	28	24	22	24	23	22	28	28	25	24	21	23	25	25	36
19							23	21	23	22	24	27	23	20	23	23	20	19	20	24	20	22	40
20								28	31	26	25	30	27	28	30	31	23	27	26	25	24	24	45
22									31	26	24	27	23	24	26	32	22	28	23	23	25	25	43
23										27	24	28	27	27	29	31	27	27	25	23	27	28	41
24											25	28	27	27	26	31	26	25	23	24	26	28	41
25												26	26	25	26	27	26	25	23	24	27	26	40
26													31	25	30	28	21	23	23	23	23	24	44
27														27	32	28	24	22	19	24	25	26	42
28															30	28	23	24	27	23	23	27	41
29																34	28	25	26	24	26	29	44
30																	31	30	26	26	30	31	50
32																		25	22	21	30	29	41
33																			23	21	27	26	37
36																				23	23	25	39
37																					23	26	38
38																						30	41
39																							51

Number of comigrating fragments, summarized for BamHI, BglII, BstEII, HindIII, and SmaI

Table 6. Reports on DNA restriction site mapping in human adenoviruses

Subgenus	Type	Reference
C	1	19
	2	19, 76[a]
	5	19
	6	75
E	4	77
B	3	17, 19
	7	17, 19
		71[a], 78[a]
	16	79
	35	80
A	12	19
D	8	81
	9, 15, 29	82
(F)	40, 41	83

a Mapping of genome types

Table 7. Reports on DNA restriction site mapping in animal adenoviruses

Host	Virus type	Reference
Monkey[a]	SAV-5 (SV20)	84
	SAV-16 (SA7)	59, 84
Cattle	BAV-2	68
	BAV-3	85
	BAV-7	40
Dog	CAV-1	69
Mouse	MAV-1	34
Tupaia	TupAV-1	86
Fowl	FAV-1	87

a Simian AV were renamed by the Simian Virus Working Team of WHO/FAO (88) restriction patterns, the degree or relatedness can be better estimated by the easier comigration analysis.

5.11 DNA-DNA hybridization

Clear and convincing results were obtained by liquid hybridization with whole DNAs of various HAV (60) and with their transforming DNA segments (89, 90). These reports provided a biochemical basis for creating subgenera A to E of HAV (see Table 3). No similar data are available on Ad40 and 41. Yet recently Takiff et al. (83), using Southern blot hybridization with cloned DNA fragments, found a strong hybridization between them and a weak one with Ad2. Results with animal AV are scarce.

Certain genome sections are homologous even between AV of the two genera (27, see above), others appear to be homologous for AV from different hosts, like mouse AV and Ad2 (map units 53-60 and 14-15 (91)), while a greater part of the genome (from 11 to 75 map units) may show hybridization between AV of a medium degree of relatedness, like Ad2 and Ad40+41 (83) or Ad2 and bovine AV3 (21). Techniques like Southern blot hybridization are prone to reveal AV relationships, which may be particularly rewarding for animal AV. Some quantification of the results is desirable, however.

Subgenus-specific hybridization may be used for recognition and identification of clinical isolates (92).

5.12 Base sequences

A comparison of HAV species on the level of nucleotide sequences has been started with genome sections from the left part of the DNA, like ITR or other parts (24, 29). Recently the whole DNA sequence of Ad2 has become known (93, and personal communication from R.Roberts). Parts of the genome of Ad5, 7, and 12 have also been sequenced and are available at the Data Library of the EMBL, Heidelberg, FRG. It remains to be seen, how the sequencing technology will spread to reveal the nucleotide sequence of more and more AV genomes or genome sections, and how a computer-aided comparison will eventually influence the subgeneric classification.

6. PROBLEMS OF SEROLOGICAL IDENTIFICATION

Identification of AV types should primarily be done by SN, although HI in addition to SN is helpful or even essential (see below). Some difficulties may arise, as in several HAV types, when cross-SN with others is found, and that in subgenus A (56), B (55), and D (94). The degree of cross-SN may even exceed the range allowed for by the above type definition. This unsatisfactory situation has historical reasons for Ad29 and Ad30, which are related to Ad15 or Ad13 respectively in SN. Ad40 and 41, although closely related in SN and identical in HI (8), have been classified as two types, because they have widely different DNA restriction patterns (83, 95). It should also be noted that Ad4 and the prototype strain of Ad16 show a bilateral cross-SN (96), although these viruses belong to different subgenera. This is not the case, however, with other Ad16 strains (55, 96).

Another problem is that AV strains may differ in their cross-reactivity with other types in SN or in HI respectively. These are the so-called intermediate or hemagglutinin variant strains (97, 98). The best known example is Ad15/H9 (H = hemagglutinin) (98), related to Ad15 in SN and to Ad9 in HI. A variety of such variant strains have been observed in subgenus D (97), and similarly in subgenus B (55, 80). These strains show one hemagglutinin type only, to which a weak cross-SN may be present in addition to the strong SN to the other type. These observations may be interpreted in that Ad15/H9 shares the hexon antigen ε with Ad15 and the fiber antigen γ with Ad9 (61). Fibers are, less than hexons, able to induce neutralizing antibodies (99). Hence, if no HI is performed, these strains may appear as doubly neutralizable, like AV "Ad3-7" described earlier (100), or may be incompletely or incorrectly identified.

It should further be noted that conventional polyclonal antisera may differ in their degree of cross-reactivity with other types. Hence, to reveal relationships or to ensure antigenic specificity, more than one antiserum of each type should be used.

Finally, weakly neutralizable strains or prime strains have been described (16, 96, 101). For all these relationships, more precise definitions and interpretations in the future shall undoubtedly be obtained by use of a series of monoclonal antibodies.

The practical identification of HAV is easily accomplished by SN in Ad1 to Ad8 which are by far the most common isolates. Isolates belonging to higher-numbered types can be classified by HA into subgenera B or D and should then be identified by SN *and* HI. DNA restriction analysis may be made instead or in addition. Subgenus A viruses are slowly growing and, because of cross-SN and -HI between the types (56), sometimes difficult to identify.

7. GENOME TYPES

It is not surprising that variant strains exist in nature with altered DNA not reflected by a change in serological properties. First reports came from Wadell and Varsanyi (71) for Ad7; the authors (48) proposed the term "genome types". Depending on the number of restriction endonucleases and the number of respective fragments, more or less genomic variation is revealed. Since each restriction site involves only few (mostly 6) base pairs, an identity of the genomes cannot be ensured for strains, even if several hundreds of restriction sites are compared.

Several observations indicate a considerable stability of AV genome types after cell culture passages. Besides our own unpublished results, HAV prototypes 1 to 39 were analyzed independently in three laboratories and showed identical restriction patterns with three enzymes, although widely different passage materials have been analyzed. It should be mentioned, however, that three plaque isolates from the same simian AV 16 virus (84) showed considerable diversity of their restriction patterns.

Results of published studies and some unpublished observations on the occurrence of genome types, shown in Table 8, may shed some light on the genomic variability of HAV. It appears that the variability is different in HAV from different subgenera. It is particularly great in Ad1, 2, and 5 of subgenus C.

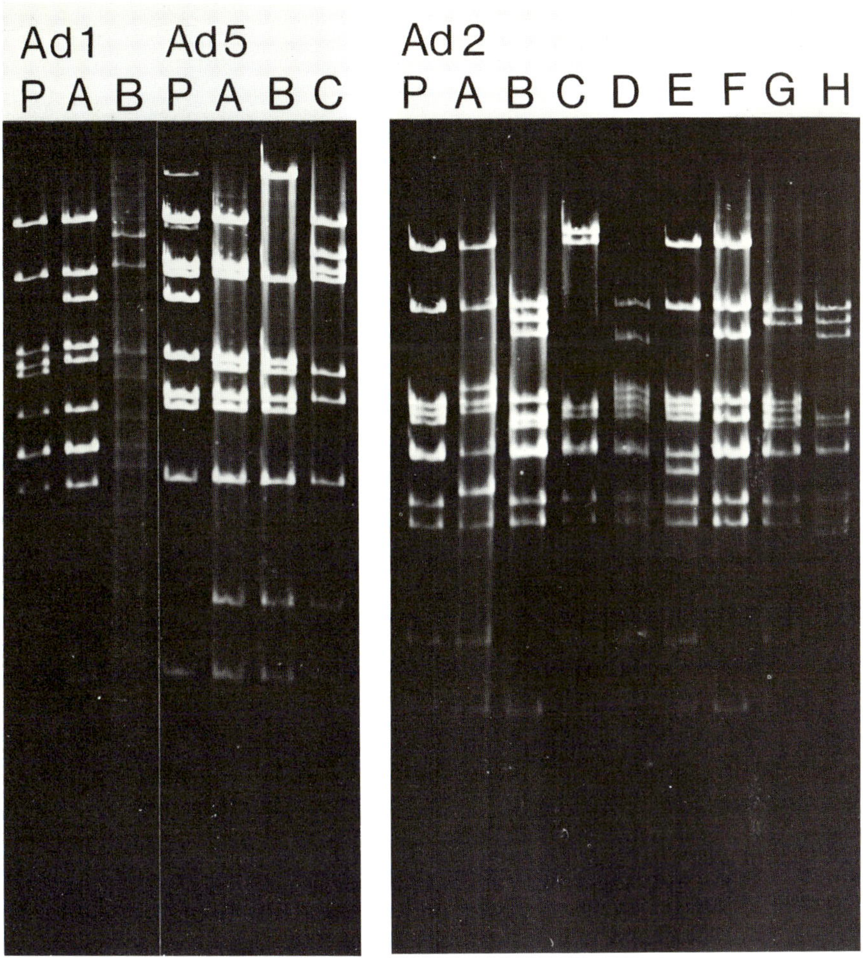

FIGURE 1. Genome type analysis: HindIII variants of subgenus C adenoviruses.
The DNA from infected HeLa cells was analyzed with HindIII and the fragments were separated in 1.2 % agarose gels. The capital letters denominate the distinct variant strains, deviating from the corresponding prototypes (P).

An example is shown in FIGURE 1. The variability is intermediate for Ad3, 7, and 8, whereas Ad4 appears to

Table 8. Recognized genome types in human adenoviruses, deviating from prototypes

Sub-	Type	Number of Enzymes used	Strains studied	Genome types	Reference
C	1	5	49	23	102, *
	2	5	94	40	102, *
	5	8	2	1	103
		2	58	7	45
		5	28	9	102, *
E	4	4	?	1	45
		7	14	1	*
B	3	?	?	1	45
		4	85	9	*
	7	5	123	4	45, 48, 71
		4	133	8	*
	11	4	1	1	73
	14	4	3	2	
	16	4	6	3	
	21	4	17	2	
	34	4	1	0	
	35	4	3	1	
D	8	4	25	2	104
		5	40	4	*
	19	3	6	1	105
(F)	40	2	11	1	106
	41	9	14	5	107

* unpublished data

have only one deviating genome type (45, and unpublished results). Only little is known on genome types in animal AV.

Field strains of Ad1, 2, and 5 have been found to be different in the size of their capsid polypeptides as compared with the respective prototypes (29). One possible mechanism for this finding could be recombination (29), as in cell cultures recombination between Ad2 and 5 has been demonstrated by means of ts mutants (108). Therefore recombination may also occur in man, although the in vitro conditions for recombination are fairly different from that what may be expected to happen in nature (109). To explain the variety of genome types, other mechanisms like deletion could also be operative. It has been shown that Ad5 has a deletion in the middle part of its hexon, as compared with Ad2 (110).

Neither in vitro recombinants nor naturally occurring strains of subgenus C have been reported up till now as serologically intermediate, and most isolates can be clearly identified by SN.

8. EPIDEMIOLOGY: HUMAN ADENOVIRUSES

8.1 General

The global incidence of AV infections in man can be estimated from the data registered by the WHO (111-113, and current yearly reports). These collective statistics have their shortcomings, however, and one of the greatest is the uneven distribution of reported cases. For example, the ratio of AV infections reported from the United Kingdom, USA and USSR were about 64:4:1, and very few data came from the developing countries.

Over the years a constant rate of about 8 % of all reported virus infections were AV infections. From the so-called respiratory viruses, the incidence of AV is second only to influenza A virus (111). The following age distribution of AV infections were derived from Assaad and Cockburn (111) plus WHO reports from several years thereafter:

Age (years)			
<1	1-4	5-14	15 or more (adults)
26	35.5	16.5	22 per cent of all cases

Hence the frequency of AV infections is highest in 1 to 4 years old children. However, this statement disregards the unequal size of the population segments as well as the habits of those submitting specimens for laboratory diagnosis.

The greatest number of AV infections has always been found associated with diseases of the respiratory tract, followed by those of the gastrointestinal tract. About twice as much

AV infections had been found in patients with respiratory than with gastrointestinal disease till 1979. Since 1980, however, the number of gastrointestinal infections approached those of the respiratory tract, which may be due to an increased recognition of Ad40 and 41, causing gastroenteritis in infants (114, 115).

The number of all respiratory infections associated with AV was about equal to those with respiratory syncytial virus, lower than influenza A, but higher than all other respiratory viruses (111). AV infections affect the upper respiratory tract more frequently than the lower.

8.2 Incidence of AV types

The greatest part of identified AV are Ad1 to 8; the decreasing order of frequency (113) generally is Ad2, 1, 7, 3, 5, 4, 6, 8. This statement, however, disregards the recently discovered enteric Ad40 and 41 (see below). All other types are more rarely isolated. The prevalence of neutralizing antibodies in persons of different age (116, 117) also confirm the early spread of Ad1, 2, 3, 5 and 7 (see below) and the rarity of the other types. These viruses, though uncommon, may be under-estimated in their prevalence, however, as many of them are slowly growing and/or difficult to identify, so that they may be reported as untyped.

The incidence of AV types can be derived from WHO data collection. The same qualifications as above are pertinent, however. An identification of the infecting type has been made in less than half of the reports to WHO; these have been evaluated for the WHO data from 1967 to 1976 by Schmitz et al. (113). The so-called epidemic types, namely Ad3, 4, 7, and 8, showed a periodicity over the years with significantly high incidence: Ad3 in 1967 and 1970, Ad7 in 1973 and 1974, Ad4 in 1970 and 1971, Ad8 in 1967, 1971, 1975 and later in 1977 and 1979. On the other hand, no such differences were found for the "endemic" types Ad1, 2, and 5. Major seasonal influences were not apparent, when data from the Northern hemisphere were considered only. Yet, a certain preponderance of subgenus C viruses in winter and of Ad3 and 7 in late summer was found. Highly significant age predilections were subgenus A for infants (less than 1 year), subgenus C for infants and small children, Ad3 for schoolchildren, Ad7 for schoolchildren and adults, Ad8, other subgenus D (118) and B AV for adults. A predilection for males was observed in all types of subgenus B and C, and in Ad4 and Ad19.

8.3 AV types and disease

The clinical syndromes and the AV types mostly found associated may be seen from Table 9. For respiratory and ocular diseases,

Table 9. Clinical syndromes caused by adenoviruses

Syndrome	Types	Remarks
Abacterial pharyngitis	1 2 3 5 7	endemic (1, 2, 5) epidemic (3, 7)
Acute respiratory infection	3 4 7 21	epidemic in army camps, sporadic among the civil population
Pneumonia	3 4 7	may take a severe course in infants
Pharyngo-conjunctivitis	3 7	epidemics
Follicular conjunctivitis	3 4 7	
(Epidemic) kerato-conjunctivitis	8 19 37 3 4 7	spreads in eye clinics
Intussusception, mesenterial lymphadenitis	1 2 3 5	infants
Haemorrhagic cystitis	11	
Gastroenteritis	40 41	infants

the causal relationship between AV infection and disease is firmly established. Ad8 is the main causative agent of epidemic kerato-conjunctivitis (119), followed by Ad19 and 37.

The association of AV with intussusception and mesenterial lymphadenitis is also probably causal (120); the same is true for Ad11 and hemorrhagic cystitis (121).

In connection with respiratory disease in children, gastro-intestinal symptoms are often seen and may be predominant.

However, a asymptomatic infection of the gut by AV is common in children (122); hence the causal relationship with gastroenteritis is often doubtful. It could recently be shown, however, that Ad40 and 41 which are often recognized by electron microscopy or by ELISA, are important pathogens and causative agents of infantile gastroenteritis (114, 115, 123), second in frequency only to rotaviruses. Ad40 and 41, which are difficult to cultivate in cell culture (8), have been found in various parts of the world (114, 124). Gastroenteritis associated with AV recognizable by electron microscopy was reported in about 5 % of US children (125).

AV have also been found associated with CNS diseases, cardiovascular diseases, or exanthems. In many cases the causal relationship is doubtful. Virus isolations from the genital tract are rare; but in recent years Ad19 and 37 have repeatedly been recovered from urethral and cervical specimens (126).

8.4 Mode of spread

Besides the respiratory spread, the fecal-oral route of transmission is important, since most AV types are excreted from the stool, some of them over an extended period (122). As AV are fairly stable viruses, indirect as well as water-borne or food-borne transmission must be envisaged. Ad8, 19, and 37 as ocular viruses are predominantly excreted in the conjunctiva; hence the mode of infection is mainly ocular. Offices of ophthalmologists were, and still are (127, 128), often the source of an epidemic outbreak with one of these viruses. Virus antigen has been found in ophthalmic solutions (127, 128). A swimming-pool with a defect chlorination has been reported as a source of an Ad3 outbreak (129). A venereal or oculo-genital transmission of AV has been suspected, but not proved (126).

Except for the spread of conjunctivitis agents, nosocomial spread of AV had been rarely reported (130). However, Ad40 and 41 may be expected to spread readily in children's wards.

8.5 Latency

Since the first discovery of AV from human tonsillar and adenoid tissue (1), Ad1, 2, and 5 have repeatedly been found there in a latent stage. It appears that these viruses are not pathogenic for their host; neither are they a likely source for further virus spread.

The mechanism of another site of latent infection, the kidney, is completely unknown. The kidney appears to be susceptible mainly for AV from subgenus B (except Ad3 and 7) which are excreted in the urine from patients under immuno-suppressive therapy (131), after renal transplantation (132) or in patients with AIDS (80, 133). Apart from Ad11 isolated in acute hemorrhagic cystitis (see Table 9) it is not known whether these viruses may cause disease of kidney or other organs in their host. A counterpart is the frequent shedding of AV in the urine by chimpanzees (134). Monkeys also harbor a variety of AV in their kidney which then appear as cytopathic "SV" viruses in kidney cultures (88, 135).

8.6 Molecular epidemiology

This term has been devised when it became possible to characterize individual virus isolates by DNA restriction analysis or other biochemical methods. Wadell and his colleagues first applied these techniques to AV (48, 71). The aim is a more detailed analysis of the distribution and spread of virus strains, either on a small scale to determine chains of infection, or on a larger basis for epidemiological or pathogenicity considerations. The emergence of a new syndrome or a changing etiology, for example in AV conjunctivitis (136), may be more fully unravelled. This approach is particularly fruitful in the frequently occurring "epidemic" Ad7 and Ad3.

Wadell (45) even presented a worldwide geographic map of the occurrence of Ad7 and Ad3 genome types (see Table 8). An example of a deduction which may be derived from these data is Japan: here many respiratory diseases by Ad3, but few by Ad7 were reported, whereas in Europe almost the opposite was observed. It turned out that the isolates in Japan were Ad7 and Ad3a, whereas in Europe mainly Ad7b, Ad7c, and Ad3 were identified. Wadell's results suggest that Ad7 and 3 are less virulent than the other genome types, which would explain the uneven distribution of AV-associated disease in both continents.

Since 1978 Ad4 has increasingly been found in patients with conjunctivitis which had been rare before (137, 138). This epidemiologic change may be due to the emergence of the Ad4a genome type (45). The same may also be true for Ad19a as a causative agent of conjunctivitis, observed since 1973 (105).

The well-known, but only partially explained mechanism of AV outbreaks in military camps (Ad4, 7, sometimes Ad21) may also be studied by molecular analysis of virus isolates.

The great variability in the genome types of subgenus C adenoviruses (see above, Table 8) cannot at present be correlated with epidemiologic or clinical data; studies on these frequently occurring infections concerning association with specific syndromes, age groups or other may be fruitful.

9. EPIDEMIOLOGY: ANIMAL ADENOVIRUSES

AV types found in various animals are listed in Table 1. Here the references are given for the more recently detected types. Generally the information on the prevalence of AV infections in domestic animals in various parts of the world is scarce. Much less is known on AV in wildlife mammals or birds, and nothing for lower vertebrates. AV infections on chimpanzees or in various monkey species have been studied only sporadically (88, 134, 135). Studies on the pathogenicity of AV for various animals, sometimes inconsistent in their results, are so patchy that they cannot be reviewed here. Other review articles (51, 139, 140) may be consulted for selected topics.

10. CONCLUDING REMARKS

The species concept, based on serology and proposed for AV in 1981 (5, 42), has been superseded by the copious growth of molecular biological techniques, knowledge and thoughts in recent years. David W. Kingsbury of the present International Committee for the Taxonomy of Viruses proposed the following (unpublished) definition of a species:

"A virus species is a population of viruses sharing a pool of genes that is normally maintained distinct from the gene pools of other viruses." This definition stresses the primary importance of the genome structure for classification from species level upwards. It may be easy to apply e.g. to myxoviruses, but not so for AV, although possibly HAV of subgenus C fit into one "species". Much more knowledge is needed on the genomes of a variety of AV types of man and animals, before this or a similar concept can be applied on a broader scale. Thereafter, supraspecies and subgeneric taxonomy has to be reconsidered.

It is intriguing to speculate on virus evolution, as has been attempted for AV (29) and on a broader scale (141). Some observations make Ad4 a candidate of a common archetype of HAV (45). A computer-aided comparison of DNA sequences may ultimately govern classification, while the use of monoclonal antibodies will help to understand serological relations between AV types and variants.

ACKNOWLEDGEMENTS

We gratefully acknowledge the valuable contribution of the following colleagues who made 'preprint' information available to us (B.M.Adair, Belfast; A.Bartha, Budapest; S.Belák, Budapest; A.D.J.Bellett, Canberra; P.Boulanger, Lille; F.Bürki, Vienna; K.Fujinaga, Sapporo; I.Nász, Budapest; U.Pettersson, Uppsala; B.W.Stillman, Cold Spring Harbor; S.E.Straus, Bethesda; L.Svensson, Stockholm; A.van der Eb, Leiden; G.Wadell, Umea). Drs J.C.de Jong, Bilthoven, J.C.Hierholzer, Atlanta, K.W.Knocke, Hannover, and G.Wadell, Umea, kindly sent us AV isolates.

The expert secretary work by Gabriele Krug is gratefully acknowledged. The experimental work in the authors' laboratory was supported by grants from the Deutsche Forschungsgemeinschaft (Wi 3/21-1, 21-2) and from the "Fördererverein der Deutschen Vereinigung zur Bekämpfung der Viruskrankheiten eV", Munich, FRG.

REFERENCES

1. Rowe, W.P., Huebner, R.J., Gilmore, L.K., Parrott, R.H. and Ward, T.G. Proc. Soc. Exper. Biol. Med. 84:570-573, 1953.
2. Hilleman, M.R. and Werner, J.H. Proc. Soc. Exper. Biol. Med. 85:183-188, 1954.
3. Enders, J.F., Dingle, J.H., Francis jr., T., Hilleman, M.R., Huebner, R.J. and Payne, A.M.-M. Science 124:119-120, 1956.
4. Norrby, E., Bartha, A., Boulanger, P., Dreizin, R.S., Ginsberg, H.S., Kalter, S.S., Kawamura, H., Rowe, W.P., Russell, W.C., Schlesinger, R.W. and Wigand, R. Intervirology 7:117-125, 1976.
5. Wigand, R., Bartha, A., Dreizin, R.S., Esche, H., Ginsberg, H.S., Green, M., Hierholzer, J.C., Kalter, S.S., McFerran, J.B., Pettersson, U., Russell, W.C. and Wadell, G. Intervirology 18:169-176, 1982.
6. De Jong, J.C., Wigand, R., Adrian, Th., Hierholzer, J.C., Kapsenberg, J.G., Muzerie, C.J. and Wermenbol, A.G. Intervirology 22:164-169, 1984.
7. Hierholzer, J.C., Kemp, M.C., Gary jr., G.W. and Spencer, H.C. J. Clin. Microbiol. 16:15-21, 1982.
8. De Jong, J.C., Wigand, R., Kidd, A.H., Wadell, G., Kapsenberg, J.G., Muzerie, C.J., Wermenbol, A.G. and Firtzlaff, R.-G. J. Med. Virol. 11:215-231, 1983.
9. Davies, D.H., Dungworth, D.L. and Mariassy, A.T. Vet. Microbiol. 6:113-128, 1981.
10. Adair, B.M., McFerran, J.B. and McKillip, E.R. Arch. Virol. 74:269-275, 1982.
11. Studdert, M.J. and Blackney, M.H. Am. J. Vet. Res. 43:543, 1982.
12. Assaf, R., Marsolais, G., Yelle, J. and Hamelin, C. Can. J. Comp. Med. 47:460-463, 1983.
13. Bodon, L. and Prohaszka, L. Acta Vet. Hung. 28:247-255, 1980.
14. Naumann, S., Kunstyr, I., Langer, I., Maess, J. and Horning, R. Lab. Anim. 15:235-242, 1981.
15. Sutjipto, S. and Simmons, D.G. Am. J. Vet. Res. 42:491-494, 1981.
16. Zsâk, L. and Kisary, J. Avian Pathology 13:253-264, 1984.
17. Tibbetts, C. J. Virol. 24:564-579, 1977.
18. Esche, H., Reuther, M. and Schughart, K. Curr. Top. Microbiol. 111:91-106, 1984.
19. Tooze, J. Molecular Biology of Tumor Viruses, Cold Spring Harbor Laboratory, 1980, p.923-949.
20. Vilnis, A.E., Loparev, V.N., Zavizion, B.A., Tarasishin, L.A., Khilko, S.N., Naroditsky, B.S., Grodnitskaya, N.A., Vikhnovich, E.M. and Manykin, A.A. Vestn. Akad. Med. Nauk. SSSR 9:88-94, 1981.
21. Hu, S.-L., Hays, W.W. and Potts, D.E. J. Virol. 49:604-608, 1984.
22. Brusca, J.S. and Chinnadurai, G. J. Virol. 39:300-305, 1981.
23. Fujinaga, K., Yoshida, Y., Yamashita, T. and Shimizu, Y. Curr. Top. Microbiol. 110:53-72, 1984.

24. Van Ormondt, H. and Galibert, F. Curr. Top. Microbiol. 110:73-142, 1984.
25. Shinagawa, M. and Padmanabhan, R. Proc. Natl. Acad. Sci. USA 77:3831-3835, 1980.
26. Stillman, B.W., Topp, W.C. and Engler, J.A. J. Virol. 44:530-537, 1982.
27. Aleström, P., Stenlund, A., Li, P. and Pettersson, U. Gene 18:193-197, 1982.
28. Shinagawa, M., Ishiyama, T., Padmanabhan, R., Fujinaga, K., Kamada, M. and Sato, G. Virology 125:491-495, 1983.
29. Sambrook, J.F., Sleigh, M., Engler, J.A. and Broker, T.R. Ann. N.Y. Acad. Sci. 354:426-452, 1980.
30. Van Ormondt, H. and Hesper, B. Gene 21:217-226, 1983.
31. Kruijer, W., van Schaik, F.M.A., Speijer, J.G. and Sussenbach, J.S. Virology 128:140-153, 1983.
32. Li, P., Bellett, A.J.D. and Parish, C.R. J. Gen. Virol. 64:1375-1379, 1983.
33. Green, M., Brackmann, K., Wold, W.S.M., Cartas, M., Thornton, H. and Elder, J.H. Proc. Natl. Acad. Sci. USA 76:4380-4384, 1979.
34. Larsen, S.H. and Nathans, D. Virology 82:182-195, 1977.
35. Gelderblom, H. and Maichle-Lauppe, I. Arch. Virol. 72: 289-298, 1982.
36. Bartha, A. Acta Vet. Acad. Sci. Hung. 19:319-321, 1969.
37. Aleström, P., Stenlund, A., Li, P., Bellett, A. and Pettersson, U. J. Virol. 42:306-310, 1982.
38. Li, P., Bellett, A.J.D. and Parish, C.R. J. Gen. Virol. 65:1803-1815, 1984.
39. Adair, B.M., McKillop, E.R., McFerran, J.B. and Todd, D. Vet. Microbiol. 8:121-128, 1983.
40. Hu, S.-L., Battles, J.K. and Potts, D.E. J. Virol. 51:880-883, 1984.
41. Zsák, L. and Kisary, J. J. Gen. Virol. 56:87-95, 1981.
42. Matthews, R.E.F. Intervirology 17:1-119, 1982.
43. Adair, B.M. and McFerran, J.B. Arch. Virol. 51:319-325, 1976.
44. Willimzik, H.-F., Kalter, S.S., Lester, T.L. and Wigand, R. Intervirology 15:28-36, 1981.
45. Wadell, G. Curr. Top. Microbiol. 110:191-220, 1984.
46. Rabson, A.S., O'Connor, G.T., Berezesky, I.K. and Paul, F.J. Proc. Soc. Exp. Biol. Med. 116:187-190, 1964.
47. Kelly, B. and Pereira, H.G. Brit. J. Exper. Path. 38:396-400, 1957.
48. Wadell, G., Hammarskjöld, M.L., Winberg, G., Varsanyi, T.W. and Sundell, G. Ann. N.Y. Acad. Sci. 354:16-42, 1980.
49. Mauss, M. Immunologische Beziehungen zwischen Adenoviren des Schimpansen und des Menschen. Dissertation, Homburg/Saar, 1982.
50. Belák, S. and Pálfi, V. Arch. Ges. Virusforsch. 46:366-369, 1974.
51. McFerran, J.B. and Adair, B.M. Avian Pathology 6:189-217, 1977.
52. Rosen, L. Virology 5:574-577, 1958.
53. Rosen, L. Am. J. Hyg. 71:120-128, 1960.

54. Rapoza, N.P. Am. J. Epidemiol. 86:736-745, 1967.
55. Wigand, R., Sehn, N., Hierholzer, J.C., de Jong, J.C. and Adrian, Th. Arch. Virol. (in press) 1985.
56. Wigand, R. and Keller, D. J. Med. Virol. 2:137-142, 1978.
57. Trentin, J.J., van Hoosier, G.L. and Samper, L. Proc. Soc. Exp. Biol. Med. 127:683-689, 1968.
58. Huebner, R.J. Perspectives in Virology 5:147-166, 1967.
59. Tikchonenko, T.I. Curr. Top. Microbiol. 110:169-189, 1984.
60. Green, M., Mackey, J.K., Wold, W.S.M. and Rigden, P. Virology 93:481-492, 1979.
61. Norrby, E. J. Gen. Virol. 5:221-236, 1969.
62. Pettersson, U. and Wadell, G. in: Neurath and van Regenmortel (ed.) Immunochemistry of Viruses and the Basis for Serodiagnosis and Vaccines (in press) 1984.
63. Johansson, M.E. and Wadell, G. J. Immunol. Meth. 19:259-267, 1978.
64. Gerna, G., Cattaneo, E., Revello, M.G. and Battaglia, M. J. Infect. Dis. 145:678-682, 1982.
65. Wadell, G. Intervirology 11:47-57, 1979.
66. Khilko, S.N., Grigoriev, V.G., Naroditsky, B.S., Zolotarskaya, E.E., Dreizin, R.S. and Tikhonenko, T.I. Vopr. Virusol. 48-51, 1981.
67. Marusyk, R.G. and Cummings, G.T. Intervirology 10:356-360, 1978.
68. Belák, S., Berencsi, Gy., Rusvay, M., Lukács, K. and Nász, I. Arch. Virol. 77, 181-194, 1983.
69. Darai, G., Rösen, A., Flügel, R.M. and Delius, H. Abstracts 6th Intern. Congress of Virology, Sendai, Sept. 1984, W-29-8.
70. Goodheart, C.R. Virology 44:645-648, 1971.
71. Wadell, G. and Varsanyi, T.M. Infect. Immun. 21:238-246, 1978.
72. Garon, C.F., Berry, K.W., Hierholzer, J.C. and Rose, J.A. Virology 54:414-426, 1973.
73. Adrian, Th., Wigand, R. and Hierholzer, J.C. Arch. Virol. (in press) 1985.
74. Zsák, L. and Kísary, J. Intervirology 22:110-114, 1984.
75. Naroditsky, B.S., Kalinina, T.J., Goldberg, E.Z., Borovik, A.S., Karamov, E.V. and Tikchonenko, T. Biochim. Biophys. Acta 606:214-227, 1980.
76. Aird, F., King, J.J. and Younghusband, H.B. Gene 22: 133-134, 1983.
77. Kitchingman, G.R. Gene 20:205-210, 1982.
78. Wadell, G., de Jong, J.C. and Wolontis, S. Infect. Immun. 34:368-372, 1981.
79. Winberg, G. and Hammarskjold, M.-L. Nucleic Acids Res. 8:253-264, 1980.
80. Horwitz, M., Valderrama, G., Korn, R. and Spigland, I. Abstracts 6th Intern. Congress of Virology, Sendai, Sept. 1984, P-17-1.
81. Takács, M., Berencsi, G., Lengyel, A. and Nász, I. Acta Virol. 27:289-298, 1983.
82. Adrian, Th., Bastian, B., Best, B. and Wigand, R. Abstracts 6th Intern. Congress of Virology, Sendai, Sept. 1984, P-31-16.

83. Takiff, H.E., Reinhold, W., Garon, C.F. and Straus, S.E. J. Virol. 51:131-136, 1984.
84. Denisova, T.S., Makhov, A.M. and Gibadulin, R.A. Vopr. Virusol. 483-488, 1982.
85. Kurokawa, T., Igarashi, K. and Sugino, Y. J. Virol. 28:212-218, 1978.
86. Matz, B., Delius, H., Flügel, R.M. and Darai, G. J. Gen. Virol. 51:421-423, 1980.
87. Shimada, K., Yoshida, T., Kuroishi, T. and Ishibashi, M. Biochim. Biophys. Acta 740:169-178, 1983.
88. Kalter, S.S., Ablashi, D., España, C., Heberling, R.L., Hull, R.N., Lennette, E.H., Malherbe, H.H., McConnell, S. and Yohn, D. Intervirology 13:317-330, 1980.
89. Mackey, J.K., Wold, W.S.M., Rigden, P. and Green, M. J. Virol. 29:1056-1064, 1979.
90. Fujinaga, K., Sawada, Y. and Sekikawa, K. Virology 93: 578-581, 1979.
91. Antoine, G., Aleström, P., Schilling, R., Pettersson, U. and Winnacker, E.-L. EMBO J. 1:453-459, 1982.
92. Suzuki, N., Ueno, T., Yamashita, T. and Fujinaga, K. Microbiol. Immunol. 25:1291-1301, 1981.
93. Gingeras, T.R., Sciaky, D., Gelinas, R.E., Bing-Dong, J., Yen, C.E., Kelly, M.M., Bullock, P.A., Parsons, B.L., O'Neill, K.E. and Roberts, R.J. J. Biol. Chem. 257: 13475-13491, 1982.
94. Wigand, R., Keller, D. and Werling, I. Arch. Virol. 72:199-209, 1982.
95. Uhnoo, I., Wadell, G., Svensson, L. and Johansson, M. Develop. Biol. Standard. 53:311-318, 1983.
96. Wigand, R. Arch. Virol. 49:323-328, 1975.
97. Wigand, R. and Fliedner, D. Arch. Virusforsch. 24:245-256, 1968.
98. Adrian, Th., Bastian, B., Benoist, W., Hierholzer, J.C. and Wigand, R. Intervirology (in press) 1984
99. Norrby, E. Virology 37:565-576, 1969.
100. Matumoto, M., Uchida, S., Hoshika, T. and Matsuyama, T. Jap. J. Exp. Med. 28:305-315, 1958.
101. Rosen, L., Hovis, J.F. and Bell, J.A. Proc. Soc. Exper. Biol. 110:710-713, 1962.
102. Adrian, Th., Müller-Stein, I., Hierholzer, J.C., Knocke, K.W. and Wigand, R. Abstracts 6th Intern. Congress of Virology, Sendai, Sept. 1984, P-31-3.
103. Bruckova, M., Wadell, G., Syrucek, L. and Kunzova, L. Acta Virol. 24:161-165, 1980.
104. Fujii, S.-I., Nakazono, N., Sawada, H., Ishii, K., Kato, M., Aoki, K., Ohtsuka, H. and Fujinaga, K. Jap. J. Med. Sci. Biol. 36:307-313, 1983.
105. Wadell, G. and de Jong, J.C. Infect. Immun. 27:292-296, 1980.
106. Brown, M., Petric, M. and Middleton, P.J. J. Clin. Microbiol. 20:334-338, 1984.
107. Kidd, A.H. J. Med. Virol. 14:49-59, 1984.
108. Ginsberg, H.S. and Young, C.S.H. in: Fraenkel-Conrat, H. and Wagner, R.R. (ed.) Genetics of Adenoviruses. Comprehensive Virology. Plenum Press, New York, London, 9:27-88, 1977.

109. Meinschad, C. and Winnacker, E.-L. J. Gen. Virol. 48: 219-224, 1980.
110. Von Bahr-Lindström, H., Jörnvall, H., Althin, S. and Philipson, L. Virology 118:353-362, 1982.
111. Assaad, F. and Cockburn, W.C. Bull. World Health Organ. 51:437-445, 1974.
112. World Health Organization Techn. Report Series 642, 1980.
113. Schmitz, H., Wigand, R. and Heinrich, W. Am. J. Epidemiol. 117:455-466, 1983.
114. Chiba, S., Nakamura, I., Urasawa, S., Nakata, S., Taniguchi, K., Fujinaga, K. and Nakao, T. Lancet II 1983, 954-957.
115. Uhnoo, I., Wadell, G., Svensson, L. and Johannson, M.E. J. Clin. Microbiol. 20:365-372, 1984.
116. D'Ambrosio, E., Del Grosso, N., Chicca, A. and Midulla, M. J. Hyg., Camb. 89:155-161, 1982.
117. Fatz, W. and Wigand, R. Immun. Infekt. (in press)
118. Jung, D. and Wigand, R. Am. J. Epid. 85:311-319, 1967.
119. Editorial Brit. Med. J. 283:629-630, 1981.
120. Nicolas, J.C., Ingrand, D., Fortier, B. and Bricourt, F. J. Med. Virol. 9:267-271, 1982.
121. Mufson, M.A., Belshe, R.B., Horrigan, T.J. and Zollar, L.M. Am. J. Dis. Child. 126:605-609, 1973.
122. Fox, J.P., Brandt, C.D., Wassermann, F.E., Hall, C.E., Spigland, I., Kogon, A. and Elveback, L.R. Am. J. Epidemiol. 89:25-50, 1969.
123. Yolken, R.H., Lawrence, F., Leister, F., Takiff, H.E. and Straus, S.E. J. Pedr. 101:21-26, 1982.
124. Dowling, J.M. and Wynne, H. Lancet II 1981, 305-306.
125. Brandt, C.D., Kim, H.W., Rodriguez, W.J., Arrobio, J.O., Jeffries, B.C., Stallings, E.P., Lewis, C., Miles, A.J., Chanock, R.M., Kapikian, A.Z. and Parrott, R.H. J. Clin. Microbiol. 18:71-78, 1983.
126. Harnett, G.B. and Newnham, W.A. Brit. J. Vener. Dis. 57:55-57, 1981.
127. D'Angelo, L.J., Hierholzer, J.C., Holman, R.C. and Smith, J.D. Am. J. Epidemiol. 113:44-49, 1981.
128. Keenlyside, R.A., Hierholzer, J.C. and D'Angelo, L.J. J. Infect. Dis. 147:191-198, 1983.
129. Martone, W.J., Hierholzer, J.C., Keenlyside, R.A., Fraser, D.W., D'Angelo, L.J. and Winkler, W.G. Am. J. Epidemiol. 111:229-237, 1980-
130. Straube, R.C., Thompson, M.A., Van Dyke, R.B., Wadell, G., Connor, J.D., Wingard, D. and Spector, S.A. J. Infect. Dis. 147:814-819, 1983.
131. Zahradnik, J.M., Spencer, M.J. and Porter, D.D. Am. J. Med. 68:725-732, 1980.
132. Stalder, H., Hierholzer, J.C. and Oxman, M.N. J. Clin. Microbiol. 6:257-265, 1977.
133. De Jong, P.J., Valderrama, G., Spigland, I. and Horwitz, M.S. Lancet I 1983, 1293-1296.
134. Asher, L.V.S., Asher, D.M., Shah, K.V., Amyx, H.L., Gibbs jr., C.J. and Gajdusek, D.C. Infect. Immun. 21:458-461, 1978.
135. Hull, R.N. Virology Monographs 2:5-14, 1968.

136. Kemp, M.C., Hierholzer, J.C., Cabradilla, C.P. and Obijeski, J.F. J. Infect. Dis. 148:24-33, 1983.
137. Tullo, A.B. and Higgins, P.G. Brit. J. Ophthalmol. 64: 489-493, 1980.
138. Aoki, K., Kato, M., Ohtsuka, H., Ishii, K., Nakazono, N. and Sawada, H. Brit. J. Ophthalmol. 66:776-780, 1982.
139. Ishibashi, M., Yasue, H. in: Ginsberg, H.S. (ed.) The Adenoviruses. Plenum Press, New York, 1984, p.497-562.
140. Kraft, V. in: Horzinek, M.C. (ed.-in-chief) Viral Infections of Vertebrates. Vol. 6: Viral Infections of Rodents and Lagomorphs, Osterhaus, A.D.M.E. (ed.). Elsevier, Amsterdam, 1983.
141. Gorman, B.M. Intervirology 20:169-180, 1983.

INDEX